Toward More Sustainable Infrastructure

Toward More Sustainable Infrastructure

Project Evaluation for Planners and Engineers

Carl D. Martland

Department of Civil and Environmental Engineering
Massachusetts Institute of Technology
Cambridge, Massachusetts

WILEY

John Wiley & Sons, Inc.

VP and Publisher	Don Fowley
Acquisition Editor	Jennifer Welter
Editorial Assistant	Alex Spicehandler
Marketing Manager	Christopher Ruel
Marketing Assistant	Diane Smith
Production Manager	Janis Soo
Assistant Production Editor	Elaine S. Chew
Cover Designer	Seng Ping Ngieng
Cover Image	© Carl D. Martland

This book was set in 9.5/12 Times Roman by Thomson Digital, NOIDA, India and printed and bound by Hamilton Printing Company. The cover was printed by Hamilton Printing Company.

This book is printed on acid free paper. ⊗

Founded in 1807, John Wiley & Sons, Inc. has been a valued source of knowledge and understanding for more than 200 years, helping people around the world meet their needs and fulfill their aspirations. Our company is built on a foundation of principles that include responsibility to the communities we serve and where we live and work. In 2008, we launched a Corporate Citizenship Initiative, a global effort to address the environmental, social, economic, and ethical challenges we face in our business. Among the issues we are addressing are carbon impact, paper specifications and procurement, ethical conduct within our business and among our vendors, and community and charitable support. For more information, please visit our website: www.wiley.com/go/citizenship.

Library of Congress Cataloging-in-Publication Data

Martland, Carl D. (Carl Douglas)
   Toward more sustainable infrastructure: project evaluation for planners and engineers / Carl D. Martland.
      p. cm.
   Includes bibliographical references and index.
   ISBN 978-0-470-44876-2 (cloth)
 1. Infrastructure (Economics) 2. Sustainable development. 3. Sustainable engineering. 4. Green technology. I. Title.
   HC79.C3M37 2011
   338.9'27–dc22

                                                                                                    2010051281

Printed in the United States of America
10 9 8 7 6 5 4 3 2 1

Contents

PART III DEVELOPING PROJECTS AND PROGRAMS TO DEAL WITH PROBLEMS AND OPPORTUNITIES

Chapter 11 Developing a Strategy to Deal with a Problem

Chapter 12 Public-Private Partnerships

Chapter 16   Final Thoughts and Further Reading   435

Preface

MOTIVATION

This textbook presents methods for evaluating projects and programs aimed at improving the performance and sustainability of infrastructure projects. It introduces system performance, concepts of sustainability, and methods of engineering economics, and it provides numerous case studies, examples, and exercises based on real-world problems. The text is designed to fill what may be a void in the education of planners and engineering students, namely an understanding of why major infrastructure projects are undertaken, how they are structured and evaluated, and how they are financed.

OBJECTIVES OF THE TEXTBOOK

The text addresses three primary objectives:

1. Provide a framework for understanding and evaluating projects, taking into account not only the financial and economic issues but also the social and environmental impacts affecting the sustainability of infrastructure. Engineers and planners need to be able to communicate with all the people involved in a project, and they must have some feeling for the financial, economic, environmental, and social aspects of a project as well as the purely technical matters.
2. Demonstrate how to apply the basic methods of engineering economics in evaluating major infrastructure projects—and also demonstrate how these same techniques can be useful with many routine business and personal decisions.
3. Promote an approach to project evaluation that is based on an appreciation of the needs of society, the potential for sustainable development, and recognition of the problems that may result from poorly conceived or poorly implemented projects and programs.

KEY CONCEPTS

The text addresses the following key concepts:

- Justification of large investments: how to determine whether future benefits justify current costs using net present value, equivalence of cash flows, internal rate of return, and benefit-cost analysis.
- Engineering-based performance functions: creating functions that treat cost, service, capacity, and safety in sufficient detail to explore major project options related to size, design, technology, and level of service.
- Probabilistic analysis: the ability to include probabilistic features when structuring cost and service functions.
- Identification of key factors: the use of financial analysis, scenarios, and sensitivity analysis to identify the most important factors affecting a project, the use of new technology, or the choice of operating or marketing strategies.
- Cost effectiveness: how to compare options for achieving nonmonetary benefits.
- Sustainability: financial, economic, environmental, and social aspects of sustainability.

- Evolution of systems in response to changes in needs, technologies, financial capabilities, and societal norms.

The text introduces analytical methodologies that can be applied to each of these concepts. However, analysis does not necessarily determine what projects are proposed, which projects are evaluated, which projects are approved, or which projects are ultimately successful. Projects may be motivated by a vision of a greater society, by an idea for addressing a specific local problem, by the prospects of making a profit while providing a needed service, or by simple greed. Some apparently excellent projects cannot be financed, while it may be easy to find financial support for some very questionable projects.

FEATURES OF THE TEXT

A Broad View of the Planning Process

This text takes a broad view of the planning process required for implementing successful projects. This process involves much more than a well-defined, logical series of analytical steps that can be undertaken, although there are methodologies that must be learned. The process involves identifying needs and opportunities, developing alternative ways to address these needs, and using creativity in finding ways to take advantage of opportunities. The process requires consideration of economic, social, and environmental factors, not just the financial issues that are naturally of paramount importance in the private sector or the engineering issues that are of great concern during design and construction. Since the evaluation of any major infrastructure project will almost certainly generate public controversies, there is a need for some way to deal with conflicts and to reach consensus about what should or should not be done. This text therefore provides many examples of infrastructure projects, the problems they encountered, political issues that had to be addressed, and qualitative approaches that have helped people refine problem statements or find better solutions.

Engineering-Based Performance Models: What Can Be Done? What Should Be Done?

For many purposes, notably many kinds of policy analysis, statistical models of past performance and economic theory provide useful insights. But in considering major infrastructure projects, engineers, planners, entrepreneurs, politicians, and the public are more interested in what can be done in the future than in what was done in the past. New technologies, new designs, changes in relative costs of inputs, and many other factors influence what will be possible or desirable to do in the future. Someone, presumably the engineers and the planners, has to figure out what can be done and convince others that it should be done. This text therefore focuses on methods and approaches to problem solving that will be useful in these endeavors, including the development of engineering-based functions that can be used to represent infrastructure cost, service, and safety as they would be affected by proposed projects.

Microeconomics Plus Space and Time

Microeconomics is an interesting and challenging field, and indeed many microeconomic concepts are used throughout this textbook. However, classes in microeconomics generally do not address some of the most central issues in designing and developing infrastructure projects, especially those concerning space and time. Should a project be built? If so, where and when should it be built? Can it be developed in phases, so that capacity can be added only when and where it is needed? What options should be preserved for the future? For engineers, planners, and entrepreneurs, these are critical questions. Those who want to be engineers, planners, or entrepreneurs must learn how to balance current versus future costs and benefits, and they must be able to understand and respond to the many factors that influence the pace and location of development. In particular, they must understand the time value of money, the equivalence of cash flows, and the effects of risk and inflation on discount rates and the attractiveness of projects. These are all central topics in engineering economics, and they form the basis for a major portion of this textbook.

Case Studies, Examples, and Realistic Exercises

It is possible to overemphasize methodologies and theories while doing little to encourage independent thought, creativity, and judgment. To avoid such a situation, this text includes case studies, references to projects in the news, examples, open-ended problem sets, and discussion questions that provide some feeling for the context within which projects are evaluated, revised, and eventually implemented. Detailed exercises based on realistic situations require creativity in analysis and judgment in reaching conclusions and making recommendations. Examples and case studies convey some of the breadth and excitement of project evaluation as applied to major infrastructure projects. The overriding goal is to provide tools and concepts that can be used throughout one's career in understanding the need for projects, the options that are available, and the kinds of methods that will help society move toward more sustainable infrastructure systems.

An Overview of Project Management

At the recommendation of some of the reviewers, this text includes a chapter that introduces critical path, budgeting, and other fundamental concepts of project management. This chapter will be useful in an introductory class that is designed to introduce basic concepts of engineering economics, project evaluation, and project management. While entire textbooks and multiple subjects can be devoted to project management, many students may be unable to fit such a class into their undergraduate program.

The Art of Estimation

As several of my students have told me years after their graduation, one of the most important things they learned is that it is seldom necessary to get a precise number in order to get a solid conclusion. The question is likely to be whether the cost is greater or less than, say, $200 per unit, so it is unnecessary to do further analysis to show that the cost is actually $17.38 or $349.32 per unit! Getting a precise answer is seldom as important as asking the right question or making a justifiable recommendation based on sufficient analysis. Now that computational technology makes it possible to produce unmanageable volumes of results at a totally unwarranted level of precision, it is essential to retain some skepticism in interpreting results and to recognize when to stop crunching numbers and when to start interpreting the results and formulating conclusions. The "art of estimation" is essential for successful completion of the open-ended problems and case studies in this text, including the Canal Case Study (Chapter 2), Skyscraper (Chapter 8), and Cammitibridge A and B (Chapters 5 and 11).

STRUCTURE OF THE TEXTBOOK

The text contains three parts. Part I provides an overview of project evaluation as a multidimensional process aimed at creating projects that meet the needs of society. This part emphasizes the need to consider economic, environmental, and social factors along with the technological and financial matters that are crucial to the success of a project.

Part II provides in-depth coverage of the engineering economic methodologies that can be used to compare cash flows or economic costs and benefits over the life of a project. This part presents the techniques that are used by investors, bankers, and entrepreneurs in deciding whether to finance projects. It also shows how public policy can use taxes and other regulations to encourage projects that have public benefits.

Part III presents sensitivity analysis, the use of scenarios, probabilistic analysis, and other methodologies that are useful in developing and evaluating projects to deal with problems and opportunities. This part provides an introduction to project and program management and ends with a discussion of some of the challenges that must be faced to create more sustainable infrastructure-based systems in the 21st century.

BACKGROUND

This textbook is based on "Project Evaluation," a subject that I designed and then taught for 10 years as one of the required subjects in the civil engineering undergraduate curriculum at MIT.[1] Like the class, this text is structured not only for civil engineering undergraduates but also for students in other departments, including economics, urban studies and planning, and management as well as the other engineering disciplines.

INTENDED AUDIENCE AND POSSIBLE SUBJECTS

It should be clear by now that this is not another textbook on engineering economics attempting to compete directly with those already on the market. Instead, this text offers a much broader approach to project evaluation, an approach that should prepare students for careers devoted to creating more sustainable infrastructure systems. Engineers and planners need to be able to communicate with all the people involved in a project, and they must have some feeling for the financial, economic, environmental, and social aspects of a project as well as purely technical matters. This textbook therefore should be of interest to multiple audiences:

- Undergraduates in civil and environmental engineering: the text could be used for a new class on Sustainable Infrastructure or Project Evaluation, or it could be used as an alternative or supplemental text for a traditional class on Engineering Economy or Engineering Practice.
- First-year graduate students taking a class related to Large Projects or to Sustainability as part of a program in civil and environmental engineering, urban studies and planning, engineering systems, or management: these students do not have the time or the need to take a traditional class on engineering economy, and they will appreciate the focus on projects and programs and the more complete treatment of nonfinancial matters. They may also appreciate the introduction to project management.
- Continuing education and independent education: students and professionals could use the text to obtain a better understanding of project evaluation via independent reading, online teaching, or access to MIT's OpenCourseWare.
- Professionals involved in project evaluation or infrastructure management: professionals may find the text a concise reference covering many facets of project evaluation. Experienced planners and engineers may recommend the text to less experienced members of their staff.

TOWARD MORE SUSTAINABLE INFRASTRUCTURE

As I indicated previously, but wish to reemphasize, the overriding goal of this textbook is to provide students with tools and concepts they can use throughout their careers in understanding the need for projects, the options that are available, and what must be done to help society move toward more sustainable infrastructure systems. With sustainability becoming more critical, it is essential that

---

[1] The syllabus, PowerPoint presentations, homework assignments, and selected readings for "1.011 Project Evaluation" are available online at MIT's OpenCourseWare (OCW) website.

students broaden their perspectives and learn how to apply their skills within the context of very complex—and very interesting—projects and programs.

ACKNOWLEDGMENTS

Many people have helped in the efforts that resulted in this textbook. First of all, I thank my colleagues at MIT—especially Rafael Bras, Eric Adams, Herbert Einstein, Patrick Jaillet, John Miller, Feniosky Peña-Mora, Fred Salvucci, Sarah Slaughter, Joseph Sussman, and Nigel Wilson—for their encouragement and support when I first proposed designing a new subject on project evaluation as part of our new undergraduate curriculum for civil engineering students. For many years, Susan Murcott inspired me and my students to become more aware of and concerned about sustainability in general and environmental and social impact assessment in particular. My TAs (Jackie Henke, Lexie Lu, Mahdi Mattar, Tom Messervy, Jodie Misiak, Carlos Mojica, and Yanni Tsipis) were all instrumental in helping students grasp the breadth of topics addressed in project evaluation and handle the complexity and excitement of open-ended problems. Many people provided detailed feedback on the structure and content of the text, including Clare Conley, Mark Hickman, Rusty Lee, Pat Little, Roger Meier, Rabi Mishalani, Joseph Olson, Bill Robert, Robert Stokes, John Wilson and several anonymous reviewers. Most recently, I was fortunate to work closely with Joseph Sussman as he became the first professor to use this text to teach Project Evaluation, the course from which this text evolved. I am very thankful to my wife Nancy for her encouragement and to my son Sam for his insights into historical trends and his knowledge about infrastructutre projects in Latin America. Finally, I must thank Jenny Welter, my editor at John Wiley, who urged me to proceed with the project, arranged for excellent feedback from a cross section of potential users, and helped with the overall structure of the text. She and her assistant Alex Spicehandler, along with copy editors Elaine Chew and Chris Thillen, enabled the publication of this book.

Carl D. Martland
Senior Research Associate and Lecturer (Retired)
Department of Civil & Environmental Engineering
Massachusetts Institute of Technology
September 2010

ABOUT THE AUTHOR

Carl D. Martland is a transportation consultant and a research affiliate at the Massachusetts Institute of Technology. He has taught undergraduate and graduate classes on project evaluation, transportation systems, and freight transportation management. His research has addressed many aspects of rail systems performance, including life-cycle costing for track, technology assessment, equipment utilization, operations planning, terminal design, and service reliability. As a consultant, he has worked with all the major railroads in North America; numerous local, state, and federal agencies; the World Bank; and railroads and transportation agencies in a dozen countries in Europe, Asia, and South America. The author of seven award-winning papers, he was honored in 1998 as the recipient of the Outstanding Researcher Award from the Transportation Research Forum in recognition of his lifetime contributions to transportation research.

Chapter 1

Introduction

Focus first on those aspects of infrastructure that provide essential services, that is, those involving drinking water, wastewater, transportation, energy, and communications. . . . Business and population growth have already outpaced the capacity of existing systems. To meet user's expectations, planners should first determine the public's expectations with respect to the levels and resiliency of such services and the amount of money that should be spent to maintain them and then determine what alternatives exist and what actions need to be taken to meet those expectations.[1]

1.1 TOWARD MORE SUSTAINABLE INFRASTRUCTURE: BETTER PROJECTS AND BETTER PROGRAMS

Modern societies depend upon vast infrastructure-based systems that support efficient transportation and communications, provide ample supplies of clean water and energy, and enable effective treatment and disposal of wastes. The performance of such systems can be measured in terms of many factors, including cost, energy consumption, resource requirements, capacity, service quality, safety, impacts on society, and impacts on the environment. Performance can also be measured in terms of sustainability, a broad concept that refers to the ability of a system to perform well over a very long period of time.

Sustainability is a particular concern for systems that rely heavily on nonrenewable resources and systems that result in severe degradation of the environment. However, troubles in any aspect of performance can limit the sustainability of an infrastructure-based system, as discussed in Section 1.2. Sustainability can be enhanced by reducing costs, improving social and economic benefits, restricting the use of fossil fuels and other nonrenewable resources, or reducing negative social and environmental impacts.

Many infrastructure projects and programs are aimed at improving some aspect of sustainability. Some are designed to ensure that the system continues to function properly. If infrastructure is inadequate or poorly managed, people may suffer from congestion, high costs, pollution, economic stagnation, or environmental degradation. To limit such problems, ongoing investments may be required in new facilities, better materials, or new management techniques, although the nature of the infrastructure may remain about the same. Highways in 2010 may have real-time information signs, better paving materials, and synchronized traffic signals, but they still look and function much as highways did 50 years ago.

---

[1] One of the conclusions of *Sustainable Critical Infrastructure—A Framework for Meeting Twenty-first Century Imperatives,* a report based upon a May 2008 workshop sponsored by the National Research Council ("Report Urges New Framework for Planning Critical Infrastructure," *Civil Engineering,* June 2009, p. 20).

Other infrastructure projects and programs are designed to replace or upgrade systems that for some reason have become obsolete or nonsustainable. Over time, as economies develop, as societal norms change, and as certain resources become less available, the demands on infrastructure systems will change along with public perceptions of infrastructure performance. If infrastructure systems fail to evolve, they may eventually be recognized as being too costly, unsafe, disruptive to society, or overly damaging to the natural environment. At that point, new systems are needed. For example, solar power and wind power can produce electricity that otherwise would have required additional power plants and more imported oil.

In short, infrastructure projects and programs are designed to improve some aspect of system performance. Better projects and better programs will lead to more sustainable infrastructure. The main objective of this text is to demonstrate how to determine which projects and which programs are better.

1.2 INFRASTRUCTURE PROJECTS AND PROGRAMS

Infrastructure projects include large-scale, multidimensional, long-term investments in transportation systems, buildings, water resources, communications, power generation, and public services. Such projects always have multiple objectives, they will be controversial, and they involve people with many different perspectives who must come together to complete the projects and make them successful. Such projects have important impacts for the public at large, because they affect the environment, our society, and our economic prosperity. A program consists of a set of related projects, such as the Interstate Highway Program or a program designed to promote investment in wind power.

Historically, there have been numerous large-scale infrastructure projects and programs, some brilliant, some misguided, and many of them quite interesting for planners and engineers. The projects discussed in this text include:

- The Erie Canal and the Panama Canal
- Skyscrapers in Manhattan
- Parks created out of wasteland in Tempe, Arizona, and parks saved from highways in Franconia Notch, New Hampshire
- Railroads and highways that bridged continents
- Bridges in Hong Kong and Chesapeake Bay
- Flood control and disaster relief
- Water projects, sewer systems, and sewage treatment facilities

Projects like these affect the way we live, they are the backbone of much of our history, and they are the pathway to our future (Figure 1.1).

This book is about understanding where projects come from, how they are evaluated, how decisions are made to proceed with them, and what separates good projects from bad projects. The book spends considerable time on methodology, especially the methods of engineering economics that can be used to understand how projects are financed; but it also provides real-world examples and case studies that convey some of the flavor, excitement, and challenge of designing, evaluating, and implementing projects.

Implementing, operating, and maintaining infrastructure requires planners and engineers to work with bankers, entrepreneurs, politicians, community leaders, and the public in order to meet society's needs more effectively. Planners and engineers must learn to deal with the social, financial, and environmental issues related to infrastructure projects, and these issues will become more important over time. Engineers are likely to start out building and designing projects, and many engineers spend their entire careers concentrating on these activities. Planners and managers are likely to start out working at a low level on projects and programs that were begun years ago. However, someone, somewhere, is trying to figure out what next to build, when and where to build it, and how to convince investors and governments to pay for it. Actually, there are many such people, and some of them are

Figure 1.1 The Panama Canal. After decades of frustration, tens of thousands of deaths from tropical disease, bankruptcy, and disgrace for the initial French Canal Company, the canal was finally was completed in 1914 and remains today a critical link in global transport and a highly profitable enterprise for Panama.

destined to become famous. These people may end up proposing projects, or they may simply define problems and convince other people to begin working on them.

Engineers, planners, and managers naturally expect to work on large-scale infrastructure projects. To succeed and to advance in their professions, they will need to understand the big picture—the needs of society—and take the lead in designing, implementing, or marketing new technologies or new systems. Leaders will need a broader outlook on problems than is ordinarily conveyed in an engineering subject or a textbook on finance. They will need to understand how projects begin, how they are sold to the public, and how they become successful. They will need to combine engineering or planning skills with marketing, financial, and communications skills. Anyone who grasps this broader outlook will have a chance to become involved in projects and programs that are increasingly complex and offer more possibilities for design and implementation strategy, less certainty regarding the outcomes, and greater need for imagination and leadership. Hopefully this text will help provide this broader perspective on infrastructure systems.[2]

Infrastructure is usually defined in terms of public systems, and constructing and maintaining infrastructure is an ongoing process and problem for local and national governments:

> *"Infrastructure" refers to the physical systems that provide transportation, water, buildings, and other public facilities that are needed to meet basic human social and economic needs. These facilities are needed by people regardless of their level of economic development. When infrastructure is not present or does not work properly, it is impossible to provide basic services such as food distribution, shelter, medical care, and safe drinking water. Maintaining*

---

[2] Not all projects are large scale, nor is this text relevant only to the most complex projects. Many of the techniques presented in this text are widely applicable in the business world and in private life. The methodologies needed to determine car payments or mortgage payments are the same as those required in decisions to expand a business, upgrade equipment, or develop a new product. Although the focus of the text is on infrastructure projects, many of the methods are commonly used in other types of economic and financial analysis.

infrastructure is a constant and expensive process which often is neglected in favor of more attractive political goals.[3]

In practice, much of the civil infrastructure may be owned and maintained by private companies or individuals. Much infrastructure was originally built by private corporations with licenses or other authorization from government; private toll roads were the norm in the United States in the early nineteenth century, and private expressways are being built today in many parts of the world. Many railroad systems are privately owned and operated. Large office buildings or apartment buildings are mostly privately owned, and they are certainly part of the basic infrastructure of a modern city.

Box 1.1　Assets and Asset Management

Infrastructure systems include many individual components, such as roads, bridges, buildings, rights-of-way, power lines, and water treatment facilities. The companies and agencies that own and operate these systems are likely to view them as a set of well-defined **assets**, each of which can be characterized in terms of its design, condition, usage, and performance. Considerable effort has been devoted to developing systems and tools for **asset management**, including such things as creating a digital inventory of assets, routinely monitoring asset condition, documenting asset utilization, and planning for maintenance and renewal. For the purposes of this text, the terms *infrastructure* and *assets* can be used interchangeably.

This text examines projects aimed at creating, maintaining, rehabilitating, or decommissioning any kind of infrastructure, whether carried out by the public sector, the private sector, or a public-private partnership. Thus building a new road, adding a lane to an existing road, or paving an existing road could all be considered infrastructure projects. For administrative convenience, a large project is often broken down into multiple smaller projects. The construction of a new road may involve construction of a dozen bridges, three major interchanges, extensive cut-and-fill operations to prepare the right-of-way, and eventually the actual paving of the road. Moreover, the road may be completed in multiple phases over a period of many years. Whether to consider each of these activities as a separate project, each phase as a separate project, or the entire road as a single project could be debated; there will certainly be a well-defined set of contracts and subcontracts so that all of the contractors have a clear perspective on their portion of the overall project. The public, however, will likely view the whole road construction as a single, multiphase project. The distinction is usually unimportant, although at times a small segment of a road, or a small portion of some other project, is proposed with the hope of gaining approval more easily later on for an extension (also known as "getting your toe in the door").

An infrastructure program may be established as a way to manage a series of projects or a way to simplify the design and approval process for multiple projects. A program may specify goals and criteria for measuring progress against those goals. It may also specify what kinds of projects will qualify to be included in the program and what kinds of incentives will be available to qualifying projects. For example, a state may establish a program aimed at attracting private investment in housing for low-income families. The program may provide subsidies, tax relief, or other benefits to projects that qualify according to the criteria specified in the legislation or regulations. A company may also have infrastructure programs; retailers such as Home Depot or Wal-Mart will have plans for expanding their network of stores and warehouses. A railroad may have a plan for upgrading its oldest bridges on certain high-density lines; each bridge renewal would become a separate project as a part of

---

[3] Neil S. Grigg, *Infrastructure Engineering and Management* (New York: John Wiley & Sons, 1988), 1.

the program. Cities and states may have programs aimed at providing housing for the elderly or for low-income residents, and they may have programs aimed at improving water supplies or sewage treatment facilities. The various interest groups and political leaders who favor or oppose a certain type of project will fight over the structure of a program, perhaps for many years, but eventually they may reach agreement about the objectives, scope, funding amounts, and funding eligibility for the program. Once a program has been established, those prolonged fights will cease, and projects can rather quickly be identified, approved, and implemented. It will be desirable from time to time to review programs to ensure that the objectives remain valid, that the funding mechanisms are adequate and fair, and that the projects as implemented under the program actually have been achieving the program's objectives.

Infrastructure projects and programs have several common and very interesting aspects:

- Infrastructure is intended to last a very long time, so it is necessary to compare what may be very large current expenses with the potential for benefits that will be gained only over a period of decades.
- Infrastructure influences and perhaps defines the location and land use of cities and regions, so the location of infrastructure will have long-term implications for local and regional land use.
- Infrastructure often involves networks of facilities that are widely dispersed, perhaps with severe consequences for the environment or for the people who live where the networks are located.
- Infrastructure benefits are frequently qualitative or difficult to measure (e.g., mobility, safety, air quality, or the availability of clean water).
- Infrastructure projects and programs will be of great concern to many different groups of people, including developers, the public, special interest groups (some of which may be public interest groups and some of which may be supporting very narrow private interests), governments (including elected officials, regulatory officials, and administrative officials), lawyers, users, abutters, construction companies, and investors.

Infrastructure is costly to build and costly to maintain. According to the American Society of Civil Engineers (ASCE), which periodically issues a report card for the major categories of infrastructure, the average condition of infrastructure in the United States in 2009 merited only a grade of D. ASCE estimated that investment of more than $2.2 trillion would be needed over a five-year period to "substantially improve conditions" so as to raise the grade to an acceptable level. The greatest needs were for roads and bridges, with a gap of $550 billion in funding; mass transit, with a gap of $190 billion; and water resources, with a combined gap of just over $100 billion for drinking water and wastewater (Table 1.1).

The ASCE issues its report card to draw attention to the need for continuing funding for infrastructure projects and programs. The long lives expected for infrastructure cannot be achieved unless funding is available for proper management, including safe operating practices, ongoing inspection and maintenance, and periodic renewal and upgrades. Without such funding, infrastructure systems will deteriorate and eventually be unable to meet the societal needs they were designed to serve. Without adequate funds for renewal and expansion, it will be impossible to meet growing needs for services or to capture the benefits of new technologies.

Adequate financing must therefore be considered an essential factor in improving the **sustainability** of infrastructure systems, where *sustainability* refers to the ability of a system to function long into the future. Poorly managed infrastructure systems that steadily deteriorate, become congested, or become unsafe clearly are not sustainable. However, adequate financing is but one of the major factors affecting the sustainability of infrastructure.

Large-scale infrastructure, even if it appears to be adequately financed, can be sustained over long periods of time only if it is supported by society and the resources it requires are available at a reasonable cost (Figure 1.2). If infrastructure requires excessive use of nonrenewable resources, if it requires too much water or energy, or if its use results in devastation of the environment, then the lack of resources, increasing costs of materials, or public outrage will force changes. If construction,

Table 1.1 ASCE's Report Card for America's Infrastructure

	Grade	Estimated Five-Year Investment Needs	Estimated Investment Shortfall
Transportation			
Roads and bridges	D– (roads) C (bridges)	$930 billion	$550 billion
Mass transit	D	$265 billion	$190 billion
Aviation	D	$ 87 billion	$ 41 billion
Rail	C–	$ 63 billion	$ 12 billion
Inland waterways	D–	$ 50 billion	$ 21 billion
Water and Environment			
Drinking water and wastewater	D–	$265 billion	$109 billion
Hazardous waste and solid waste	D (hazardous) C+ (solid)	$ 77 billion	$ 43 billion
Levees	D–	$ 50 billion	$ 49 billion
Dams	D	$ 13 billion	$ 8 billion
Public Facilities			
Schools	D	$160 billion	$ 35 billion
Public parks and recreation	C–	$ 85 billion	$ 48 billion
Energy			
National power grid	D+	$ 75 billion	$ 30 billion
Total	D	$ 2.2 trillion	$ 1.2 trillion

Source: ASCE, *2009 Report Card for America's Infrastructure* (Reston, VA: ASCE, 2009).

maintenance, and operations continually disrupt neighborhoods, cause human suffering, or expose people to potentially catastrophic risks, then society will be reluctant to support further expansion of that kind of infrastructure.

Over time, social norms may change, the costs of resources may vary, and new technologies may emerge. What one generation viewed as highly beneficial investments may be viewed as dubious achievements or even disasters by following generations. Infrastructure systems must evolve along with society, and rising concerns about public safety, public health, pollution, and environmental decline mean that society will require more sustainable infrastructure. Water shortages, highway fatalities, urban congestion, overdependence on fossil fuels, toxic chemicals associated with large-scale agriculture, acid rain, oil spills, and excessive amounts of solid waste are all symptoms of problems that reflect a need for more sustainable infrastructure and a more sustainable way of life.

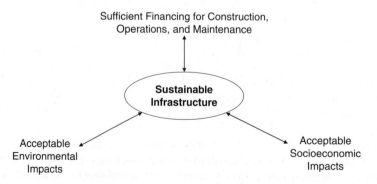

Figure 1.2 Three Requirements for Sustainable Infrastructure

Box 1.2   Societal Expectations Regarding the Environment: From Exploitation toward Stewardship

In the eighteenth and early nineteenth centuries, as European settlers in North America pushed development westward, they cleared land for agriculture, cut down forests for lumber, and dug mines for coal and iron ore, often with little or no concern for their impact on nature. When the population was small, and people relied on hand tools, the environmental impact was limited. By the end of the nineteenth century, however, population growth, better transportation, and power tools made large-scale devastation possible. Fearful of destruction of even the most beautiful and unique natural areas, John Muir and others pushed for the creation of national parks. After wildfires devastated much of the recently logged slopes of the White Mountains in New Hampshire, public outrage—and the leadership of Senator Weeks—led to the Weeks Act and the creation of the National Forest System. During the Depression, following the tragic effects of the "dust bowl" on small farmers, major federal programs were introduced to promote soil and water conservation. In the latter half of the twentieth century, better understanding of environmental science led to bans on pesticides such as DDT and aerosols. The National Environmental Policy Act, passed in 1969, required an environmental impact statement (EIS) for any major federal program or legislation (as discussed in Section 4.4). Soon thereafter, major federal legislation led to improvements in air and water quality by such things as limiting automobile emissions, limiting discharges of raw sewage, and restricting development along waterways. By the early twenty-first century, many public agencies were regarding their role as stewards of the environment and taking a more active stance toward conservation and enhancement of the environment instead of just regulating development. Challenges such as those posed by climate change, oil depletion, collapse of fisheries, and large numbers of endangered species combine to make stewardship of the environment and sustainable development greater concerns for society.

Achieving more sustainable infrastructure will require thought, innovation, planning, financing, regulation, and leadership. There clearly is a continuing need for large investments in infrastructure, and there will be many opportunities for evaluating projects and programs related to all types of infrastructure. Evaluating projects and programs will require methodologies for comparing current and future impacts, for considering multiple objectives, for assessing both quantitative and qualitative information, and for communicating and negotiating with diverse groups of people.

1.3   EVALUATING INFRASTRUCTURE PROJECTS

The main goal of project evaluation is to help in identifying and implementing successful projects and programs. From an overall perspective, a project is successful if:

1. It was built, which means proved that construction was feasible from engineering, financial, and social perspectives.
2. The benefits were indeed greater than the costs.
3. The project as built was an effective way to achieve those benefits.
4. The project was built in an efficient and effective manner:
   a. There were no clearly better options.
   b. There were no significant negative externalities.
5. Building this project did not foreclose other, even better projects.

Different participants might have far narrower definitions of success. Did the engineers design a building that was safe? Did the contractors get paid? Did clean water actually come to the neighborhoods? Did the mayor get reelected? These different perspectives must of course be considered in evaluating projects, but it is useful for students, consultants, concerned citizens, honorable developers, and honest politicians to pay some attention to the overall issues.

Project evaluation is a qualitative process as much as it is a quantitative one. A critical step is to create a "story" for the project that can be used to explain why the project is needed, what it will do, what the benefits and costs will be, and why this is the best way to proceed. There will certainly be quantitative aspects to the process, although estimates of costs and benefits may be rather ill-defined and subject to debate.

Implementing and maintaining a project over a long period of time will require:

- Financing: sufficient income to cover expenses, whether the income comes from user fees, investors, subsidies, or contractual payments
- Government approvals: licensing and periodic inspections to ensure compliance with safety, environmental, and other regulatory matters
- Engineering skills: sufficient knowledge and skilled manpower to conduct the maintenance and rehabilitation necessary to perform at an acceptable level of service
- Resources: people and materials as required for maintenance and operations and whatever additional resources are needed by users (e.g., asphalt for highway maintenance plus gasoline for drivers)
- Public support (or tolerable opposition and interference)

The financing issue is different from the economic issue. Financing provides the cash necessary to construct, operate, and maintain a project. The ability of a project to be financed depends upon the availability of money—not upon the actual economic benefits of the projects. Economic issues concern the costs and benefits associated with a project, the distribution of those costs and benefits, and whether the benefits are sufficient to justify the costs. Economic benefits may include creation of jobs, congestion relief, reduction in accidents, or improved productivity for those affected by the project. Some of these benefits may be easily described in monetary terms, and some may be very difficult to quantify in monetary or any other terms. They are economic benefits because they allow more efficient and more effective use of resources, even if the benefits do not translate directly into cash for the project or for investors.

It may be helpful for a project to have economic benefits in order to attract public or private financing. For example, governments may choose to subsidize transit operations, housing for low-income or elderly residents, or agriculture. The cash provided by those subsidies can in fact attract investors, who will create commuter rail services, apartment buildings, and more productive farms. Whether or not these projects are really worth the subsidies they receive is important for legislative bodies and elected officials to consider, but not necessarily something that will concern investors.

Government approval will be needed for any almost any project. A building permit will be needed for constructing a screen house in your backyard or for constructing a 100-story office building. Governments may establish regulations concerning land use, protection of the environment, the siting and size of buildings, construction materials and methods, the use of union or local labor, and many other factors that may affect the feasibility, cost, and ultimate success or failure of the project. Whether or not government agencies approve proposals or provide the necessary permits may depend upon legislation, regulations, the whim of administrators, and/or feedback from the public. Large projects tend to generate large criticism, so developers must always be concerned with public perceptions of their projects, and they must be aware of ways to make their projects more attractive to the public.

People with the necessary skills are needed in designing projects, in constructing them, and in ensuring they continue to function. It is one thing to build a road. It is another thing to enforce weight limits to ensure that overloaded trucks do not destroy the pavement within a few years, to enforce speed limits so as to promote safe driving conditions, and to establish periodic inspections, maintenance, and rehabilitation to keep the road in safe condition.

Projects and the people who use them or depend upon them will need resources for operations and maintenance over what may be a very long lifetime. Projects may fail because the resources needed to sustain them become too costly or unavailable. Some of the most pressing issues of the twenty-first century relate to the continued availability of fossil fuels for transportation, electrical power generation, and home heating, and to the availability of water for irrigation, household consumption, and industrial use. Many projects and infrastructure choices were justified based upon usually unstated assumptions that unlimited supplies of cheap oil and water would always be available. Fossil fuels, however, will not last forever, and prices will rise as reserves of oil, coal, and natural gas are used up. With cheap oil, automobiles and airlines prosper; with expensive oil, transit and rail transportation become more competitive. With abundant water supplies, crops can be grown in irrigated deserts, people can compete for the greenest lawns, and industries can use processes that consume vast amounts of water. Eventually, however, as population growth and other demands for water increase, the supply of water is no longer sufficient for all the possible uses, so the use of water will be regulated and the price of water will rise. Moreover, water supplies may diminish. Regions that are heavily dependent upon well water may find that their aquifers are drying up. In other regions, changes in climate may diminish the amount of water available. Since drainage and river basins follow geographical rather than political boundaries, rival demands for the use of water have sparked, and will continue to spark, political battles between neighboring states and countries. A populous region, such as the Los Angeles metropolitan area, will seek to divert water from distant regions in order to support its needs, while perhaps limiting the growth and productivity of the regions from which the water is diverted. Disputes over oil reserves have already sparked conflicts in the Middle East, and the potential for future conflict will continue as long as so much of the world's transportation, power generation, and industrial production is fueled by oil.

Public support, or at least tolerable opposition, is the final factor necessary for the long-term success of a project. The public normally does not have a direct role in decisions regarding major projects, because most decisions regarding projects are made by elected officials, appointed officials, and legislative bodies. However, the public can provide input into the decision process, whether by participating in a process established to promote public involvement, by writing to newspapers or elected officials, or by organizing groups to support or oppose projects. Public opposition can prevent particular projects, it can lead to new regulations or legislation, and it can change programs and policy. In the late 1960s and early 1970s, public opposition was the major factor in halting construction of major urban portions of the Interstate Highway System, including the so-called Inner Belt and the Southwest Expressway in Boston and the Embarcadero in San Francisco. Public concerns over the safety of nuclear power plants had led to stringent regulation of the construction of such plants in the United States by the 1970s; public outrage after a rather minor leakage incident at the Three Mile Island Nuclear Power Plant effectively ended construction of these plants in the United States.

1.4 INFRASTRUCTURE, CITIES, AND CIVILIZATION

It can be argued that infrastructure projects are the key to urbanization, which is perhaps the chief characteristic of civilization. If people are to be able to congregate in cities, then they will need access to large amounts of clean water, and they will need some system for treating or isolating wastes. They will need to import food, building materials, and energy resources. They will need facilities and materials to support various kinds of manufacturing and trade. They will want to create facilities for education, sports, and worship as well as for communications and entertainment. In short, people will have to construct the infrastructure necessary to support all of the normal functioning of a densely populated society.

The benefits of urbanization can be great for people's lifestyles and for efficient use of resources. Higher populations can support a diversity of lifestyles and greater opportunities for jobs and recreations. There can be a greater frequency and higher quality for social events. When people no longer have to spend all of their time eking out a living, whether on a farm or in isolated rural areas, they will have sufficient time to enjoy the fruits of civilization. From a systems standpoint, having large numbers of people living in a small area allows more efficient use of resources in constructing and operating transportation networks, creating housing, supplying water, and treating waste (Figure 1.3). As activities are differentiated, complementary activities can be concentrated within special districts of the city. When people are concentrated in well-situated cities with sound infrastructure, they can be protected from natural disasters, and it is possible to manage development so as to reduce the consequences of man-made disasters.

> *Your imagination, your initiative, and your indignation will determine whether we build a society where progress is the servant of our needs or a society where old values and new visions are buried under unbridled growth. For, in your time, we have the opportunity to move not only toward the rich society and the powerful society but toward the Great Society. The Great Society rests on abundance and liberty for all. It demands an end to poverty and racial injustice . . . It is a place where the city of man serves not only the needs of the body and the demands of commerce but the desire for beauty and the hunger for community. . . .*
>
> *Our society will never be great until our cities are great. . . .*
>
> Lyndon B. Johnson,
> President of the United States, excerpts from the
> "Great Society Speech" delivered at the University of Michigan,
> Ann Arbor, May 22, 1964.

Of course, as Freud pointed out in his book, *Civilization and Its Discontents*, crowding vast numbers of people into cities may not be good for everyone. The more we protect ourselves from natural disasters, and the more contact we are forced to have with each other, the more difficult it may be for us to live together. There is not only the loss of self-sufficiency that may be achievable on a farm, but there is also the possibility of extreme poverty. A city is dependent upon its infrastructure—and transportation or water resource systems may fail. If diseases break out, thousands may die, and pollution and the inability to absorb wastes may become continuous drains on health and happiness. As cities grow ever larger, congestion is likely to limit mobility, and it may become ever more difficult to limit pollution, to provide open space and to ensure adequate housing for everyone.

Whether cities evolve into safe, livable, aesthetically pleasing places or degenerate into over-crowded dens of despair depends greatly upon the ability of the people of those cities to undertake the projects that will enable them to meet the needs of human life and challenges of urban life. Anticipating and responding to challenges is the driving force for successful civil and environmental projects. And there will always be new challenges.

Tomorrow's challenges may be quite different from yesterday's; but there will always be basic needs to be met, and there will always be a need for evaluating and choosing the best ways to meet

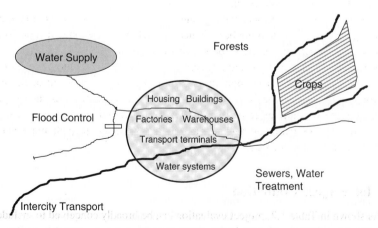

Figure 1.3 Infrastructure Projects Make Cities Possible and More Livable

those needs. Even with tremendous advances in communications and computers, with automated factories and computer-controlled highway networks, with cheap transportation for freight, a global economy, and ever-improving medical care, there will still be plenty to do. After all, only about half of the world's population has access to clean running water, hundreds of thousands of people die each year in transportation accidents, earthquakes and other natural disasters cause thousands of fatalities, billions of people live in substandard housing, and nearly everyone who lives in a large city spends a large portion of their life stuck in traffic and breathing bad air.

> *For it is absurd to suppose, given the amount of ugliness and suffering the world still affords, that we have run out of work: what we have run out of is a set of organized connections between purposes crying to be achieved and people crying out for purpose.*
>
> Eric Larrabee, former associate editor of *Harper's*,
> "Time to Kill: Automation, Leisure, and Jobs," *Nation*,
> 100th Anniversary Issue, September 20, 1965.

1.5   OVERVIEW OF PROJECT EVALUATION

1.5.1   Where Do Projects Come From?

A project begins long before the groundbreaking, long before the first contract is signed, and long before a specific plan is identified and agreed upon by people with the resources and political power to make something happen. A project begins with an idea, with a vision of what is wrong or what is needed or what is possible. Initial ideas quickly evolve into whole families of ideas and possibilities, and soon different, competing options begin to emerge. Long before the time for computer analysis and project planning, strong-minded, imaginative, entrepreneurial, and political individuals are vying to promote their concepts for the future. The players might include engineers, politicians, charlatans, financiers, developers, or dreamers. There are no bounds to how they might think or talk about the project, or how they conceive the project fitting into what is already in place or what could be put in place. Their creative processes can be slow or rapid, rational or chaotic, cooperative or acrimonious— there are no rules, and there are no limits to how hard people will push.

This undisciplined, often unmannerly process eventually leads to a specific project that will be constructed to finely drawn plans with a well-defined scheme for paying for it all. At this point, and not before this point, project management skills are needed; and there will be plenty of work for those with specialized software, algorithms, and risk management techniques that can lead to more efficient designs and timely completion of the project. But those skills are not much use in the early stages of project design and evaluation.

In these early stages there is the greatest uncertainty, the most excitement, the widest opportunities for egregious errors, and the best chances for achieving elegance in a project. It is difficult to teach how to conduct this process for which there are no rules and few guideposts. By the time the processes are well enough defined to create guidebooks for planners, the damage of poorly conceived projects will be only too apparent. We built highways straight through cities for decades before stopping to think seriously about the effects on the neighborhoods and the possibility for justifying less disruptive, more effective approaches. We need to think before we leap; we need to appreciate the creative, political, and entrepreneurial efforts that are needed; and we need to avoid the pitfalls that can catch the unwary.

1.5.2   A Framework for Project Evaluation

As shown in Table 1.2, project evaluation can be broadly conceived to include five phases that cover the entire life-cycle of a project.

Table 1.2   *A Framework for Project Evaluation.* Five stages of project evaluation span the entire life-cycle of a project. The first three phases may require much iteration before a final project is approved. The final phase should continue over the entire life of a project.

I	**Project Identification**
	• Identify needs and opportunities that could be addressed by a project.
	• Define objectives to be met by a project.
	• Identify alternatives for meeting these objectives.
II	**Analysis of Alternatives**
	• Conduct a preliminary technical analysis.
	• Analyze economic issues and impacts.
	• Analyze environmental issues and impacts.
	• Analyze social issues and impacts.
III	**Assessing and Comparing Alternatives**
	• Assess results of analysis.
	• Identify the most promising alternatives.
	• Identify need for further analysis (more detailed study, more options, or different criteria).
	• Select the best alternative.
IV	**Implementation**
	• Undertake value engineering to reduce costs and enhance benefits of the preliminary design.
	• Determine how best to mitigate negative social and environmental impacts.
	• Complete the final engineering design.
	• Manage the construction of the project.
V	**Ongoing Evaluation**
	• Conduct post-audit of construction.
	• Assess performance relative to original objectives.
	• Identify needs for modification or expansion.
	• Provide input to evaluation of similar projects.

Phase I—Project Identification

The first phase is the least well-defined and yet the most important for the ultimate success of a project or a program. Many ideas for projects arise in response to perceived problems and the needs of society. Congestion leads to ideas for new roads or new transit systems. Rising populations require new schools, housing, and drinking water. If problems and needs are understood, and if there is a process for examining possible ways to deal with them, then it should be possible to develop effective projects and programs that result in a better society. However, there will not necessarily be any process for determining and responding to societal needs. The ideas for many projects may originate when someone senses an opportunity to make some money or to create some sort of monument. Ideas for projects might well come from someone—an entrepreneur, a company, or a public official—who spots an opportunity for using a new technology, for developing a particular plot of land, or for expanding an existing network of facilities. It may well happen that project proponents first identify the project and then address the problems or needs that would be addressed by this project.

Nevertheless, it is useful to have a framework in which the first step examines problems or needs. For an infrastructure-based system, problems are likely to relate to cost, capacity, service quality, or safety. A problem may exist if some aspect of performance is believed to restrict the efficiency or effectiveness of the system. System operators will likely be aware of ways to improve performance, based on their own insight into operations or based on comparisons with similar systems in other locations. A need for better performance may be evident from user complaints, media reports, or scientific studies. Needs may be expressed in terms quite different from those used to define problems. For example, transportation needs might be expressed in terms of mobility and accessibility, whereas

Table 1.3   Selected Problems and Needs Related to America's Infrastructure

Bridges	*"More than 26%, or one in four, of the nation's bridges are either structurally deficient or functionally obsolete."*
Drinking water	*"Aging facilities are near the end of the useful life; . . . leaking pipes lose an estimated seven billion gallons of clean drinking water a day."*
Inland waterways	*"Of the 257 locks still in use on the nation's inland waterways, 30 were built in the 1800s and another 92 are more than 60 years old . . . well past their design life of 50 years."*
Rail	*"Growth and changes in demand create bottlenecks that constrain traffic in critical areas."*
Roads	*"Americans spend 4.2 billion hours a year stuck in traffic at a cost of $78.2 billion to the economy."*
Wastewater	*"Aging systems discharge billions of untreated wastewater into U.S. surface waters each year."*
Aviation	*"Travelers are faced with increasing delays and inadequate conditions as a result of the long overdue need to modernize the outdated air traffic control system."*
Public parks and recreation	*"The acreage of parkland per resident is declining."*
Solid waste	*"The increasing volumes of electronic waste and lack of uniform regulations for disposal create the potential for high levels of hazardous materials and heavy metals in the nation's landfills, posing a significant threat to public safety."*
Mass transit	*"Nearly half of American households do not have access to bus or rail transit, and only 25% have what they consider to be a 'good' option."*

Source: "ASCE's Infrastructure Report Card Gives Nation a D, Estimates Cost at $2.2 Trillion," *ASCE News*, February 2009.

transportation problems might be expressed in terms of travel delays and maintenance costs. Table 1.3 indicates some of the problems and needs highlighted by ASCE in its 2009 review of infrastructure in the United States.

The objectives of the project need to be clear and well-defined, but they can be modified based on feedback and assessments concerning completed projects or new information related to needs and opportunities. The need for flexibility may lead to certain challenges in the overall decision-making and implementation process. Sometimes strategic objectives are too narrowly defined and remain fixed despite changing conditions and acquisition of new information. Sometimes objectives are in conflict with objectives of other programs, particularly in the public sector, so that projects can be developed only after due consideration of related programs.

The next step is to generate alternatives for addressing the problems and needs that have been identified. Problems and needs should be considered in general terms, so that different kinds of alternatives can be considered. For example, many systems must deal with potential capacity problems related to growth in population. If so, then alternatives could consider not only expanding capacity to keep pace with population growth but also increasing prices in order to limit demand, or increasing efficiency of operations in order to allow more effective use of existing capacity.

The project identification phase concludes with a clear statement of needs, a set of objectives and specific assessment criteria, and an initial list of alternatives for achieving the objectives. Key results from this stage of project evaluation include clear statements of needs and objectives, the establishment of criteria, and the selection of alternatives for further study.

Phase II—Analysis of Alternatives

The process then enters the analysis phase, in which studies provide information that will help in assessing and comparing the various alternatives being evaluated. Various studies will be necessary to assess the viability of each alternative with respect to technical, financial, operational, social, economic, environmental, or other objectives. Considerable discussion and thought will be devoted to identifying performance measures and evaluation criteria for each major objective. Preliminary studies may give an early indication of the viability of an alternative, along with the risk involved. The most promising alternatives will be studied in greater depth. Analysis may include market demand studies, cost-benefit analysis, environmental impact assessment, and social assessment. Very detailed analyses involving multiple groups of people with backgrounds in engineering, economics, environmental science, or other disciplines may be required. Important planning decisions during this phase of project evaluation include the allocation of resources to the different types of studies and the extent to which the process allows refinement and modification of alternatives.

Phase III—Assessing and Comparing Alternatives

Assessing the result of the analysis is a separate stage from analysis, because many different kinds of results need to be considered. During this phase, it will be necessary to compare alternatives with respect to how well they satisfy the objectives that were previously established. Assessment will involve consideration of financial, economic, environmental, and social factors. To what extent does each alternative meet the needs that are being addressed? What are the costs and benefits of each alternative? Are costs and benefits measured properly? To what extent does each alternative lead to positive or negative externalities—that is, to broader impacts on the environment or the community or the region that would result from implementing a particular alternative?

Whereas analysis requires specialists and may include many independent studies, assessment requires generalists. For public projects and for large private projects that require public approval, there will have to be opportunities for input from potential users, abutters, and the general public. Users may push for a bigger and better system. Abutters, those who live next to the construction sites, may like the concept of the project, but oppose the proposed location. This type of opposition is so common that it is known by an acronym—**NIMBY**—that means "not in my backyard." The general public, to the extent that it is informed about the issues, is likely to be more receptive to a more balanced approach that recognizes the potential benefits of the project while acknowledging the importance of externalities.

The goal at this stage is not necessarily to define the exact, best option, but to determine the general approach that is best. The outcome from this stage could be one of three broad conclusions:

1. One alternative clearly is the best.
2. Further study is necessary to determine which alternative is best.
3. None of the alternatives is worth pursuing.

If one alternative is clearly the best, then it is possible to proceed to the next phase. If no alternative is clearly the best, then more detailed analyses may be needed that focus on what are believed to be the most promising alternatives. It may also be desirable to revise some of the alternatives, or to suggest new alternatives or different kinds of analysis. This phase of the evaluation process requires the consideration of multiple objectives as well as risk assessment in order to compare what could be markedly different alternatives. It also requires some mechanism for ensuring there are no other, better alternatives that should have been studied, as well as a mechanism for determining that the preferred alternative in fact is a cost-effective way of meeting the needs identified at the outset. Table 1.4 suggests some guidelines for this phase of project evaluation.

Phase IV—Implementation

Project identification, analysis, and assessment are iterative processes that may continue for years or decades without finding an alternative that is technically, financially, and politically feasible. Eventually, it may be possible to agree upon a particular alternative. The fine-tuning of a particular

Table 1.4   Guidelines for Assessing the Projects

1. Address the grand issues.
   • Economic viability—is there a clear case for supporting the project?
   • Engineering—what are the options regarding capacity, staging, and flexibility?
   • Financial feasibility—is there a way to cover investment and operating costs?
   • Environmental impacts—can the project be done with less negative impact on the environment? Can it result in improvements to the environment?
   • Political feasibility—who is likely to support or oppose the project? How can negative social impacts be mitigated?
   • Organizational structure—is the project best done as a public project, a private project, or a public-private partnership?
   • Size—would a larger or smaller project be better than what is proposed?
2. Consider comparable projects to get a quick, though rough estimate of the viability of the project.
3. Consider the possibility that the benefits are so great that there is more danger from doing too little than from doing too much.
4. Be prepared to think at all scales: local, regional, and national.
5. Think about aesthetics, and plan with an eye to style.

alternative may involve mitigation of environmental or social impacts, it may involve modifications aimed at reducing costs or increasing benefits (a process known as **value engineering**), and it may involve modifications to incorporate recommendations resulting from public input or the various studies that were conducted. At some point, detailed engineering design can be completed, and a construction management program can be initiated. A strategy for construction must be developed. How soon should construction begin? How quickly should construction proceed? What are the possibilities for implementing the project in stages? Once these questions have been answered, a project management team will be in charge of the actual construction process, and there will be innumerable decisions related to the best construction techniques, logistics, coordination of subcontractors, communications and cooperation with relevant public authorities, and maintaining the safety and security of the site. Before construction is complete, it will be necessary to begin the transition to operation. Eventually the construction phase ends, and the project is up and operating: the bridge is open, the tenants are in the building, the water is flowing, or the park is opened to the public.

Phase V—Ongoing Evaluation

Few projects are so well-planned and so carefully executed that everything goes perfectly on day one of the transition. Some period of time will be needed to identify and correct minor problems. After operations have settled down, it will be possible to compare the actual performance to what was intended. Was the project completed as planned? Was it completed on time and on budget? Most importantly, how effectively has the project addressed the original problems and needs? Answers to questions such as these will help in planning the next project and perhaps help in creating criteria for a program for constructing many similar projects.

   In summary, the process of defining a project can be viewed as a logical sequence of well-defined steps:

• Phase I: identification of needs, objectives, and alternatives.
• Phase II: analysis of alternatives with respect to technical, financial, economic, environmental, and social criteria.
• Phase III: assessing alternatives with respect to technical, financial, economic, environmental, and social objectives; selecting the best alternatives for refinement and further analysis.

- Phase IV: implementing the project (engineering design, construction management, and transition to operations).
- Phase V: monitoring performance, revising the project as needed, and identifying better approaches for evaluating similar projects and programs.

While it is useful to have a framework such as this for thinking about projects and project evaluation, it is important to recognize two fundamental aspects of the process of defining and selecting projects. *First, the process is iterative.* It may begin with identifying needs, with technological opportunities, or with an idea for a specific project. Once assessment begins, new ideas may emerge or people may find serious problems with all of the proposals, so it will be necessary to reconsider the needs and the opportunities. *Second, the process may not necessarily be logical or rational.* Suggestions for projects may come from those who want to build them or from those who want to operate them—whether or not the projects they propose are the best projects or the projects that respond to the most pressing needs of society. Companies that build roads and bridges want to build more roads and bridges, just as highway authorities may respond to all transportation problems by recommending construction of more highways. New technologies quickly lead to ideas for new projects, but it may be years or decades or longer before those projects can be justified. With many new technologies, the new capabilities create new needs, or at least perceived needs (e.g., continuous, instantaneous connections to the Internet; high-definition TV). With advertising, suppliers can create needs that drive construction of new plants and distribution facilities (bottled water is a good example, especially when the water is obtained directly from a region's public water supply). It is a mistake to expect the process to be completely rational. On the other hand, it is also a mistake not to try to impose a rational process on defining needs, identifying alternatives, and assessing, selecting, and modifying alternatives.

1.6 STRUCTURE OF THE TEXTBOOK

The first major goal of the text is to provide a *framework for understanding and evaluating projects*, taking into account not only financial and economic issues but also social and environmental factors. Examples and case studies illustrate the complexity of major projects and demonstrate the role for and the limits of analysis in clarifying and resolving issues.

The second major goal of this text is to show how to apply the *basic methods of engineering economics* in evaluating major infrastructure projects. Examples and exercises indicate how to develop and apply models for estimating the costs of resources required for such projects and how to estimate their life-cycle costs.

The third major goal of the text is to promote an approach to project evaluation that is based on an appreciation of *the needs of society, the potential for sustainable development, and the recognition of problems that may result from poorly conceived or poorly implemented projects and programs.*

Key concepts that will be covered include the following:

- *Justification of large investments*: how to determine whether future benefits justify current costs.
- *Technology-based performance functions*: creating functions with sufficient detail to explore how cost, service, capacity, and safety vary with major project options related to size, design, and technology.
- *Cost effectiveness*: how to compare options for achieving nonmonetary benefits.
- Sustainability: environmental, financial, economic, and social aspects of sustainability.
- *Evolution of systems*: understanding how systems evolve in response to changes in needs, technologies, and financial capabilities.

Analytical methodologies can be applied to each of these concepts. However, it is critical to recognize that analysis does not necessarily determine what projects are considered, what projects are proposed, which of these projects are approved, or which projects are ultimately successful.

Projects may be motivated by a vision of a greater society, by an idea for addressing a specific local problem, by the prospects of making a profit while providing a needed service, or by simple greed. Some apparently excellent projects cannot be financed, but it may be easy to fund some very questionable projects. Lackluster projects may prevent outstanding projects, and highly acclaimed projects may prevent dozens of less showy, but more effective projects. Financially successful projects may be terrible in terms of their consequences for the environment, and projects sold as being good for the environment may turn out to be overly expensive or socially unacceptable.

Project evaluation is not a hard science, since there are so many factors to consider, so many unknowns, and so many different perspectives concerning what is good or bad. Nevertheless, there is a role for analysis, if only to help people recognize and agree upon the likely magnitude of the most important costs and benefits. Experience, a coherent framework for analysis, and a concern for sustainability will provide a sound basis for evaluating projects, whether you are the developer, the consultant, the banker, the neighbor, the user, or the politician.

The text contains three parts.

Part I: Building Infrastructure to Serve the Needs of Society
Part II: Comparing Economic and Financial Impacts over the Life of Proposed Projects
Part III: Developing Projects and Programs to Deal with Problems and Opportunities

Part I provides an overview of project evaluation as a multidimensional process aimed at creating projects that meet the needs of society. This part emphasizes the need to consider economic, environmental, and social factors along with the technological and financial matters that are crucial to the success of a project. Part II provides in-depth coverage of the engineering economic methodologies that can be used to compare cash flows or economic costs and benefits over the life of a project. This part presents the techniques that are used by investors, bankers, and entrepreneurs in deciding whether or not to finance projects. It also shows how public policy can use taxes and other regulations to encourage projects that have public benefits. Part III presents methodologies that are useful in developing and evaluating projects to deal with problems and opportunities. This part includes an introduction to project and program management, and it concludes with a chapter that considers the evolution of infrastructure-based systems and the need for more sustainable infrastructure in the coming decades.

DISCUSSION AND ESSAY QUESTIONS

1.1 What are the key characteristics of infrastructure projects? What are the challenges inherent in evaluating infrastructure projects?

1.2 What are the major phases associated with an infrastructure project? What are the key issues addressed in each phase?

1.3 What is the difference between a project and a program? Why would a public agency or a private company consider undertaking an infrastructure program?

1.4 Everyone is familiar with many, many projects, ranging from the Seven Wonders of the Ancient World to the highway or power plant or mall that was constructed in their neighborhood. Identify one project that you really like, and explain why you like it. Identify another project that you really dislike, and explain why you dislike it.

1.5 Identify a project that you believe would be very beneficial for your school, your city, or your state. Why do you think this project would be useful? How would you describe this project if you had a chance to make a presentation to school or city officials?

1.6 Identify a proposed project that is currently generating controversy in your city or region. Who favors the project and who is against it? What are the major issues? How well do you think the issues are presented in the news accounts you have read?

1.7 Global warming and overdependence on fossil fuels are strategic concerns facing the world today. Both problems are exacerbated by our love affair with the automobile, yet we continue to buy automobiles; and our governments in 2008 acted swiftly to provide loans to auto companies that faced bankruptcy because of the international credit crisis. What kinds of projects might you want to consider or recommend that would reduce the oil consumed in transportation and the carbon dioxide (CO_2) emissions produced from transportation?

PROBLEMS

1.1 Restate the following statements of problems so as to encourage a better selection of alternatives:

a. The city's population is expected to increase by 25% by 2020. The capacity of the reservoir needs to be increased to provide sufficient supplies of drinking water to the city.

b. The city should expand highway capacity so as to reduce congestion by at least 10% within the next 10 years.

c. All bridges that are more than 50 years old should be replaced by the year 2525.

1.2 What sounds like a great idea may, upon reflection, turn out to be much less. For example, consider the following hypothetical but entirely realistic situation: a coalition of highway users has advocated investment in railroads in order to move more freight by rail and thereby remove trucks from the highways and reduce highway congestion. The coalition has lobbied the state legislature, arguing that the proposed investment will remove "thousands of trucks per year" from the highways. Public reception to the program was initially good, but an intern working for a state representative pointed out that removing thousands of trucks per year from the highways might amount to removing only a few trucks per day from a single interstate—and the politicians concluded that the proposal was being oversold. Indicate why you believe each of the following proposals would or would not be worth pursuing:

a. Installing new plumbing in a school is expected to save hundreds of gallons of water per year.

b. An airport authority proposes building a new runway to increase capacity and save an average of three minutes per flight.

c. A proposed 30-mile link in the rural interstate system is expected to reduce travel time by two to three minutes for people driving between the state capital and the rural regions in the southwest portion of the state.

d. Following an earthquake that caused widespread devastation in a poor country, government officials propose changing building codes to require new construction to be built that would be likely to survive similar earthquakes in the future.

1.3 For one of the major classes of infrastructure (see Table 1.3), identify at least three projects that were recently completed. Compare and contrast the goals, engineering approaches, and expected costs and benefits of these projects. Which of these projects do you expect to be most effective? Why?

1.4 Identify a major infrastructure proposal that was recently delayed, rejected, or dramatically modified because of concerns about social or environmental impacts. Describe the proposed project, the major objections to the project, and the actions taken in response to the social or environmental concerns. To what extent do you believe that the objections were justified? To what extent do you agree with the actions that were taken?

Part I

Building Infrastructure to Serve the Needs of Society

Part I of the text is an overview of many of the processes needed to assess the performance of infrastructure projects and to evaluate alternatives for improving performance. Inevitably, there will be many aspects of performance to consider and many possible impacts on society or the environment that must be minimized or mitigated. Part II shows how engineering economics provides the methodologies needed to compare projects in terms of financial and economic impacts, including the effects of taxes and depreciation. Part III provides greater detail concerning the various topics relevant to projects and programs aimed at improving infrastructure performance, including dealing with uncertainty, public-private partnerships, program management, and the evolution of infrastructure systems.

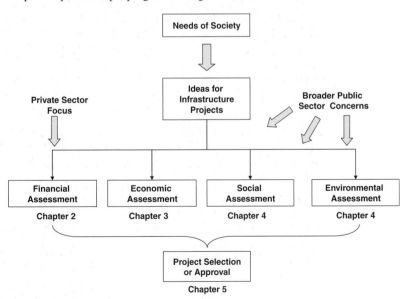

The first three chapters of Part I deal with the different types of impacts that must be considered in designing and evaluating infrastructure. Chapter 2 considers the measures of performance that are most relevant to the owners and users of a system, namely cost, capacity, service, and safety. These measures all ultimately relate to the attractiveness of the system to potential users, the willingness of users to pay for the use of the system, and the ability of the owners to finance the system. This chapter therefore addresses the perspective of the private sector. The economic impact of a project extends beyond the costs and revenues directly associated with it. Benefits to users exceed what they pay to use infrastructure, and there are benefits to society related to the money that is spent to construct and maintain a project. Chapter 3 introduces the various economic concepts relevant to determining the need for projects and estimating the impacts of a project on the economy. Chapters 2 and 3 both consider impacts that can be measured in monetary terms. While such impacts are important, they are just part of what must be considered, especially for large infrastructure projects. Chapter 4 introduces the social and environmental impacts that affect public projects and also public decisions concerning the approvals necessary for private projects. With so many different types of impacts, it is often impossible to condense evaluation of a major to a single metric. Multiple measures are needed, and different options are likely to be preferred depending upon which measures are deemed most important. As presented in Chapter 5, a reasonable process is needed to select the best alternative, and there could be considerable debate as to how to structure that process.

Part I concludes with a case study of the Panama Canal, which actually covers a series of transportation projects that include roads and railroads, a disastrous and unsuccessful attempt to build a sea-level canal, and the ultimately successful construction of the existing canal. Some of these projects were sponsored by ambitious entrepreneurs while others were undertaken by governments. The history of the canal illustrates many of the concepts introduced in Part I.

Chapter 2

System Performance

As it is always easier and in the end less costly to be accurate than inaccurate, the good engineer will always be accurate in all essentials, but he will not waste time in attempting unnecessary precision which does not add appreciably to the final value of his work.

Arthur M. Wellington, *The Economic Location of Railways*
(John Wiley & Sons, 1911)

CHAPTER CONCEPTS

Section 2.1 Introduction to Performance of Infrastructure-Based Systems
Performance cannot be captured by a single measure
Performance depends upon social, political, and business issues as well as engineering issues
Performance can be documented, modeled, and managed

Section 2.2 System Cost
Total, average, marginal, and incremental costs
Fixed and variable costs
Break-even volume for preferring an option with higher fixed costs
Resource utilization: economic versus financial costs
Life-cycle costs
Engineering-based cost functions

Section 2.3 Profitability, Break-Even Volume, and Return on Investment
Demand curves
Break-even demand for achieving a profit
Return on investment as an incentive for further investment

Section 2.4 Service
Service characteristics of transportation and water resource systems
Engineering-based service functions

Section 2.5 Capacity
Difficulties in measuring infrastructure capacity
Deterioration in service as demand approaches capacity
Maximum, operating, and sustainable capacity

Section 2.6 Safety, Security, and Risk
Differences between safety and security
Probabilistic risk assessment
Cost effectiveness of strategies for improving safety

2.1   INTRODUCTION TO PERFORMANCE OF INFRASTRUCTURE-BASED SYSTEMS

Infrastructure supports the needs and activities of society. Transportation systems provide mobility for people within and between regions, and they support the distribution of food, lumber, clothing, and everything else that people and society need and use. Water resource systems provide water for consumption in household, for agriculture, and for manufacturing while reducing the risks of flooding and generating electricity. Vast infrastructure systems support the generation and distribution of electricity, cable TV, and the Internet. Cities are composed of interrelated systems of infrastructure for housing, fire protection, offices, retail outlets, manufacturing, recreation, religion, heath care, and the many other aspects of social life.

　　The performance of any of these infrastructure systems cannot be captured by one or two measures. Cost, quality of service, capacity, safety, environmental impacts, and sustainability are all important, while the extent of coverage, accessibility, equity, and appearance can also be critical. Moreover, performance depends upon one's perspective, as owners and managers seek financial rewards, users seek good service at reasonable prices, and the public worries about such things as the need for subsidies, environmental impacts, safety, land use, economic development, and aesthetics (Figure 2.1). Managing these systems thus involves trade-offs among multiple factors, often with

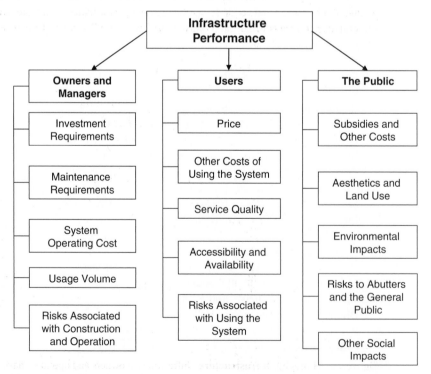

Figure 2.1 Infrastructure Performance. Owners, managers, users, and the public will have differing priorities for measuring and monitoring infrastructure performance.

no obvious way of determining which factors are most important. Evaluating proposals for creating new systems or for expanding or modifying existing systems will always require judgment and often requires some sort of political process to determine which aspects of performance should be emphasized.

System performance can be documented, studied, modeled, predicted, and managed. If system performance is well-understood, and if there is a consensus about the relative importance of the various aspects of performance, then it is possible to provide a clear, objective basis for evaluating new projects and programs. If performance is poorly understood, then research and analysis may be able to clarify the potential trade-offs for various options. If there is no consensus as to which aspects of performance are most important, then research and analysis of performance can at least provide an objective framework for evaluating proposals.

Performance of infrastructure systems depends upon engineering and managerial issues, such as the nature, condition, and deterioration rates of structures and equipment. Infrastructure managers develop plans and policies that they use to guide operations and maintenance, usually with the hope of increasing profitability or other financial goals. The management structure often has separate departments responsible for operations, maintenance, and marketing, and each department has its own concerns and ideas about what types of projects are most beneficial in improving performance. Each department's objectives ideally reflect strategic plans, which may include financial goals, goals for expanding or shrinking the system, plans for new services or markets, or goals related to service quality, risk management, or interactions with the public (Figure 2.2).

Performance also depends upon many other factors, including the regulatory structure, operating capabilities, and business strategies pursued by the organizations and individuals that make use of the infrastructure and the existence and performance of competing infrastructure systems.

Most importantly, perhaps, performance reflects the demand for the service, especially when usage approaches system capacity. A highway that allows motorists to drive through the city at high speeds in the middle of the night can be more like a parking lot during rush hour. Predicting the average speed on a highway during rush hour is complex, for it requires some way to estimate demand while taking into account the fact that the level of demand will affect performance. It is more straightforward to develop a **performance function** for the highway—that is, to predict the performance of the highway for any particular level of demand.

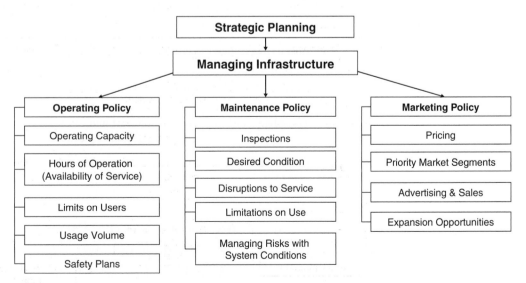

Figure 2.2 Managing Infrastructure. Infrastructure owners and operators have separate departments for operations, maintenance and construction, and marketing. Departmental activities ideally are coordinated to reflect the strategic goals of the company or agency.

Performance functions can be developed for any infrastructure system based upon the engineering characteristics of the infrastructure and the ways it is used. Developing **engineering-based performance functions** enables planners and entrepreneurs to understand the options for developing or improving systems, to evaluate the impact of proposed projects, to understand the potential benefits of new technologies, and to select and design better projects. A major purpose of this chapter is to demonstrate how engineering-based performance functions can be designed to provide a mathematical basis that is helpful for investigating trade-offs. Performance functions may or may not be precise, depending upon the state of knowledge of the system and also upon the context for their use. The key point is that it is often possible to construct a set of equations that predict system performance under specified conditions based upon scientific and engineering knowledge, regulatory constraints, unit costs, pricing strategy, and/or operating characteristics.

The remaining sections of this chapter introduce basic concepts related to infrastructure performance, with a focus on issues most relevant to owners, managers, and users; namely, cost (Section 2.2), profitability (Section 2.3), service (Section 2.4), capacity (Section 2.5), safety and security (Section 2.6), and cost effectiveness (Section 2.7). Each section presents basic microeconomics and systems concepts along with examples of how these concepts might be used in measuring or analyzing infrastructure performance. Broader social, economic, and environmental aspects of performance are covered in Chapters 3 and 4.

2.2 SYSTEM COST

Owners and investors are concerned with the initial cost to construct infrastructure as well as the continuing cost for operating and maintaining the infrastructure. Users are concerned with their costs of using the system, which include the prices they are charged plus their own time and expense associated with using the system. Basic cost concepts, which are useful whether considering costs from the owner's or the user's perspective, are introduced in the next three subsections. Section 2.2.4 illustrates how to develop an engineering-based cost function. Additional costing methodology is presented in Part III. Section 14.3 presents various techniques that can be used to estimate project costs. Section 15.6 presents a framework for incorporating risk into the determination of the lives and costs of infrastructure components.

2.2.1 Total, Average, Marginal, and Incremental Costs

Cost is a fundamental consideration for any system, and there are several important cost questions to consider:

- Total cost: what is the total cost of the system for a given level of output?
- Average cost: what is the average cost per unit of output?
- Marginal cost: what is the cost for one additional unit of output?
- Incremental cost: what is the cost for an increment of output?

These concepts can be illustrated using an extremely simple equation expressing cost as a linear function of volume (V):

$$\text{Cost} = a + bV \tag{Eq. 2.1}$$

$$\text{Average cost} = (a + bV)/V = a/V + b \tag{Eq. 2.2}$$

$$\text{Marginal cost} = (a + b(V + 1)) - (a + bV) = b \tag{Eq. 2.3}$$

The fixed cost "a" is incurred whatever the volume, and "b" is the marginal cost per additional unit. The average cost per unit declines as the fixed cost is spread over additional volume. In any particular situation, these cost functions are more complex. Total cost will be a function of the resources used and their unit costs, and the resources used will depend upon the quality of service as well as the level of demand.

$$Cost = C(resources,\ unit\ costs,\ quality\ of\ service,\ volume) \qquad (Eq.\ 2.4)$$

If, for simplicity, we use C(V) to denote the total cost for this system, then:

$$Average\ cost = C(V)/V \qquad (Eq.\ 2.5)$$

$$Marginal\ cost = C(V+1) - C(V) \qquad (Eq.\ 2.6)$$

If C(V) is known and differentiable, then the marginal cost will be dC/dV. In practice, the problem in getting a derivative will be that nobody really knows the cost function, at least not in enough detail for valid engineering analysis. Instead, costs can be based upon analysis of existing systems or simulation of proposed systems, either of which produces a series of cost estimates for specific conditions or assumptions. The relevant cost concept that is used in these situations would be the incremental cost for increasing volume from V_0 to V_1 rather than the marginal cost of increasing output by 1 unit:

$$Incremental\ cost = C(V_1) - C(V_0) \qquad (Eq.\ 2.7)$$

In many analyses, the question of interest is actually the cost of a project that provides a rather large increment to capacity. For example, adding a lane to a road will add capacity of approximately 1000 cars per hour, while constructing a hotel may add hundreds of rooms to a city's ability to absorb tourists, conventions, and other visitors. The marginal cost can be estimated as the incremental cost per unit:

$$Incremental\ cost/unit = (C(V_1) - C(V_2))/(V_1 - V_2) \qquad (Eq.\ 2.8)$$

It is often possible to represent a complex cost function by a simple function of the type shown in Eq. 2.1 by using the concepts of fixed and variable cost:

- Fixed costs: costs that are likely to remain unchanged over the range of volumes under consideration
- Variable costs: costs that are expected to change in proportion to changes in volume

The fixed cost (a) could well be the sum of many different types of costs that would not vary with changes in volume, such as the salaries of the senior administrative staff, the costs of establishing a maintenance facility, and the costs of acquiring basic equipment and machinery. Variable costs would likely include the labor and energy costs associated with operations. Which costs are fixed and which costs are variable depends upon the level of volume and the time frame under consideration. In the short run, many more costs are fixed; in the long run, most costs can be affected by restructuring the system. Whatever the time frame, the simple cost function shown above as Eq. 2.1 may (for a reasonable and meaningful range of volumes) actually be a useful approximation of a much more complex cost function. Hence, it is worth taking a closer look at this cost function (Example 2.1).

EXAMPLE 2.1   Graphing a Linear Cost Function

Graph the total, fixed, variable, average, and marginal cost for the following cost function, which is believed to be approximately valid for volumes (V) ranging from 10 to 100 units per day:

$$Total\ Cost = \$50 + \$1\ (V)$$

The fixed cost is $50, which plots as a straight horizontal line in Figure 2.3, while the variable cost plots as a straight line with a slope of 1. If this line were extended beyond the range of interest, it would intersect at the origin of the graph.

Figure 2.4 shows the average and variable cost for this cost function. The marginal cost is equal to $1 over the entire range, while the average cost declines from $6 when volume is 10 units per day to $1.50 when demand is 100 units per day. The average cost function is nonlinear, and the average cost approaches the marginal cost for high volumes.

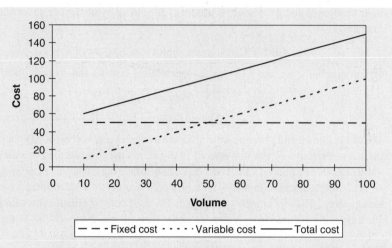

Figure 2.3 Fixed, Variable, and Total Cost for a Linear Cost Function of C = 50 + V

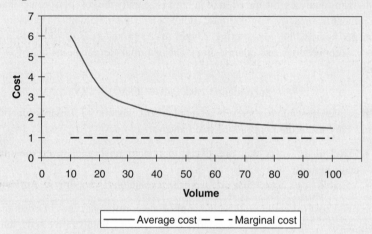

Figure 2.4 Average Cost and Marginal Cost for the Linear Cost Function C = 50 + V

It is common to find a trade-off between two options, one of which has higher fixed cost but lower marginal cost. Figure 2.5 shows a second cost curve in which the fixed cost has increased from $50 to $95 per day, but the variable cost has dropped from $1 to $0.50. The break-even point is at 90 units per day: above this level, option 2 is preferred; below this level, option 1 is preferred.

For linear functions of the type $TC = a + bV$, the break-even point (V_b) can readily be calculated:

$$\text{If} \quad TC_1 = a_1 + b_1 V$$

$$\text{and} \quad TC_2 = a_2 + b_2 V$$

then, at the point where the costs are equal for the two technologies, the following equation will hold:

$$a_1 + b_1 V_b = a_2 + b_2 V_b$$

Solving for V_b yields:

$$V_b = (a_2 - a_1)/(b_1 - b_2) \quad \text{(Eq. 2.9)}$$

In words, the break-even volume is calculated as the increase in fixed cost divided by the savings in variable cost per unit. Major civil and environmental engineering (CEE) projects require investments that are aimed at reducing marginal cost. Larger projects typically provide an opportunity for greater

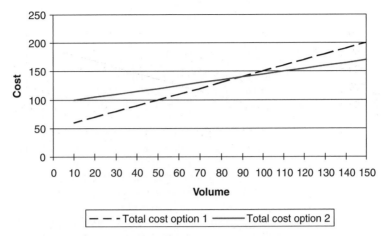

Figure 2.5 The Option with the Higher Fixed Cost Is Preferred if the Volume Is Greater than 90 (Cost of Option 1 = 50 + V; Cost of Option 2 = 95 + V/2)

reductions in marginal cost. A key question is whether there is enough demand to cover the extra costs associated with the larger project. In general, smaller projects are better until congestion and capacity concerns make a larger project desirable.

L. F. Loree, past president of the Delaware & Hudson Railway and Chair of the Kansas City Southern, described how railroads worked hard to avoid investing unnecessarily in capacity:

We hang back and postpone as long as possible work to increase facilities the use of which may be increased by ingenuity and method.
Source: L. F. Loree, *Railroad Freight Transportation* (New York: D. Appleton & Co., 1922), 61.

EXAMPLE 2.2   Break-Even Volume for Using an Alternative Technology

Suppose the company referred to in the previous example had an alternative process with a higher fixed cost equal to $600,000, but a variable cost that would be $5,000 lower. What would the break-even volume be for using this alternative technology?

Solution

The break-even volume could be determined by plotting the two curves and identifying where they cross, as shown in Figure 2.5. It is easier and quicker to use Eq. 2.9:

$$V_b = (a_2 - a_1)/(b_1 - b_2)$$
$$V_b = (\$600,000 - \$500,000)/(\$20,000/\text{unit} - \$15,000/\text{unit})$$
$$V_b = \$100,000/\$5,000 \text{ per unit} = 20 \text{ units}$$

Nonlinear Cost Functions

Linear cost functions are easy to draw, but more complex functions often are necessary. The same logic applies to plotting total costs, average cost, or marginal costs, and the same approach can be used to determine break-even volumes. Example 2.3 considers a situation where three technological options are available, and each of them can be represented by a nonlinear cost function.

EXAMPLE 2.3 Nonlinear Cost Functions

Plot the average costs for three options with the following cost functions:

$$\text{Total Cost Option 1} = 50 + V + 0.03V^2$$
$$\text{Total Cost Option 2} = 100 + 0.5V + 0.02V^2$$
$$\text{Total Cost Option 3} = 300 + 0.25V + 0.01V^2$$

The third term in these equations means that the average costs eventually begin to increase, as the contribution of the squared term eventually offsets the savings from spreading the fixed costs over a larger volume. Figure 2.6 shows the plots of the average costs of the three technologies for the ranges of volumes for which they are feasible options. Option 1, the lowest-cost option, is shown to have costs that drop from more than $6 for the minimum volume of 10 units to a low of less than $4 for volumes of 30 to 50 units, then rise steadily as volume increases to 100 units, which is the maximum that can be handled by this technology. Option 3 is much too expensive for or ill suited to low volumes, so the costs for that technology are shown only for volumes of at least 40 units. Figure 2.6 indicates that Option 1 is favored for volumes less than about 50, Option 2 is cheapest for volumes between 50 and 130, and Option 3 is favored for greater volumes.

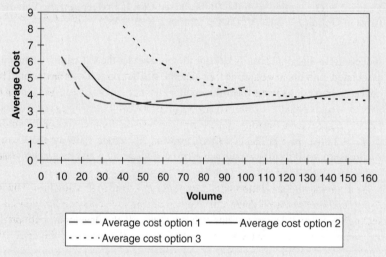

Figure 2.6 Long-Run Average Costs

The marginal costs for these three cost options can be obtained by differentiating the cost functions with respect to V. If

$$C = a + bV + c\,V^2$$

Then $dC/dV = b + 2cV$, and the marginal costs for the three options are

$$\text{Marginal Cost Option 1} = 1 + 0.06V$$
$$\text{Marginal Cost Option 2} = 0.5 + 0.04V$$
$$\text{Marginal Cost Option 3} = 0.25 + 0.02V$$

Short-Run versus Long-Run Average Costs

The three cost curves shown above in Figure 2.6 each depict **short-run average costs** for a particular technology. Each curve shows the average costs that would result from using a particular technology for a specified range of volumes. They are called short-run average costs because they do not allow a shift to a more efficient technology.

It is often useful to understand how average cost of a particular type of process or production would change with volume, assuming that the best technology is used for each level of volume. This cost is called the **long-run average cost**, assuming that in the long run it will be possible to adjust the

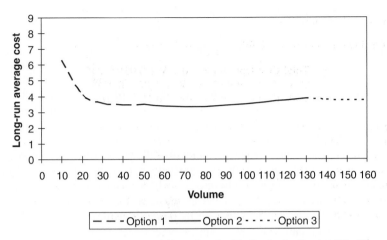

Figure 2.7 Long-Run Average Costs for the Technologies Depicted in Figure 2.4

production process to what is best for the actual volume served or produced. The long-run average cost curve can be constructed from the applicable short-run cost curves: for each volume, the long-run average cost is equal to the short-run cost of the technological option with the lowest short-run average cost. In other words, the long-run average cost will be the envelope of the minimum short-run costs, as depicted in Figure 2.7. This is a single cost function based upon the three cost functions shown in the previous figure. The long-run average cost in Figure 2.7 could be described as a little less than $4 for volumes ranging from 20 to 160.

You (the developer, entrepreneur, or planner) know, or should know, your costs and technologies. You therefore should be able to develop an algebraic expression for your costs that accounts for the technological and design options you may have. It the developer, entrepreneur, or planner doesn't know his or her costs or technologies, then you (the consultant, the researcher, or the smart young analyst) may be able to do some analysis and create relevant cost functions.

2.2.2 Resource Requirements

Costs ultimately are linked to resources: people, land, materials, energy, and capital. There will always be many different ways to use resources in constructing a particular project, and choices will probably be made by developers, engineers, and planners seeking to minimize the cost of the project. From their perspective, minimizing cost is purely a financial exercise aimed at minimizing the cash required to complete the project (this topic is discussed in detail as an important part of project management in Section 14.3). At this point, it is sufficient to recognize that resource requirements can be an important part of project evaluation, because the availability and prices of resources will differ geographically and could well change over the long life of an infrastructure project.

The design of a project and the resources required for projects are influenced by both the current and the expected future availability and prices of people, land, materials, energy sources, and capital. If average salaries and wages are expected to rise relative to the cost of capital, then the tendency is to use fewer people and more machines. If the cost of energy is expected to rise relative to the cost of building materials, then there is an incentive to construct buildings that require less energy and to use materials that are less energy intensive. If longer-lasting materials are created, they will be used where their longevity provides benefits to a project. Standard methodologies of engineering economics, which are presented in Part II, make it possible to decide what system design is best for a project, which materials to use, and which construction techniques to follow, taking into account the initial investment cost and costs to the owners and users over the life of the project.

> *There is more than one way to skin a cat.*
>
> American Proverb

From an economic perspective, however, the price of a resource may not reflect the true cost of using that resource. There are many reasons why the economic cost might differ from the financial cost and why the economic cost should be considered in evaluating projects:

- The cost of using a resource is an **opportunity cost**; the use of resources on one project means that they are not being used on other projects or are being reserved for future use.
- **Negative environmental impacts** may be associated with using a resource, such as the paving over of wetland, emission of greenhouse gases in the manufacture of construction materials, or the devastation of remote wild areas by mining activities.
- **Nonsustainable use** of renewable resources, such as groundwater or timber, will lead to declines in the amount of such resources that can be used in the future.
- Extensive use of nonrenewable resources can eventually lead to **resource depletion**.
- Exposure of workers to unsafe conditions or hazardous environments may lead to **risks of injury or serious illness**; unscrupulous companies may be able to avoid covering the costs associated with these problems, and governments may or may not regulate workplace conditions.
- **Government subsidies** may be available for certain types of workers, or government regulations may require the use of excessive numbers of workers.

Some of these costs may be reflected in the market prices for certain resources, such as the cost of capital and salaries and wages in developed countries. Other costs are notably absent from the prices charged, such as the effects of emissions on global warming or the environmental damages related to strip mining. If a company or a country already owns land that could be used for a project, it may not even consider the cost of using that land in the analysis, yet developing that land may forestall even better opportunities in the future.

If the prices for resources actually approximate the costs associated with using those resources, then financial analysis will perhaps produce a reasonable result. If the prices for some resources are markedly above or below their full economic costs, then the financial analysis could result in poor decisions from the perspective of society, even when those decisions do appear to increase the profitability of the project. Because of the discrepancy between financial and economic costs, public policy must enforce rules and regulations that promote or enforce consideration of the true economic costs of a project. Such regulations include:

- Restrictions on development of wetlands and other sensitive environments
- Regulations related to occupational safety and health
- Minimum wages and laws allowing the formation of unions
- Licensing of engineers and others involved in project design and implementation
- Regulations concerning the technologies and methods used in mining and other extractive activities
- Regulations requiring assessment of environmental and social impacts

For most privately sponsored projects in developed countries, project evaluation is based strictly upon a financial analysis. If government regulations are effective, then the private decisions will be reasonably good. If government regulations fail to capture significant aspects of economic costs—as has been the case with many environmental impacts—then private decisions could lead to increasing problems for society.

When evaluating projects, or when estimating costs, it will always be worthwhile to (1) detail the actual resources that will be used in addition to how much money is needed, and (2) document the methodologies and logic used in reaching decisions about design, choice of materials, and choice of construction methods. As prices of resources change relative to each other, as new technologies become available, and as more economic aspects of costs are recognized, different types of projects and different approaches to constructing those projects will be needed. However, the original methodologies and logic may still be suitable for evaluating future projects.

EXAMPLE 2.4   How Resource Costs Can Affect Operating Practice and Performance[1]

A team of North American railroad experts visited a rail terminal in China and compared labor productivity with similar terminals in the United States and Canada. They found that productivity in the Chinese terminal was less than half what it was in North America. They identified what they felt to be glaring examples of "wasteful use of labor": (1) trains were short, so that two to three times as many engineers and trainmen were needed to move a given amount of freight, and (2) a team of ten people was used to inspect a train rather than just a pair of inspectors. On the other hand, the Chinese terminal was vastly superior in terms of the time that freight cars spent in the yard. Rather than averaging more than 24 hours, as in North America, the Chinese were able to move cars through the terminal in an average of less than 8 hours. How was a railroad that was so wasteful in its use of people able to achieve such a high level of performance in moving freight through the terminal?

Solution

The differences between the Chinese and the North American terminals could be explained by differences in relative resource costs. Labor costs for North American railroads were many times higher than labor costs in China, but the costs of freight cars were the same. Moreover, the North American railroads and their customers were able to acquire freight cars by borrowing money at a much lower rate of interest than was possible for the Chinese railways. Thus it made sense for the Chinese to use a lot of people while maximizing equipment utilization, just as it made sense for the North American railroads to use as few people as possible while moving freight more slowly through terminals. In both rail systems, management was looking for ways to reduce total costs while providing a level of service that was acceptable to their customers.

2.2.3   Life-Cycle Costs

Since infrastructure lasts a long time, it is important to consider costs over the entire life of the infrastructure. Figure 2.8 shows the costs that might be incurred over the life of a project. At the beginning are the design costs, which are likely to be modest compared to the construction costs that

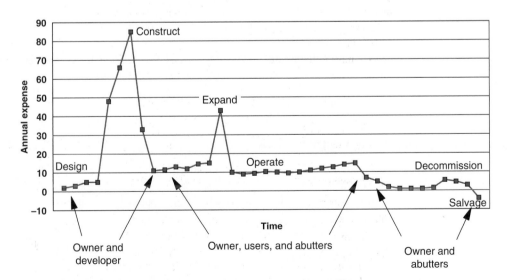

Figure 2.8 Life-Cycle Costs for Infrastructure

---

[1] This example is based upon a discussion with U.S. railroads experts who had worked in China in the mid-1990s.

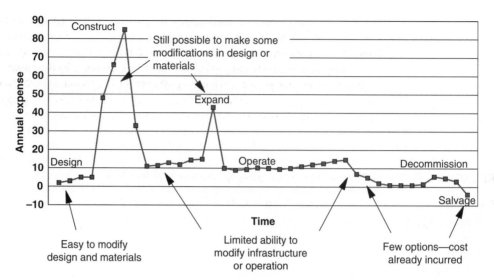

Figure 2.9 Opportunities for Reducing Cost Are Greatest at the Outset

follow. Once the project is completed, the costs of operation may be borne by the owner, users, or abutters. As the facility ages, it may be desirable to expand or rehabilitate the facilities. Eventually, because of declining demand or excessive costs of maintenance, it will be necessary to decommission the facilities and to salvage metal or whatever is left of value. Notice that the original owner and developer incur the initial costs, while users and abutters and subsequent owners are left with whatever the long-term costs turn out to be.

The design phase offers the greatest opportunity to affect the life-cycle costs of a project (Figure 2.9). At this point, while all things may not be possible, many things will be. A major design consideration is the extent to which the owner or developer considers the costs to users and abutters. Small changes in design conceivably produce more efficient operations or limit the negative effects on third parties, but only to the extent that such costs are even considered in the design. Mistakes regarding the size of the project—too big, too small, too soon, too little flexibility, too difficult to rehabilitate, too much impact on neighbors—conceivably can be rectified with little or no additional construction time or expense.

The opportunities for savings are clearest when the owner is responsible for operations for the indefinite future. The opportunities for misguided design (or fraud) are greatest when the developer is interested only in minimizing the construction cost, since operations and maintenance will be the responsibility of others.

2.2.4   Engineering-Based Cost Functions

There are various approaches to estimating cost functions: accounting, econometric, and engineering based. Each approach has its usefulness, but for project evaluation, an engineering-based approach is essential. This section describes each of these approaches and provides examples illustrating how the approaches differ.

Accounting Approach

For an existing system, whether an apartment building, a transportation terminal, or a wind farm, it should be possible to identify all of the expenses related to the system's construction and operation. If the accounting system is accurate and complete, then the costs associated with each phase of the design, construction, and operation of the project should be evident. Complications will arise if there are many different owners and users, or if it is difficult to allocate specific costs to specific purposes or users. Special studies may be required to support a realistic allocation of costs or to identify user costs. However, the companies or agencies that manage infrastructure are likely to have good information concerning their own costs and good estimates of the costs borne by their users. Using accounting costs for project evaluation presents two main problems. First, accounting costs do not readily relate to the costs of new projects, unless the new projects are very similar to existing projects. Second, accounting systems can provide rich detail concerning existing operations, but they do not show how costs vary with demand, quality of service, capacity, or technology.

EXAMPLE 2.5   Cost of Upgrading a Rail Line—the Accounting Approach

A railroad plans to upgrade 30 miles of its track, which will entail several activities:

- Dismantling and taking away the old track
- Placing new ties on the roadbed
- Attaching new rail to the ties
- Adding sufficient ballast to ensure a safe, smooth ride

Costs could be estimated based upon experience in a recent project that resulted in upgrading a similar segment that was 20 miles long, as summarized in Table 2.1. If the new project also costs $290,000 per mile, then we can estimate that the new 30-mile project will cost $8.7 million.

Table 2.1   Costs of Upgrading 20 Miles of Railway Line

Activity	Total Cost	Cost/Mile
Dismantle and take away old rail and ties	$600,000	$30,000
Install new ties on the roadbed	$1,200,000	$60,000
Purchase new rail	$2,000,000	$100,000
Install new rail	$1,000,000	$50,000
Add ballast	$1,000,000	$50,000
Total cost of upgrading the line	**$5,800,000**	**$290,000**

Econometric Approach

If cost data are available for a variety of completed projects, it may be possible to discern trends in cost as a function of the type or size or location of the project. If so, then it will be possible to predict cost of a new project as a function of its design, size, and location.

EXAMPLE 2.6   Cost of Upgrading a Rail Line—the Econometric Approach

Continuing the railroad example, perhaps the company has access to the costs of dozens of projects that were carried out over the past several years. This scenario is summarized in Table 2.2, which shows the length and cost per mile for ten projects carried out over the past 10 years. The cost per mile clearly seems to be increasing over time, so it might be prudent to assume that costs would continue to increase in the future.

Figure 2.10 plots the actual cost per mile and an equation that estimates cost per mile as a linear function of time, T:

$$Cost/mile = a + bT$$

where

a = $209,000 cost/mile in 1999
b = $9,200 per year after 1999

Table 2.2   Costs of Upgrading Railway Lines over the Past 10 Years

Project	Year	Cost/Mile
1	1999	$200
2	2000	$233
3	2001	$225
4	2002	$237
5	2003	$240
6	2004	$250
7	2005	$260
8	2006	$300
9	2007	$270
10	2008	$290

The coefficients "a" and "b" in this model were estimated using regression analysis, a statistical technique designed to provide the best fit between a mathematical model and observed data. The difference between the model's estimate of cost per mile and the observed cost per mile is called the error of the estimate. Regression analysis chooses "a" and "b" so as to minimize the sum of the squares of the errors for all of the observations. The values were obtained from an Excel spreadsheet using the functions for Slope and Intercept. Both of these functions require the user to specify a range of values for the independent and the dependent variables, which in this case are the years since the base year (i.e., 1999 is 0 and 2008 is 9) and the observed cost per mile.

The estimated value for a project in 2009 would be $209,000/mile plus 10 years of cost increases (at $9,200/mile per year) for a total of $301,000 per mile. This estimate might be perceived as more reliable than simply using the average cost of the most recent project, which was $290,000 per mile. A 30-mile project would therefore be expected to cost $9.33 million if completed in 2009.

However, experienced railroad engineers would want a more detailed estimate that takes into account the number of miles of track that were upgraded each year, since there is a considerable mobilization expense associated with assembling the work gangs and the machines necessary to upgrade the track. Hence, short-distance projects would cost more than long-distance projects. Table 2.3 provides additional information concerning the length and costs of the projects. Mobilization costs consistently increased by $1,000 per year, a trend that was expected to continue. A better cost estimate would therefore be obtained by analyzing the construction cost per mile and then adding the expected cost of mobilization. Using the Slope and Intercept functions in Excel, the construction cost per mile was estimated to be

Construction cost per mile = $204,000 + $8,900 per year since 1999

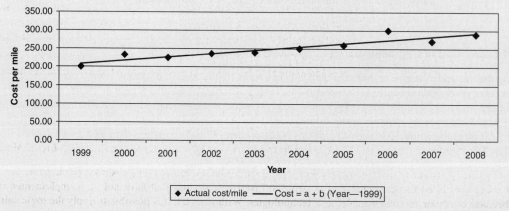

Figure 2.10 Actual and Estimated Cost per Mile for Railroad Rehabilitation

Table 2.3   Year, Length, and Cost per Mile of Rail Rehabilitation Projects

Project	Year	Miles of Track	Total Cost/ Mile	Mobilization Cost/ Project	Construction Cost/ Mile
1	1999	30	$200	$55	$198
2	2000	6	$233	$56	$224
3	2001	6	$225	$57	$216
4	2002	30	$237	$58	$235
5	2003	10	$240	$59	$234
6	2004	20	$250	$60	$247
7	2005	20	$260	$61	$257
8	2006	3	$300	$62	$279
9	2007	20	$270	$63	$267
10	2008	10	$300	$64	$294

The predicted construction cost per mile for a project in 2009 would therefore be $204,000 plus $89,000 for a total of $293,000 per mile. The total project cost for a 30-mile project planned for 2009 would therefore be 30 miles multiplied by $293,000 per mile plus the expected mobilization cost of $65,000 for a total of $9.44 million.

Econometric analysis can be useful for public policy. For example, if data were available on the cost of constructing apartment buildings in a major city, it would be possible to determine the cost per apartment or the cost per square foot of living space for each project. This data could then be plotted to determine if the cost/apartment and the cost/square foot vary with the size of the project. If large projects involving 100 or more apartments are markedly cheaper to construct than small projects, then public policy perhaps should be slanted toward facilitating larger projects.

The econometric approach can be useful in determining some basic trends in cost, but it is less useful in estimating costs for a particular project. It is very ill suited to estimating the costs of projects that use new designs or technologies—and these of course are among the characteristics of many large-scale infrastructure projects.

Engineering-Based Approach

The preferred method for estimating infrastructure costs is to break a project into well-defined pieces for which it is feasible to estimate unit costs. For each piece of the project, unit costs can be developed based upon past experience, expert judgment, or special studies. These unit costs can reflect new technologies or new designs, so that this approach is not restricted to past experience. Like the previous two approaches, the engineering-based approach can provide estimates of the cost/square foot for a hotel or the cost per mile for upgrading a rail line, but only the engineering approach provides a valid, detailed estimate based upon current or future technologies, resource prices, and designs.

EXAMPLE 2.7   Cost to Upgrade a Rail Line—What Can Be Gained by Using an Engineering-Based Model?

What can be gained by making the effort to develop an engineering-based model of the cost to upgrade a rail line? The main benefit of using a detailed model is that it is possible to investigate options that have not been implemented in the past, possibly because of resource costs or inadequate technologies. With a model, it is possible to apply the logic suitable to the decision rather than relying on costs estimated based upon past experience:

- Cost of the rail could vary with the size and quality of the rail that is selected.
- Cost of each activity can depend upon the size of the gang, the wage rates of the gang, and the productivity of the gang.
- Different types of machinery may available for each activity. Each type of machine will have its own production rate, gang size, and daily cost. Over a period of 10 years, there may well have been 20–50% improvement in productivity as a result of the use of better technology.

Summary of Rail Relay Examples

This series of examples shows how better cost models can produce better estimates of the costs of a project. If a project is similar to one that was recently constructed, then it may be reasonable to assume that costs will be similar. Examples in this section showed three approaches to estimating the cost of a proposed rail replacement project. The first approach simply used the cost per mile of the most recent rail project to obtain an estimate of $8.7 million for the new project. The second approach used regression analysis to determine the trend in costs, which were found to be increasing by nearly $10,000 per year. With this approach, the cost for the proposed project jumped to $9.3 million. The third approach took into account the fixed cost for a rail relay project; namely, the mobilization cost of getting the crews, machines, and materials to the site at the beginning of the project. A somewhat more complex regression analysis resulted in a slightly higher predicted cost of $9.4 million, which is 8.5% above the initial rough estimate. All of these approaches assumed there would be little or no change in the technologies used for replacing the rail. If different technologies are used, or if different materials are required for the track, or if different-sized gangs are used, then an engineering-based cost model will have to be developed.

A case study at the end of this chapter guides you through the steps required to develop engineering-based cost models for the cost, capacity, and service of a canal. The canal case study addresses many of the cost concepts raised in this section, including fixed and variable costs, long-run cost functions, and break-even volumes. It also considers costs from the perspectives of both the owner and the user. A major issue facing canal developers was how wide and how deep to make their canals. A canal that allows larger boats would increase construction costs but reduce costs to users, so the evaluation of canal projects requires cost functions for the user as well as estimates of construction and operating costs for the canal owner.

This section has of necessity just provided an introduction to cost concepts. Part II of this text presents techniques for understanding the costs associated with infrastructure systems, taking into account both investment costs and future operating and maintenance costs. Once these techniques are available, it will be possible to develop more complex engineering-based cost functions that take into account nuances of investment strategies and long-term decisions concerning design and operations of infrastructure. In the initial stages of project evaluation, rough cost estimates are satisfactory; but eventually more accurate estimates will be needed, as discussed in Chapter 14, "Managing Projects and Programs." Sections 14.3 and 15.6 cover additional techniques that can be useful in estimating the costs of a project and the costs of infrastructure components.

2.3   PROFITABILITY, BREAK-EVEN VOLUME, AND RETURN ON INVESTMENT

2.3.1   Profit

Entrepreneurs and owners are concerned with the financial success of a project. Three main questions are of interest. First, will the project be profitable? Second, will the profit be sufficient to justify the investment required? Third, once the project is completed, will it be worth more than it cost to build it?

The project will be profitable if the revenues received from the project are sufficient to cover its costs. The revenue could include subsidies from government agencies as well as revenues from users of the project. Revenue from users will depend upon the price that is charged and the value of the project to potential users. For projects that add capacity within a competitive market, such as most real estate projects, the prices that can be charged will rise and fall with market forces. For projects where competition is difficult, such as new bridges or toll roads, the prices can to some extent be established by the owner. In the competitive situation, the question is whether the project can be constructed and operated so that it achieves a profit given expected market prices. In a monopolistic situation, the challenge is to choose the prices that will maximize profits, assuming that there is in fact a range of prices that could be profitable.

Let's begin with the simple situation we discussed in the previous section:

$$\text{Total cost} = a + bV$$

If this is a project that will add capacity to a competitive market, then the price (P) will be determined by the market, and the total revenue can be expressed as

$$\text{Revenue} = PV$$

Profit will be the difference between revenue and cost:

$$\text{Profit} = PV - (a + bV) \qquad \text{(Eq. 2.10)}$$

2.3.2   Break-Even Volume

If a company has a linear cost function of the type shown in Figures 2.1 and 2.2, then it must sell enough units so that the average cost of production is less than the average sale price. If the sale price is less than the variable cost, then the company will never make a profit. If the sale price is higher than the variable cost, then the company will receive enough cash from the sale to cover the variable cost and have something left over that could go toward covering fixed costs; once sales volume is sufficient to cover fixed costs, then each sale adds to profit. The difference between the sale price and variable cost is called the **contribution to fixed costs or profit** or the **contribution to overhead or profit**:

$$
\begin{aligned}
&\text{If} &&\text{Total Cost} = a + bV \\
&\text{and} &&\text{Price} = P \\
&\text{Then} &&\text{Contribution} = P - b \\
&\text{and} &&\text{Profit} = PV - (a + bV) = (P - b)V - a
\end{aligned}
$$

Sales will be enough to generate a profit if the total contribution is enough to cover the fixed costs. The point at which contribution equals fixed cost is known as the break-even volume (V_p) relative to making a profit:

$$V_p = a/(P - b) \qquad \text{(Eq. 2.11)}$$

There are two conditions for profitable operation in this simple situation: price must be above variable cost, and the volume must be above the break-even volume. V_p is the volume where the incremental contribution to profit from each unit produced is sufficient to cover its share of the fixed cost.

EXAMPLE 2.8   Break-Even Volume for Making a Profit (using Eq. 2.11)

The fixed cost related to the production of a product is $500,000, the variable cost is $20,000 per unit, and the selling price is $30,000 per unit. What is the minimum volume that must be sold in order to make a profit?

Solution

Each unit sold provides a contribution of $10,000 toward the fixed costs or profit. To cover the $500,000 in fixed costs, it will be necessary to sell at least 50 units, because $500,000/$10,000 per unit = 50 units. After 50 units are sold, all of the contribution will add to profits. Note that the solution could also be found by plotting the average cost curve (which will have the shape shown in Figure 2.4) and finding the volume for which the average cost equals $30,000. The next example illustrates the graphical solution for a similar problem.

EXAMPLE 2.9   Break-Even Volume for Profit (Graphic Solution)

Calculate the break-even volume for a situation with a linear cost function and a fixed price. Figure 2.11 plots total revenue and total cost for a case where the price per unit is 1.5 and the total cost is $TC = 50 + V$. The total revenue equals the total cost, and the two lines intersect when the sales volume is 100 units. As in the previous example, Eq. 2.11 could also be used to arrive at the same answer:

$$V_p = a/(P - b) = 50/(1.5 - 1) = 100$$

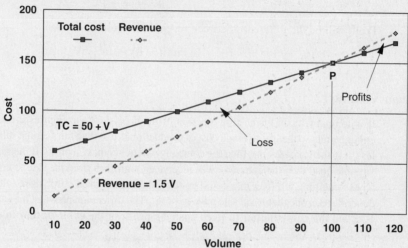

Figure 2.11 Break-Even Volume, Linear Cost Function, and Competitive Markets

Volume Varies with Price—the Demand Curve

More generally, usage volume will vary with the price that is charged. The higher the price, the lower the volume. This is traditionally represented as a demand function, as shown in Figure 2.12. Note that price, presumably the independent variable, is shown on the y-axis while volume, presumably the dependent variable, is shown on the x-axis.

When faced with a downward-sloping demand function, someone trying to decide upon a price for their services faces this dilemma: set the price too high and no one will show up; set the price too low and there won't be any profit, no matter how high the volume. The expression for profitability will be the same as shown above (Eq. 2.11), and it will still be necessary for price to exceed variable cost to make a profit.

The best price—at least from the owner's purely financial perspective—is the price that maximizes profit. Figure 2.13 plots profit as a function of volume, where the price is defined by the demand curve shown above. The first point on each curve corresponds to a price of $225 per unit and a volume of 1 unit per day. The third point corresponds to a price of $140 and revenue of $420 for sales of 3 units per day; this is the price that maximizes profit. If price is further lowered to $110, the volume increases to 4 units per day and revenue rises to $440 per day, but profit declines. With further price reductions,

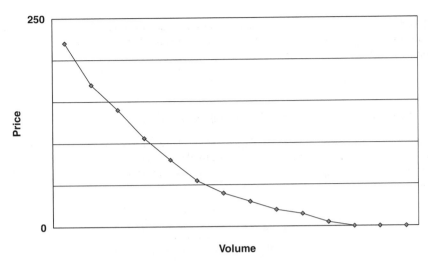

Figure 2.12 A Typical Demand Curve

the total revenue declines. When price drops to \$15, the volume is much higher, but total revenue equals total cost. For even lower prices, the owner loses money. The project therefore would be profitable for any price between \$15 and \$225/unit. Note that the owner would rationally choose to keep prices high, even if there were additional benefits to society to be gained by attracting more users.

There is not necessarily any price at which the owner could make a profit, as illustrated by Figure 2.14. If total cost always exceeds total revenue, then private companies will not provide the service. If the service is deemed to be desirable for society, then the public may decide either to provide the service via a public agency (e.g., public transit) or to subsidize the service so that the private sector will continue in the business (e.g., subsidized housing for low-income families).

Prices people are willing to pay for use of infrastructure, for renting space in an office building, or for any other service or product will depend upon many factors beyond the control of the company or agency that is trying to make a profit. As shown in "The Rise and Fall of Office Rental Rates," general economic conditions and the overall supply of office space sometimes lead to dramatic changes in rental rates. This excerpt from the *Boston Globe* highlights the surprising peak in rental rates in October 2000. Of course, when the bottom fell out of the economy in 2007, rental rates plummeted and vacancy rates again increased.

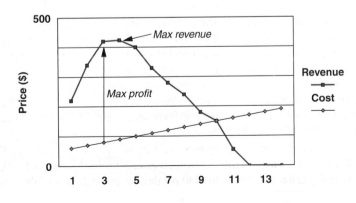

Figure 2.13 Maximizing Profit Is Not the Same as Maximizing Revenue. As prices fall, volumes rise, but total revenue eventually drops below total cost.

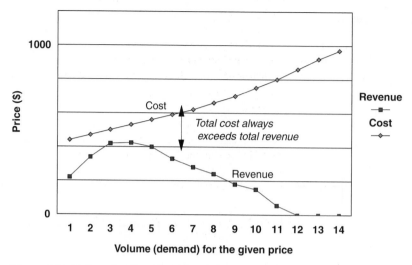

Figure 2.14 If Costs Are Too High, then There May Be No Price for Which a Profit Can Be Achieved

---

Box 2.1   Projects in the News

The Rise and Fall of Office Rental Rates

Office rental rates fall when vacancy rates are high, and they rise when vacancy rates are low. Rents for office space are generally specified as the annual rent per square foot, including operating expenses, taxes, and insurance. In 1980, Boston had essentially no vacant office space and annual rents were about $20/sq ft. Rates rose steadily to $40/sq ft in 1988, while vacancy rates fluctuated between 3 and 12%, as a building boom created new office towers. Following the collapse of the stock market in 1987 and the recession in 1990–91, vacancy rates rose to nearly 20% in 1991 while rents fell to about $27. For the next 12 years, new construction subsided, and demand slowly overtook supply; by 2000, vacancy rates were down to 1.5%, the lowest since 1980, while rents averaged $60. And over the summer, rents for prime locations skyrocketed:

Boston office rents, after increasing steadily through most of the 1990s, have exploded this year, stunning long-time real estate professionals with the ferocity of their rise. A new threshold was pierced last month when a rent of $100 a square foot—four times what landlords could get for the best space when the market hit bottom in 1992—was posted on the top two floors of One Post Office Square.

James Hooper, a principal and downtown broker, was shocked at the jump in prices when he returned to town after being away this summer. "This market is standing on its ear," he said. "I can't believe what is going on there." With new space being developed at a fraction of the pace of the 1980s building boom, no one sees a big jump in available space anytime soon.

Even though many companies want to be downtown, seeing the location as a valuable recruiting tool in a tight job market, some are beginning to buckle. In addition to making the Back Bay a mainstream corporate address, companies are venturing into the still-evolving South Boston Waterfront in hopes for a break on rent. Neighboring cities such as Chelsea, Somerville, and Medford are beckoning tenants.

Source: Richard Kindleberger, "Through the Roof:
Explosive Rise of Hub Office Rents Stunning
Real Estate Observers," *Boston Globe*, 25 October 2000.

2.3.3   Return on Investment

The financial success of a project is often measured by the return on investment, which is the annual profit divided by the initial investment in a project. If someone invests $100,000 in a project that provides a profit of $15,000 per year, then the return on investment is $15,000/$100,000 or 15%. Usually the return on the investment varies from year to year because prices and usage volume vary with the economy, competitive conditions, and other factors that affect supply and demand. If return on investment is high, then investors will consider additional investments in similar projects; if returns are low or negative, then investors turn to other types of projects, leave their money in the bank, or perhaps invest in other companies by buying their stocks and bonds. As seen in the above example, spikes in demand for office space can lead to intensive development of new buildings in a city, but eventually there is a glut of office space, rents decline, and many years pass before any further construction is considered.

Profitability and return on investment are covered in much greater detail in Part II, which provides the methodologies needed for comparing cash flows that occur at different times over the long life of an infrastructure project. Section 8.2 provides more detail concerning return on investment, while Section 9.5 introduces the concept of the **internal rate of return**, which is a method of calculating the average expected return on investment over the life of a project.

2.4   SERVICE

Potential customers for a project are concerned about the quality of service as well as the cost. Moreover, many different aspects of service quality may be important to certain customers. Section 2.4.1 identifies some of the service measures that might be used in conjunction with projects related to various types of infrastructure projects. Section 2.4.2 shows that the quality of service provided by infrastructure relates to its design, its physical condition, the way it is operated, and the level of demand. As with cost, engineering-based performance functions can be developed to represent the service that will be provided, and these equations are very useful in comparing and evaluating alternative projects. As usage increases, a point is eventually reached where congestion, queues, or service outages emerge as symptoms of a capacity problem, which is discussed in Section 2.5.

2.4.1   Measures of Service Quality for Various Infrastructure Systems

Transportation Systems

For passengers, key service measures include the average trip time, trip time reliability, the probability of excessive delays, comfort, and convenience. The relative importance of these measures varies with the type of trip, the type of traveler, and the travel options available. For the movement of freight, key measures include the size of the shipment that can be carried and the probability that the shipment will be lost or damaged in addition to trip times and reliability. During the nineteenth century, the introduction of railroads revolutionized intercity travel because it was so much faster than walking, riding a horse, or going down a canal in a barge. During the twentieth century, the introduction of automobiles and air travel drastically reduced the role of railroads and led to massive investments in infrastructure for highways and airports. During the twenty-first century, high oil prices, congestion, and concerns about emissions may result in another major realignment of transportation systems and investments in transportation infrastructure.

Water Resource Systems

In developed countries, the main concern is likely to be the price, quality, and reliability of the water supply. Potable water is needed for every household, and sufficient quantities of water must be provided for all uses and all users at a reasonable price. In many regions of the world, where running water is a luxury, other measures are useful, such as the percentage of homes with running water, the average distance to the nearest clean water in rural villages, and the probability of disruptions in supply. In some regions, the chemical content of the water is a major concern for public health. For example, arsenic has been found in rural water supplies in Nepal, resulting in serious health problems in many villages.

Historically, failure to keep drinking water separate from wastewater and sewage resulted in outbreaks of cholera and other diseases that periodically killed tens of thousands of people in major cities. Contamination of drinking water remains a critical problem today in situations where earthquakes or tsunamis endanger water supplies and in regions where wars or rumors of war cause thousands of people to move to refugee camps with inadequate infrastructure for either drinking water or for sewage.

Navigation, flood control, and the use of water power to generate electricity are other matters related to the construction of dams and levees. Riverbanks and lakeshores are wonderful locations for parks, but also prime spots for industrial uses. The conflict between the natural and the man-made environment is very apparent as wetlands are filled in, whether to provide additional space for ports, seaside housing, or industrial uses. In many rural areas, vast expanses of wetlands have been drained (via extensive networks of drainage pipes, ditches, and channels) to provide fertile lands for agriculture.

Energy Systems

The use of electricity is pervasive throughout the developed world. Normally, customers can access electricity simply by flipping a switch, and their concerns about service relate to the time the electric company requires when responding to infrequent outages caused by bad weather. Difficulties in finding an electrician to install a new appliance or disputes with the power company about billing are likely to cause more concerns than the availability of power.

In the rest of the world, the availability of electricity can be a dominant concern. Many regions have either no access to the power grid or access that is limited by time of day or day of week due to limitations in capacity. Poorly maintained systems may have frequent outages, and some systems are prone to disruption related to civil unrest.

Access to energy, limitations on the use of energy, likelihood of planned and unplanned disruptions, and the time required to restore service are aspects of service related to other types of energy used directly by consumers. These include gasoline, diesel fuel, natural gas, and firewood.

Energy companies have options for the kind of energy resources they use and the origins of those resources. From their perspective, a critical measure of service quality is the reliability of the supply chain, which is the process by which coal, oil, or another energy resource is mined, processed, and transported to their facility. To meet their customers' needs, energy companies prefer reliable supply chains for their fuel. If the supply chains are unreliable, companies have several options. They could maintain large inventories of fuel that will enable them to continue operations despite periodic breaks in deliveries; they could have multiple sources of fuel; or they could have redundant networks for providing their services, so that failure in one portion of the system does not shut down the entire system.

2.4.2 Engineering-Based Service Functions

As was the case with cost, engineering-based functions can be developed for each aspect of service. However, service is more complex than cost, because it is not possible to have a single measure (such as "dollars") that is readily accepted by owners and users and everyone else as a valid measure of performance. Thus there is a multistep process in developing service functions. First, and probably most important, it is important to understand what aspects of service are most important to potential customers. Second, it is necessary to understand how design relates to the quality of service that can be offered. Example 2.9 shows how different measures of service might be considered in evaluating options for improving urban transit.

EXAMPLE 2.10   An Engineering-Based Service Function for Bus Operations

Commuters in large cities usually have a choice of driving or taking transit. Public officials advocate greater use of transit as a way to reduce highway congestion, improve air quality, and reduce emissions of greenhouse gases. However, commuters will take transit only if they perceive the cost and service levels to be acceptable in comparison to driving. Transit agencies therefore need to understand how long it will take commuters to get from home to work if they decide to take the bus; if the transit agency can provide better service, then perhaps it can attract more riders. What is needed is an engineering-based service function that can predict trip time for a commuter based upon existing or potential transit schedules.

Any commuter's journey to work can be divided into individual segments. For example, one commuter's transit journey might include the following:

- Walk to bus stop (5 minutes)
- Wait for bus, which operates every 10 minutes during rush hour (0–10 minutes)
- Ride bus 2 miles to subway station (5–10 minutes depending upon traffic, number of stops, and weather)
- Transfer from bus to subway platform (3 minutes)
- Wait for subway train, which operates every 5 minutes during rush hour (0–5 minutes)
- Ride train 3 miles to destination, stopping at 5 intermediate stations (12–15 minutes)
- Exit station and walk to destination (7 minutes)

The total trip could take as little as 36 minutes, if the connections are perfect and there are no delays on the bus or the subway. On the other hand, if the commuter just misses the bus and also just misses the train, and if there are delays for both the bus and the train, then the journey could take 55 minutes. The average trip is likely to be about 45 minutes.

After a couple of weeks taking this route, the commuter knows when to leave home in order to be on time for a meeting and when to leave home in order to experience the least delay. The commuter relies on experience and does not require an engineering-based service function.

The transit agency, however, has designed the bus routes, built the subway system, determined how many stops to make along the bus routes, and established schedule frequencies. It also has extensive experience in travel time along the bus routes and the time required at bus stops and at subway stations. The agency therefore can develop equations that can predict service for any commuter under any set of operating conditions and any assumptions about future bus routes and subway extensions. The agency also knows, from census data and surveys, where people live and where the jobs are. It can therefore select a representative sample of commuters and calculate the following for each of them:

- Access to the system
  - If distance from home to subway is less than 0.25 miles, assume the commuter walks to the subway at an average speed of 3 miles per hour.
  - If distance from home to subway is more than 0.25 miles, assume that the commuter will take a bus if there is a bus stop within 0.25 miles; the time to walk can be estimated assuming an average speed of 3 miles per hour.
  - If there is neither a subway station nor a bus stop within 0.25 miles, assume that the commuter will drive to work.
- Wait for bus (assume time is between 0 and the average time between buses during rush hour).
- Ride bus to subway station (estimate time based upon observed travel times, scheduled frequency of stops, and time required for stops along route).
- Transfer from bus to subway platform (estimate based upon distance from bus to platform, expected congestion at stairways or escalators, and time required to purchase ticket).
- Wait for subway train (assume time is between 0 and the average time between trains during rush hour).
- Ride train to destination (estimate time based upon distance, speed limits, train acceleration and braking capabilities, number of stops, and time required for loading and unloading at each station).
- Exit station and walk to destination (estimate based upon distance and walking speed).
- Sum all of the times to determine the minimum, expected, and maximum travel times for each user.

This is an engineering-based service function because it provides a way to predict trip times and reliability as functions of engineering parameters, operating strategies, and network design. This process could be used to estimate the changes in service, for particular groups of people, that would result from increasing the frequency of bus service, adding bus routes, creating an express-bus lane, or expanding the subway system.

2.5   CAPACITY

Capacity is another important aspect of the performance of infrastructure-based systems. Measurement of capacity is complicated by a number of factors, not the least of which is the difficulty in defining output. Infrastructure-based systems typically consist of networks of facilities that are managed or used by multiple organizations using labor, equipment, energy, and other resources to provide what may be a rather large array of multidimensional services. To understand capacity, it is best to start with a small piece of the system, where it is possible to enumerate all of the key inputs, where volume is well-defined, and where it is straightforward to estimate what happens to performance as volume increases. As with both cost and service, it is useful to develop engineering-based functions that relate capacity to the characteristics of the infrastructure and its users.

EXAMPLE 2.11   Capacity of a Highway Intersection

Consider a typical intersection of two arterial streets in an urban area. One street carries eastbound and westbound traffic, the other street carries northbound and southbound traffic. To simplify the situation, assume that (1) all of the vehicles are automobiles, (2) none of the vehicles are making any turns, and (3) there are no pedestrians. The traffic signal has a cycle time of 120 seconds, and it provides equal time (60 seconds each) for the east-west and the north-south flows. When the signal turns green, it takes 2 seconds for the first car to move through the intersection; if there is a queue of cars, then additional cars can pass through every 2 seconds. Thus the maximum number of cars in one lane that can move through the intersection in each 2-minute cycle is

$$(60 \text{ seconds green per cycle for each direction})/(2 \text{ seconds per car}) = 30 \text{ cars per cycle}$$

If both roads are two-lane roads, then a total of 4 lanes would approach the intersection. As many as 4 lanes multiplied by 30 vehicles per lane (120 vehicles) could be moving through the intersection during each 2-minute cycle, and on average 1 vehicle per second could be moving through the intersection.

Is the capacity of this intersection therefore 1 vehicle per second or 60 vehicles per minute? If so, is this the same as saying that the capacity is 60 vehicles per minute or 3600 vehicles per hour or 24 hours/day (3600 vehicles/hr) = 86,400 per day?

Given the assumptions, these may appear to be reasonable statements. However, we probably should be somewhat suspicious of the nice round numbers used in the above analysis. If you really want to know the capacity of the intersection, you could stand on the corner and count cars. If you had a stopwatch and a clipboard, you could record the actual number of cars moving through on each cycle; and perhaps you would find that sometimes the first car takes 4–5 seconds (until the driver behind honks), while subsequent cars take a little more than 2 seconds. With better information about how cars move through the intersection, you would be able to say something like this: the maximum number of cars passing through the intersection at rush hour during one week in October was 56; the average number of cars passing through the intersection was 52. Moreover, you noted that there was never enough time for all of the waiting cars to get through in one cycle, so that these observations were directly related to the capacity of the interchange. Armed with this knowledge, you might conclude that capacity was somewhat less than 60 cars per minute. And you might wonder, should the capacity be stated as 60 (which is what was calculated using the textbook approach described above), 56 (the maximum that you yourself saw), or 52 (the average that you observed)?

Before answering this question, let's think about how this measure of capacity might be used. If we are traffic engineers or planners, we are likely comparing the capacity of the intersection to the volume of traffic that is expected to be trying to get through the intersection; if capacity is insufficient, we would then consider whether it is worthwhile to expand capacity or to reduce demand (e.g., by establishing tolls or by promoting the use of transit). If we are commuters, we are likely worried about excessive delays if there is insufficient capacity. Engineers, planners, and commuters would all be thinking about the extent of the delays, and they would all be concerned with what happens during the morning and afternoon rush hours. Thus they would be thinking about cars per hour rather than cars per minute. While all might agree that the theoretical capacity is somewhere between 56 and 60, they would all prefer to use no more than the observed average of 52 in their comparison of capacity to demand.

Is there any reason to consider something less than 52 cars per minute as the estimate of capacity? Yes, indeed. Any experienced commuter would have some questions about the study that you completed:

- Were there any accidents during that week?
- What was the weather like (were there any heavy rains or snowstorms)?
- Were there any emergency vehicles trying to move through the intersection?
- Was there any maintenance in or near the intersection?
- Was the intersection ever gridlocked (i.e., did cars ever enter the intersection without being able to continue all the way through, thereby blocking the intersection when the light turned green for the other traffic)?

Commuters ask these questions because they know that accidents, bad weather, emergency vehicles, and aggressive drivers all disrupt normal flows, not necessarily every day but often enough to be a problem. A more extensive study that included winter conditions and periods of maintenance might show that the average flow through the intersection was actually only 48 cars per minute during rush hour.

So is the capacity of the interchange 48 cars per minute? Well, the commuters and the planners would ask another question: how bad are the delays at this intersection? This might be a notorious intersection where cars back up for a half mile or more and often take 10–15 minutes to get through. Anyone regularly driving that route would say that such delays are unacceptable, and planners would likely agree that the intersection was experiencing traffic volumes that were higher than its capacity. This may seem to suggest that the capacity is less than 48 cars per minute, but actually it doesn't. The capacity of the intersection is in fact about 48 cars per minute, but traffic volumes during rush hour are greater than this; the long queues build up because cars arrive faster than cars can get through the intersection. When an incident happens—whether related to an accident, a snowstorm, or anything else—fewer cars move through the intersection, and queues build up to an unacceptable level. A traffic engineer might then ask the following question: if the average throughput of the intersection is 48 cars per minute during rush hour, what is the maximum traffic volume for which the performance of the intersection is acceptable? This question could be phrased in terms of the structure of the street network: what is the maximum traffic volume for which the queues at this intersection will interfere with the performance of neighboring intersections no more than once or twice a month? This question could be answered by using algebraic models or by simulating traffic conditions. If a delay of a few minutes is acceptable, then it will certainly be possible for more than 48 cars per minute to be arriving during rush hour, even though only 48 cars per minute will get through the intersection.

EXAMPLE 2.12   Hourly, Daily, Monthly, and Annual Capacity of Highway Intersections

The discussion in the previous example considered only rush hour, and the estimates of capacity all related to what happens during rush hour. In designing systems, it is essential to consider capacity during the peak period; but estimates of demand may be based on daily, weekly, monthly, or annual usage. It is therefore necessary to be able to translate peak period capacity into capacity for these other periods. This can be done by considering the pattern of demand over longer periods. For highways, the following relationships can be estimated based upon observations of traffic volumes:

- Average vehicles per day on weekdays over the course of an entire year
- Percentage of traffic on weekdays
- Percentage of traffic during peak periods of the day (e.g., 2 hours in the morning plus 2 hours in the afternoon)

For example, if the capacity of an intersection is 50 vehicles per minute during rush hour, then the capacity for the four peak hours will be 12,000 vehicles during four peak hours. If the rush hour periods account for 40% of weekday traffic, then the intersection can be considered to have capacity for 12,000/0.40 = 30,000 vehicles per weekday. If weekdays have 80% of the average weekly traffic, then the capacity is 30,000 vehicles per weekday × 5 weekdays/week ÷ 0.80 = 187,500 vehicles per week, or 9.375 million per year, assuming that the current traffic patterns remain the same. This statement can be interpreted as follows: If traffic along these two roads rises to 9.4 million vehicles per year, then this particular intersection is likely to experience unacceptable delays on a regular basis. The intersection might actually handle more traffic, but the delays would increase and the congested rush hour periods would be longer. In other words, this intersection is expected to act as a bottleneck once traffic along these two roads rises to 9.4 million vehicles per year.

The capacity of the network will be harder to define because of the many possible bottlenecks. Severe problems may occur at other points in the network long before our intersection experiences significant delays. If it is easy to add capacity, then it may be possible to add capacity as demand grows, thereby avoiding the development of bottlenecks.

If the network has many potential bottlenecks, then fixing one of them will not necessarily affect system performance. Increasing the capacity of one intersection by building an overpass may simply shift the delays to the next intersection.

The examples in this section have illustrated several key concepts regarding capacity of a well-defined element of an infrastructure-based system.

- **Maximum capacity**—the maximum flow through the system when everything is operating properly
- **Operating capacity**—the average flow through the system under normal operating conditions
- **Sustainable capacity**—the maximum flow through the system that allows sufficient time for maintenance and recovery from accidents or other incidents

Since demand may vary substantially on a daily, weekly, or seasonal basis, it is important to consider how well the system performs during periods of peak demand. There could well be delays to users or denials of service during peak periods, but the system may be able to recover soon after peak demands subside. Delays during peak periods do not mean that capacity has been reached; it is only when delays become serious impediments to users that capacity has been reached. The solution to peak-period problems could be expansions to the infrastructure, changes in operations, changes in pricing, or other attempts to limit peak-period demands.

2.6   SAFETY, SECURITY, AND RISK

Safety and security are important aspects of performance for infrastructure systems. Both refer to the likelihood of injuries or fatalities to employees, customers, abutters, and others along with the probability of damages to the infrastructure or to the property of users or abutters.

Security generally refers to the measures that might be taken to prevent deliberate attacks on people or property, whether those attacks involve pickpockets, thieves, or terrorists.

Safety is generally used in reference to accidents or problems that occur in the course of normal operations of the system, although it may also be used with respect to the possibility of deliberate attacks. Safety records for a transportation company show such things as the number of accidents, the number of fatalities, and the number of fatalities per million miles traveled.

Risk is a broader concept that can be used to describe the potential for future accidents or incidents.[2] A methodology known as probabilistic risk assessment can be used to measure risk in a way that is useful in evaluating system performance. Risk is the product of two factors:

- The probability of an accident or incident
- The expected consequences if an accident or incident occurs

Consequences could include fatalities, injuries, disruption of service, release of toxic chemicals, and inconvenience to people living or moving through the neighborhood of the accident. A weighting scheme is needed to compare the severity of the different types of accidents. Strategies for reducing risks could focus either on reducing the probability of accidents or on reducing the expected consequences if an accident occurs.

The design of infrastructure systems always involves consideration of risks, including risks to those involved in construction of the facility as well as risks to users. Construction standards and choice of materials will depend in part upon the potential for natural disasters, such as hurricanes or earthquakes, and the need to provide a safe operating environment. It is never possible to eliminate all risks, and judgment is required to determine which risks are worth worrying about.

EXAMPLE 2.13   What Are the Risks at Grade Crossings?

Hundreds of people are killed each year in grade-crossing accidents. A grade crossing (also known as a level crossing in some countries) is where a road crosses a railroad at grade, so that it is possible for a train and a vehicle to collide. Accidents may be caused by people who are too sleepy or too drunk to notice that they are approaching a crossing, by people whose car is stuck in traffic while trying to get across the tracks, or by people whose car breaks down on the crossing. A few accidents are caused by malfunctioning of the signals, and a great many are caused by people who ignore the warnings and try to beat the train across the intersection.

---

[2] The term *risk* is used in two different ways by safety engineers and financiers. To safety engineers, risk refers to accidents or incidents that result in property damage, injuries, or fatalities; that is the sense in which the term is used in this section. In finance, the term is used to include any of the many factors that may affect the success of a project, including such things as the risk of an economic downturn, the risk of new competition, and the risk of political upheaval as well as the risks related to safety and security. Both meanings are used in this text; which sense of the term is being used should be clear from the context.

The probability of such an accident occurring varies primarily with the density of rail traffic and with the type of protection available. At a crossing equipped with flashing lights, the probability of an accident is on the order of 2 per million trains. If there are 20 trains per day, then the expected number of accidents per year would be (20 trains/day × 365 days per year/1 million trains) × (2 accidents per million trains) = 0.0146 accidents per year for such a crossing. For a rail route with 100 such crossings, the expected number of accidents per year would be 0.0146 (100) = 1.46. Although the probability of an accident at any crossing is very low, the likelihood of an accident somewhere along this route is quite high. In fact, such accidents are common. In the United States, more than a quarter million grade crossings and thousands of grade-crossing accidents occur each year.

The second step in estimating risk is to determine the expected consequences if an accident occurs. Even a small passenger train weighs hundreds of tons, so the consequences of a collision between a train and an automobile are very predictable. The car will be destroyed, any people who fail to get out of the car are likely to be killed or severely injured, and the locomotive may sustain some minor damage. In addition, the engineer and anyone else in the cab of the train will be suffering an emotional shock after (1) knowing that the accident was about to happen and (2) being completely unable to stop a train in time to avoid hitting someone trying to sneak across before the train arrives. If the train is a passenger train, the major consequence for most passengers will be a delay to the train; passengers might not even notice the impact and will simply wonder why the train stopped.

Table 2.4 shows the consequences of a typical grade-crossing accident in Japan and also of the worst grade-crossing accident. The typical grade-crossing accident was based upon the average consequences for all crossing accidents that had occurred on the East Japan Railways (JRE) over the period 1988 to 1995. In many cases, the accident involved a vehicle that was broken down, stalled, or caught on the track; so the occupants had a chance to escape. On the average, there were 0.2 fatalities and 0.4 injuries for the occupants of the highway vehicle, and no fatalities or injuries for the train crew or passengers. The typical accident resulted in JRE delaying 8.5 trains and canceling 0.2 trains. A more catastrophic accident can occur if a train hits a very heavy highway vehicle. The worst accident occurred when a train derailed after hitting a cement truck that was stalled on the crossing; the derailed cars struck and injured 90 JRE maintenance workers who happened to be standing next to the track near the crossing, waiting to resume their work on the track after the train passed.

Table 2.4   Consequences of Grade-Crossing Accidents on the East Japan Railway, 1988 to 1995

Category	Typical Accident	Worst Accident	Risk Conversion Factor	Weighted Consequences of Typical Accident	Weighted Consequences of Worst Accident
Fatalities, occupant of auto	0.2	1	100	20	100
Fatalities, JRE employees			100	0	0
Fatalities, train passengers			100	0	0
Injuries, occupants of auto	0.4		1	0.4	0
Injuries, JRE employees			1	0	0
Injuries, JRE passengers			1	0	0
Injuries, bystanders		90	1	0	90
Delayed trains	8.5	26	1	8.5	26
Canceled trains	0.2	39	10	2	390
Total				**30.9**	**606**

Source: Sudhir Anandarao and Carl Martland, "Risk Assessment of Level Crossing Safety in the JR East System," JRE/MIT Report 4, December 1996.

The risk conversion factors shown in Table 2.4 were used informally by the JRE Safety Research Lab to compare the different categories of consequences. Each unit represented an amount equal to 1 million yen, or approximately $10,000. The Safety Research Lab established these measures as a basis for determining cost-effective means of reducing risks. As shown in the table, the weighted average consequences associated with the average incident totaled 30.9 units, or approximately $300,000; the consequences of the worst single accident totaled 606 units, or approximately $6 million.

In addition to grade-crossing risks, JRE also considered the risks associated with train collisions, derailments caused by track or equipment failure, and derailments caused by earthquakes and other natural hazards. These other types of accidents, although far less common than grade-crossing accidents, could have catastrophic consequences with hundreds of fatalities or injuries. Therefore JRE focused more of its safety research and investments on strategies that would prevent potentially catastrophic accidents, such as reinforcing the infrastructure for high-speed rail to reduce the risk of failure during an earthquake.

Some adjustments in predicted risks may be necessary to reflect the perceived importance of certain types of accidents or consequences to various stakeholders. Quantifying perceived risks requires answers to questions like these: "Who is at fault?" "Is it a catastrophic accident?" and "Is new technology involved?" Public perceptions of risk will be greater for situations where that risk could result in catastrophic accidents with hundreds of fatalities or accidents with dreadful consequences, as could be the case with the release of toxic chemicals or radiation after an accident at a chemical factory or at a nuclear power plant. The public is more concerned with unknown risks that might be associated with new technologies than with well-known risks. People know that automobile accidents kill tens of thousands of people per year, but they still drive cars. Fears of accidents involving nuclear power plants hampered development of such projects in many countries, despite an excellent safety record.

Chapter 13 deals with these issues in more detail. For now, it is important to understand the basic concepts:

- Risk is an important aspect of infrastructure performance.
- Risk reflects both the probability of an accident and the consequences of the accident.
- It is often possible to use past experience as a guide for estimating accident probabilities and expected consequences.
- Weights can be devised for comparing different types of consequences.
- Strategies for reducing risks can be evaluated by comparing the costs of those strategies to the expected reduction in risk.

Companies, operating agencies, and agencies that create, use, or regulate infrastructure are continually seeking cost-effective ways to reduce risks. Since resources are limited and risks can never be entirely eliminated, it is important to allocate those resources effectively. In practice, this means that the expected reduction in the consequences of accidents should be high enough to justify the costs.

2.7 COST EFFECTIVENESS

The previous sections have shown that there are many ways to measure the performance of a system, including cost, financial performance, service quality, capacity, and risk. When dealing with costs and finances, it is natural to measure everything in terms of money. When dealing with the other matters, it is more difficult to use money as a performance measure. Thus, if an investment is proposed to improve infrastructure performance, it is usually difficult if not impossible to determine whether the non-monetary benefits justify the financial cost of the investment. Ultimately, as discussed in Chapter 5, some kind of political process is necessary to weigh all of the costs and benefits to determine whether a proposed project is justified. This section introduces the concept of cost effectiveness, which is a valuable technique that can be used in evaluating potential investments aimed at achieving non-monetary benefits.

Cost effectiveness is the ratio of the benefit to the cost of achieving that benefit. Thus, as long as it is possible to measure a benefit, it is possible to calculate cost effectiveness. If there are multiple ways of achieving the same type of benefit, then the most cost-effective approach is the one that costs the least per unit of improvement. Just because an option is the most cost effective does not mean that it is worth pursuing—it may be better than any of the others, but it may still be judged as too costly.

Cost effectiveness is therefore a concept that is most useful in eliminating projects from consideration and in identifying ones that deserve further consideration. A project is not implemented simply because it is the most cost effective. Example 2.9 showed how a transit agency might develop an engineering-based service function that it could use to evaluate options for improving transit service. Using such a service function, the agency could determine how much service would improve for various operating and investment strategies. The cost effectiveness could be measured by comparing the improvements in service to the cost of the investments. The service improvements might be stated as the percentage reduction in average travel time for current users, which could be achieved by operating strategies that increased bus frequency as well as by investment strategies that allowed faster trains or extensions of the subway system. By considering cost effectiveness, the transit agency could determine whether it would be best to improve the existing services or to extend these services. The agency might conclude that the most cost-effective approach is to increase bus frequency—but it might also eventually decide that it could not afford to do so.

The transit agency might also be under pressure to expand the system, so that more people would have access to transit. The main question would be whether to expand the bus routes or extend the subway lines. Cost effectiveness would favor one over the other, but cost effectiveness alone would not be enough to convince the transit authority to buy more buses or to invest in a new subway line.

Example 2.13 shows how cost effectiveness can be used in conjunction with probabilistic risk assessment to determine the best ways to reduce the risks of grade-crossing accidents.

EXAMPLE 2.14   What Are the Most Cost-Effective Strategies for Reducing Risks Associated with Grade Crossings?

Accident rates at grade crossings can be reduced by installing flashing lights, putting in crossing gates (arms that automatically come down and block the travel lane when a train approaches), installing four-quadrant gates (four arms block the entire road, so that motorists cannot run around the gate), or building a bridge. It is even possible to block the entire road, while uniformed personnel ensure that pedestrians do not try to skip across in front of a train (Figure 2.15); but this expensive solution can be justified only in very unusual circumstances.

Figure 2.15 The Highest Level of Crossing Protection. At this extremely busy intersection in Tokyo, the configuration of grades and buildings makes it impossible to put in either a highway or a railway bridge. Guards are stationed at both sides of the tracks, a wire "gate" drops across the entire width of the road, and a signal (visible at left in the picture) shows whether trains are coming from the left, the right, or both. The intersection also has an obstacle detector that alerts trains in time for them to stop if a vehicle is stuck in the crossing.

Table 2.5 shows the cost of installing (or upgrading) to each level of protection along with typical accident rates achieved with this type of protection. Note that the accident rate is driven (in this simplified model, but also in reality) by train traffic, not by highway traffic.

Table 2.5 Grade-Crossing Accident Rates

Protection	Cost per Crossing	Accident Rate (per million trains)
Signs only	$500	10
Flashing lights	$20,000	2
Gates	$100,000	1
Four-quadrant gates	$200,000	0.2
Bridge	$2,000,000	0

1. Assume that you are the safety officer in a state Department of Transportation, and you have a budget for improving highway safety. You would like to use some of this budget to reduce crossing accidents. You have categorized crossings into the groups shown in Table 2.5. Given the information shown in the table, calculate the cost effectiveness for each strategy in reducing accidents, and identify the three most cost-effective strategies to pursue. Give the equation you are using for cost effectiveness, calculate measures of cost effectiveness for each strategy, and indicate the three most cost-effective upgrades among those listed in Table 2.6.

Table 2.6 Possible Upgrades

Highway Traffic per Year (in millions)	Trains per Year	Base Acc. per Year	Current Protection	Possible Upgrade
20	100,000	0.1	Gates	Bridge
20	100,000	0.1	Gates	Four-quadrant
2	50,000	0.05	Gates	Four-quadrant
0.2	50,000	0.1	Flashing lights	Gates
20	5000	0.01	Flashing lights	Gates
0.02	2000	0.02	Signs	Flashing lights
0.02	200	0.002	Signs	Flashing lights

2. How do you decide how much to invest in improving grade-crossing safety as opposed to other projects that could improve highway safety?

Solution

The first step is to estimate the effect of the upgrade on accident rates and the number of accidents per year for each category of crossing. Table 2.7 has two new columns added to the information in Table 2.6. The new accident rate per million trains comes directly from Table 2.5. The expected accidents per year is the product of the new accident rate per million trains multiplied by the number of trains per year. For example, if four-quadrant gates are installed for crossings with 20 million highway vehicles and 100,000 trains per year, then we can expect the accident rate to drop to 0.2 per million trains, while the expected number of accidents per year will be 0.2 accidents per million trains multiplied by 0.1 million trains per year or 0.02 accidents per year, as shown in the final column of the table.

Table 2.7 New Accident Rates and Annual Accidents if Grade Crossings Are Upgraded

Highway Traffic per Year (in millions)	Trains per Year	Base Acc. per Year	Current Protection	Possible Upgrade	New Acc. Rate	New Acc. per Year
20	100,000	0.1	Gates	Bridge	0	0
20	100,000	0.1	Gates	Four-quadrant	0.2	0.02
2	50,000	0.05	Gates	Four-quadrant	0.2	0.01
0.2	50,000	0.1	Flashing lights	Gates	1	0.05
20	5000	0.01	Flashing lights	Gates	1	0.005
0.02	2000	0.02	Signs	Flashing lights	2	0.004
0.02	200	0.002	Signs	Flashing lights	2	0.0004

Table 2.8   Cost per Accident Avoided per Year if Upgrades Are Implemented

Highway Traffic per Year (in millions)	Trains per Year	Current Protection	Possible Upgrade	Accidents Avoided per Year	Cost of Upgrade ($)	Cost per Accident Avoided/ Year ($Millions)
20	100,000	Gates	Bridge	0.1	2,000,000	20.0
20	100,000	Gates	Four-quadrant	0.08	200,000	2.5
2	50,000	Gates	Four-quadrant	0.04	200,000	5.0
0.2	50,000	Flashing lights	Gates	0.05	100,000	2.0
20	5000	Flashing lights	Gates	0.005	100,000	20.0
0.02	2000	Signs	Flashing lights	0.016	20,000	1.25
0.02	200	Signs	Flashing lights	0.0016	20,000	12.5

The next step is to compute the cost effectiveness, which is the cost per annual reduction in accidents. This can readily be calculated as the cost of the upgrade divided by the number of accidents avoided, as summarized in Table 2.8. The accidents avoided per year is calculated as the difference between the base and the new number of accidents per year (shown above in Table 2.6). The cost of the upgrade is shown above in Table 2.5. The result is the cost per accident avoided, as shown in the last column of Table 2.8.

The most cost-effective measures therefore are as follows:

1. Install flashing lights at crossings with 2000 trains per year that are currently protected only by signs ($1.25 million per accident avoided per year).
2. Install gates at crossings with 50,000 trains per year that are currently protected only by flashing lights ($2 million per accident avoided per year).
3. Install four-quadrant gates at the very busy crossings with 100,000 trains per year ($2.5 million per accident avoided per year).

As the safety officer for the state department of transportation, you would still have to determine whether there are more cost-effective strategies to pursue in terms of reducing risks within your state. To do this, you would need an estimate of the consequences of grade-crossing accidents, so that you could calculate cost effectiveness in terms of risk reduction rather than in terms of accident reduction. You could then compare strategies for reducing risks at grade crossings to strategies such as adding more policemen to enforce speed limits, requiring seat belts, or upgrading dangerous highway intersections.

2.8   DEVELOPING COST AND REVENUE FUNCTIONS FOR A NINETEENTH CENTURY CANAL

This section integrates many of the concepts presented in this chapter by considering investment decisions concerning canals. Canals provide a useful case for studying infrastructure investment, because it is straightforward to develop engineering-based functions for cost, service, and capacity. Thus it is possible to understand the performance issues that had to be considered and why decisions were made to invest in canals.

Canals are also interesting in that their history plays an important role in the evolution of transportation systems. They were long the most efficient means available for transporting grain, other agricultural products, coal, ore, and general freight. Two hundred years ago, canals were the dominant focus of transportation infrastructure investments in the United States, but changes in technology and demand ultimately led to their demise. Infrastructure systems are long lasting, but they do change; and historical perspective concerning one type of infrastructure is informative when looking at other types of infrastructure.

The examples in this section also illustrate a principle related to analysis of complex systems. When there are multiple measures of performance, it is always worthwhile to begin with a broad rather than a narrow approach. By obtaining a rough idea about service, cost, and capacity, it may be possible to determine which factors or design features must be studied in detail and which can safely be treated very simply. Depending upon the situation, cost, service, or capacity might be the most important. Before delving into a complicated methodology for measuring any of these aspects of performance, it is well worth determining whether such an extensive effort is required.

The analytical methods used in this section can easily be implemented in a spreadsheet. Many of the assignments in this textbook require the use of spreadsheets, which provide an effective way to structure problems, conduct analysis, and present results. Once you understand the basic design and performance factors associated with canals, you will be able to develop a spreadsheet model and use it to explore the issues presented in this section. An expanded version of this case is therefore included as the case study assignment at the end of this chapter.

2.8.1   Background on Nineteenth Century Canals

Canals were among the first major civil engineering projects in the United States, as in many other countries. In the era before railroad or truck transportation, land transportation was cumbersome, slow, and expensive. Water transportation—when available—was much cheaper, more reliable, and provided the only means of handling large volumes of freight, which of course was why cities grew up at the best harbors and along the major navigable rivers. To avoid costly transshipment of goods, the first canals simply bypassed rapids. Later canals linked major cities to their hinterlands. The early canals were designed with a towpath on either side so that a horse or a team of horses could pull the canal boats along the canal (Figure 2.16). The average speed of canal boats was only 2–3 miles per hour, and the distance traveled per day was only 20–30 miles. More ambitious projects, such as the Erie Canals, sought to open up western regions and thereby promote development (not to mention the importance of the port city whose citizens promoted the project).

Canals must be close to level to allow safe, easy movement in both directions. To create a level route in uneven topography, it is necessary to construct locks. A lock is a chamber with two sets of gates. The water level is higher on one side of the lock than it is on the other side. By opening one gate at a time, the water level within the chamber can be adjusted to match either the high side or the low side. The depth of the lock must be sufficient to accommodate canal boats when the water is low; the height of the gates above the low water level limits the extent of the lift that can be achieved by one lock. Several locks can be operated next to each other, or in close proximity to each other, if the terrain requires more lift than can be achieved with a single lock. The time required for a move through a lock includes the time to position the canal boat (or boats) in the lock, the time required to raise or lower the water, and the time required for the boat(s) to move out of the lock.

The width and depth of a canal and the size of the locks determine the size of the boats that can use the canal. Figure 2.17 shows a lock on the Chesapeake and Ohio (C&O) Canal in the Georgetown neighborhood of Washington, D.C. Note that the lock is much narrower than the canal, and it has a lift of less than 10 feet. This location has only one channel but if more capacity were required, a second lock could have been built right next to this one.

The deeper and wider the canal, the more material must be excavated, and the more expensive the project (Figure 2.18). The larger the locks, the more expensive they become, and the more water is

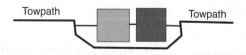

Figure 2.16 Cross Section of a Typical Nineteenth-Century Canal. The canal was barely wide and deep enough to allow two small boats to pass. The boats were towed by a horse that walked along the towpath next to the canal.

Figure 2.17 Remnants of the C&O Canal Can Still Be Seen in Georgetown in the District of Columbia. The long, narrow boat is sized to squeeze into locks such as the one at the back of this picture.

required to operate the system. Hence, there are fundamental design issues concerning the size of the canal and the type of boats to be accommodated.

The channel—at the bottom of the canal—must be wide enough for two boats to pass (or there must at least be periodic basins where opposing boats can pass) and deep enough to provide the required draft for the largest boats that are allowed to use the canal. Some of the early canals could handle only small boats with a capacity on the order of 15 tons; these boats required a draft of only 12 inches when loaded. Larger canals could handle larger boats, for example, boats that could carry 75 tons along canals providing more than 4 feet of draft. As suggested by Figure 2.18, increasing the width and depth of a canal can require massive excavations in hilly regions.

The number of locks required is a function of the route and the size of the locks. The topography of the route determines the minimum lift that will be required, which is the difference in elevation between the beginning and end of the canal. If a canal is constructed to connect two river basins, then the canal must either cut through the intervening hills or move up, across, and down the other side in a series of locks (Figure 2.19). Excavating a level route reduces the number of locks required, which also reduces the time required to move along the canal. However, it may be infeasible or extremely expensive to create a level route.

Each time a boat moves upstream through a lock, the lock fills with water; when a boat moves downstream, the lock is emptied. If a canal is to move from one watershed to another, there must be a reliable water supply to support the functioning of the locks (Figure 2.20).

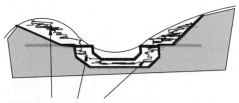

Doubling the width and depth of the canal can lead to major increases in excavation

Figure 2.18 Topography Has a Large Effect on the Costs of Constructing a Canal

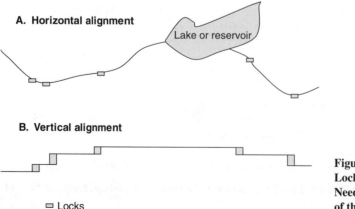

Figure 2.19 Locks Enable Canal Boats to Move Up and Over Hills

2.8.2   Service and Capacity of a Canal

For canal service, the main concerns are the size of the boats and the time it takes to move along the canal. First let's consider the time it would take to complete a 120-mile trip along a nineteenth century canal. We can assume that boats could tie up for the night at either end of the canal or at frequent locations along the length of the canal (as shown in the photo of Regent's Canal in Figure 2.21). At 2 miles per hour, a 120-mile trip would require 60 hours of travel time, which would be six 10-hour days; at 3 miles per hour, the same trip would require only 40 hours of travel, or four 10-hour days. Thus the travel time would be at least 4–6 days, and a reasonable estimate might be stated as "five days plus or minus a day" or "less than a week." This may strike you as a fairly imprecise estimate; but before worrying too much about how to refine it, think about what else goes into the travel time. First of all, there are likely to be locks located every few miles along the canal, and it takes an hour or two to get through each lock, perhaps much longer if the canal is very busy and queues of boats are waiting to get through the busiest locks. If the 120-mile trip has 40 locks, then the time spent in the locks could well be 40–80 hours, which is as long as and more variable than the travel time along the canal. There is also the possibility that bad weather, high or low water, lock maintenance, or other problems will limit or prohibit travel along the canal. Thus the estimate that it will take about 50 hours to travel along the canal between locks is probably the most reliable portion of the overall estimate of travel time. Adding in locks and considering the possibility of major weather-related delays, the trip would likely be estimated as a journey of 2 to 3 weeks.

The capacity of a canal (maximum throughput measured in tons of cargo) can be estimated for various times and conditions:

- A peak day with 12 hours of operation
- A peak summer month with 12 hours of operation, 7 days/week
- A year, with operations ceasing during the winter and during major storms

The canal's capacity is a function of the characteristics of the canal, the boats, and the operating characteristics. The maximum capacity of the lock is the inverse of the function of the cycle time for

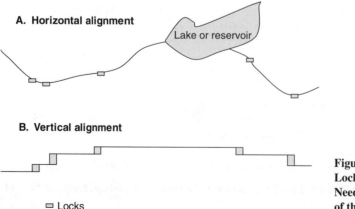

Figure 2.20 To Operate the Locks, a Water Supply Is Needed Above the Highest Point of the Canal

Figure 2.21 Regent's Canal, London. England's canals remain busy, but the old canal boats have been converted to houseboats used as mobile homes or for weekend getaways. Dozens of boats may tie up for the night at a convenient, picturesque spot like this.

the lock. If the cycle time is 30 minutes per boat, then the maximum capacity is 2 boats per hour. The operating capacity, expressed in boats per day, is limited by the hours of operation. If the lock operates 12 hours per day, then the operating capacity is 24 boats per day. The sustainable monthly capacity is further reduced to allow for lock maintenance, for periods of slower or interrupted operations during bad weather, and for smooth functioning of the canal despite the normal variations in traffic and routine delays that might occur. For example, the lock might be closed for 1 day per month for routine maintenance, heavy rains or winds might reduce capacity by 50% for several days per month, and miscellaneous delays might amount to an hour or so each day. This would reduce sustainable monthly capacity as follows:

- Days per month: 30
- Days required for maintenance: 1
- Expected days of operating at 50% capacity: 3–4 days, which is equivalent to 1–2 days of lost operation
- Miscellaneous delays: about 1 hour per day or 30 hours per month, which is equivalent to about 4 days per month
- Net days available for normal operation per month: $30 - 1 - 2 - 4 = 23$ days

Thus the sustainable capacity would be no more than 75% of the operating capacity. A further reduction in sustainable capacity might be necessary based upon traffic patterns and service requirements. For example, traffic volumes might be much higher during the middle of the day and much lower on weekends, and users of the canal might expect at most modest delays during normal operations (i.e., delays related to maintenance or bad weather may be acceptable, but extensive delays related to congestion may be viewed as unacceptable). These considerations would reduce sustainable monthly capacity to less than 70% of operating capacity:

- Weekend days per month: 8–10 days, with perhaps traffic at 50% of normal volume, which is equivalent to 4–5 days of normal operation or about 15% of monthly capacity.

- Peak patterns of traffic: the lock must be functioning reasonably well during these peak periods, so there will be unused capacity the rest of the day. If peak period totals only 6 hours per day, then there will be idle capacity the rest of the day. If the off-peak volume is 50% of the peak volume, then the maximum daily capacity under normal operations and normal service levels will be only 75% of the operating capacity (100% for 6 hours and 50% for 6 hours produces an average utilization of 75% for the entire 12 hours).

If you consider these two factors together, there will be a further reduction in capacity:

- Operating capacity: 24 boats per day through the lock
- Sustainable capacity, without considering traffic patterns: 70% of operating capacity
- Adjustment for traffic patterns:
  - Weekends: 15% reduction
  - Weekday peaks: 25% reduction
- Sustainable capacity, taking into account traffic patterns: If these factors are considered to be independent, the sustainable capacity will drop to less than half of the operating capacity: (0.70)(0.85)(0.75) = 45% of operating capacity, or 11 boats per day.

However, there is certainly some overlap among these various factors. Storms may occur on weekends; and if they do, they will not disrupt operations as much as if they occur on weekdays. Providing capacity for acceptable service during peak periods may also make it easier to schedule routine maintenance during slow periods of the day or the weekends. Miscellaneous delays that occur during off-peak times will not seriously affect capacity. The estimate of sustainable capacity is therefore perhaps 50–60% of the operating capacity, or 12–15 boats per day, or 360 to 450 boats per month.

Is this an acceptable estimate? Do we really believe all of these assumptions? Wasn't that just hand waving and magical thinking when the sustainable capacity was increased from the calculated 45% to a rather broad range of 50–60% of the operating capacity? The answer to all of these questions may be "maybe"! We could study canals in much greater detail to refine the assumptions, and we could develop simulation models to look much more closely at the effects of traffic patterns on service levels. However, rather than spending a lot of effort trying to answer these questions about methodology, let's look at several more aspects of capacity—namely, the loads carried on each canal boat, the ability to operate throughout the year, and the ability to adjust operations as needed to keep up with demand. Perhaps the estimate of sustainable capacity is acceptable as it is.

First of all, the size of the average load is very important. The maximum capacity measured in tons per day could be based upon the size of the largest boat that can use the canal. So, if the canal can be used by boats that carry 15 tons, then the maximum capacity is 24 boats/day × 15 tons/boat = 360 tons/day. If the largest boat can carry 50 tons/day, then the maximum capacity is 1200 tons/day. In either case, the average load—and therefore the operational capacity—is considerably less than the maximum load, especially if some of the boats carry loads in only one direction. If the average load were estimated to be 11–12 tons/boat, that would be a reduction of 20–25% in capacity measured in terms of tons/day—and another reason for being somewhat imprecise in our measure of capacity.

Weather is an even larger factor. Canals are unusable when the water levels are too high or too low, or when they are frozen over. Depending upon the climate, periods of inoperability may last for months or half the year, and these periods can vary greatly from one year to the next. If weather shuts down the canal for 3 months, then annual capacity is reduced by 25%; if weather shuts down the canal for 4 months, then annual capacity is reduced by 33%. There could easily be a 5–10% variation in annual capacity depending upon the weather.

Finally, the users and operator perhaps have considerable options regarding operating and pricing policies. All of the above estimates assumed operations of 12 hours per day. Is this a credible constraint? Would it not be possible to add some extra hours of operation during the peak season? And couldn't some of the off-peak capacity be utilized by reducing lock fees for these periods in order to promote somewhat different usage patterns?

Hold on! These are too many questions for this stage of the analysis! What can we conclude with some certainty? We think that the locks on the canal can probably handle close to 24 boats per day when things are going well, but that they may not be able to handle even half that much on a sustainable basis due to a variety of potential problems. We also think that the canal will be able to operate for about 8 months of the year. If we take 10 boats per day as the sustainable capacity, that would be 150 tons/day and 36,000 tons in 8 months. We know this is a rather fuzzy number, but perhaps it will be sufficient. Now let's turn to costs and competition. Will there be a market for the canal?

2.8.3   Canal Competitiveness

The early canals competed with horse-drawn wagons. The canals had a marked advantage in cost, as illustrated by the immediate success of the Middlesex Canal, which opened for operations between New Hampshire and Boston in 1803:

> *The advantages of canal travel over wagon transport were obvious at once. One horse, for example, could easily draw 25 tons of coal on the canal. On land the same horse could pull only 1 ton. One team of oxen could pull 100 tons, an amount that would take eighty teams on land. In the first eight months of the canal's operation, 9,405 tons were carried at a cost of $13,371. The cost for such a shipment by land would have been $53,484.*
>
> *Source*: Daniel L. Schodek, *Landmarks in American Civil Engineering*
> (Cambridge, MA: MIT Press, 1987), 12.

These estimates of operating cost for this 27-mile canal can easily be converted to the cost per ton-mile of transporting freight by canal boat or by wagon. Expressing cost as the cost per ton-mile is useful because it allows a normalized comparison among different modes of transportation, different lengths of haul, and different time periods. Assuming that all of the 9405 tons were transported the entire length of the canal, the cost per ton-mile (tm) of using the canal was $13,371/(27 miles $\times$ 9405 tons) = $0.056/tm. If the distance by road was also 27 miles, then the cost of using a wagon was four times as large: $53,484/(27 miles $\times$ 9405 tons) = $0.21/tm. These numbers are very interesting because they are considerably larger than the costs of transporting freight along the inland waterways and highways in the twenty-first century! Transporting coal on the inland waterways or on railroads now costs less than $0.02 per ton-mile, while the cost of truck transportation is well under $0.10 per ton-mile. (And a penny in 1803 was worth a whole lot more than a penny is worth some 200 years later!)

2.8.4   Engineering-Based Cost Model for a Canal

Knowing the cost per ton-mile is interesting historically, and it is a useful indicator that customers or managers might use. Being able to estimate the cost per ton-mile as a function of design factors and operating conditions is essential to planning and evaluating a transportation project. It is mildly interesting to know that the Middlesex Canal cut costs by 75% relative to horse-drawn wagons, but a canal designer and his financial supporters will want to understand the potential costs and benefits related to constructing a specific canal that would attract traffic from wagons. Given assumptions about unit costs and productivity, it is possible to create an engineering-based cost model for a canal. Let's assume that the operating costs and productivity parameters for a canal are based upon typical values for the early nineteenth century:

- Cost for the two-person boat crew ($1/day each, for 10 working hours)
- Cost for the teamster and the horse ($1/day each, or $2/day total)
- Cost for the boat ($2/day for a boat with a capacity of 15 tons)
- Cost for lock operations ($2/day for an operator and routine maintenance)
- Cost for canal and embankment maintenance ($40/year per mile)

- Average speed (3 miles per hour along the towpath)
- Average time per lock (12 minutes for a 15-ton boat)
- Annual operations: 225 days per year
- Annual tonnage along the canal: 10,000 tons

Using these assumptions, we can estimate the variable costs of operation for a canal of any length and with any number of locks. For example, let's calculate the variable cost per ton-mile for a trip by a 15-ton canal boat along a 30-mile canal that has 10 locks. First we need to know how long the trip will take:

- Travel time along the canal: 30 miles/3 mph = 10 hours
- Time in the locks: 10 locks × 0.2 hours per lock = 12 hours

The boat operates 10 hours per day, so the trip will take 1.2 days. Perhaps the crew finishes unloading in morning at one end of the canal and hopes to reach the other end in time to start unloading the following afternoon. We will assume that the cost of the trip includes 1.2 days for the crew, the boat, the teamster and the horse:

- Boat Crew: 2 people × $1/day × 1.2 days = $2.40
- Teamster and horse: one team × $2/day × 1.2 days = $2.40
- Boat cost: $2 per day × 1.2 days = $2.40
- Total variable cost = $7.20
- Total ton-miles = 15 tons × 30 miles = 450 tm
- Variable cost per ton-mile = $7.20/450 = $0.016/tm
- Variable cost per ton = $7.20/15 = $0.48/ton

The fixed costs include the cost for the lock operators (who are assumed to be available whether or not there is any traffic) and the cost for canal maintenance.

- Lock operations: ($2/lock per day) × (10 locks) × (225 days per year) = $4,500 per year
- Maintenance: $40/mile/year × 30 miles = $1,200 per year
- Total fixed cost: $5,700 per year
- Total fixed cost per ton: ($5,700/year)/(10,000 tons/year) = $0.57/ton
- Total fixed cost per 15-ton load: $8.55

The total, fully allocated cost for the trip is the sum of the variable cost and the allocated portion of the fixed cost: $7.20 + $8.55 = $15.75. The total, fully allocated cost per ton-mile is $15.75 per trip/ 450 ton-miles per trip = $0.035. This cost is well below the estimated cost of $0.21 per ton-mile of using a horse and wagon, so the canal likely could attract traffic and earn profits for the investors.

By incorporating the cost and capacity calculations used in this example into a spreadsheet, we can explore how costs and capacity will vary with differing assumptions concerning the structure of the canal, the size of the canal boats, the operating parameters, and the unit costs. Such a model can also be used to see how costs per ton-mile will vary with the annual tonnage on the canal and the size of the boats.

2.8.5 Evaluating the Canal Project

From the designer's perspective, critical questions would concern the size of the canal: should it be built to accommodate only 15-ton boats, or should it be built to handle larger boats? To handle larger boats, the canal had to be a little wider and deeper, and the locks had to be larger, so the costs of construction would rise. On the other hand, with larger boats, the variable costs per ton-mile were expected to decline, since the same crew could handle a larger boat.

From the owner's perspective (or the perspective of the banks and investors who were providing the funds for the project), the key question was whether they could charge tolls sufficiently large to cover their fixed costs of $5,700 per year and yield sufficient profit to justify their investment. To figure out how much they would need to charge for a toll, we need an estimate of what it would cost to build the canal. Canal costs were on the order of $20,000 per mile in the early nineteenth century, so that a 30-mile canal would cost about $600,000 to construct. This money had to be raised from investors who would expect (or at least hope) to receive a substantial annual dividend. A 10% dividend amounts to $60,000 per year, so the total amount of money raised by the toll would have to be $65,700 per year. If this were to be raised by a toll based upon tonnage, the average toll would have to be $6.57 for the expected 10,000 tons/year.

Is this a realistic toll? To answer this, we need to consider the perspective of the user. By using the canal rather than a horse and wagon, the user transports freight at a variable cost of $0.48 per ton. The toll would raise this cost to more than $7/ton or $0.25/tm for the 28-mile canal—which is more than the cost of using a horse and wagon, which was estimated above to be $0.21/tm! With such a large toll, the canal would have difficulty attracting any traffic at all.

Upon hearing this sad news, the investors would have several options. They could cancel the project as unprofitable. They could settle for a smaller return on their investment; cutting the toll to $3.57 per ton would provide a 5% return, and it would keep the cost per ton-mile well below the cost of the competition. They might also conduct a more careful study of demand, including an assessment of the potential for growth in canal traffic over a 10- to 20-year horizon. They could also consider building the canal for larger boats, a strategy that would further reduce operating costs for users, possibly for a fairly modest increase in construction costs.

Notice how the degree of precision has softened as we have progressed through this example. We started with some concern over the many assumptions we were making about capacity and operating costs. By the time we got to the end, however, we discovered that the largest cost by far was the return on investment that would be required to attract investors. Given the cost of the canal, the required toll would be an order of magnitude larger than the users' operating costs, and the major problem with the project appears to be that there is not enough traffic to justify the investment.

To reach this conclusion, it doesn't matter much—if at all—whether the canal boats move along at 2 or at 3 mph, and it doesn't matter whether the locks take 12 minutes or 15 minutes. It also doesn't matter whether the capacity of the canal is 50% more or less than our preliminary estimate of 36,000 tons per year. Why? Because our projected demand is barely 25% of that amount; adequate capacity seems to be assured. The lack of precise estimates of operating costs and capacity doesn't matter nearly as much as what we have discovered to be much more important considerations:

- How much traffic will use the canal—when it first opens and over the next 10 to 20 years? Perhaps the investors can defer dividends for a few years in order to secure very attractive profits in the future.
- What will it really cost to build the canal, and how great a return on their investment will investors require?
- Should the canal be redesigned to handle larger boats, which might attract more traffic?

2.8.6   Examples of Canals

China constructed its Grand Canal more than 1300 years ago.[3] Linking Beijing with the country's major river systems and ultimately the coast, this canal provided a means of transporting a steady supply of grain from the south to the north of the country. During the seventh century, 300,000 tons per year of grain were transported along this route. The canal was an enormous undertaking: 5.5 million laborers worked 6 years on one 1500-mile stretch of the canal (20 person-years per mile).

England's industrial revolution got a strong push when canals were constructed that provided cheap transportation for coal, agricultural products, and everything else.[4] The Bridgewater Canal, built in

---

[3] E. L. Newhouse, ed., *The Builders* (Washington, DC: National Geographic Society, 1992), 29.

[4] Newhouse, *The Builders*, 30.

1761 to link Manchester with coal mines, halved the price of coal in Manchester and helped Manchester become England's leading industrial center. By the 1840s, the country had a network of 5000 miles of canals and navigable rivers, and nearly every city or town was within 15 miles of a canal. As the country prospered, canals were built to be straighter, wider, and deeper; aqueducts were constructed to allow canals to cross rivers.

The Potowmack Canal was the first extensive system of river navigation in the United States.[5] Championed by none other than George Washington, the canal was designed to open up the area west of the Appalachian Mountains by providing a water route to the Potomac River (current-day Washington, D.C.). The canal allowed boats with a 16- to 20-ton payload to make the 185-mile trip in 3 days at half the cost of transporting the same freight by horse and wagon. The canal ran into problems because of recessions, lack of skilled workers, and bad weather. The route was navigable for only 3 months of the year, the canal tended to fill up with sediments, and the wooden locks decayed. The canal did help spur investment in and development of the western region, but it was a financial failure. After investing $750,000 between 1785 and 1802, the canal company was $175,000 in debt by 1816.

The Middlesex Canal, a similar project, was aimed at providing a better link between Boston, Massachusetts, and the farms and forests of New Hampshire.[6] The 27-mile canal required the construction of 50 bridges, 8 aqueducts, and 27 locks. The investment of $528,000 ($20,000 per mile) was, for the time, an enormous amount equal to 3% of the assessed value of all property in Boston. The canal suffered because the freight—what little there was—was mostly southbound. Despite the small volume of freight, political disputes erupted because people in New Hampshire did not appreciate a company chartered by Massachusetts diverting freight from New Hampshire's major port.

The Erie Canal was the most ambitious canal project undertaken in the United States during the nineteenth century.[7] The project was first proposed in 1724, and it was widely discussed for nearly 100 years as a means of linking New York City and the Hudson River with the Great Lakes at a point to upstream of Niagara Falls. Due to the geography of the eastern United States, the Erie Canal route was the easiest way to get from Atlantic ports across the Appalachian Mountains. Thomas Jefferson called the Erie Canal "a splendid project—for the twentieth century!"

Gaining support for construction of the canal required a major political effort, and not only because of the difficulty in financing such a large project. There was uncertainty about the route; an inland route would be expensive, but it would avoid exposing trade to attacks from Canada if the route used Lake Ontario. Merchants who used ground transport were against the project because it would expose them to competition. There was also a lack of skilled engineers for carrying out the project—and the project therefore led to the creation of civil engineering schools at Rensselaer Polytechnic Institute and Union College.

DeWitt Clinton was a member of the commission formed in 1810 to consider the construction of the canal. As a former mayor of New York City, a U.S. senator, and eventually as governor of New York, Clinton was the foremost champion of the project, and it was finally approved by the state legislature. The 363-mile canal, with 83 locks and 18 major aqueducts between Albany and Buffalo, was constructed between 1817 and 1825 at a cost of $8 million plus the loss of 1000 lives from malaria and pneumonia.

To limit costs of construction, the canal was built just wide enough (40 feet) and deep enough (4 feet) to handle medium-sized boats. When the canal opened, demand and revenue exceeded all expectations, and it was possible to finance projects that increased the canal to a width of 70 feet and a depth of 7 feet.

The long-term impact of the canal was immense. According to architectural historian Daniel Schodek, opening up Lake Erie was the "decisive impetus" for commerce in the eastern United States to move east-west rather than north-south. Rochester and Buffalo became boom towns, and population in New York increased all along the route of the canal. The success of the Erie Canal sparked development of a system of canals in Ohio, as population and economic development in the country pushed further westward.

---

[5] Daniel L. Schodek, *Landmarks in American Civil Engineering* (Cambridge, MA: MIT Press, 1987), 3–6.

[6] Ibid., 5–12.

[7] Ibid., 13–19.

2.8.7 Epilogue: Canals versus Railroads

Table 2.9 summarizes the costs and operating characteristics for the main transport options in the early nineteenth century. Turnpikes provided a way to achieve substantial improvements over rough roads, and they could easily be financed by tolls. Canals cost two to four times as much to construct, but they cut freight transport costs to a third of the costs of using turnpikes. Canals enjoyed only a brief period of supremacy in the United States, because they were restricted by topography and their service was much slower than what was possible once railroads were introduced. Railroads were similar to canals in terms of construction cost and operating cost, but they were much faster and therefore much more attractive for passengers and for most kinds of freight.

2.8.8 Lessons for Other Infrastructure Projects

This brief review of a few major canal projects in the United States provides some useful lessons regarding projects:

- Ideas and concepts for major infrastructure projects may abound long before the means to build the infrastructure are available.
- Important public figures may become champions for particular projects.
- Major projects can, like the Erie Canal, be decisive in directing development and population growth; but it is also possible to spend major resources on projects like the Patowmack and Middlesex Canals, which have only modest potential.
- Changes in technology can kill projects (railroads quickly put these canals out of business by the mid-nineteenth century).
- Financing is a major concern.

The discussion of canals also offers some insight into the different perspectives of the various participants in evaluating projects. If potential users' costs are lower, they will use the facility. In deciding whether to use a new canal, potential users had to ask whether they could lower their freight costs by using canals rather than horse and wagon, or by using large canal boats rather than smaller boats. Potential users therefore had to compare their costs for equipment and operations for their current and newly available options.

Owners or entrepreneurs have a different question: should they build the facility? They have to compare annual revenues to annual costs, taking into account the costs of construction and the continuing costs of operating and maintenance. If they are going to borrow money, they must be able to pay back the interest. If they are going to charge tolls, they have to compare the amount of the toll to what users would actually save by using their facility. Set the toll too high, and nobody will use the facility.

Potential investors have a simpler and more direct question: if they put their money into the project, can they recover their investment plus a reasonable return? They should be worried about the feasibility of the project, the time it could take to complete the project, and the ability of the project to

Table 2.9 Comparison of Transportation Costs, First Half of the Nineteenth Century

Rough Road $1–2,000/mile to construct	1 ton/wagon 12 miles/day 12 tm/day/vehicle	$0.20 to $0.40/tm for freight rates
Turnpike $5–10,000/mile	1.5 tons/wagon 18 miles/day 27 tm/d/v	$0.15 to $0.20/tm
Canal >$20,000/mile	10–100 tons/boat 20–30 miles/day 200-3000 tm/d/v	$0.05/tm
Railroad $15–50,000/mile	500 tons/train 200 miles/day 100,000 tm/d/v	<$0.05/tm

actually generate revenue. They do not necessarily have any interest in the details of the construction or the operation, and they will be comparing this project with completely different options for making money.

Contractors may not care at all about what the project ultimately accomplishes, as long as they can complete their portion of the project on time, safely, and on budget. They are very interested in trying to predict construction costs, in choosing the most effective technologies and materials, and in planning and managing the process. They must determine whether the potential profits from the project are worth the risks that they perceive to be associated with the project.

2.9   SUMMARY

Transportation, water resource, and other infrastructure-based systems serve various needs of society. One aspect of the performance of such systems is therefore how well they actually serve those needs, as measured by the cost to the user and the quality of the service they receive.

Average, Marginal, and Incremental Costs

Total, average, marginal, and incremental costs are all important aspects of system performance. The average cost is the total cost divided by the total volume of usage; the units could be the cost per vehicle for a highway or the cost per gallon for a municipal water system. The marginal cost is the cost per additional user, which would be the cost for one more car on a highway or for an additional gallon of water. Pricing and operating decisions are often based upon marginal costs rather than average costs. The incremental cost is the added cost for a larger increment of users (e.g., a 10% increase in vehicles or water usage). Investment and longer-term operating decisions are often based upon consideration of incremental costs: what is the best way to handle the expected increase in usage over the next 5 years? It is possible to construct engineering-based cost functions that can be used to estimate costs based upon the engineering and operating characteristics of the system. A case study showed how the costs of constructing and using a canal could vary with the width, depth, and length of the canal; the number of locks; and the maximum size of boats that can be used. The canal case study also showed how building a larger canal could reduce travel times and operating costs for the users.

Fixed versus Variable Costs

For many systems, it is useful to consider the distinction between fixed and variable costs. Fixed costs are those associated with making the system available, while variable costs are those that vary with the level of usage. For investments in infrastructure, there are often options that can provide better service or higher capacity, but that require higher fixed costs. A common question is therefore to decide whether the demand will be sufficient to justify the alternative with the higher fixed costs.

Life-Cycle Costs

Since infrastructure lasts a long time, it is important to consider costs over the entire life of the infrastructure. Design costs are likely to be modest compared to the construction costs that follow. Once the project is completed, the costs of operation may be borne by the owner, users, or abutters. As the facility ages, it may be desirable to expand or to rehabilitate the facilities. Eventually, as a result of declining demand or excessive costs of maintenance, it will be necessary to decommission the facilities and to salvage metal or whatever is left of value. Ideally, the design of infrastructure-based systems will address total life-cycle costs, rather than simply construction costs. Adding room for expansion, allowing more efficient operations, using designs that facilitate maintenance, and ensuring safer operations may require additional investment, but these measures can result in lower costs over the life of the infrastructure.

The design phase offers the greatest opportunity to affect the life-cycle costs of a project. At this point, while all things are not possible, many things will be. A major design consideration is the extent to which the owner or developer considers the costs to users and abutters. Small changes in design conceivably produce more efficient operations or limit the negative effects on third parties, but only to the extent that such costs are even considered in the design. Mistakes regarding the size of the project—too big, too small, too soon, too little flexibility, too difficult to rehabilitate, too much impact on neighbors— conceivably can be rectified with little or no additional time or expense related to construction.

Financial Performance

Another aspect of performance that is critical for investors and for accountants is the financial performance—that is, the profitability and the return on investment for the system. The return on investment is the annual profit divided by the total amount of the investment. While financial performance is not the only measure—or not even the most important measure—if financial performance is deemed to be unacceptable, then it will be difficult or impossible to find investors (or taxpayers) willing to invest more money in improving or expanding the system. The interaction between supply and demand is an important factor determining the prices that can be charged for a service.

Capacity

A third aspect of performance is the capacity of the system to serve growing needs and the level of deterioration in service as demand approaches capacity. The maximum capacity of a system is limited by the initial design, engineering factors, and operating constraints. The maximum capacity may be useful in design, but it cannot be achieved except for brief periods. Operating capacity recognizes the importance of considering such things as the normal variations in volume, weather, and the need of routine maintenance. The operating capacity is what is achievable on most days when the system operates pretty much as planned. The limit to operating capacity is likely to be what users view as acceptable delays or restrictions on use during peak periods. The sustainable capacity is somewhat lower, because time must eventually be allowed for more maintenance and there will be periods when severe weather or accidents disrupt operations for hours, days, or weeks. Sustainable capacity for most systems will be on the order of 70% of maximum capacity.

Safety and Security

A fourth aspect of performance is the safety and security of the system—that is, the likelihood of accidents or disruptions to the system based upon system problems or attacks upon the system. Risk is a useful concept for considering safety and security issues, since risk is the product of two key factors: the likelihood of an accident or incident and the expected fatalities, injuries, and other consequences if there is an accident or incident. Probabilistic risk assessment is a methodology that can be used to determine the cost effectiveness of various options for reducing risk.

ESSAY AND DISCUSSION QUESTIONS

2.1 "In the United States, railroads put the canals out of business; but in Panama, a canal put the railroads out of business." How can this be? Discuss the logic underlying this apparent paradox.

2.2 If urban sprawl is so bad, why do we have so much of it? When you fly across the country, why is so much of it empty? What is the marginal value of farmland, why are farmers going broke, and what happens to all the chemicals in that fertilizer? Is suburbia worse than a forest or another cornfield? Comment on the nature of urban/suburban development and the types of projects that might produce better places for us to live—with each other and with nature.

2.3 The twenty-first century is viewed by some as the century when we begin to fight wars over access to water. Hundreds of millions of people worldwide lack easy access to clean water. Los Angeles and many other large cities seek farther and farther for water, with severe consequences for the environment.

2.4 In the latter part of the twentieth century, Boston Harbor suffered from pollution caused by the release of untreated or poorly treated sewage and by stormwater runoff. The harbor's potential for swimming and fishing was limited by the poor water quality. Boston completed a multibillion-dollar cleanup of the harbor that featured the construction of a modern sewage treatment plant and improvements in the sewer system. The project was paid for in part by increasing the rates that consumers paid for water. When water and sewer rates quintupled in the 1990s, there was great public consternation. With these new rates, a typical Boston family's water bill averaged about 25 cents per person per day. Is this a reasonable price to pay for cleaning up Boston Harbor? How do you think people responded to the higher rates?

2.5 What is the difference between safety, security, and risk?

2.6 Bad weather can lead to unsafe conditions on highways. Many accidents are caused by cars that lose traction when roads are snowy or icy. Discuss ways to reduce the risks associated with driving during winter storms.

PROBLEMS

2.1 Define the following terms:

- Total cost
- Fixed cost
- Variable cost
- Average cost
- Marginal cost
- Incremental cost
- Opportunity cost
- Life-cycle cost
- Profit
- Return on investment

2.2 The cost of constructing an office building is estimated as follows:

- Acquisition of land: $2 million
- Site preparation and landscaping: $1 million
- Foundation for a building of up to 25 stories: $5 million
- Lobby and ground-level shops: $3 million
- Each floor of office space: $2 million
- Roofing: $5 million

a. If each story has 10,000 square feet of rentable space, what is the total cost and the average cost per square foot for a 20-story building?
b. What is the marginal cost of adding one story to the building?
c. Write an equation for the total cost, average cost per story, and marginal cost per story of an office building that could be constructed on this site; the building can have from 10 to 25 stories.
d. Plot the total, average, and marginal costs of an office building as a function of the number of stories in the building.

2.3 The cost of constructing a football stadium is estimated to be $100 million for a facility that will include the field, dressing rooms for the teams, offices for management, space for the press, and seating and amenities for 50,000 fans. More seating can be added either by making the stadium larger, to accommodate up to 10 additional rows of seats, and/or by adding a second tier of seating for some or all of the stadium. Each added row of seats would increase capacity by 1000; the cost for each additional row is estimated to be $2 million. A second tier covering a quarter of the stadium would provide seating for 5000 fans, and three such tiers could be constructed. The cost for each such tier is estimated to be $20 million.

a. What is the most cost-effective way to increase the seating capacity of the stadium from 50,000 to 55,000? From 50,000 to 65,000?
b. Plot the total cost of constructing a stadium as a function of the capacity of the stadium for 50,000 to 75,000 fans.
c. Which concept—incremental cost or marginal cost—is more useful for discussing the size of the proposed stadium?

2.4 *Long-run average costs*: You have three options for constructing a transportation network, as shown in the table. Assume that the investment will last indefinitely. Assume that the money needed for the project will be raised by selling 30-year bonds with an annual interest rate of 10% (i.e., the $10 million project must pay $1 million per year in interest).

a. Sketch the *long-run average cost curve* that includes all three categories of cost (show the long-run average cost per vehicle for annual vehicles from 100,000 to 1 million per year).
b. Indicate the range of volumes for which the standard technology is the best choice.

Problem 2.4 Options for a Transportation Network

Technology	Investment	Annual Fixed Costs	Variable Cost/ Vehicle	Capacity Vehicles/ Year
Low cost, low capacity	$10 million	$1 million	$20	400,000
Standard	$20 million	$2 million	$7	500,000
High capacity	$30 million	$3 million	$2	2,000,000

2.5 A communications company is considering providing high-speed Internet access/cable TV to rural regions. The estimated cost of providing the basic cable network is estimated at $200 million, with annual maintenance and fixed operating costs of $20 million. Rural customers will be expected to pay a modest one-time fee to cover the costs of laying cable to connect their house or business to the regional network. Variable operating costs are expected to be $5/month per customer. If the monthly connection fee is $100, how many customers must sign up for the service before the company starts making a profit?

a. 16,666
b. 17,544
c. 20,000
d. 210,000

How many customers must sign up for the service for the company to make a 10% return on its $200 million investment?

a. 20,000
b. 35,000
c. 200,000
d. 420,000

2.6 One way to reduce the risks associated with driving in bad weather is to purchase a vehicle with four-wheel drive (4WD). With 4WD, the probability of an accident is reduced because the car has better traction and is less likely to spin out of control. For example, suppose that someone driving a car with 4WD is only half as likely to have an accident or a fatal accident as someone driving the same model car without 4WD. Now consider whether it is worth spending an extra 10% ($2,000) to purchase the optional 4WD for your new car. You live in a region that gets a lot of snow, and you must commute a long distance to work. You estimate that the likelihood of having a severe snow-related accident sometime during the life of your car is on the order of 50%, with a 0.5% chance that the accident would result in a fatality. Would the extra cost of 4WD be justified by the reduction in risk?

2.7 You are continuing to debate whether to buy a car equipped with four-wheel drive. One of your friends says that drivers of such cars are involved in more accidents, because they become overconfident and drive too fast when road conditions deteriorate. The table below shows the risks that might be associated with a particular trip as a function of whether the car has 4WD, the speed, and the road conditions.

Problem 2.8 Risks Associated with Driving Conditions

Four-Wheel Drive?	Speed	Conditions	Accident Probability	Probability of Fatal Accident
No	60 mph	Poor	.01	.0001
No	60 mph	Terrible	.05	.0005
No	40 mph	Poor	.001	.0001
No	30 mph	Terrible	.01	.00005
No	20 mph	Terrible	.005	.000005
Yes	65 mph	Poor	.005	.00005
Yes	65 mph	Terrible	.01	.0001
Yes	50 mph	Poor	.001	.00001
Yes	40 mph	Terrible	.005	.00005

a. Does having 4WD reduce the risks of driving in poor or terrible conditions? If not, is there any benefit of having 4WD?

b. What difference does it make in deciding whether to buy a car with 4WD if you have flexibility in what trips you make or in when you make trips?

c. What difference does it make if you plan to move from a place with severe winters to a place with mild winters?

2.8 A company is considering whether to build a warehouse next to one of its existing manufacturing plants. The warehouse would store products from five different manufacturing facilities that are scattered across the country. Several factors make this the ideal site compared to several locations that were considered in other states. First, the company has plenty of vacant land, so it would not have to purchase a site for the building. Second, the warehouse operation could be managed effectively by people who currently work at the existing site; no one would have to be relocated if the warehouse is built at this location. Third, the company has recently experienced a drop in orders, which may necessitate layoffs for several dozen employees; if the warehouse were built, many of these people could still be employed. Fourth, the other sites might allow a reduction in truck miles traveled to and from the warehouse, but the company has its own trucks and currently has a surplus of truck drivers, so no additional trucks or drivers would be needed. How should these four factors be considered in deciding whether to locate the warehouse at this site?

2.9 A consultant has been hired to recommend whether to construct a new manufacturing plant for a state-owned cement manufacturing company in Africa. The company currently has an old facility that requires more workers per unit of output and much more energy than would be required in the more modern facilities used in Europe and Japan. The production process also generates a lot of pollution, but that is not perceived as a problem since the plant is located far from the wealthier neighborhoods of the city. As a state-owned enterprise, the company is required to provide employment for people who otherwise cannot find a job; about 10% of the workforce currently consists of poorly paid people who sweep the halls, serve coffee, greet visitors, and work on landscaping for the factory. Although the facility uses a lot of energy, that is not a problem; the country produces oil and makes fuel available to government-owned enterprises at a price that covers the cost of production, but that is less than half the world price for oil. It is clear that a company in a developed country would upgrade the plant, but company management so far has been reluctant to invest in what they perceive to be a very expensive project. How should the consultant deal with the issues related to low labor productivity, the make-work workforce, the pollution, and energy consumption? Which of these factors, if any, might tilt the recommendation toward building a more modern plant?

2.10 Coal is the fuel most commonly used in the United States to produce electricity. Burning coal is also one of the main causes of air pollution and emission of greenhouse gases, and therefore a big target for those interested in reducing emissions to slow global warming. The cost of mining coal is on the order of $5 to $10/ton, while the cost of transporting coal from the mine to a power plant is on the order of $20 to $25/ton; the delivered cost of coal is therefore about $30/ton. Go online to find out the amount of (a) carbon dioxide produced by burning a ton of coal and (b) the amount that would likely be charged per ton of carbon dioxide if some sort of carbon tax were imposed. How much would the cost of coal increase?

2.11 Each gallon of gasoline used by an automobile produces 22 pounds of carbon dioxide. How much would a carbon tax increase the cost of gasoline for someone who drives 20,000 miles per year in a car that gets 20 miles per gallon? For someone who drives 7500 miles per year in a car that gets 35 miles per gallon? (As in the previous problem, use online sources to obtain an estimate of what the tax per ton of carbon dioxide might be if such a tax were imposed.)

CASE STUDY: DEVELOPING COST AND REVENUE FUNCTIONS FOR A NINETEENTH CENTURY CANAL

This assignment examines the costs, capacity, and service capabilities of a canal that is similar to the Middlesex Canal described in this chapter. You will be asked to develop cost and capacity functions that you can use to address questions about the design of the canal.

1. Capacity Model
The capacity of a canal (maximum throughput measured in tons of cargo) could be estimated for various time periods, for example, a peak day with 12 hours of operation; a peak summer month with 12 hours operation, 7 days/week; for a year, with operations ceasing during the winter and during major storms. Boats can tie up for the night at either end of the canal or at frequent locations along the length of the canal; as a result, boats will likely be dispersed all along the canal at the beginning and end of the day.

Questions
a. Develop a spreadsheet model of a canal's capacity as a function of the characteristics of the canal, the boats, the operating characteristics (e.g., speed through the canal, cycle time of the locks [the time required for one boat to pass through the lock], the hours available for operation per day and week, and the months available for operation per year).
b. Use this model to estimate the capacity (tons/month) of a canal that is 30 miles long, has 10 locks (1 lock at each of 10 locations), and is wide enough for two boats to pass assuming that it is designed for (i) 15-ton, (ii) 25-ton, or (iii) 75-ton boats.
c. What are the most important factors affecting the capacity of the canal?

2. Operating Cost Model
Estimate the benefits (cost savings) from constructing a canal that would attract traffic from wagons. Assume that the cost per ton-mile for wagon transport is constant at $0.056, as described in Section 2.8.3.

Questions
a. Develop a spreadsheet model for estimating the cost per ton and cost per ton-mile of moving freight along a canal similar to the Middlesex Canal. Use the unit costs given in Section 2.8.4 for freight moving in boats with a capacity of 15 tons. If freight moves in 25-ton boats, the cost of the boat increases from $2 to $2.50 per day and the time per lock increases from 12 to 15 minutes. If freight moves in 75-ton boats, the cost of the boat increases from $5 per day and the time per lock increases to 20 minutes. The traffic volume (tons/year), length of the canal, number of locks, hours of operation per day, days of operation per week, and months available for operation should all be variables in your model. Show results for a canal that is 30 miles long and has 10 locks.
b. Use your model to estimate the operating cost for 15-, 25-, and 75-ton boats, assuming that annual traffic varies from 10,000 to 100,000 tons per year.
c. What are the most important factors affecting the operating cost of the canal?

NOTE: Question 2 requires you to estimate the fixed and variable costs of operations; to get the total cost per ton at a particular level of traffic, you must allocate the fixed cost to the tonnage that is handled.

3. What Size Boats?
Assume that the state legislature (which has allowed you to build the 30-mile canal) has authorized you to charge no more than 50% of the wagon costs per ton-mile, no matter what size boat you use. Further assume that you can raise money for investment that equals 10 times the annual operating profits (i.e., if projected operating profits were $10,000 per year, you could raise $100,000 to construct the canal). The Middlesex Canal cost $528,000, which is just under $20,000 per mile. To keep the calculations simple, assume that the construction costs per mile would be $20,000, $25,000, and $30,000 to handle the three sizes of boats (thus operating profits would have to be at least $2,000, $2,500, and $3,000 per mile respectively before the investors would be willing to provide enough funds to construct the canal). How much annual tonnage would be needed to justify building a canal that handled (a) 15-, (b) 25-, or (c) 75-ton boats?

NOTE: You can approach this problem as a break-even analysis. The contribution per ton (i.e., revenue per ton – variable cost per ton) multiplied by total tonnage must be sufficient to cover the fixed operating costs of the canal and leave enough operating profit to justify the investment.

4. What Role for Analysis?
Explain how analysis related to one of the following disciplines could be used to make *meaningful* improvements to your model of the performance of the canal: fluid mechanics; materials science; structural mechanics; soil mechanics; physics; probability and statistics; biology.

Chapter 3

Basic Economic Concepts

The quality of a nation's infrastructure is a critical index of its economic vitality. Reliable transportation, clean water, and safe disposal of wastes are basic elements of a civilized society and a productive economy. Their absence or failure introduces an intolerable dimension of risk and hardship to everyday life, and a major obstacle to growth and competitiveness.

National Council on Public Works Improvement,
Fragile Foundations: A Report on America's Public Works,
Final Report to the President and Congress (February 1988), p. 1

CHAPTER CONCEPTS

3.1 INTRODUCTION

This chapter introduces various economic concepts that are useful in understanding infrastructure systems and in identifying and evaluating potential projects for improving their performance. The chapter begins with a discussion of how equilibrium prices result from the interaction of supply and demand. If prices are high, say for office space or for energy, then a great deal of investment in new buildings or oil drilling or wind power is justifiable. If too many buildings are built or if too much oil is available on the world market, then prices fall and investments based upon continuing high prices may well fail. The success or failure of any major project depends in part upon the future interactions between supply and demand.

Costs, prices, and values are distinct concepts that should not be confused. The cost of providing a service or of manufacturing a product depends upon such things as resource requirements, capacity requirements, and unit costs associated with operations. Chapter 2 included many examples of cost and capacity functions. While owners surely desire that prices be higher than costs, prices are usually determined by market forces that may have little or no relationship to cost. The value of a product or a service is something that can be determined only by potential purchasers: if they perceive the value of a product or service to be higher than the price, then they will go ahead and make the purchase. The difference between what they were willing to pay and what they actually paid is an economic benefit known as consumer surplus, which is in fact an economic benefit even though it does not result in any revenue to the supplier. Large infrastructure projects are often justified in part by increases in consumer surplus, so this is an important concept for evaluating such projects.

From an economic perspective, a major goal of any project is to increase productivity, which is defined as the ratio of system output to system input. If productivity improves, then more output can be obtained using the same or fewer resources, resulting in an overall benefit for society. If a company can produce more without increasing its labor force, then it may be able to pay higher wages to its employees. Companies and agencies that manage infrastructure are continually seeking ways to make more productive uses of their resources, and productivity improvement motivates many infrastructure projects and programs. Section 3.2 defines various concepts related to productivity and identifies several common productivity problems that may hurt infrastructure performance. In most infrastructure systems, there are economies of scale, scope, or density that allow larger, more complex systems to offer more benefits at a lower cost.

Lower cost would seem to be a clear benefit to society, but project evaluation must consider who will capture the benefits of lower cost, the supplier or the customers? To answer this question, Section 3.3 considers pricing policies in competitive and monopolistic conditions. If there are many potential suppliers, then there will be competition for customers, prices will fall to marginal costs (marginal cost pricing), and customers will benefit from any productivity improvements. However, if a single supplier has no competition or very limited competition, then it will be able to charge prices that are well above marginal costs. The threat of monopoly pricing is therefore present whenever there are strong economies of scale. To achieve public benefits from scale economies related to essential infrastructure, it may be necessary to have public ownership or some sort of price regulation.

As noted in previous chapters, there are multiple reasons why infrastructure performance and major infrastructure projects will always be of interest to the public:

- The public uses the infrastructure, and the performance of the infrastructure affects everyone's daily life.
- Much of the public infrastructure is owned or regulated by public agencies, so there is a direct public interest in managing and investing in that infrastructure.
- Infrastructure projects are large projects with long-lasting impacts on society and the environment, and the public has a justifiable interest in questioning whether these impacts are positive or negative and whether the costs and benefits of a major project are equitably shared.
- Investment in infrastructure projects can provide a boost to the region in terms of jobs, income, and economic growth through what is called the multiplier effect.

Section 3.5 discusses the regional economic impacts related to infrastructure investment, while Chapter 4 focuses on the social and environmental impacts.

Infrastructure needs depend in part upon the economic forces that drive regional, national, and international development. Where goods are produced depends in part upon where raw materials can be found, where it is most efficient to produce the goods, where labor and other resources are cheapest, and the cost of transportation. As transportation costs decline, because of improvements in technology and expansions of transportation infrastructure, distance ceases to be an impediment to consolidation of agriculture, manufacturing, mining, and other industrial activities. Cheap transportation has enabled the rise of a global economy, and regions in one country now compete with regions in other countries for all sorts of economic activities. Section 3.6 presents two concepts that are directly relevant to understanding the global economy: spatial price equilibrium and comparative advantage. As patterns of trade and production shift, the needs for industrial facilities and transportation infrastructure also shift. In the less developed parts of the world, investments in infrastructure may be required for the economic growth. In the developed parts of the world, existing infrastructure that was designed for the economy of the 19th or 20th centuries may need to be redeveloped or replaced by infrastructure relevant to the 21st, with greater emphasis on major ports and continental distribution systems and less emphasis on local production.

Most of this chapter addresses economic issues in rather broad terms; it considers such topics as globalization of the economy, the need for regulatory policy, corporate decision making, and the importance of regional economic impacts. However, it is also worth considering the perspective of the individuals who ultimately make the decisions that determine which types of infrastructure are used, how much revenue is gained, and whether infrastructure projects prove to be successful. Individuals decide such things as how much living space they need, whether to live in the city or a suburb, whether to water the lawn regularly, whether to drive a car or take the bus to work, whether to switch from oil to natural gas for home heating, and where to go on vacation. Section 3.7 introduces utility, an abstract concept that can be used to understand how these decisions are made. The basic idea is that individuals are assumed to make decisions that maximize their utility based upon personal constraints related to time and money. An example in Part III of this text (in Section 13.4.2) uses the concept of utility to address how intercity travelers might choose between flying, driving, taking a train, or taking a bus.

3.2   SUPPLY AND DEMAND

3.2.1   Supply, Demand, and Equilibrium

Supply, demand, and equilibrium are central issues in economics. At the most basic level, both supply and demand are described as functions of price, and the **equilibrium price** is the price at which supply equals demand. The **supply function** shows the quantity of goods or services that will be produced for each price. Under normal circumstances, the supply of goods and services is expected to increase as the price increases. If the price is higher, then existing suppliers will be willing to produce more, and new suppliers may be enticed to enter the market. The **demand function** shows the quantity of goods or services that will be purchased for each price. Under normal circumstances, the demand declines as the price increases. Some people may be willing to pay a high price, but more people are willing to pay when the price is lowered.

The interaction between supply and demand can conveniently be expressed in a chart (Figure 3.1). Note the convention that price is shown on the y-axis, although that is assumed to be the independent variable, while the volume or quantity of supply and demand are shown on the x-axis. The point where the supply and demand functions intersect is the equilibrium price. The most important idea is that this equilibrium price reflects both supply and demand: under competitive market conditions, prices will adjust to changes in supply and demand, and there is a tendency for supply to match demand.

Over time, factors that affect both supply and demand are subject to change. First consider changes in supply. Investing in new technologies or in more efficient production facilities or simply adopting

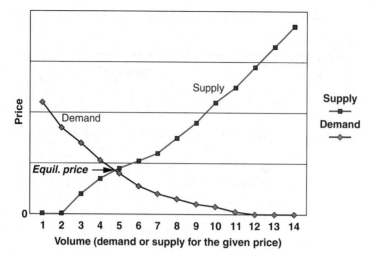

Figure 3.1 Prices Are Determined by the Interaction of Supply and Demand

better management techniques may make it possible for suppliers to offer greater quantities for any given price. Graphically, this results in a shift in the supply curve to the right and leads to a new—lower—equilibrium price, as shown in Figure 3.2. Note that the demand curve has not changed at all: with the lower prices, people are willing to buy more, which is what is described by the demand curve.

The demand curve may also change. For example, growth in population or increases in family income may result in an increase in cars purchased, attendance at movie theaters, or use of air transportation. These changes appear on the graph as an upward shift in the demand curve: at each price level, a greater quantity of goods and services are purchased or used.

If demand increases, then prices will rise; if demand declines, then prices will fall. How much prices rise or fall will depend upon the shapes of the supply and demand curves. How quickly prices rise or fall will also depend upon the nature of the goods and services being sold. Outside a sold-out baseball stadium, the prices that scalpers charge for tickets will react within minutes to changes in demand. Achieving an equilibrium in the prices of new homes is something that may take years, as evidenced by the steady decline in home prices that began in 2007 and continued through 2009 (not to mention the steady increase in prices for many years before that). Adjusting transportation networks to changes in oil prices or new technologies is a process that takes decades—and may never reach equilibrium,

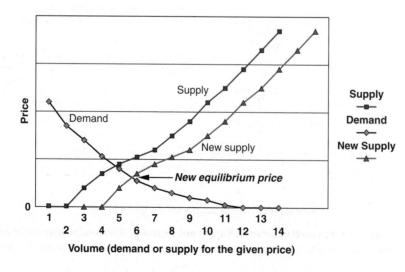

Figure 3.2 Investment May Allow Suppliers to Be More Productive and Lower Their Prices. This will result in a shift in the supply curve and a lower equilibrium price (resulting from a movement along the demand curve).

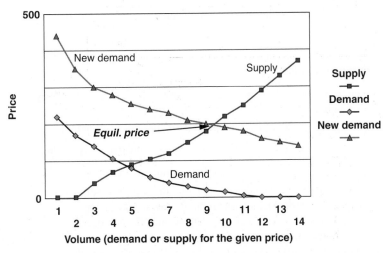

Figure 3.3 Population Growth, Advertising, Higher Incomes, or Other Factors May Increase Demand and Raise Equilibrium Price.

because only a small portion of the transportation network can ever be changed within just a few years.

In many circumstances, changes in demand result not in a change in price, but in poor service, congestion, or long lines as too many people try to buy something or to use something at the same time. The time spent in line can be viewed as part of the price of the service being sold: some people will come intending to buy but depart as soon as they see the line.

Over time, suppliers will react to changes in demand by adjusting their levels of production. New companies may emerge in response to increases in demand; companies may go out of business in response to decreases in demand.

A lot of time can be spent in trying to understand the supply and demand curves, and there are some ingenious methods for estimating these curves based upon past experience. However, it is important to retain some humility, for we probably know only a little about how supply and demand vary within a fairly small range of prices and existing conditions (Figure 3.4). When new projects are being considered, the quantity or quality of services provided may be far different from what is currently

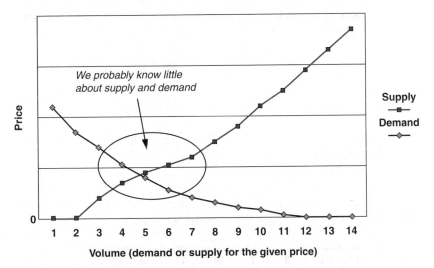

Figure 3.4 Interaction between Supply and Demand. This relation may not be readily understood, especially if the factors affecting supply and demand have been fairly stable for a long time.

available. Special studies can be undertaken to estimate the effects of the new projects on demand, but such studies will never be exact.

Consumer Surplus

Given the shape of the demand curve, it is clear that the equilibrium price is lower than the price that many would be willing to pay. The difference between what someone is willing to pay and the equilibrium price is called **consumer surplus**. For each individual:

$$\text{Consumer Surplus} = (\text{Willingness to Pay}) - (\text{Equilibrium Price}) \qquad \text{(Eq. 3.1)}$$

Consumer surplus is greatest for those willing to pay the highest prices. For someone willing to pay only the actual price and not a penny more, the consumer surplus is zero. The total consumer surplus in principle could be obtained by summing the surpluses for everyone using a product or service. In practice, this is infeasible because data is collected and decisions are made based upon actual prices. Unless special studies are undertaken, little is known about how much more people are willing to pay for things that they now buy or for services that they now use. For this reason, it is easier to focus on the changes in consumer surplus that may result from changes in equilibrium prices.

Consider the change in supply illustrated above in Figure 3.2. The shift in the supply curve increased consumer surplus by (1) lowering the price for those who previously were willing to pay a higher price and (2) allowing more people to purchase the product. The increase in consumer surplus can be estimated just by looking at prices and volumes before and after the change in supply:

$$\text{Increase in Consumer Surplus} = V_0(P_0 - P_1) + \tfrac{1}{2}(V_1 - V_0)(P_0 - P_1) \qquad \text{(Eq. 3.2)}$$

The first term in this equation is the benefit to existing users from the reduction in price; the second term represents the benefits gained by new users. The full decline in price is not a benefit for new users, since they were unwilling to pay the old price. If the relevant portion of the demand curve is assumed to be a straight line, then the consumer surplus for the new users will be the area of a triangle whose base (B) is the difference in volume and whose height (H) is the difference in price, and whose area is ½ BH. In effect, this assumption—sometimes called the **rule of ½**—provides a simple way to estimate the area under the demand curve without needing to estimate an equation for that curve.

Note that consumer surplus is an economic rather than a financial concept. Price is a financial measure, as is manufacturing cost; these are things that can be measured in dollars and cents, and these are things that can and will be recorded in checkbooks and accounting systems. Consumer surplus is not related to any such accounting, but it is still an important matter for evaluating the economic impact of projects. A public benefit results from projects that increase consumer surplus, because people will still have the money they otherwise were willing to pay for the product or the service. They can save that money or use it to buy something else. Either way, there is a benefit for the individuals and for the local economy. Thus change in consumer surplus, though not a direct concern for the private sector or for investors, is an important consideration in evaluating the public economic impacts of proposed infrastructure projects.

EXAMPLE 3.1   Supply, Demand, Equilibrium, and Consumer Surplus

A consultant has been asked to look at the possibility of producing and selling a new product that would be manufactured locally and marketed throughout the state. Tables 3.1 and 3.2 describe the results of the consultant's survey of potential customers and suppliers for this product. Table 3.1 shows the results of a survey of eight groups of people who were asked the question, "Would you purchase this product at the stated price?" (The stated price ranges from $100 down to $50.) This survey was believed to represent a 1% sample of the target population. Multiplying the survey results by 100 therefore provided an estimate of the regional demand for this product. Table 3.2 shows the results of the consultant's interviews with the four companies that were deemed most likely to manufacture this produce. Suppliers S1 and S2 had relatively high fixed costs and were uninterested in producing any units unless they reached a break-even volume. Suppliers S3 and S4 were smaller companies with lower fixed cost; they had the ability to produce a smaller number of units, but limited ability to expand their production.

Table 3.1   Results of a Survey of Eight Groups of People Showing Willingness to Purchase the New Product at the Stated Price

Group	$100	$90	$80	$70	$60	$50
A	3	4	5	7	8	8
B	2	2	3	6	7	8
C	6	7	7	10	11	12
D	4	5	6	7	7	7
E	8	10	11	15	16	17
F	7	9	11	14	14	15
G	3	4	6	9	10	11
H	5	5	6	7	7	7
Sample Total	38	46	55	75	80	85
Regional Total	3800	4600	5500	7500	8000	8500

Table 3.2   Results of a Study of Potential Suppliers: Units Supplied at the Stated Price

Supplier	$60	$65	$70	$75	$80	$85
S1	None	None	None	2000	5000	10,000
S2	None	None	None	None	10,000	20,000
S3	None	1000	2500	2500	2500	3000
S4	3000	5000	5000	5000	5000	5000
Total	**3000**	**6000**	**7500**	**9000**	**17,500**	**38,000**

Using the information in these two tables, the consultants created the plots of supply and demand as shown in Figure 3.5. Based upon this analysis, the consultant estimated that the equilibrium price would be about $67, and the equilibrium demand would be 7500 units. The consultant therefore concluded that the two large companies S1 and S2 would not enter the market, but the two small suppliers could produce enough to satisfy the market.

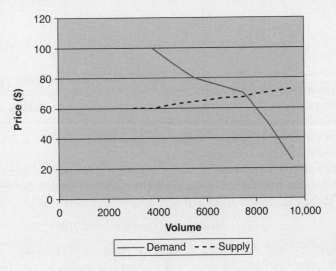

Figure 3.5 Plots of Supply and Demand for the New Product

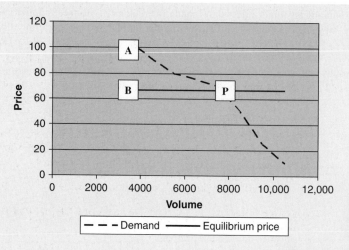

Price (y-axis), Volume (x-axis)

- - - Demand ——— Equilibrium price

Figure 3.6 Consumer Surplus is the Area ABP between the Demand Curve and the Equilibrium Price

This example can also be used to illustrate the concept of consumer surplus, which is the difference between what people are willing to pay and the actual price they are charged. In Figure 3.6, the consumer surplus is defined by the area between points A, B, and P. The consumer surplus can also be estimated by using the information from the consumer survey (see Table 3.1). The consultant estimated that 3800 people would be willing to pay at least $100 for the product. For them, the consumer surplus would be at least $33 each. Another 800 people would be willing to pay $90, and their consumer surplus would be at least $23 each (Table 3.3).

Table 3.3 Estimating the Consumer Surplus

Willingness to Pay	Equilibrium Price	Difference (consumer surplus per person)	Number of Customers	Total Consumer Surplus
$100	$67	$33	3800	$125,400
$90	$67	$23	800	$ 18,400
$80	$67	$13	900	$ 11,700
$70	$67	$ 3	2000	$ 6,000
Total			**7500**	**$161,500**

3.2.2 Elasticity of Demand

In the hypothetical case of Example 3.1, a consultant conducted surveys and interviews to obtain the information needed to plot supply and demand curves. In practice, it is difficult to obtain the detailed information needed to plot supply and demand, so a more abstract approach is often used. Consultants or marketing managers may use past experience in trying to answer questions such as, "How much will demand change for a given change in price?" or "Will total revenue go up, down, or stay the same if the price is changed?"

These questions can be answered by using the concept of **elasticity of demand**, which is a measure of how sensitive demand is to changes in price. Elasticity of demand is defined as the negative of the derivative of the quantity demanded (Q) with respect to price (P).

$$\text{Elasticity of Demand} = -dQ/dP \qquad \text{(Eq. 3.3)}$$

This measure may also be referred to as **price elasticity**. The minus sign in this equation is conventionally used because the quantity demanded is expected to vary inversely with the price charged. Elasticity of demand can also be estimated by looking at the change in demand that occurs after a change in price:

$$\text{Elasticity of Demand} = -((Q_1 - Q_0)/Q_0)/((P_1 - P_0)/P_0) \qquad \text{(Eq. 3.4)}$$
$$= -((Q_1 - Q_0)/(P_1 - P_0))(P_0/Q_0)$$

In this equation, the changes in quantity and price are both normalized by dividing by their values before the price change. The equation therefore can be interpreted as the percentage change in quantity divided by the percentage change in price.

To understand the importance of price elasticity, consider the two effects of a price decrease from P_0 to P_1 on total revenue (PQ). Existing customers will pay less because the price is lowered, and there will some loss of revenue:

$$\text{Reduced Revenue from Original Customers} = Q_0(P_0 - P_1) \qquad \text{(Eq. 3.5)}$$

However, the lower price will attract new customers, so the quantity demanded will increase from Q_0 to Q_1, providing some additional revenue:

$$\text{Additional Revenue from New Customers} = P_1(Q_1 - Q_0) \qquad \text{(Eq. 3.6)}$$

If elasticity is greater than one, then the percentage increase in Q will be greater than the percentage decrease in P, and the added revenue from Eq. 3.4 will be greater than the loss of revenue from Eq. 3.3. If this is the case, demand is said to be elastic, because there is a large response to changes in price. If elasticity is less than one, then the opposite is true: total revenue will decrease if prices are lowered, since the added revenue from new customers will be insufficient to offset the loss of revenue from existing customers. When elasticity of demand equals one, there is no change in total revenue PQ, since the effect on the change in price is exactly offset by the change in demand.

Price elasticity is an important factor in infrastructure systems, because these systems tend to have high fixed costs and low variable costs. Maximizing revenue may therefore seem to be a reasonable goal, since the greatest obstacle to making a profit is having enough revenue to cover the fixed costs of the system. In the canal case study in Chapter 2, the nuances of operations and the details of design were found to be surprisingly insignificant: the big question was whether there would be enough demand to pay for the very high fixed costs related to the construction of the canal. During the early portion of the twenty-first century, tremendous investments in satellite-based communications were justified in part by the expectation that creating a very high-capacity system with very low prices would lead to extraordinary increases in demand—which is exactly what happened as technological advances lowered the costs of e-mail, cell phones, and wireless Internet access.

In the short run, demand tends to be less elastic than in the long run. For example, when the price of oil rose dramatically in 2007 and 2008, people initially had to pay the higher price and perhaps cut back on nonessential driving. Over a period of a year or two, however, people were able to adjust in part by buying more fuel-efficient cars and in part by figuring out how to combine errands, share rides, and use public transportation. Over a period of a decade or longer, the automobile companies will develop cars that use alternative energy sources, thus allowing people to drive more while using less oil.

Elasticity of demand is an important concept to keep in mind when evaluating infrastructure projects, because demand forecasts will drive decisions related to the size and therefore the cost and capacity of infrastructure. Forecasts based upon continuation of low prices or free access will lead to extravagant statements of infrastructure needs. Such has long been the case with urban road networks: except for a few toll roads, there is no charge for using highways; and to no one's surprise, these roads have become quite congested. Where tolls have been introduced, as in London and Singapore, it has proved possible to reduce traffic volumes and thereby limit congestion to reasonable levels. Water supply is another area where unrealistically low prices have in many locations led to unnecessarily high rates of consumption; future "needs" for water should consider the effect of more rational pricing on consumption.

Based upon the concepts of equilibrium prices and demand elasticity, future demand for infrastructure clearly will depend to a greater or lesser extent on the prices charged. If demand is elastic, then pricing could dramatically affect demand, and raising prices could be viewed as a way to reduce or avoid investments that increase infrastructure capacity. If demand is highly inelastic, then pricing will probably not be an effective means of limiting demand, and failure to expand capacity could lead to extremely high equilibrium prices, extremely poor service, or a need for regulating use or access. Elasticity of demand is therefore an important factor both in pricing infrastructure services and in forecasting demand for infrastructure. The next section continues the discussion of pricing in the context of the degree of competition among suppliers.

3.3   PRICING

As seen in the previous sections, prices for goods and services are expected to move toward an equilibrium that balances forces related to both supply and demand. If demand is elastic, then suppliers have an opportunity to sell more and increase their revenue by lowering their prices. If demand is inelastic, then suppliers have an opportunity to increase their revenue by raising their prices.

This section introduces two markedly different pricing regimes: competitive markets (Section 3.3.1) and monopolistic pricing (Section 3.3.3). A competitive market includes many suppliers and many potential customers, none of whom have the power to set prices. Instead, as described above, prices reflect an equilibrium between supply and demand. Not all markets are competitive, and geography, politics, or economic factors may encourage the development of companies or agencies that have a monopoly for particular goods or services. With no regulation, a monopoly can set prices, and customers have little power. Monopolies are not necessarily evil, because in many situations a single large supplier can produce goods or services at the lowest possible cost. Moreover, the danger of monopolistic pricing can be controlled by government regulation, so that the benefits of low-cost production are passed on to society and not simply captured as excessive profits by the owners of the monopoly. Section 3.3.2 discusses economies of scale, scope, and density, three economic factors that tend to make monopolies possible and regulated monopolies desirable.

3.3.1   Marginal Cost Pricing in a Competitive Environment

In a competitive environment, prices will fall to marginal cost. A competitive environment includes many suppliers, all of whom have access to the same or similar technologies, and they are serving customers who are able to purchase goods or services from any of the suppliers. Under these conditions, a supplier who tries to raise prices above marginal costs will have a problem: another supplier will be willing to offer a slightly lower price and thereby capture the business. So long as the price is above the marginal cost, each sale will give the supplier some **contribution to overhead and profit**. Under perfect competition, no supplier has any pricing power; prices are determined by the cost structure, technologies available, and level of demand.

Marginal cost pricing is efficient in the sense that prices reflect the actual cost of the product or service. All of those who purchase the product or service are in fact willing to pay—and do pay—the marginal cost of production. Others may desire the product or service, but they are unwilling to pay enough to make it worthwhile to any of the suppliers. Any supplier who can provide the product or service at a lower price is free to enter the market and make a profit by selling at, or somewhat less than, the prevailing price.

Situations where prices differ from marginal cost are likely to be inefficient in economic terms. If prices are too low, as in the highway and water systems mentioned earlier, then demand will be too high, and some users will incur costs that they are not willing to pay for. If prices are too high, then many who would have been willing to pay a reasonable amount for the service will be unable to afford the product or service.

There is a major difficulty with marginal cost pricing in situations where marginal costs drop below average costs, as is commonly the case with infrastructure systems. In such situations, the marginal cost pricing does not provide sufficient revenue to cover costs, and all the suppliers will face bankruptcy. Technological advances and increasingly efficient production may help some suppliers stave off bankruptcy, but only the most efficient suppliers will survive. In these situations, some kind of government regulation or subsidies may be needed to enable suppliers to remain in business. Regulation could take the form of limiting entry into the market or establishing prices at a level that allows suppliers to make a profit. Examples of limiting entry include

- Issuing taxi medallions in an attempt to limit the number of taxis to a level that will be efficiently utilized
- Requiring railroads to seek regulatory approval before constructing new lines
- Creating public utilities for communications or energy services

Generally, when entry is limited, prices must be regulated to ensure that prices are reasonable.

3.3.2 Economies of Scale, Scope, and Density

A competitive market requires multiple suppliers who are free to determine whether to enter the market based upon the prevailing prices. If there are many companies, and it is easy to enter and exit the market, then supply and demand can quickly approach equilibrium. However, the equilibrating process will be hampered if there are barriers to entry, such as the need to make large investments in order to compete. For infrastructure-based systems, this is certainly an issue because these systems by definition require substantial investments, and it takes time and effort to construct a competing system. Moreover, there are very likely to be economies in creating large facilities that can serve multiple purposes for many different users. Having competition among a great many—or even a few—smaller companies may be less efficient than having a single supplier. Larger systems may have three types of advantages over smaller systems: economies of scale, scope, and density.

Economies of scale exist when an increase in the size of the system results in reductions in cost. If $C(Q)$ is the total cost of providing infrastructure adequate for usage Q, then there are economies of scale if

$$C(Q_1 + Q_2) < C(Q_1) + C(Q_2) \qquad \text{(Eq. 3.7)}$$

For transportation, water resources, electric power grids, and other network-based systems, there will often be economies of scale because

- A single management team can manage a larger system using the same basic information technology.
- The same advertising can be used for a wider audience.
- A larger network allows a company to provide single-company service to more customers, and direct service may be cheaper than service that requires cooperation among multiple suppliers.
- A larger network provides direct links between more locations, which in transportation or communication systems can be a major benefit for potential customers.
- Consolidated maintenance facilities can serve a wider area.
- The costs of energy and materials can be reduced because a larger company can negotiate lower prices from suppliers.

Economies of scope exist when it is more efficient to use facilities for two or more types of service than it is to use them for a single service. If $C(Q_i, S_i)$ is the cost of serving Q_i customers of type S_i and $C(Q_1, S_1, Q_2, S_2)$ is the cost of serving two groups of customers, then there are economies of scope if

$$C(Q_1, S_1, Q_2, S_2) < C(Q_1, S_1) + C(Q_2, S_2) \qquad \text{(Eq. 3.8)}$$

A situation that clearly has economies of scope includes highways, which serve commuters and intercity travelers moving in automobiles or buses along with local trucking and intercity trucking. Another situation is a dam that is constructed for flood control but also can be used to generate electricity and support irrigation. If there are potential economies of scope, then there will be advantages to society from building joint facilities.

On the other hand, in some situations it does not make sense to have a single facility for multiple services. Because of potential safety problems, pedestrians and cyclists are not allowed on high-speed, limited-access highways. High-speed passenger trains cannot operate on tracks designed for freight trains, because high-speed trains cannot be safely operated on routes with sharp curves and frequent grade crossings. Swimming is not allowed in reservoirs, because of possible public health problems.

Economies of density exist when average costs decline as a result of adding more volume to an existing system:

$$C(Q_1 + Q_2)/(Q_1 + Q_2) < C(Q_1)/Q_1 \qquad \text{(Eq. 3.9)}$$

If applied to a single facility, economies of density would exist wherever scale economies exist. The distinction, however, between economies of scale and economies of density is critical in transportation and other networks that involve many facilities and an extensive route structure. Costs in these networks relate to both the links and nodes of the system, and two major strategies are employed for capturing more business; namely, expanding the network or adding more volume to the existing network. In network systems, economies of scale refer to situations in which the network expands in proportion to the increase in demand, whereas economies of density refer to the effects of adding more traffic to existing facilities.

Even if there are no economies of scale, there could be strong economies of density. Much of the investment in transportation systems has been attempting to capture economies of density (e.g., by concentrating more cars on existing roads and more flights at existing airports). Most transport networks have strong economies of density up to the point where added traffic causes extreme congestion.

The distinction between economies of scale and economies of density can also be seen in restaurants and retail sales. Large retail outlets, such as Staples or Home Depot, are able to achieve lower fixed costs per unit of sales by having large, efficient buildings with managers and employees who are more specialized and more productive than they would be at smaller stores. The larger workforce makes it easier for these stores to adjust up or down for peak periods or slack periods. Big-box retail stores therefore capture economies of density.

Fast-food outlets such as McDonalds and Burger King, which have thousands of restaurants all over the country and around the world, are able to achieve scale economies. Although some of their facilities are larger than others, these companies have vast numbers of smaller restaurants that benefit from brand recognition, common procurement, common design, and standard management. These restaurants have lower unit costs than those of the individual restaurants and smaller chains they compete with, and they use extensive marketing to convince us that their food is not only cheaper, but just as tasty. People know what to expect when they walk into one of these restaurants, and so people are likely to go to these restaurants not just when they are near home but also when traveling or vacationing in another state or another country. These companies clearly profit from scale economies.

3.3.3 Monopoly Pricing

Where there are possibilities for economies of scale, scope, and density, there can be strong forces leading to supply-side consolidation. The motivation initially is to save costs or expand markets; but if competition is reduced, then a single company could achieve monopoly pricing power. If demand is inelastic, this can lead to extremely high prices, not to mention extremely high profits.

If there are economies related to size, a larger company can always underprice smaller companies and still make a profit. Hence, a large firm can drive competitors out of business and then, when no one else is left, it may have the opportunity to raise prices so as to maximize profits. Naturally, the public and public officials are against monopolistic pricing, but they also are likely to favor lower prices. Therefore, in situations with strong economies of scale, public agencies often allow one or a few companies to exist, but regulate their prices and perhaps their services. Examples of publicly owned or regulated monopolies include most transit systems in the United States, most agencies that provide water and sewage treatment, and most public utilities.

3.4   PRODUCTIVITY

3.4.1   Measuring Productivity

Productivity, a classical consideration in economics, is defined as output divided by input. Productivity can be increased either by increasing the outputs obtained from the same level of inputs or by reducing the inputs required to obtain the same level of output. Improving productivity allows a company, an industry, or a society to produce more and/or to consume fewer resources. Improving productivity thus is generally viewed by most everyone as an important goal. Officials in the private sector believe that productivity improvements will lead to higher profits, while those in the public sector believe that higher productivity will lead to higher income for workers, lower prices for consumers, and better opportunities for growth in the economy. Achieving higher productivity motivates many infrastructure projects.

Measuring productivity is complicated because there are usually multiple types of outputs and inputs. Thus, to get the ratio of outputs to inputs, some kind of weighting system is needed for measuring both outputs and inputs. For example, in looking at the productivity of the air transportation system, output cannot simply be measured as the number of passengers or the number of passenger miles; it is necessary to consider the differences between short- and long-distance flights for business and nonbusiness travelers. With water supply systems, it is necessary to distinguish among water supplied to residential, industrial, and agricultural users. With electricity companies, it is necessary to consider peak loads versus off-peak loads. In all of these systems, inputs will include many factors that can be summarized under broad headings of labor, capital, land, materials, and energy.

One simplifying approach is to consider just a single measure of output and a single factor of production, leading to measures such as the total number of air passenger trips per airline employee or the amount of electricity generated per unit of investment. These single-factor measures tend to be somewhat arbitrary since production really does require multiple factors, but such measures may be reasonable for measuring productivity changes for a single, fairly stable operation.

Another possible approach is to weigh outputs by their prices and to weigh inputs by their costs, perhaps using information from a base year for both prices and cost. If this is done, then the aggregate measure of output will be something close to "total revenue," the aggregate measure of input will be something close to "total cost," and the ratio of output to input will be the ratio of revenue to cost.

A variant of this approach is to assume that prices reflect marginal cost, in which case price can be assumed to be a measure of cost. Since price information is more readily available than cost information, this can be a useful assumption. If there is a meaningful measure of output, then the inverse of the price per unit may be a reasonable measure of productivity. Over time, changes in the price per unit can therefore be viewed as an indicator of changes in productivity, as suggested in Example 3.2.

EXAMPLE 3.2   Productivity Improvements in Freight Transportation, 1800 to 2000

Table 3.4 illustrates the tremendous productivity gains achieved in freight transportation over the past 200 years. This table shows three factors that contribute to freight productivity: the cost per mile for construction, the tons carried per vehicle, and the miles traveled per day. The most common measure of output in freight transportation is the ton-mile, which is 1 ton carried 1 mile. Typical costs and prices are shown in this table for two periods, the early nineteenth century and the early twenty-first century. The costs are current costs, unadjusted for inflation.

Table 3.4   Increasing the Productivity of Freight Transportation

Technology	Cost per Mile to Construct ($)	Tons/Vehicle	Miles/Day	Ton-miles per Vehicle Day	Typical Prices (cents/ton-mile)
Early Nineteenth Century					
Rough road	1–2000	1 per wagon	12	12	20 to 40
Turnpike	5–10,000	1.5 per wagon	18	27	15 to 20
Canal	> 20,000	10 to 100 per canal boat	20 to 30	200 to 3000	5 to 10
Railroad	15–50,000	500 per train	200	100,000 per train	3 to 5
Early Twenty-first Century					
Arterial roads	1–5 million	10 per truck	100	1000	10 to 50
Interstate Highway	5–100 million	20 per trailer	500	10,000	10 to 15
Heavy-haul railroad	1–5 million	5–15,000 per train	500	5 million per train	2
Inland waterway	Highly variable	1500 per barge; up to 40 barges per tow	50–200	6 million per tow	1

Before 1800, only two types of freight transportation were generally available: wagons moving over bad roads and somewhat larger wagons moving over improved roads, which were usually limited to a few turnpikes radiating out from major cities. At that time, a turnpike was often just a dirt road maintained to allow slightly heavier vehicles to travel the straightest possible path between two towns. When taking the turnpike, people could use a larger wagon and go a little faster simply because the road was smoother and a bit wider. Even with the turnpikes, transport was slow and expensive, and typical prices exceeded 15 cents per ton-mile. Rivers and canals allowed larger loads and longer daily hauls, since it is much easier to pull a canal boat along a river than to drive a horse and wagon up and down the hills. As documented in Chapter 2, canals were built in the early nineteenth century for upward of $20,000 per mile; the expense was justified by the increased productivity for the freight carriers, and typical freight rates fell below 10 cents per ton-mile. Canals were limited by geography, so rail technology had a great advantage as soon as it became available. Even with only 500 tons per train, a railroad allowed much more productive freight operations than was possible with small canal boats, and typical freight prices dropped below 10 cents per ton-mile. Technological improvements continued throughout the nineteenth and twentieth centuries, so that today the prices for freight transportation are actually lower than they were 200 years ago. The lowest prices are achieved for fully loaded vehicles traveling at the maximum speed on the main routes—tractor-trailer combination trucks on the Interstate Highway System, heavy coal trains on high-density, well-maintained rail lines (Figure 3.7), and tows of 40 barges moving along the major rivers.

Figure 3.7 Coal Train. Trains carrying up to 15,000 tons of coal operate on narrow rights-of-way through difficult territory like this canyon in Colorado. A double-track route like this can handle more than 100 million tons of freight per year, plus a pair of 79-mph passenger trains. The interstate highway is visible on the other side of the river. To minimize its footprint, the highway was double-decked for a portion of the route. There is also a bike path beside the river below the highway.

Photo: S. J. Martland, 2000.

This example uses the ton-mile as a simple measure of output for freight transportation, even though the costs and benefits of transporting different commodities different distances can vary widely. For example, it is easier to move coal in single shipments of 10,000 tons than it is to move 10,000 tons of general merchandise as 200 separate shipments. Nevertheless, even though the ton-mile is far from a perfect measure of rail output, the cost per ton-mile remains useful in highlighting the dramatic productivity improvements achieved in freight transportation.

The intricacies of productivity measurement are beyond the scope of this text, but some intuition concerning productivity will be helpful in understanding why projects are undertaken and how they are evaluated. Perceived productivity problems often suggest the types of projects that need to be undertaken:

- Peak demands may cause delays at bottlenecks in transportation or systems (so consider investing to relieve bottlenecks).
- Engineering constraints, such as weight limits on bridges or bandwidth limitations in communications networks, may restrict the usage of the system (so consider investing to increase the ability of the infrastructure to handle larger or heavier loads or higher volumes of usage).
- Lack of communication and control may inhibit efficient use of resources (so consider investing in communications and control systems).
- Facilities that were designed and built many years ago may no longer match what is needed today or in the future (so consider rehabilitating, expanding, or redesigning facilities or networks).

Sections 3.4.2 to 3.4.6 discuss these four causes of low productivity in infrastructure systems.

3.4.2   Causes of Low Productivity: Peak Demands May Cause Delays at Bottlenecks

Figure 3.8 shows a common occurrence at Clark Junction on Chicago's transit system. This is a busy junction where the double-track Brown Line joins the four-track Red Line at a point midway between two closely spaced stations. The picture was taken looking north from a control tower at one station, and two southbound trains are visible waiting at the next station. The first train on the left is moving southbound and will soon stop at the station. The next train is a northbound train moving across the southbound Red Line tracks and heading out on the Brown Line. The two distant trains must wait until

Figure 3.8 Clark Junction, Chicago Transit Authority.
This busy intersection limits the capacity of two major transit lines operating north out of downtown Chicago. The infrastructure was originally designed for shorter trains.

Photo: C. D. Martland, January 2003.

these two trains clear the tracks before they can proceed. Since they are several hundred yards from the nearby station, each train will take another minute or two to reach the platform once it is able to proceed. Because of the extra time to bring each train into the station, this location becomes a bottleneck for both the Brown Line and the Red Line during rush hour.

3.4.3 Causes of Low Productivity: Lack of Demand

Being labeled unproductive does not mean that a system is outmoded or obsolete. The system may have been built with too much capacity, it may have too many unnecessary features, or it may have been built all at once without considering opportunities for staging the construction over a period of years or decades. Vast portions of the Interstate Highway System in rural areas of the United States still have very low traffic volumes half a century after they were designed. The productivity of these highways, which can be measured as the vehicle miles per year per lane-mile of highway, is very low compared to any urban highway. Nevertheless, the minimum design standard for the Interstate System was a four-lane, limited-access, divided highway, and thousands of miles of the interstate were constructed in the 1960s in locations where the capcity of a four-lane highway is still unnecessary for current traffic volumes. The option of a staged development was not pursued, probably because the cost of constructing highways through rural areas was usually very low. In Europe, where land values were much higher and space more limited because of prior development, modern highways were often implemented incrementally by expanding existing roads rather than by creating vast networks of new, four-lane, divided highways across entire countries.

A brand-new system may be reasonably sized for the forecast demand, but that demand may have been overestimated. Many transit systems have been built in anticipation of demand that failed to materialize; although these systems can be very advanced technologically, they will be unproductive if usage is well below forecasts. Figure 3.9 shows the nearly empty escalators serving a modern subway station on the London Metro. Unless people actually use the subway, a facility like this is unproductive.

The recession of 2007–09 was initiated by the collapse of the bubble in real estate that had driven prices to unsustainable levels. The effects of the real estate debacle were greatest in areas like southern Florida, where rapidly rising prices were fueled by speculators hoping to make easy money by building condominiums for what was apparently perceived as a never-ending stream of buyers. When the bubble broke, many of the condos remained unsold, and many of those that did sell were foreclosed because the owners could not afford to pay their mortgages (and may never have intended to pay those mortgages!). At many other locations in the country, the bursting real estate bubble exposed extensive overbuilding of condominiums and other housing.

The case study at the end of this chapter illustrates some of the factors that lead to bubbles in real estate. Working on that case might show you how to make a million in real estate—and warn you that you could lose everything when the bubble bursts.

Figure 3.9 Tate Modern Subway Station, London (shown at a lull between trains). This is a station on a highly used new line built south of the Thames to relieve pressure on other lines and promote growth along that side of the river.

Photo: C. D. Martland, October 2002.

3.4.4 Causes of Low Productivity: Limits on Usage

Size and weight limits place constraints on the productivity of many infrastructure systems. As shown above in Table 3.4, the ability to use larger vehicles carrying heavier loads has been a driver of productivity improvements in freight transportation for 200 years. However, the largest, heaviest vehicles can be used only where the route structure is adequate. Elsewhere, smaller vehicles must be used, at higher costs per ton-mile, and larger vehicles will be forced to detour around restricted areas.

The same constraints show up in other systems, including the power grid and the Internet. The Internet offers high-speed transmission of data, but only if you have the necessary access. Extending high-speed Internet services to rural areas could become a major public policy objective in many countries, just as extending electricity and postal services to rural areas was a major priority in the early twentieth century.

3.4.5 Causes of Low Productivity: Lack of Communications and Control

Improving communications and control systems can at times be an effective way to improve system performance. Many system problems are exacerbated when demand is either too high or too low. If capacity can be adjusted, then delays will be shorter during peak periods and costs will be lower during off-peak periods. Improvements in communications and control have enabled system operators to improve performance in many types of systems over the past 50 years.

In air transportation, for example, air traffic controllers can extrapolate from the current situation to what will likely be happening in several hours. They therefore can take action to relieve congestion by rerouting some flights, and they can allow planes to absorb necessary delays while still on the ground rather than having them take off and be forced to circle the airport at their destination. The capacity and efficiency of the electric power system have been enhanced by shifting power generation to the most efficient or the cleanest power plants, thereby allowing older, less-efficient plants to be shut down during periods of slack demand. At the Panama Canal, increasing tolls has helped the Canal Commission to manage congestion and capacity; ships willing to pay higher tolls can move to the head of the queue and avoid delays. In London and Singapore, as mentioned above, strict regulations and high fees have limited highway traffic and relieved congestion.

3.4.6 Causes of Low Productivity: Facilities That No Longer Match Demand

Facilities can be too large or too small. If a facility is too large for the demand, then it will be difficult to cover fixed costs. If a facility is too small, then it will tend to be crowded, uncomfortable, and less successful than it could have been.

Figure 3.8 above illustrates a key bottleneck in Chicago's transit system. The operation of trains through that junction may seem very inefficient, and this junction may appear to be a poorly designed bit of infrastructure. However, it was designed decades ago when transit operations were much different from what they are today. The line actually does have space for two trains to stop just before the nearby station, so this type of bottleneck was not supposed to occur. However, to increase capacity of its service, the Chicago Transit Authority (CTA) decided to lengthen station platforms to allow for longer trains—and the longer trains no longer fit in the spots that were available at this particular junction. Instead, the longer trains must wait, as shown in the picture.

Bottlenecks like this can be found in any network, especially networks that were designed long ago. While identifying the problems is straightforward, rectifying them can be far from it. In this instance, it would severely disrupt two very busy transit lines to adjust the track and signals to allow greater capacity through this junction.

Productivity problems can of course arise from poor design, not just from outmoded design. A new facility can be too large or too small for the actual demand. A facility can even be both too large and too small. For example, during the 1970s, a number of cities constructed new stadiums for their major league baseball and football teams. With some ingenuity in design, both sports could be accommodated; but the question was, how much seating should be provided? Some settled on seats for 70,000 or so. Over time, it became apparent that the baseball teams could seldom fill up the stadiums, so much of the capacity costs were wasted for baseball; but the same stadium proved to be too small for football, where 100,000 fans might regularly come to see a game. Once this lesson was learned, separate facilities were constructed for baseball and for football. The baseball facilities were smaller, and they were designed to resemble the old-time ball fields of the early 1900s (e.g., Camden Yards in Baltimore). The football fields were much larger, and they were designed to facilitate large numbers of fans arriving and departing within a short time. Both types of stadiums installed luxury box seats, since owners realized that corporations and wealthy individuals were willing to pay a great deal for season tickets in plush surroundings.

Facilities that were originally sized appropriately may become unsuitable over time. Competitors may have impinged on the market, leaving companies with too much capacity and the wrong-sized facilities. Some examples include

- Discount airlines offered low-priced services that pulled traffic away from the larger airlines, whose fixed costs were much higher.
- Apartments with too few bathrooms and kitchens that are too small have reduced rents.
- Movie theaters with only a single large room have to close or be redesigned because they cannot compete with large theaters that can offer a choice of films.

3.5   MEASURING AND IMPROVING THE ECONOMY

Infrastructure projects have impacts that go well beyond the financial affairs of owners and users. Infrastructure allows and supports other economic activity, and the greatest benefits of investment in infrastructure may be the new opportunities made available to society. In evaluating large infrastructure projects, two types of economic impacts are commonly considered:

1. The short-term boost to the local economy resulting from the planning and construction of the project
2. The long-term impact of the project on the region's economy once the project is completed

The short-term impact occurs as a result of all the jobs created and all of the expenditures for materials and services that are incurred as the project is implemented. The multiplier effect of all this activity is discussed in Section 3.5.1.

The long-term impact occurs as a result of the benefits to users of the new infrastructure as well as the permanent jobs directly linked to the project. Constructing a light-rail line to the airport to relieve highway congestion and improve access does much more than provide jobs for operating the trains and maintaining the tracks; it will saves time for air travelers, airport employees, commuters who use the new line, and highway commuters who experience less congestion. It also creates opportunities for developing real estate near the light-rail stations. The light-rail project could therefore have an impact on the overall economic activity of the region, as discussed in Section 3.5.2.

3.5.1   Short-Term Economic Impacts: The Multiplier Effect of New Investment

The design and construction of a major project boosts the economy, because of what is known as the **multiplier effect**. Consider the construction of a new building in a city. Much of the investment cost will be made up of wages and salaries paid to local construction workers and payments to local merchants for materials and services. These workers may save some of their wages, but they are likely to spend most of it; likewise, the local suppliers will spend most of what they receive. The proportion of the new income that they consume is called the **marginal propensity to consume**.

Let's say that the construction of the building resulted in wage payments of $1 million to local workers and companies. This $1 million is in itself an addition to the regional economy, but that is just part of the story. If the marginal propensity to consume is 0.5, then workers and companies will spend another $0.5 million—another addition to the local economy. And that $0.5 million will go to other workers and companies who will save some and spend some. If they also save half and spend the rest, then another $0.25 million is added to the regional economy. And some of that money will also be spent. If half of the money is saved at each step, then the total addition to the regional economy will be $1 million $(1 + 0.5 + 0.25 + 0.125 + \ldots)$, which will converge to $2 million dollars. In this case the multiplier is 2, since each dollar invested leads to an increase of $2 in the regional economy.

In general, the total addition to the economy can be expressed as a function of the marginal propensity to consume (MPC):

$$\text{Addition to GRP} = \text{Investment in Region} \left(1 + \text{MPC} + \text{MPC}^2 + \text{MPC}^3 + \ldots\right) \qquad \text{(Eq. 3.10)}$$

So long as MPC is less than 1, this sequence converges to $1/(1 - \text{MPC})$. The factor $(1 - \text{MPC})$ is the marginal propensity to save, so the multiplier effect increases inversely with the marginal propensity to save:

$$\text{Multiplier Effect} = 1/(1 - \text{MPC}) \qquad \text{(Eq. 3.11)}$$

For example, if the marginal propensity to consume increases from 0.5 to 0.75, then the marginal propensity to save drops from 0.5 to 0.25. If so, then more money goes into the economy. The total addition to the regional economy is $1 million $(1 + 0.75 + 0.75(0.75) + (0.75)(0.75)(0.75) \ldots) =$ $1\text{million}/(0.25) = $4 million. With less money going into savings, the multiplier effect jumps from 2 to 4.

The multiplier effect applies to both the construction phase and the operations phase of a project. For infrastructure projects, however, since investment costs are so much higher than continuing costs, the greatest interest is in the multiplier effect from the investment. Multipliers are typically between 2 and 3 for transportation and other infrastructure projects. Note that the multiplier effect relates only to the money spent within the region, so a project that imported costly materials and used highly automated equipment would have a much lower regional impact than a more labor-intensive project that used local labor and materials.

Note that the multiplier effect does not enter into a purely financial analysis of a private or public project, because these economic benefits are dispersed throughout society and are not captured by the owners of the project. Also, a public agency comparing the benefits and costs of a project may first want to ensure that the benefits directly related to the project exceed the costs directly related to the project. The multiplier effect can then be used as further justification for a project whose direct benefits exceed its direct costs.

On the other hand, the multiplier effect can at times drive projects. This effect is what motivates governments to initiate stimulus programs during a recession. In the short run, the stimulus will be most effective in reviving the economy if it is directed toward projects and programs that direct money toward people who are likely to spend most of what they receive. The long-run economic benefit depends upon the success of the project in providing permanent jobs, making society more productive, or enabling other economic benefits to society.

3.5.2   Long-Term Economic Impacts: Gross Domestic Product

The most common economic measure used to monitor the health of the economy is the **gross domestic product (GDP)**, which equals the sum of private consumption (C), investment (I), and government expenditures on goods and services (G) plus exports (E) minus imports (M):

$$\text{GDP} = C + I + G + E - M \qquad \text{(Eq. 3.12)}$$

Growth in the economy is measured as the change in GDP, and growth in GDP is generally viewed as a critical objective for a nation. A growing economy provides opportunities for more jobs, higher wages, more goods and services, and higher profits for companies. If GDP declines for two successive quarters, then the economy is said to be in a recession. Unemployment rises during recessions, wages may fall, and company profits decline. Thus maintaining GDP is an important economic objective for a nation.

GDP is not a perfect measure, in part because life involves more than economics. Even in the realm of economics, however, the major problem with GDP is that it fails to account for the losses associated with depreciation of the capital stock of the country. Machines wear out, buildings age, infrastructure deteriorates, and these losses from depreciation are not captured until and unless repairs are made or facilities are replaced. The net domestic product is calculated by subtracting total depreciation from GDP. Net domestic product is less commonly used because it is difficult to estimate depreciation of assets but relatively easy to monitor consumer purchases, investments, government expenditures, and foreign trade. Since the two measures usually rise and fall in tandem, the GDP measure is most frequently used.

GDP is an aggregate measure that does not reflect conditions for particular regions, groups of people, or sectors of the economy. However, similar measures can be estimated for each region of a country. The **gross regional product (GRP)** can be defined in the same manner, except the various factors defined will apply to the region, not to the nation. As with the national economy, growth in GRP is a major economic objective for any region.

Jobs and average income are other important aspects of the regional economy. Adding jobs to the regional economy is always viewed positively, but especially so during a period of high unemployment. Higher-paying jobs are preferred, and local governments may provide tax breaks and other incentives to attract or to retain companies that have such jobs (as discussed in Chapter 10).

Economic models can be constructed to predict the impact of infrastructure investment on the regional economy. Such models may be able to show that transportation investments will make the region more attractive to new businesses or that investing in dams and irrigation will make local agriculture more profitable and lead to growth in all activities related to agriculture. Analysis may also show that investment in infrastructure is expected to have a measurable impact on congestion, public safety, or public health. Savings in time, reductions in risk, and improvements in health can be translated into economic benefits by using the average value of time for commuters, the expected savings in accident costs, and the expected reduction in health care costs. While such benefits do not result in cash flows that help pay for infrastructure investments, they are quantifiable factors that can help justify (and gain public approval for) public investments.

It is beyond the scope of this textbook to discuss these models in detail, but it is important to recognize that predicted economic benefits will indeed help increase public support for infrastructure projects.

3.6   TRADE

A great deal of infrastructure investment is based upon projections for population growth and growth in regional economies. Over the long term, both types of growth depend to a large extent upon forces that act on a national or international scale, such as technological change and trade. **Technological change** results in new products, new materials, new development opportunities, and new processes for manufacturing and distribution. Over time, there can be marked changes in the types of things produced, how and where they are produced, and how they are distributed around the world. These changes influence and respond to changes in **economic geography**—the location of people and economic activities throughout a region, a nation, or the world. And these changes in economic geography require and motivate many investments in infrastructure.

Trade is the exchange of goods among regions or countries. Transportation makes trade possible. Differences in regional resources, economies of scale in production, and differences in costs and capabilities make trade desirable. The ability to exchange currencies of different countries and transfer monies between countries makes trade feasible financially. The ability of wealthy countries to buy vast amounts of goods and materials makes trade flourish.

Two key concepts are helpful in understanding how trade works and why trade is important. First, if it is possible to produce something for a lower cost in one region than in another, then there is an opportunity for trade, but only if transport costs are sufficiently low. Section 3.6.1 describes **spatial price equilibrium**, the process by which transport costs and manufacturing costs together determine prices for products that can be produced in one area and sold in another.

Section 3.6.2 discusses comparative advantage, the reason why it may be sensible for different regions to concentrate on what they do best. Understanding these concepts is essential to understanding the global economy, and recognizing the existence of a global economy is essential for understanding the kinds of infrastructure investment that will be needed to support industrial production, trade, and population growth.

3.6.1 Spatial Price Equilibrium

To begin, consider a product that can be made in two locations. At location "a," production costs are lower than at location "b." The producers at "a" and "b" compete with each other for business in their region. Although costs are lower at location "a," there are two major reasons why they might not capture all of the business. First of all, costs are not prices. Manufacturer A may decide to charge whatever Manufacturer B charges, so that they will share whatever market is available—and Manufacturer A will have higher profits. Second, transportation costs must be added to the manufacturing costs, and the prices for the products must cover the total cost.

Figure 3.10 illustrates how transport costs will affect the markets captured by each manufacturer. The x-axis of this chart represents distance. Manufacturer A is located at point "a," and Manufacturer B is located at point "b." The y-axis represents the sum of manufacturing and distribution cost. The costs of manufacturing (M_a and M_b) are indicated by the vertical lines at "a" and "b"; costs are much lower at "a."

Assume that transport costs (T_{ax}) from Manufacturer A to any potential location (x) are proportional to the distance (D_{ax}) from the manufacturer:

$$T_{ax} = k_{ax}D_{ax} \qquad \text{(Eq. 3.13)}$$

Thus, if Manufacturer A decides to market its product at location "x," then the total cost will be

$$C_a = M_a + T_{ax} = M_a + k_{ax}D_{ax} \qquad \text{(Eq. 3.14)}$$

A similar equation would apply for Manufacturer B:

$$C_b = M_b + k_{bx}D_b \qquad \text{(Eq. 3.15)}$$

The total costs (C_a and C_b) are both plotted in Figure 3.10. The costs for each manufacturer are Y-shaped, since total costs rise linearly by moving in either direction in this two-dimensional figure.

The slope of the total cost line is the cost per mile for transporting the product, which is k_{ax} for Manufacturer A and k_{bx} for Manufacturer B. The slope is steeper for Manufacturer B, indicating that

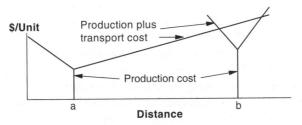

$/Unit

Production plus transport cost

Production cost

a

b

Distance

Figure 3.10 Spatial Price Equilibrium

transportation costs are more expensive for that manufacturer. Manufacturer A will presumably capture the market for all locations where its total costs are less than those of Manufacturer B. At points where they are equal, the manufacturers will share the market. In the two-dimensional world depicted in Figure 3.10, the total cost lines intersect in two places—one just to the left of Manufacturer B's location on the graph and one a little farther to the right of that location. This indicates that Manufacturer B will be serving only its local market, while Manufacturer A will capture all of the rest.

In general, if two manufacturers of identical products are competitive in a market, then they must be charging the same price at that location. For a manufacture to compete and make a profit, the sum of its production cost plus transportation cost must be less than or equal to the prevailing price at that location. Prices will vary from one location to another, reflecting the differences in production and transportation cost. Over time, if transportation and production technologies are stable, spatial price equilibrium will be achieved, and the markets served by each manufacturer will be well-defined.

Of course, neither production nor transportation technologies remain stable for very long. As we have already seen in this chapter, transportation costs have declined dramatically over the past 200 years, enabling today's manufacturers to compete globally. With low-cost global distribution feasible, it makes sense (1) to have large manufacturing facilities that take advantage of whatever economies of scale can be found in manufacturing and (2) to have those manufacturers located in regions where costs are lowest, whether because of local labor rates, local energy costs, or the geographical position relative to sources of inputs and major markets. Because of the tremendous improvements in freight transportation productivity, it is now feasible to manufacture many consumer goods in Asia, where labor and production costs are very low, and to ship those goods on large container ships to major ports for distribution throughout the rest of the world.

3.6.2 Comparative Advantage

The unequal distribution of resources, including capital and skilled labor as well as natural resources, is another force promoting trade. Due to accidents of location or history, one region may be able to produce certain products at a lower cost or higher quality than other regions. If this region makes an excess amount of such products, then it can sell them to other regions and use the proceeds to purchase other types of goods from those regions. For example, one country may be very good at making automobiles, while another country is very good at agriculture. Opportunities for trade thus seem pretty clear: trade automobiles for food, and allow the country to concentrate on its best products.

The potential benefits of specialization were first highlighted nearly 200 years ago by David Ricardo, who developed the theory of **comparative advantage** in 1817. The economist Paul Samuelson summarized this theory as follows:

> *Whether or not one of two regions is absolutely more efficient in the production of every good than is the other, if each specializes in the products in which it has a comparative advantage (greatest relative efficiency), trade will be mutually profitable to both regions. Real wages of productive workers will rise in both places.*[1]

This theory is the basis for reducing tariffs and other barriers to trade. Tariffs are taxes charged on inputs as a means of protecting local manufacturing. However, the theory of comparative advantage indicates that it is better to allow imports, so that local workers and local capital can be put to work more productively in areas where the region enjoys its greatest efficiency relative to other regions. Example 3.3 shows how comparative advantage might work today.

---

[1] Paul Samuelson, *Economics: An Introductory Analysis*, 6th ed. (New York: McGraw-Hill, 1964), 665.

EXAMPLE 3.3   Comparative Advantage and Protective Tariffs

Tables 3.5 and 3.6 show the costs of production and distribution for two countries that both could produce grain and automobiles. In Table 3.5, transport costs and tariffs are high. It costs $2,000 to transport an automobile between the two countries, so it would be possible for auto manufacturers in A to deliver cars to B at a cost of $22,000 for manufacturing and transport. Since Country B's automobiles cost $25,000 to produce, it would seem that Country A could sell cars in Country B. However, to protect its own manufacturers, Country B has established a $4,000 fee per auto for importing a car, so it would cost $6,000 per auto to export an automobile from A to B. Nothing is exported, because the total of production cost plus transport cost plus tariff is now $26,000, which is greater than the cost in B. For grain, the opposite happens. Transportation costs are $20 a ton, but Country A charges a tariff of $50 to protect its farmers. Under this system, both countries are spending $45 billion per year on automobiles and grain.

Table 3.5   Transport Costs and Protective Tariffs Inhibit Trade between Country A and Country B

	Country A			Country B		
	Cost per Unit ($)	Demand per Year	Cost per Year ($)	Cost per Unit ($)	Demand per Year	Cost per Year ($)
Domestic auto production	20,000	1 million autos	20 billion	25,000	1 million autos	25 billion
Grain production	250/ton	100 million tons	25 billion	200/ton	100 million tons	20 billion
Export auto production	6,000/auto	0	0	2,000/auto	0	0
Export grain production	250 + 20 per ton	0	0	200 + 20 + 50 per ton	0	0
Total			**45 billion**			**45 billion**

In Table 3.6, the tariffs have been eliminated, which allows both autos and grain to be exported. Country A produces less grain and more automobiles, while country B produces more grain and fewer automobiles. With trade, both countries benefit since the total spending, including the costs of transportation for the exported goods, is reduced to $42 billion in each country.

Table 3.6   Elimination of Tariffs Promotes Trade between Country A and Country B

	Country A			Country B		
	Cost per Unit ($)	Demand per Year	Cost per Year ($)	Cost per Unit ($)	Demand per Year	Cost per Year ($)
Domestic auto production	20,000	1 million	20 billion	25,000	0	0
Grain production	250/ton	0	0	200/ton	100 million	20 billion
Export auto production	20,000 + 2,000	1 million	22 billion	25,000 + 2,000	0	0
Export grain production	250 + 20	0	0	200 + 20	100 million	22 billion
Total			**42 billion**			**42 billion**

In this example, Country A was better at producing automobiles and Country B was better at producing grain, so it is not surprising that there is an opportunity for trade. However, the benefits of trade remain even if one country is better at producing both types of goods. Suppose that Country A could produce grain not at a cost of $250 per ton, but at a cost of $190 per ton. In that case, it could produce 100 million tons of grain for $190 billion, which is much less than the $22 billion it would cost to import the grain. If there were no trade with Country B, then Country A would produce 1 million automobiles and 100 million tons of grain at a cost of $39 billion. However, if trade were allowed, then perhaps Country A could make enough money selling automobiles to justify paying more for grain. Suppose that Country A exports autos to Country B, where they are sold for $24,000 apiece—a price below the cost of local manufacturing. Now Country A could make a profit of $4 billion by selling cars, which is enough to justify spending an extra $3 billion to buy grain. At the same time, Country B saves $1 billion on auto purchases and sells more grain. In this situation, Country A has a comparative advantage in producing automobiles and Country B has a comparative advantage in producing grain.

3.6.3   Currency Exchange and International Banking

International trade depends upon a banking system that can do two things. First, there must be a common medium of exchange, so that the money used in one country can be used or converted into an amount of another country's money that has the same value. Second, it is necessary to have some system of credit, so that a person in one country can borrow the funds needed to buy goods that will be exported from a second country and perhaps transported and sold in a third country. The people transporting the automobiles and the grain in Example 3.3 do not carry huge bags of cash back and forth after making their deliveries; someone has agreed to pay for the automobiles or the grain once the deliveries are made, and a bank somewhere provides a way for a payment drawn on a bank in Country A to be transferred to a bank in Country B.

Today, when it is easy to stick a credit card into an ATM and get local currency when traveling abroad, it is difficult to imagine how important—and how difficult—it was for traders to have access to a banking system so they could carry out their business without personally carrying vast sums wherever they went.

The exchange rates between currencies can be based upon market forces or regulatory forces. Major newspapers provide daily reports on the exchange rates for the major currencies. In early 2009, for example, $1.00 was worth about 0.77 euros or, to put it the other way around, one euro was worth about $1.30.

At times, exchange rates are quite volatile. On September 20, 2008, just before the announcement that the world's financial system was on the brink of collapsing, the euro was worth $1.60. Two weeks later, as the magnitude of the crisis spread and the credit crunch was found to be even more severe in Europe, the value of the euro dropped to about $1.35. This abrupt 15% drop in the value of the euro made European exports cheaper in the United States, thus making U.S. exports more expensive in Europe. This change affected everyone traveling overseas. Suddenly, everything priced in euros was 15% cheaper for anyone who had dollars to spend—and everything priced in dollars was 15% more expensive for anyone who had euros to spend. A change of this magnitude is equivalent to putting a 15% tariff on everything exported from the United States into Europe and having a 15% sale on everything imported to the United States from Europe. Changes of this magnitude have broad repercussions on international trade and travel, even without a credit crisis.

A credit crisis can be devastating to trade and economic growth. Without credit, it is hard for businesses to get the loans they need to expand production, and it is hard for consumers to borrow money to buy houses, cars, and other items. The credit crisis in late 2008 and 2009 resulted in stock markets plunging, the auto industry teetering at the edge of bankruptcy, and many banks and investment banks collapsing. Without credit, trade declined abruptly and the world economy slipped into a serious recession. In later chapters of this text, we consider how risks like these affect project planning and evaluation.

3.7   MAKING DECISIONS

3.7.1   Discrete Choice

Economists use the concept of **utility** as a way to understand how individuals make decisions. It is assumed that people act rationally so as to maximize their utility, subject to budgets for both time and money. Utility is a useful concept, even though few of us can say why we do or don't do something, or why we prefer one product over another. While people don't always act rationally, they do so often enough that utility maximization provides a meaningful way to help understand many aspects of human behavior.

Utility can be studied by documenting the choices that people make or by conducting surveys—that is, by considering what are called **observed preferences** or **stated preferences**. Analysis of actual choices is likely to provide better insight into behavior, but it is much easier to obtain detailed information by using surveys and documenting stated preferences. Earlier in this chapter, Example 3.1 illustrated the results of a survey that asked people whether they would buy a particular product for various prices. That survey was an example of using stated preferences to estimate how potential consumers would make decisions. More complex surveys can be devised to explore quality of service, timing, and other factors that might be important in addition to price.

For example, consider the journey to work. A person may have the choice of driving alone, riding with a friend, or taking the bus. Direct observation may show that this person drives alone 60% of the time, rides with a friend 20% of the time, and takes the bus 20% of the time. An interview can probe further, seeking to understand how cost, travel time, work schedules, weather, errands, and shopping factor into the decision. Utility models may be constructed based upon the actual decisions or the stated preferences. These models typically include variables that reflect cost, service quality, and convenience. An example of how utility models can be used to understand infrastructure design is given in Chapter 13, Section 13.4.

For now, it is enough to know that utility is a concept used to understand how people make choices, that utility can be studied by considering actual or stated preferences for different groups of people, and that the result of a utility analysis will be something like "People seem to value the time spent commuting at something close to their average wage."

People make many decisions based upon time and money, and it is possible to consider budgets for both time and money. In theory, people may be viewed as trying to maximize their utility subject to budget constraints for time and money. Therefore, projects that save users time or money are liable to attract customers. These savings in time and cost can also be viewed as economic benefits of the projects.

3.7.2 Sunk Costs

Economic decisions concern future costs and benefits. Money spent in the past should not affect what we decide to do today, and such costs can be viewed as **sunk costs**. For instance, if you are about to buy a new car, and you plan to trade in your old car, it does not matter—to you or the car dealer—what you paid for that car. What matters is the value of the car, today, which depends upon the condition of the car and the demand for used cars. On a larger scale, when you are trying to decide whether to buy or sell a building, it does not matter what that building cost to build. Only the market value of the building will affect the price. Of course, if you have yourself put a great deal of money into buying a car or your house, you may well perceive that the car or house is worth a lot more than anyone else does, but that is a factor only in deciding whether you are willing to part with it. The current market value is what should enter your economic analysis.

3.8 SUMMARY

This chapter has introduced supply and demand, pricing, productivity, utility, and other economic concepts relevant to project evaluation. These concepts provide a framework for thinking about needs, projects, and project evaluation. Examples in this chapter illustrated how the economic concepts might be encountered in specific situations.

Supply and Demand

The supply function describes the amount of output that will be provided as a function of the price per unit sold. The demand function describes the amount of output that will be purchased as a function of the price per unit. The equilibrium price is the price at which supply equals demand. In most complex systems, both supply and demand are continually changing. Suppliers will be adopting new management techniques and new technologies, and they may expand or contract their facilities. The level of demand for a service or product will vary with changes in population, average income, competing products and services, and other opportunities for spending money. With supply and demand in continuous flux, it is more realistic to think about systems moving toward equilibrium rather than always being in equilibrium. This is especially true for infrastructure-based systems because it is costly and time-consuming to adjust their capacity. Congestion, delays, and poor quality are likely when demand exceeds supply, while underutilization of equipment and reductions in the workforce of suppliers are common when supply exceeds demand.

While it is difficult to quantify functions for supply and demand, past experience can be used to predict how demand will respond to changes in price. The elasticity of demand with respect to price can be estimated by observing the effects of price changes on demand. If demand is elastic, then demand will be more responsive to price changes, and an increase in price will lead to a decrease in total revenue. Demand tends to be more elastic in the long run than in the short run since people and businesses will generally find ways to reduce their dependence on higher-priced goods and service.

Pricing

In a competitive environment, no individual supplier has the power to set prices, and prices will fall to marginal cost. In infrastructure-based systems, marginal cost pricing is generally well below average cost as long as the system is operating below its design capacity. Thus there may be a need for price regulation or subsidies to ensure that revenues are sufficient to cover total costs of operation. If demand approaches capacity, then marginal costs for both users and operators will rise as delays and high utilization levels make it difficult to use and maintain the system.

If a supplier has a monopoly, it can set prices well above marginal costs so as to maximize profits. Monopolies may also be slow in adopting new technology or expanding capacity to meet demand, and they may display little concern for service quality. However, many infrastructure-based systems achieve tremendous economies of scale and density, so that the cost of service can be greatly reduced by limiting competition. Thus many transportation companies and public utilities are allowed to operate as monopolies to achieve cost savings while also being subjected to regulation to ensure reasonable prices and service.

Productivity

Productivity is defined as the ratio of output to input. Improving productivity is an important objective, because productivity improvements make it possible to produce more goods and services using fewer people and resources. Many projects eliminate productivity problems related to bottlenecks, constraints on usage, inadequate control, or outmoded facilities.

Productivity may also be improved by changing the structure, design, and size of networks or facilities so as to achieve economies of scale, scope, or density. Scale economies exist when expanding the size of the system leads to reductions in average cost. Scope economies exist when it is cheaper to use facilities for multiple uses. Density economies occur in a network when more volume is concentrated on each route.

Measuring and Improving the Economy

Public officials and the general public are naturally interested in expanding economic output, which can be measured as either the gross or the net domestic product. The GDP is the sum of all private consumption, private investment, government expenditures, and net exports. The net national product is GDP minus an allowance for the depreciation of factories, buildings, and other parts of the capital stock of the country. While the net national product might seem to be the more relevant measure, GDP is much easier to monitor and is therefore the more commonly used measure. Similar measures can be developed for the gross or net regional product. Other measures of the economy include total jobs, unemployment levels, and average income per person or per family. All of these measures can be developed for a region as well as for a country.

A major project affects the regional economy both directly and indirectly. The direct effect is related to the jobs created and the expenditures required to complete and subsequently operate and maintain the project. For a private company, these expenditures are viewed as a cost that must be recovered by revenues that will eventually be obtained over the life of the project. For the residents of the region, job opportunities and purchases of local goods and services are viewed as an immediate economic benefit. In addition, there will be a multiplier effect, because the people who work on the project and the companies that sell materials to the project will spend much of what they earn, whether on food or cars or housing, and other people and other companies will enjoy some added income. Because of the multiplier effect, pro-development officials and organizations will support endeavors that lead to more construction projects.

Trade

Trade allows regions to specialize in economic activities that give them a comparative advantage relative to other regions. By producing more of what they need for some types of products, regions are able to trade for other things they need or desire. The ability to trade depends upon the ability to transport goods efficiently between regions, because of spatial price equilibrium. For trade to make sense, the cost of producing something in one location plus the cost of transporting the product to another location must be less than the price that can be charged in that location. Investments in transportation systems have produced dramatic reductions in transport costs, thereby enabling the shift of manufacturing, mining, agricultural production, and other activities to the regions of the world with the lowest costs. The global economy reflects low transportation costs and generally high economies of scale in production.

Banking and currency exchange are another essential aspect of global trade. Exchange rates between currencies of different countries may be determined by market forces or by regulations. Over time, exchange rates may vary substantially; this tends to change the patterns of trade by making some countries relatively cheaper and others relatively more expensive. Growth in trade and changes in trade routes are important considerations in many infrastructure projects.

Making Decisions

Although people are often less than rational in their decisions, economists assume that individuals generally make decisions that maximize their own utility. Utility is a rather vague—but thoroughly useful—concept that can incorporate disparate factors such as cost, convenience, reliability, safety, or aesthetics that might affect our choice of a new car, a new house, or where to eat dinner. By observing what decisions people make (revealed preferences) or by asking people about hypothetical choices (stated preferences), analysts can infer what factors they consider in making choices. Those who plan projects must, at some level, consider how many people will use the completed project (road, water system, park, or office building) and how much they will be willing to pay for their use of it.

When evaluating proposals, project planners need consider only future costs and benefits. Money spent in the past is a sunk cost. Whether that money was well spent or wasted does not and should not affect decisions about what to do in the future.

ESSAY AND DISCUSSION QUESTIONS

3.1 What factors might cause a change in the demand function for an infrastructure-based system (e.g., communications network, office space, air travel, or sewage treatment facilities)? What is the difference between a change in the volume using a particular system and a change in the demand function for that system?

3.2 What is the role of price regulation in a competitive environment? What is the role of price regulation in a situation where there are very large economies of scale or density, and a very large company will be able to provide services at a much lower cost?

3.3 Consider a particular infrastructure-based system (e.g., transportation, water supply, communications, or office space in a major city). What factors are likely to affect the extent to which people use the system? What decisions will people make, and what aspects of the system's performance are most likely to affect people's utility?

3.4 What pricing options would you consider (a) if you were the secretary of transportation for a state with congested highways; (b) if you were in charge of water supply for a region experiencing water shortages due to increasing development and irrigation at a time when its climate is becoming drier; (c) if you were an Internet service provider and your capacity was being chewed up by extensive downloading of images, videos, music, and spam?

3.5 Clearly, there are economies of scale in construction and operation of infrastructure. The larger the building, the lower the cost per square foot for both construction and operation. What limits the size of a new office building in Boston, where buildings are larger than those in Cambridge but smaller than those in New York City?

3.6 Clearly, there are economies of scale in construction and operation of infrastructure. The more seats in a football stadium, the lower the cost per seat for both construction and operation. What limits the size of a football stadium?

PROBLEMS

3.1 Define the following terms:

a. Supply function
b. Demand function
c. Equilibrium price
d. Elasticity of demand

3.2 Define the following terms:

a. Productivity
b. Economies of scale
c. Economies of scope
d. Economies of density

3.3 Define the following terms:

a. Marginal cost pricing
b. Consumer surplus
c. Comparative advantage
d. Spatial price equilibrium
e. Exchange rates
f. Sunk cost

3.4 A consultant has estimated that the demand (D) for a particular product can be approximated by the following equation for prices between \$200 and \$800:

$$D = 1000 - P$$

The supply (S) for this product can be estimated as follows:

$$S = 500 + 0.5P$$

a. What will be the equilibrium price?
b. Suppose that demand is expected to increase in proportion to population growth. What will the equilibrium price be in 10 years if population grows by 10%?
c. Suppose a new technology allows suppliers to cut their marginal costs by \$50/unit. What would be the new equilibrium price,

and what would be the new equilibrium volume? (Assume that demand will grow by 10%, as in part b.)

3.5 A transit system in a medium-sized city operates at a loss, as do essentially all transit systems in the United States. The transit agency is under pressure from its advisory board to increase fares in order to cover a larger proportion of operating costs. The agency is also under pressure from various groups representing riders and from others who argue that fares should be lowered to attract more riders, thereby reducing highway congestion, air pollution, and gasoline consumption. Research conducted at a local university indicates that the price elasticity for bus service in similar-sized cities is on the order of 1.2. If so, how much would revenue change if bus fares were increased from \$1.25 to \$1.50? If fares were reduced to \$1.00? Given that the vast majority of bus service costs would remain the same, since no one advocates reducing scheduled service, explain why you would recommend raising or lowering fares.

3.6 An energy company has recently enhanced its capabilities for generating electricity, resulting in a 10% reduction in the cost per kilowatt-hour (kWh). The company, which is regulated by the state, is required to pass the savings on to consumers via a 10% reduction in price per kWh. Assume that the company has 100,000 residential customers who paid an average of \$60/month in the year before the new system was implemented; the same 100,000 customers paid \$58/month following the shift to the new system with its lower prices. The total revenue therefore dropped from \$6 million per month to \$5.8 million per month. Based upon these facts, answer these questions:

a. What was the increase in volume of electricity generated?
b. Was there a change in the demand function? If so, what is the new demand function? If not, why did the power company have to produce more electricity?

c. What was the price elasticity in the demand for electricity?

d. What was the increase in consumer surplus as a result of the reduction in prices?

3.7 Which of these factors should be considered in deciding whether to buy a new car? Why?

a. Your current car is 8 years old.

b. Your car cost $4,000 when you purchased it.

c. You recently spent $1,000 on a new muffler and a new set of tires.

d. The car gets 20 mpg.

e. Your mechanic said you can expect maintenance expenses to increase by 20% or more over the remaining life of the car.

f. A friend has offered to sell you her 6-year-old car for $5,000.

g. You could buy a new car with better gas mileage for $15,000.

h. The car broke down twice last year on the way to work, and you are afraid it will break down again.

i. You never did like the sickly green color of the car.

3.8 Last year you took a foreign vacation when the exchange rate between the dollar and the euro was 1.60 ($1.60 to purchase one euro). That vacation cost you $2,000 not including transportation to Europe. This year, you want to go again; and now the exchange rate is 1.30. Assuming that prices are the same in Europe as they were last year, how much will the European portion of the trip cost this year?

3.9 Your company has two divisions, one operating in the United States and one operating in Europe. Each division was started with an initial investment of $1 million. The accounting for the U.S. company is stated in dollars ($), while the accounting for the European company is stated in euros (€). Based upon the exchange rates at the time of the investment, the $1 million invested in Europe was equal to €800,000. Both companies earned 10% on investment in two recent years. The average exchange rate ($ per €) was 1.50 in the first year and 1.20 in the second year. If the profits were expressed in $, what was the return for the combined operations in these two years? Discuss the implications of this result for evaluating projects that involve foreign investment.

3.10 Two regions with extensive apple orchards attempt to sell their apples to grocery stores throughout the country. Farmers in the western region have very large orchards, and they can obtain rates of $1.40/truckload/mile for shipping their products to market in the middle of the country. Due to the large size of their farms, these farmers use large trucks that can transport 20 tons of apples. Their costs of growing, picking, and packing the apples average $200/ton. Farmers in the eastern market have smaller farms, and they must use smaller trucks. Their costs are $250/ton for growing, picking, and packing the apples and $1.60/truck-mile for transporting a 16-ton truckload. The distance between the eastern and western areas is 2500 miles. Assuming that the apples are in fact identical in size and taste, what will be the dividing line between the markets served by the two regions? What will be the price/ton of apples at this midpoint?

CASE STUDY: BOOM AND BUST IN REAL ESTATE DEVELOPMENT

1. How to Make a Million in Real Estate

Until the 1960s, there was nobody involved in assembling sites for office developments, arranging loan finance from the banks, overseeing construction and selling on completed schemes as long-term investments for insurance companies or pension fund.
Frank McDonald and Kathy Sheridan, *The Builders: How a Small Group of Property Developers Fuelled the Building Boom and Transformed Ireland*
(London: Penguin, 2008)

In *The Builders*, McDonald and Sheridan describe the intriguing process whereby a handful of developers went from building houses to building apartment buildings to building malls to building small towns in Eastern Europe—and in the process took advantage of tax loopholes, land deals, shady dealings with zoning officials, and other nefarious activities in order to become billionaires. Many of their friends and relatives had a chance to become millionaires along the way. Here's how they did it, according to one developer interviewed for their book.

A developer has plans for an apartment building as part of a major development outside of Dublin. He shows these plans to a couple of friends and convinces them to buy an apartment for €190,000 (approximately $300,000) with a deposit of €5,000. The building was built in 18 months, at which time the apartments were selling for €290,000. The developer then allowed his friends to sell the apartment to someone else, without ever occupying it (this is known as "flipping" the apartments). At that point, the developer received €185,000, which along with the deposit made up the agreed-upon purchase price of €190,000. The two friends took the rest of the €290,000 purchase price as a profit. Not bad for an 18-month investment of €5,000. But not nearly enough!

The two friends used all of their profits to buy more units in another of the developer's projects; they put down €10,000 per unit and agreed to pay the rest of the €270,000 price when the apartment complex was completed. After another 18 months, they were able to flip all of those units for an average of €340,000, giving them a profit of €70,000 on each unit.

Flipping appears to be a highly profitable activity—and it is easy to believe there will always be a buyer when the real estate market is soaring.

You might ask how the developer was able to build the units if buyers only had to put down €5,000 or €10,000 as a deposit. No problem—this was a time of easy credit. For every €5,000 deposit, the bank would loan the developer €100,000, knowing that there was already a buyer waiting for the unit to be built.

Figure 3.11 New Construction in Dublin, September 2008

Questions

a. Assume that the two friends put all of their profits from the first deal into buying more apartments as part of the second deal. How many apartments did they buy, and how much did they gain when they flipped those apartments?

   NOTE: Don't worry about interest payments on mortgages—just use the numbers as presented in the background or in the questions.

b. Suppose the friends each pocketed €100,000 and invested the rest of their profits in yet another scheme. How many units could they buy if the required deposit was €20,000 for apartments priced at €400,000? If those apartments ultimately sold for €500,000, how much would their profits be on this transaction?

c. Assume that the developer had 100 apartments in each of the developments referred to above, all of them assigned to investors who were given the same deals described above. Assume that his cost to construct the first set of apartments was €150,000 per unit, while his cost for the second set of apartments was €200,000 per unit. How much did he profit from these two developments?

2. How to Lose a Million in Real Estate

When banks started to tighten credit, and when there were no new buyers to be found, it became impossible for flippers to do their tricks, since the developers could find no buyers. The flippers and the developers were stuck with mortgages they had always assumed would be paid off immediately when they sold their properties. Since they had these mortgages, but no income, their credit was no longer any good; and banks wouldn't loan them any more money to start new projects. And since they and their friends were no longer buying new units, the prices of real estate began to fall. Soon, real estate speculators found themselves owning property that was worth less than the outstanding balance on their mortgages, and many developers and others in the construction business went into bankruptcy.

Questions

a. Suppose our investors had decided to try flipping once again, putting €2 million into one more "can't lose" development. This time the deposit was €40,000 per unit and the sale price was €500,000. They assume that prices will continue to climb 10% per year, so they expect to flip the units for €600,000 in two years. What will their profit be if this works?

b. What happens if housing prices fall 2%, so that the sales prices actually turn out to be only €490,000?

c. What happens if housing prices fall 10% overall—and they can sell only 80% of the units?

d. What happens to the developer if housing prices fall 10% and he can sell only 80% of the units? Assume the developer's cost per unit was €400,000 and assume that the developer had 1000 units under construction.

Chapter 4

Public Perspective: Economic, Environmental, and Social Concerns

Happiness lies not in the mere possession of money; it lies in the joy of achievement, in the thrill of creative effort. The joy and stimulation of work no longer must be forgotten in the mad chase of evanescent profits. These dark days will be worth all they cost us if they teach us that our true destiny is not to be ministered unto but to minister to ourselves and to our fellowmen. . . . Our greatest primary task is to put people to work. This is no unsolvable problem if we face it wisely and courageously. It can be accomplished in part by direct recruiting by the government itself, treating the task as we would treat the emergence of a war, but, at the same time, through this employment, accomplishing greatly needed projects to stimulate and reorganize the use of our natural resources.

Franklin D. Roosevelt, President of the United States,
First Inaugural Address, March 4, 1933

CHAPTER CONCEPTS

4.1 GENERAL OVERVIEW

This chapter introduces the basic public policy concerns that are related to the evaluation of major infrastructure projects. It also provides some examples that illustrate how the public perspective differs from the private perspective.

As discussed in Chapter 2, the private sector undertakes projects primarily in the hopes of financial rewards. Building a road or a canal will be viewed as a good thing if the tolls that are expected will be more than sufficient to pay off the bonds that were sold to pay for the construction. Building a new office building will be viewed as worthwhile if the rents that are expected will be more than sufficient to pay off the mortgage. The analysis is concentrated on financial matters: what will it cost to build the project? What revenues will it produce? What are the risks that costs will be higher or that revenues will be lower than expected? How can we raise funds to pay for the construction of our project? Would it be better to invest in financial securities rather than trying to make money constructing our own projects?

Chapter 3 introduced a broader economic framework that also includes the short- and long-term impacts of a project on the local economy. Since any major project will require some sort of public approvals, the nature and magnitude of these economic impacts could determine the willingness of public officials to support or grant approvals for the project. Still, the primary motivation for the private sector projects will be the financial returns to the owners, not the broader effects on the economy.

The public sector, which is responsible for many kinds of infrastructure systems (Table 4.1), takes an entirely different perspective in identifying needs and evaluating potential projects. The motivation is not to earn money, but to satisfy public needs or to promote growth in the economy. Financial issues are important, but not necessarily dominant. Social and environmental impacts are central to the public evaluation process, and equity in the distribution of costs and benefits will be critical. In dealing with nonmonetary objectives, cost effectiveness will be a more relevant concept than return on investment: which of the proposed alternatives is the best way to achieve the desired objectives?

Table 4.1 Examples of Public Infrastructure: Multiple Purposes and Multiple Measures

Type of Infrastructure	Purpose	Measures
Transportation	Mobility Accessibility Regional competitiveness	Service levels (travel time, congestion) Cost of transportation Fuel consumption Safety Emissions
Dams	Flood control Irrigation Hydropower Recreation (boating, swimming, camping, picnic sites)	Risks associated with floods Volume of water available for irrigation Land area to be irrigated Electricity production (cost and revenue) Impact on wildlife
Water and sewage	Clean water for consumption Water for industry and irrigation	Volume of water available for each type of use Cleanliness (risk of disease) Cost per unit
Public housing	Housing for elderly Housing for low-income residents Housing for homeless	Number of units Size and quality of buildings Cost per unit (construction and operation)Safety and security Aesthetics
Parks and recreation	Open space for residents Protect environment Aesthetics	Open space as a percentage of total space Visitors per year Diversity of wildlife Safety

The time frame of the analysis will be much longer for the public than for the private sector, since the public entity is presumed to endure indefinitely. The long time frame requires the consideration of sustainability—are projects or programs sustainable over long periods of time, taking into account economic, financial, social, and environmental factors?

For many kinds of public infrastructure projects, the tolls, fees, and other direct revenues from the project are insufficient to cover the costs of the investment. However, the nonmonetary benefits can be considerable. A public transportation project may relieve congestion, improve air quality, and promote mobility for those without access to an automobile. While these benefits can at times be quantified, using concepts like consumer surplus or the multiplier effect of new investment, such benefits do not produce cash flows that cover the interest on bonds or the operating costs of the transit agency. If the benefits are clear, and if the public generally believes these benefits are worthwhile, then public agencies may be able to use tax revenues to supplement the direct cash flows from the project. If taxes are used to finance a project, then that project is competing not just against similar projects, but against all of the other projects undertaken by that city, state, or country. Transportation projects compete with housing for the elderly, and water projects compete with health care projects. Decisions for or against projects are political decisions, and the relative importance of various kinds of costs and benefits are subject to considerable debate.

To complicate the situation further, a project also is evaluated in terms of its impact on the community:

- Economic impacts, including employment, regional economic growth, regional competitiveness
- Environmental impacts, including air quality, water quality, noise, loss of wetlands, and impact on ecosystems
- Equity, including the distribution of costs and benefits across regions and groups of the population and the relative impact on current and future generations
- Aesthetics, including the appearance of the new infrastructure, its effect on neighboring areas, and its effect on long-term changes in land use
- Other social impacts, including such things as effects on communities during construction, displacement of residents, and long-term changes in population distribution

Multiple objectives and multiple measures mean that these projects are inherently complex, and many conflicts are possible among different objectives. The decision makers ultimately will include the public, who may have a chance to vote for or against the funding sources proposed for a project; and the politicians or appointed officials, who must justify their decisions to the public in order to be reelected or to retain their jobs. Large projects are politically sensitive, and all of the conflicts must be considered and balanced. Real and apparent conflicts of interest occur among those who are supposed to be proposing, evaluating, and approving projects. It is not possible to satisfy everybody, and there will likely be determined opposition to almost any major project. People commonly do not want anything built too close to them, even if they are going to be major beneficiaries of the project. This phenomenon, which can lead to intense community opposition, is known as the NIMBY response: "not in my backyard."

Thus a major public project is evaluated by many different groups of people, from many perspectives. These groups have varying concerns for the relative importance of various features of the projects, and potential disputes can arise over how to measure or estimate costs and benefits.

4.2 BENEFIT-COST ANALYSIS

Public projects require an evaluation process that includes, but is much broader than, financial analysis. A simple dictum is mandated both by law and by common sense: for any public project, the total benefits should exceed the total costs. This does not mean that every project with more benefits than costs is a good project; it simply means that projects whose costs exceed their benefits are bad projects that should not be funded by the public. This principle may seem rather obvious; but it surely is necessary due to the many instances of projects being built, at taxpayer expense, whose costs far exceeded their benefits. Such projects even have names: "gold plated" projects could have been constructed for far less money; "pork barrel" projects were approved in order to get a crucial politician's support for some larger political scheme; and "white elephants" are constructed at great expense, but afford no greater benefits than ordinary elephants! The political process can provide a means to fund many different projects, and many projects are earmarked (i.e., specifically authorized in the legislation) rather than subjected to a rigorous examination of their costs and benefits. Requiring that the benefits of every project exceed the costs is therefore a step toward a more rational allocation of public funds and a defense against mismanagement, stupidity, and corruption.

Measurement is a major problem in determining whether costs exceed benefits: How can different types of noncash costs and benefits be converted to monetary terms? How can important benefits such as savings in travel time or reductions in risk of accidents be converted into monetary terms? What about aesthetics? In some cases the monetization is straightforward, in other cases it is convoluted and controversial, and in still other cases it is essentially impossible.

EXAMPLE 4.1   Construction of a New Highway

Consider a proposal to construct a new highway that is intended to provide a safer, more attractive route around the congested core of a city. The basic question is whether the savings in travel time, the expected reduction in fatalities, and the prettier route justify the cost and the environmental impacts of constructing the highway through the surrounding region.

TRAVEL TIME

Traffic engineers can model how commuters, truckers, and others will use the new facility, and they can predict traffic flows on the new facility along with the changes in traffic flows on other facilities. Based upon the changes in traffic flows, they can predict travel times on the new road and changes in average travel time on each segment of the existing network. The overall effect can be summarized as a reduction in travel time measured as vehicle-hours per day or per year, with details for commuters, local delivery trucks, long-distance trucks traveling through the region, and any other group of interest. The value of these time savings is commonly estimated by making a series of assumptions. For instance, the time saved by commuters could be valued by using the average hourly wage for workers in the region, and the value of time saved by truckers could be valued by using the average hourly wage for truck drivers, the hourly cost of truck ownership, and the hourly value of the contents of the truck. Some might argue that something less than the average hourly wage should be used, and others might challenge the methodology used to estimate the hourly cost of truck ownership; but these estimates of the value of time are commonly accepted, and the benefits are clear and verifiable.

SAFETY

Estimating the value of the safety benefits is trickier. Traffic engineers can predict the number and severity of accidents based upon traffic flows and highway geometry, and safety analysts can use past history to quantify the expected damage to vehicles and the highway. However, no one can readily place an economic value on the most important safety benefits—namely, a reduction in the expected number of injuries or fatalities. Instead, departments of transportation in some countries will consider the value to society of reducing fatalities and serious injuries resulting from automobile and other transportation accidents. In the United States, the U.S. Department of Transportation uses a value in excess of $2.5 million in its risk analysis. This amount represents the benefit to society of eliminating a single, future fatality. It does not represent the value of a human life, for it is impossible to say who would have been hurt or killed. In effect, the $2.5 million can be viewed as an aggregate benefit to all users of the system of a slight reduction in the probability of a fatal accident. Every user benefits because the probability of an accident is reduced.

AESTHETICS

Now we are close to the "impossible" in trying to quantify the benefits of the new highway. Whether aesthetics is viewed as an important component of the decision will depend upon the political situation in the region. It could be viewed as an afterthought, which might mean planting some flowers along the right-of-way, or it could be a major design consideration, as in the construction of roads in national parks or the construction of parkways into major cities. The argument may well boil down to someone showing artist's renditions of the options (or photos of similar projects elsewhere) and saying something like, "Isn't it worth spending an extra $5 million to get a nice facility?" If people agree that $5 million is a small amount, then they will choose the more aesthetic option. If people note that $5 million is equivalent to the park budget for 20 years, then they will probably be vocal in their opposition!

4.3   ECONOMIC IMPACTS: MEASURES RELATED TO THE REGIONAL OR NATIONAL ECONOMY

Governments and public agencies are concerned with the effects of projects on the local, regional, or national economy. Primary measures will include gross regional or national product, jobs created or lost, average income, and personal and industrial productivity. The economic benefits could come from several types of benefits, as has already been described in Chapter 3. The benefits directly related to the project include

- Continuing productivity benefits resulting to citizens, users, industries, or public agencies as a result of the project
- Growth in the economy related to the productivity benefits provided by the project
- Construction jobs and income
- Jobs and income related to the eventual operation of the new project

The following are some additional economic impacts related to the construction and continued operation of the project:

- The multiplier effect of construction
- The multiplier effect of operation of the new project

As discussed in Chapter 2, these multiplier effects are not considered in a financial analysis, and they may be viewed as a secondary aspect of the economic analysis.

For example, consider the construction of a toll road. The initial construction may take 2 years, provide hundreds of jobs, and increase sales of construction materials within the region. The direct expenditure of several hundred million dollars would have a multiplier effect that more than doubles the economic benefit to the region during the period of construction. Once the toll road is opened, there could be long-term jobs for toll collectors (and for those who maintain any electronic toll collection devices) and for highway maintenance forces, thus having both a direct and a multiplier effect on the regional economy. The toll road presumably offers benefits to the public in terms of higher capacity for rush-hour traffic, reduced risks of accidents, and perhaps reduced travel time. With less congested highways, the region may be able to continue to attract new businesses and to absorb additional population growth. Land near interchanges is likely to increase in value and attract hotels, restaurants, trucking terminals, warehouses, and other businesses that depend on highway access or serve highway users.

These benefits could be offset by the impacts of both the construction and the continued operation of the highway. Disruption of normal activities can be a major economic cost of a highway project. Although construction of a new highway interchange will ultimately relieve congestion, it may cause increased delays for a year or two. Once the highway is built, it may act as a barrier that limits access between different parts of the region. Over time, land use will adjust to the existence of the highway, which could result in rapid growth in some areas and equally rapid declines in other areas.

4.4   ENVIRONMENTAL IMPACTS

Any project will alter the complex relationships between what might be thought of as the natural world and the man-made world. Construction activities convert more space from the natural to the man-made world. Projects require construction materials such as wood, steel, and concrete, which ultimately depend upon activities such as forestry, mining, and manufacturing that certainly disrupt and may at times destroy the natural world. Continued operation and maintenance of infrastructure requires energy and other materials that ultimately come from the natural world. Normal operations, accidents, and decay may release toxic substances that can affect air quality, water quality, and soil composition and limit or destroy the ability of plants and animals to survive near project sites. Constructed facilities will cast shadows; they may be noisy; and they might just be ugly or interfere with people's day-to-day lives. Whether the benefits of the project are worth the environmental costs will always be a relevant question, especially when those receiving the benefits are not those who bear the costs. The extent to which this question is considered will depend upon the social, cultural, and political institutions. In many countries, developers must prepare an environmental impact statement (EIS) that at least states the goals of a project, presents major alternatives for achieving these goals, identifies the major environmental impacts, and suggests ways to mitigate the most negative impacts. Preparing an EIS ensures that information is made available to the public and to public officials who must approve a project; the extent to which environmental considerations affect decisions about a project may well depend upon legal and political battles.

Courts and legislative bodies are well-structured for dealing with controversial trade-offs between environmental and economic issues and with the extent to which developers must deal with environmental concerns. Legislation has limited the development of wetlands, promoted soil and water conservation, required more fuel-efficient automobiles, and limited land use via zoning and other restrictive matters.

However, courts and legislative bodies are not well-structured for dealing with the underlying science, as evidenced by the controversies related to the extent of, the causes of, and the possibilities of responding to global warming. Global warming is one aspect of a very large and ultimately very important issue facing humankind. To what extent can we continue to modify the natural environment without running into fundamental constraints that will affect not only human life and human health, but the continued existence of all species? This question is extremely difficult to respond to, and it will affect all of our major decisions concerning energy, transportation, housing, and other aspect of infrastructure.

Humans have certainly transformed the world. Over a period of many thousands of years, we have converted vast portions of the land area to agriculture, drained innumerable wetlands, developed much of land near the oceans, seas, and major rivers, and cut down vast areas of forest. These activities have changed the chemistry of the atmosphere, altered the natural flows of fresh water, and restricted the natural habitats crucial for many species of plants and animals. These activities have also allowed humans to prosper by helping to ensure adequate food supply, clean water, housing, abundant energy sources, and protection against floods and other natural disasters. In the future, we will still eat, drink, use energy, and improve the way we live—but we will have to pay more attention to our impact on the environment.

A whole of range of environmental issues will have to be addressed in evaluating any major project, and the primary objective of many major projects will be to improve the environment. Environmental issues will range from very local debates about what gets built in whose backyard to regional and national questions related to using resources to international questions concerning the future of the planet. Since we can't expect to answer all of these questions every time we want to build a new hotel or new segment of a highway, we need to provide a reasonable structure for addressing these issues within the project evaluation process.

If we do not establish intelligent controls over land use, most Americans will be spending 95 percent of their lives in huge blighted zones, hundreds of miles long, that are neither city nor country, but one vast, dispiriting Nowhere. . . . We have not only the governmental tools, but the technology and the wealth to complete the unfinished business of America [which is to address the problems of] public safety, noise, pollution, eyesores (junkyards, utility poles, billboards), mass transit, cities, regions, . . .

Edmund K. Faltermayer, editor, *Fortune* Magazine, expanding upon the prospects raised by President Johnson in his Great Society speech (*Fortune*, March 1965)

4.4.1   Key Environmental Concerns

The following are some of the key environmental concerns:

- Ecosystems
- Pollution
- Wetlands, aquifers, and drainage
- Wildlife habitat
- Renewable versus nonrenewable resources
- Climate change

Soil, water, sunlight, and temperature are among the factors that determine what plants can grow in any location. Plants that are well-adapted to local conditions will prosper, those that are poorly adapted will struggle, and those that cannot survive the extremes of temperature or hydrological conditions will never gain more than a short-term foothold. Insects, birds, and small mammals are necessary to the propagation of many plant species, and worms, amphibians, and insects make soil into a complex, living community. Animals may feed on plants or other animals, and they prosper in locations where there is an abundance of food along with sufficient cover for their own safety and appropriate places to raise a family, whether in trees, burrows, rotten logs, stream banks, or wetlands.

Left undisturbed, any location eventually develops a characteristic set of plants and animals that can survive or flourish within the constraints posed by soil conditions, sunlight, and climate. Biologists have identified distinct **ecosystems** that can be characterized by the kinds of plants and animals that are found there. Any healthy ecosystem supports a diversity of species, each of them somehow related to the health of the overall system. Pileated woodpeckers make holes in dead trees as they look for insects, and chickadees later use these holes as nesting sites. When the dead tree finally collapses, it will provide cover for mice and other small rodents, as well as an ideal place for fungi to grow or for grouse and hares to hide.

In many cases, some species are found only in certain ecosystems, so they are considered to be **indicator species**. Indicator species are useful in documenting the existence of unusual ecosystems. For instance, wood frogs lay their eggs in vernal pools—small pools that are formed in rainy seasons or in spring as the snow melts. Vernal pools dry up for part of the year, so they cannot support fish; this means that eggs deposited in a vernal pool will be safe from predation by fish. Wood frogs are an indicator species for vernal pools. A single female wood frog lays hundreds of eggs in the early spring, so if the pool retains water long enough for the eggs to turn into tadpoles and for tadpoles to grow into tiny frogs, then the wood frogs will prosper. If the vernal pools are filled as part of the process of building a parking lot or a suburban subdivision, then the wood frogs will die off.

In most regions, a few types of ecosystem will dominate the landscape, while a dozen or more other types will be commonly scattered throughout, and some will be found only in a few locations. Preserving the rare ecosystems may be essential for preserving biodiversity, since certain plants and animals can be found only in those locations. Preserving a good distribution of the more common ecosystems will prevent populations of plants and animals from becoming too isolated. Preserving large tracts of the dominant ecosystems will ensure healthy conditions for all of the region's most common species.

Ecosystems can be harmed in several ways. **Pollution**—the introduction of foreign elements into the air, water, or soil—may lead to the death of certain plants or insects and of the animals that depend upon them for food. Pollution could be in the form of toxic chemicals that are poisonous to certain species, but it could also be just the introduction of sediment into a pristine stream, thereby making the water quality unsuitable for certain types of fish. Pollution can also refer to the heated water that is discharged from a nuclear power plant because the heated water is lethal to some species, but it attracts others that may be alien to the previously existing ecosystem.

Disruptions to the flow and retention of water can have devastating effects on ecosystems. Draining wetlands to increase the land available for highways, housing, or agriculture will lower the

Figure 4.1 Wildlife Crossing over the Trans-Canada Highway in Banff National Park. By installing fencing along the highway and constructing crossings like this at regular intervals, highway officials allow animals to cross the highway safely, thereby avoiding fragmentation of habitat and reducing highway collisions with wildlife.

water table and make the remaining wetlands more susceptible to drought and fire. Extensive development in Florida, for example, has changed the flow of water through the Everglades, threatening the future of what were once the seemingly endless wetlands of southern Florida. More rapid runoff of water means that both floods and droughts are more likely, which means that certain species of plants and animals will have greater difficulty surviving.

Fragmentation of an ecosystem eventually creates areas too small to support the wildlife that formerly flourished there. A black bear requires a range of 10 to 100 square miles; if a region that formerly supported large populations of black bears is crisscrossed by roads and disrupted by housing developments and malls, then the habitat is no longer large enough for the bears to survive.

For species requiring less extensive ranges, it is not so much the fragmentation of the habitat as the total **loss of habitat** that will be decisive. As agricultural land is turned into housing developments or malls, the birds that once fed on the insects and seeds will have to go somewhere else, and the deer that once fed on the leftover corncobs will be hit by cars as they try to feed on the shrubs and gardens of the new developments. Colonies of butterflies and dragonflies will be lost, along with vast numbers of mice, voles, and moles as well as the hawks, owls, and weasels that feed on them. For migratory birds, the loss of habitat is especially problematical because they need places to feed and to breed, perhaps on two continents, and they need extensive areas for resting and feeding along their migration routes.

A final threat to ecosystems comes from the introduction of alien species. In a well-functioning ecosystem, everything is in balance. Insects or other animals eat some but not all of the seeds, none of the animals eat all of any of the species of plants, and none of the plants grows so rapidly that it crowds out all of the other plants: it is a complex system of natural checks and balances. An **alien species** is one that originated in a distant ecosystem where it had adapted to competition with the other plants and animals that comprised that ecosystem. It undoubtedly served to control some of the other species, and other species controlled it. However, when an alien species is introduced into a new ecosystem, there may be no controls and balances, and that species may prosper until it outcompetes and eventually crowds out the native species. Purple loosestrife is a tall, tough wildflower with a large, woody ball of roots; it has numerous flowers on a spike, and it grows profusely in wetlands. When introduced to wetlands in the United States, it faces only modest competition from less aggressive plants, and it has no natural insect competitors. As a result, it can, within a few years, fill the wetlands and create a beautiful purple covering—but also a barren wetland. Because no native insects eat the loosestrife stalks or flowers, no native birds are attracted to the plant; and no hawks are circling to catch any of those birds off guard. The weeds grow so close together that muskrats or beavers must struggle to keep their channels open, and there is too little space between plants to support families of ducks. Alien species often get their start when ground is disturbed for some sort of construction project. If these species are not dealt with—which often requires people who search for the first aliens and then pull them out by hand—then they can rapidly spread and destroy many acres of land. The key point to remember is these alien species overflow their niche, eliminating the chance for native species to prosper and also eliminating the niche that was occupied by insects and animals that depended upon the native plants and animals. Alien species may be beautiful, but they tend to limit biodiversity.

Maintaining the health of ecosystems requires local, regional, and national strategies. One useful concept is **green infrastructure**, which refers to the network of natural areas that is necessary to support the diverse populations of native plants and animals living within a region. This term does not refer to man-made infrastructure that is constructed in an environmentally friendly manner. Rather, it refers to the connected natural system of open spaces, forests, waterways, and wetlands that allow plant and animal species to prosper. Green infrastructure includes the following kinds of components:

- Very large areas of undeveloped land that are able to support and protect habitat for the widest-ranging animals and ensure the continued existence of diverse ecosystems
- Small areas or undeveloped land that protect uncommon or rare ecosystems
- Numerous small or medium-sized natural areas that are large enough and close enough together to avoid isolation of plant and animal species
- Connecting corridors of open spaces that can be used by animals to move between the larger open areas

By acknowledging the existence of and the need for green infrastructure, government agencies and conservation groups can develop plans that preserve and protect suitable green infrastructure. National parks, state parks, public conservation lands (e.g., national forests or wildlife management areas) can provide the critical large areas. Smaller parks, wetlands, and private landholdings can protect enough smaller areas to ensure diversity and density of ecosystems. The hardest part is ensuring that wide enough corridors are maintained between and among all of the open spaces so that wildlife can in fact move throughout the region. The corridors need to be wide enough to be perceived as safe routes for animals to travel. For the largest mammals, 100- to 200-foot side corridors are needed. For smaller mammals and amphibians, narrower corridors will suffice. Land adjacent to waterways and wetlands is ideal for use as connecting corridors, as is land next to railroads, power lines, or other infrastructure networks.

Pollution can be controlled by limiting emissions, by confining emissions, or by cleaning up emissions. The cheapest control strategy is to prevent emissions, but that may or may not be feasible depending upon the nature of the process that causes the pollution. Some pollutants are extremely toxic, and even a small release can be hazardous to anyone living nearby. Hence, special consideration is necessary in dealing with the most toxic chemicals and spent nuclear fuels or other radioactive substances. Finding a safe means of sequestering nuclear waste is one of the main challenges facing the nuclear power industry.

Climate change caused by excessive emissions of carbon dioxide, methane, and other so-called greenhouse gases is a major challenge for the world in the twenty-first century. Scientists believe that increasing concentrations of these gases in the atmosphere will trap heat, thereby leading to warmer temperatures. With warmer temperatures, more energy will be available to power hurricanes, tornadoes, tsunamis, and other extreme weather conditions. Warmer temperatures will also accelerate the melting of glaciers and the ice caps, which will raise the level of the oceans and threaten flooding of the many cities and developed regions along the coasts. Changes in climate could also include changes in precipitation, which could have major implications for agriculture and the natural environment.

4.4.2 Environmental Impacts of Projects

Several levels of concern are evident:

1. Use of materials in construction and operation
2. Pollution: impacts on air quality, water quality, and soil toxicity
3. Loss of habitat and disruption of ecosystems: impacts on plants and wildlife
4. Impacts on the local environment (noise, shade, aesthetics)
5. Sustainability

It will not be possible or necessary to consider all possible levels of impacts for every project that is considered. Regulations can be developed that govern the use of materials and that establish acceptable limits for pollution. Regional plans can help identify the necessary green infrastructure, and zoning can be used to direct development away from the most critical natural areas. Local impacts on noise, aesthetics, and land use are of course a concern for nearly any project, and some sort of community involvement can be helpful in anticipating and responding to potential problems.

The next section provides some background on how environmental impact statements are prepared in the United States. A process has been created to ensure that environmental impacts are considered along with the economic and social impacts of any major project or program involving federal funding or approvals. This process emphasizes the need for determining and disclosing environmental impacts, and it requires developers to consider how to mitigate negative impacts, but it does not indicate what can or cannot be done. That determination is left to the legislatures and the courts.

4.4.3   Environmental Impact Statements in the United States

An EIS is required for any major federal legislation or action "significantly affecting the quality of the human environment."[1] The federal agency proposing the changes must prepare the EIS, which must include "a detailed statement of these environmental effects."

> *The National Environmental Policy Act of 1969 (NEPA), as amended, (42 U.S.C. 4321 et seq., Public Law 91–190, 83 Stat. 852), requires that all Federal agencies proposing legislation and other major actions significantly affecting the quality of the human environment consult with other agencies having jurisdiction by law or special expertise over such environmental considerations, and thereafter prepare a detailed statement of these environmental effects. The Council on Environmental Quality (CEQ) has published regulations and associated guidance to implement NEPA (40 C.F.R. Parts 1500–1508).[2]*

The Environmental Protection Agency (EPA) is responsible for reviewing the draft EIS and rating it according to two criteria. First, the EPA must decide whether the EIS is acceptable in terms of the depth of its analysis and the completeness of its findings. Second, EPA rates the environmental impact according to one of four categories:

1. *Lack of Objections (LO):* "The review has not identified any potential environmental impacts requiring substantive changes to the preferred alternative. The review may have disclosed opportunities for application of mitigation measures that could be accomplished with no more than minor changes to the proposed action."
2. *Environmental Concerns (EC):* "The review has identified environmental impacts that should be avoided in order to fully protect the environment. Corrective measures may require changes to the preferred alternative or application of mitigation measures that can reduce the environmental impact."
3. *Environmental Objections (EO):* "The review has identified significant environmental impacts that should be avoided in order to adequately protect the environment. Corrective measures may require substantial changes to the preferred alternative or consideration of some other project alternative (including the no action alternative or a new alternative)."
4. *Environmentally Unsatisfactory (EU):* "The review has identified adverse environmental impacts that are of sufficient magnitude that EPA believes the proposed action must not proceed as proposed."

The EPA review is a major hurdle for any project involving the federal government or requiring federal approval. The draft EIS is published, and time is allowed for public comments concerning what the draft includes or fails to include. The draft EIS and all of the comments and procedural rulings are

---

[1] The material in this subsection is based on EPA's "Policy and Procedures for the Review of Federal Actions Impacting the Environment" (October 3, 1984), 4, 19–20. The document is available at the EPA website (www.epa.gov).
[2] Ibid., 4.

available to the public online. If EPA finds environmental concerns, it may require substantial changes in the proposed actions or prevent the project from proceeding as proposed. Moreover, if it finds the EIS to be inadequate, EPA may require it to be revised or redone, an action that could delay a project for a year or more. The conditions that would allow EPA to raise environmental objections are specified by government regulations. Objections can be raised in five situations:

1. *"Where an action might violate or be inconsistent with achievement or maintenance of a national environmental standard;*
2. *"Where the Federal agency violates its own substantive environmental requirements that relate to EPA's areas of jurisdiction or expertise;*
3. *"Where there is a violation of an EPA policy declaration;*
4. *"Where there are no applicable standards or where applicable standards will not be violated but there is potential for significant environmental degradation that could be corrected by project modification or other feasible alternatives; or*
5. *"Where proceeding with the proposed action would set a precedent for future actions that collectively could result in significant environmental impacts."*

In other words, EPA must have a clear reason for raising objections, and other guidelines and policies will be used to determine whether proposed actions are acceptable or not. More stringent guidelines are in place for finding a proposal with environmental objections to be environmentally unsatisfactory:

1. *"The potential violation of or inconsistency with a national environmental standard is substantive and/or will occur on a long-term basis;*
2. *"There are no applicable standards but the severity, duration, or geographical scope of the impacts associated with the proposed action warrant special attention; or*
3. *"The potential environmental impacts resulting from the proposed action are of national importance because of the threat to national environmental resources or to environmental policies."*

Thus the environmental review process places the onus on the proposing agency to identify the potential impacts, while establishing an agency with the necessary skills and responsibility to review and interpret the EIS. The criteria cited above could be quite qualitative, leaving approval up to the judgment of the EPA. Since the whole process is open to the public, groups opposed to any action can make their objections to EPA.

Figure 4.2 summarizes the review process used by EPA. If a major project or action is proposed, the first step is to determine whether there will be significant impact on the environment. If EPA believes there will be no such impact, then it can allow the project to proceed without an EIS. If EPA finds that there will be significant impact on the environment, then the proponents of the project must prepare a draft EIS, which will be available for public comment and review by EPA before submission of the final EIS. EPA will then make its decision, as described above. If EPA finds that the environmental impact is unknown, then an initial environmental assessment can be required, which could lead to a finding of no significant impact or to the preparation, review, and revision of an EIS.

4.4.4 Reviewing the Draft Environmental Impact Statement

A great deal of judgment is involved in preparing and reviewing environmental impact statements. EPA has developed two sets of checklists of questions that might be asked in order to guide reviewers as they evaluate an EIS.[3] The first set of checklists address general areas of environmental concern that might apply to any proposed project: energy management, habitat preservation, landscaping, water use, and pest management. In each area, EPA has summarized the relevant scientific factors and posed a series of questions that could and probably should be asked when reviewing any project. The feature box on

---

[3] Science Applications International Corporation (SAIC), *Pollution Prevention—Environmental Impact Reduction Checklists for NEPA/309 Reviewers*, Final Report, EPA Contract No. 68-W2-0026 (McLean, VA: Author, January 1995).

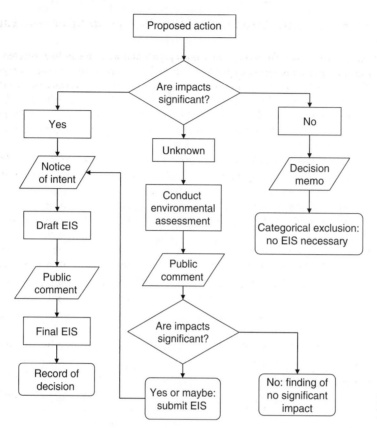

Figure 4.2 Summary of NEPA Decision Process

page 109 describes the major ecosystem issues and the threats to ecosystems that might be posed by a project. Projects can destroy, divide, or diminish habitats, thereby potentially affecting the plants and animals that live there. Projects can also introduce alien species, either by chance when the site is disturbed or by choice, as in providing plants for landscaping. Table 4.2 poses some relevant questions.

Table 4.2   Selected Questions to Consider Regarding a Project's Impact on Ecosystems

Area of Concern	Questions
Habitat fragmentation	Have other sites been considered as an alternative to encroaching on the existing habitat?
	Has the critical area necessary for survival of the ecosystem been determined? Can the area of the habitat that will be altered be minimized?
	Has the project been designed to avoid the fragmentation of existing habitats into a number of smaller areas?
	Does the project establish a system of natural corridors (that take into consideration the behavior of the species in question) to link habitat areas?
Habitat alteration	Does the project include mitigation measures, such as restoration of damaged habitats or the creation of new habitats?
	Does the project or development include adequate buffer zones between the developed area and wetlands or other habitats?
	Has project planning considered sources of water and controls of water flow to wetlands or other habitats?
Introduction of exotic species	Will landscaping activities avoid (or at least minimize) the use of exotic species?

Source: SAIC (1995), 19–22.

Box 4.1   How Can Ecosystem Preservation and Protection Affect the Environment?

In the face of development activities, populations of indigenous plants and wildlife can be protected only through the protection and preservation of ecosystems necessary for their survival. Ecosystem requirements are species-specific and can include a variety of factors, such as soil type, water regime, climate, and plant and animal associations. Ecosystems are defined by the structure and function of plant and animal communities and by the habitats they utilize. The protection and preservation of ecosystems are important for a number of reasons, which include the protection of wildlife, climate control, maintenance of biodiversity sources, pollutant detoxification, erosion control, and CO_2 sequestration.

Wetlands are ecosystems necessary for the survival of a host of aquatic and terrestrial species. In addition, wetlands are integral parts of the hydrological system and are necessary for the maintenance of water supplies and water quality.

Ecosystems face a number of threats that reduce the area available for wildlife, change the character of the species that inhabit particular habitats, or change their form through the alternation of features, including topography or water regime. Ecosystem preservation efforts are generally directed at protecting particular species, such as endangered or threatened species, recreationally or aesthetically important species or commercially important species. It should be noted, however, that habitat preservation (or creation or enhancement) for one species can adversely affect other species.

Source: Science Applications International Corporation (SAIC), *Pollution Prevention—Environmental Impact Reduction Checklists for NEPA/ 309 Reviewers*, Final Report, EPA Contract No. 68-W2-0026 (McLean, VA: Author, January 1995). pp. 18–19.

The other checklists provide questions concerning 23 types of activities (Table 4.3). In each special area, the guidebook includes a brief summary of the likely problems and then provides several dozen questions that could be asked to determine how well an EIS has addressed issues that might be anticipated. For example, the guidebook provides a one-paragraph description of the possible environmental effects of flood control projects:

Table 4.3   Checklists Have Been Developed to Guide Review of the Environmental Impacts of 23 Types of Activities

Agriculture and natural resources	Agricultural irrigation
	Forestry activities
	Grazing
	Mining projects
Building projects	Building/housing construction
Energy projects	Coal-fired power plants
	Nuclear decommissioning
	Oil and gas projects
Military projects	Chemical demilitarization
	Defense testing and related activities
	Military base closure and reutilization
	Rocketry/missile projects
Transportation projects	Airports
	Dredging
	Highways and bridges
	Natural gas pipelines
Water resources	Dams, hydropower, and water supply reservoirs
	Flood control projects
Waste projects	Hazardous waste incinerators
	Hazardous waste storage and treatment facilities
	Solid waste landfills
	Waste site cleanup activities
Other	Recreation and tourism

Source: SAIC (1995).

Table 4.4 Selected Questions to Consider in Reviewing the EIS for Flood Control Projects

Area of Concern	Questions
Ecosystem concerns	Has the use of alternatives involving levee setbacks or the use of floodways been considered?
	Will the project lead to land use changes in the watershed, particularly those that result in increased surface water runoff and nonpoint source pollution?
	Have modifications to existing flood control structures been evaluated to determine if they can eliminate the need for the new channelization or channel modification project?
	Does the plan include native plant revegetation of areas disturbed by construction to minimize erosion and sedimentation?
Project design and planning	Have alternatives, such as upstream watershed management and floodplain widening, been considered?
	Will building be prohibited within a defined distance from the streambed to protect the stream bank?
	Are channel slopes graded so animals can crawl or climb out?
Construction	Will construction take place during the dry season?
	Will site access routes and equipment storage areas be planned and located to minimize erosion potential?
	Will construction and storage areas be sited away from critical habitats?
Maintenance	Will vegetation removal methods that use chemicals, grazing, or burning be prohibited? (Chemical herbicide residuals and animal wastes can be washed into waterways during rainy periods.)

Source: SAIC (1995).

> *Flood control projects can include channelization and channel modification activities and levee construction. Such activities can change the ability of natural systems to filter pollutants from surface waters; alter the rates and paths of sediment erosion, transport, and deposition; increase the movement of pollutants from the upper reaches of watersheds into coastal waters; lower dissolved oxygen levels; increase salinity in marshes; reduce freshwater availability; and accelerate the delivery rate of pollutants to downstream sites. Pollution prevention techniques can reduce or eliminate some environmental effects.*

The guidebook then offers various questions to consider in each of four areas of concern, a few of which are included in Table 4.3.

EXAMPLE 4.2 The Franconia Notch Parkway

Environmental issues can be easy to understand or very complex, and they can generate dramatic controversy. Franconia Notch, illustrated in Figures 4.3 and 4.4, is one of the most beautiful spots in the White Mountains of New Hampshire. Most of the area visible in the photos is part of Franconia Notch State Park, which includes a swimming beach, an aerial tramway to the summit of Cannon Mountain, ski trails, and hiking trails. The notch is also the historical route for traveling between southern New Hampshire and northern Vermont—and therefore the natural route for an interstate highway. The route was indeed part of the original plan for Interstate 93 (I-93), but controversies arose over how to fit the standard four-lane divided interstate highway into the little room that is available between the Lafayette Range on the east (the left side of the photos) and the Kinsman Range to the west. One approach developed by highway planners was to blast a tunnel through Mt. Skookumchuck (the rocky outcrop at the head of the notch that is clearly visible near the center of Figure 4.3), so that the four-lane divided highway could plow straight through the notch. This plan would have obliterated two small ponds, cut off much of Echo Lake (the lovely lake visible in the photo that is used extensively for swimming and fishing), and taken up nearly all of the flatland available in the upper portions of this narrow valley.

Local opposition prevented this option and all other attempts to build a highway to interstate specifications through the notch. The public's interest in the project was represented by the Appalachian Mountain Club, whose members maintained and hiked the many trails in these mountains, and by the Society for Preservation of New Hampshire Forests, which had purchased this land more than 80 years ago and contributed it to New Hampshire for use as a state park. These two organizations worked with NH Department of Transportation to find a less-intrusive solution that preserved more of the park

Figure 4.3 Franconia Notch, with Mount Lafayette in the Distance, Skookumchuck at Left Center, and Echo Lake at Lower Right. A road built to interstate standards would have destroyed much of the scenic sites within Franconia Notch State Park. However, a two-lane road provides sufficient capacity for this rural portion of the interstate, and it has little impact on the state park. This road, completed in 1983, is an early example of context-sensitive design.

Figure 4.4 Franconia Notch Parkway at the Edge of Echo Lake. This narrow but highly scenic two-lane road is now part of Interstate 93. This photo, taken on the same day as the photo in Figure 4.3, shows how little space the highway takes up as it skirts the shoreline of a lake and a valley that are famous for their recreational opportunities.

yet allowed better transportation. Instead of the typical interstate, the existing road was enhanced but not widened, and a better link was thereby made between the sections of I-93 that were completed north and south of Franconia Notch. Since the new road had only a 45-mph speed limit and narrowed to a single lane in each direction for several miles, it was clearly not up to interstate standards. Moreover, it had frequent exits for access to the notch's attractions, including the viewing spot for the much loved but recently collapsed Old Man of the Mountain—a striking rock formation that is now visible only in photos and on the obverse of the New Hampshire quarter (the only U.S. coin with two heads). For many years, the Federal Highway Administration refused to classify the road as an interstate, and it was officially known as the Franconia Notch Parkway. FHWA eventually relented and, early in the twenty-first century, the Franconia Notch Parkway officially became part of the interstate system, despite its obvious deficiencies. Today, the stretch of road visible in these photographs is the only two-lane portion of the entire Interstate Highway System. The compact between NH DOT and the two conservation groups remains in place. These three parties worked together in 2009 to develop plans for resurfacing the roadway, repairing drainage, replacing guardrails, improving the landscaping, and keeping the small footprint of the highway unchanged. The compromise decision to allow a slimmed-down interstate highway to be constructed through this scenic region was an early example of what is now referred to as **context-sensitive design**. Considerable environmental disruption can be avoided by tailoring infrastructure to the local geography and environmental conditions instead of insisting on using standard procedures.

4.5   SOCIAL IMPACTS

Almost any project will have social impacts, which may be related to the users of the project, people who live near the project, people who are displaced or competitively disadvantaged because of the project, or people who are hurt or whose lives are hindered due to the construction or operation of the project.

Social impacts could be positive as well as negative, but it is the negative impacts that must be considered most carefully. Positive social impacts will help make a project more attractive, whereas negative impacts may be sufficient to arouse intense public opposition that prevents or markedly restricts a project. Anticipating negative impacts is therefore something that should be done early in the evaluation process, so that there will be an opportunity to adjust plans and thus reduce the negative impacts or provide means for mitigating them.

By social impacts, we mean the consequences to human populations of any public or private actions that alter the ways in which people live, work, play, relate to one another, organize to meet their needs and generally cope as members of society. The term also includes cultural impacts involving changes to the norms, values, and beliefs that guide and rationalize their cognition of themselves and their society.

Interorganizational Committee on Principles
and Guidelines for Social Impact Assessment

Major projects may have far-reaching consequences that are difficult or impossible to quantify or comprehend. In some cases, projects that appear at first to be wholly desirable turn out to have unexpected consequences that some people view very unfavorably. In their famous study of Middletown, Robert and Helen Lynd reported that some residents recognized the social benefits of having automobiles for commuting but were outraged by the dreadful social impacts of auto ownership (see the feature box below).[4]

No one questions the use of the auto for transporting groceries, getting to one's place of work or to the golf course, or in place of the porch for "cooling off after supper" on a hot summer evening; however much the activities concerned with getting a living may be altered by the fact that a factory can draw from workmen within a radius of forty-five miles, or however much old labor union men resent the intrusion of this new alternate way of spending an evening, these things are hardly major issues. But when auto riding tends to replace the traditional call in the family parlor as a way of approach between the unmarried, "the home is endangered," and all-day Sunday motor trips are a "threat against the church," it is in the activities concerned with the home and religion that the automobile occasions the greatest emotional conflicts.

Robert S. and Helen Merrell Lynd,
Middletown: A Study in Contemporary American Culture
(New York: Harcourt, Brace, & Company, 1929)

---

[4] The authors spent more than a year in Muncie, Indiana, interviewing residents about all facets of life in that small city.

Table 4.5   Principles for Social Impact Assessment

1   *Achieve extensive understanding of local and regional populations and settings to be affected by the proposed action, program, or policy.*

2   *Focus on the key elements of the human environment related to the proposed action, program, or policy.*

3   *The SIA is based upon sound and replicable scientific research concepts and methods.*
- *Research methods should be holistic in scope.*
- *Research methods must describe secondary and cumulative social effects related to the action or policy.*
- *Ensure that methods and assumptions are transparent and replicable.*
- *Select forms and levels of data collection and analysis that are appropriate to the significance of the action or policy.*

4   *Provide quality information for use in decision making:*
- *Collect qualitative and quantitative social, economic, and cultural data sufficient to usefully describe and analyze all reasonable alternatives to the action.*
- *Ensure that the data collection methods and forms of analysis are scientifically robust.*
- *Ensure the integrity of collected data.*

5   *Ensure that any environmental justice issues are fully described and analyzed:*
- *Ensure that research methods, data, and analysis consider underrepresented and vulnerable stakeholders and populations.*
- *Clearly identify who will win and who will lose, and emphasize vulnerability of underrepresented and disadvantaged populations.*

6   *Undertake project, program, or policy monitoring and evaluation and propose mitigation measures if needed:*
- *Establish mechanisms for evaluation/monitoring of the proposed action that involve agency and stakeholders and/or communities.*
- *Where mitigation of impacts is required, provide analyses and assessments of alternatives.*
- *Identify data gaps and assess data needs.*

Source: ICPGSIA, "Principles & Guidelines for Social Impact Assessment in the USA," *Impact Assessment & Project Appraisal* 21, no. 3 (2003): 231–50.

In the United States, social impacts must be considered as part of the process required for environmental impact assessment. A set of principles and guidelines for social impact assessment was developed by the Interorganizational Committee on Principles and Guidelines for Social Impact Assessment (ICPGSIA), a group of social scientists who sought to help public agencies and private organizations in carrying out responsible social impact assessment (SIA). Their motivation was that "SIAs help the affected community or communities and the agencies plan for social change resulting from a proposed action or bring forward information leading to the reasons not to carry out the proposal."[5] Like the environmental impact assessment, a major purpose for the SIA is to provide a mechanism for understanding and responding to the potential negative impacts of proposed policies, programs, or projects.

This group defined social impacts (see the feature box on page 112) and identified six principles for social impact assessment (Table 4.5). The first principle calls for identifying the people who will be affected by the proposed action and collecting information about their social conditions so as to establish a baseline for evaluating changes to those conditions. The second principle is that the analysis should be focused on the most important social and cultural issues that are likely to be affected. The SIA need not address every possible social or cultural impact that might be imagined. The third principle emphasizes the need for using proper methods and input from the public in identifying and quantifying problems that might be encountered. In other words, social scientists know how to do this kind of analysis, and they should be involved early in the design and evaluation process for major projects. The fourth principle establishes the role of the SIA as providing information to be used by

---

[5] Interorganizational Committee on Principles and Guidelines for Social Impact Assessment (ICPGSIA), "Principles and Guidelines for Social Impact Assessment in the USA," *Impact Assessment & Project Appraisal* 21, no. 3 (September 2003).

decision makers and the public; the SIA and the people conducting the SIA are not the ones who ultimately make decisions about whether to go ahead with the project.

The fifth principle deals with environmental justice, which refers to the sometimes commonly used approach to locating or structuring projects: "Locate them in the poorest neighborhoods and don't worry about how the disadvantaged will be hurt by the project." Who benefits and who pays are important considerations in SIA and in project evaluation in general. The final principal indicates that SIA doesn't end when the project or program or policy is implemented. It is necessary to monitor what happens to ensure that mitigation measures are actually implemented and to ensure that unforeseen social impacts will be recognized.

Predicted social impacts may be temporary or long-lived, and minor impacts may affect a lot of people or intense impacts may affect only a few people. To understand the importance of social impacts, it will help to consider the kinds of social impacts that might be encountered in typical infrastructure projects. Table 4.6 lists some of the notable types of impacts.

It might be even more helpful to consider some projects where the social impacts, whether foreseen or unforeseen, turned out to be devastating or reprehensible. If the leaders of the French company that set out to build the Panama Canal knew that tens of thousands would die in their failed attempt, would they ever have begun the project? If city officials in New Orleans had long ago understood the risks posed by hurricanes, would they have allowed housing to be built in the lowest-lying areas of the city?

Table 4.6   Examples of Negative Social Impacts of Projects

Type of Impact	Examples
Relocation of people	• Entire villages displaced for the construction of a dam • Hundreds of people and small businesses relocated to allow the construction of a highway through a city
Deaths and injury during construction	• Deaths of more than 20,000 from tropical disease in the various efforts that eventually led to the Panama Canal • Deaths resulting from workers falling off bridges or buildings in situations where safety nets were not installed
Deaths, injury, or illnesses resulting during normal operation of infrastructure	• Millions of people severely injured or killed in highway accidents • Bridges and tall buildings serving as jump-off points for suicides • Asthma and other illnesses resulting from air pollution caused by emissions from power plants, automobiles, or home heating • Tens of thousands of people injured or killed annually worldwide in grade-crossing accidents between highway vehicles and trains
Deaths and injuries resulting from infrastructure failure	• Thousands of deaths and destruction of cities resulting from dam failures • Loss of life from buildings and structures that collapse in earthquakes
Disruption of neighborhoods	• Limited-access highways serving as barriers when they are constructed to divide urban neighborhoods • Loss of property values following construction of large, noisy, or ugly buildings or infrastructure • Creation of suburbs and decline of central cities following construction of better highways and policies that encouraged home ownership
Loss of livelihood caused by negative environmental aspects of a project	• Destruction of fishing and shell-fishing areas following construction of bridges, port facilities, or oil spills • Decline in use of informal taxis and buses following opening of new subway lines in large cities in Latin America and Asia
Loss of livelihood related to projects that help competitors	• Bankruptcy of canal companies following construction of railroads. • Bankruptcy of railroads following construction of highways and invention of cars, trucks, and airplanes • Decline in newspapers following widespread use of the Internet
Loss of privacy	• Disruption of the lives of native peoples following construction of roads or railroads through their previously remote homelands
Reduced quality of life	• Noise and dust resulting from construction of a highway • Shade resulting from construction of tall buildings

If automobile manufacturers, highway engineers, and government officials truly understood the dangers of automobiles (hundreds of thousands killed worldwide each year), would we have the system in use today? These questions are worth some discussion. Hindsight may suggest that we would have done things differently if we had only known—but maybe the drive for a route to the Pacific, the need for more housing, and our great love affair with the automobile would have led us exactly to where we ended up.

4.5.1 A Framework for Assessing Social Impacts

A framework for assessing social impacts can be structured as shown in Table 4.7. This matrix indicates that various categories of social impacts can be considered over the major stages in the life-cycle of a project. The categories of social impacts shown in this matrix are those suggested by ICPGSIA, but other categories could be used. The social impact assessment would consider possible important social impacts during each phase of the life-cycle. We have earlier seen (in Chapter 1) that the preliminary planning phase may be the most important in developing a project; this is when it is easiest to adjust the design to be more efficient and more effective in meeting the needs of a project. This is also the phase when the greatest flexibility is available for modifying the project so as to mitigate the most important social impacts. It is also the time to think about the entire life-cycle—what is to be done with the bridge when it is no longer suitable for traffic, and what is to be done with the power plant when it is no longer safe to operate?

Various researchers and agencies have used different categories of social impacts, and elaborate topologies of social impacts can be constructed. Whatever the categories, the main concern for the

Table 4.7 A Framework for Social Impact Assessment. Consider potential impacts at different stages in the life-cycle of the project or program.

Category of Social Impacts	Preliminary Evaluation of Needs and Alternatives	Detailed Planning	Construction	Operation and Maintenance	Decommissioning or Abandonment
Population change • Size, density, and rate of growth • Relocation • Influx of people for construction					
Community and institutional structures • Size and structure of local government • Diversity of economic base • Voluntary associations • Employment levels • Level of poverty					
Political and social resources • Distribution of power • Stakeholders • Leadership capabilities					
Community and family changes • Lifestyle • Attitudes, beliefs, and values					
Community resources • Infrastructure availability and needs • Land use patterns					

Source: Adapted from ICPGSIA (2003), Table 1.

SIA will be to determine who is going to be hurt by the project or program, when problems will arise, and what can be done about them. As with environmental impact assessment, with an SIA much is to be gained simply by requiring these questions to be asked and then publicizing the answers and supporting information. If the problem is understood, then it may be possible to take action.

Three broad categories of actions can be identified:

1. Adjust the design so as to avoid or reduce the social impacts.
2. Require mitigation as a condition for approval of the project.
3. Compensate those who are hurt by the project.

Whether any or all of these actions are necessary is ultimately decided by those who are threatened, and by local governments, developers, other stakeholders, and the courts.

EXAMPLE 4.3   Rationalizing the Freight Rail System in Chicago

Chicago is the busiest freight rail center in North America. All of the major railroads run to or through this hub, exchanging freight cars in a bewildering network of yards and lines that were laid out more than 100 years ago. Moving trains through Chicago may take a day or more, and this terminal clearly needs substantial infrastructure investment so that it can continue to handle the growth expected over the next 25 years. In 2008, Canadian National Railway (CN) sought to acquire the Elgin, Joliet and Eastern Railway (EJ&E). CN is the largest railroad in Canada, and it has several important lines in the United States that converge—but do not meet—in Chicago. CN must route most of its traffic across various terminal railroads, which is a costly and time-consuming process. The EJ&E is a terminal railroad that in effect provides a bypass around metropolitan Chicago, just as Interstate 495 provides a bypass around Boston. By acquiring the EJ&E, CN could route most of its traffic along the underutilized EJ&E—which of course would increase the crossing delays and rail-related noise and emissions in the small towns along that route. The draft environmental impact statement documented the extent to which the new operations would disrupt traffic in these communities. Despite strident protests from those communities (see "Projects in the News" on page 000), the Surface Transportation Board approved the acquisition, but only if CN provided most of the funds required to build bridges over a number of the highways that would be most affected by grade-crossing delays.

Box 4.2   Projects in the News

Editorial: (Almost) All Aboard

Although CN's proposed acquisition of the EJ&E Railway would have broad economic benefits for the region, some local communities were vehemently opposed to the deal. An editorial in the Chicago Tribune acknowledged the grade crossing issue for some communities, but justified their support for the project based upon consideration of the impacts on the entire region:

Even though it's a gain for the region, the deal hasn't been very popular. That's because 34 communities along the EJ&E route, a 198-mile arc that runs from Waukegan through Barrington, Aurora and Joliet and east into Indiana, will see more local train traffic as a result of it. Those towns have been accustomed to dealing with just a handful of trains a day. There are, though, many local winners. Eighty communities will see a significant reduction in the number of trains passing through. If all this were about creating local winners and losers at the crossing gates, we'd probably argue to give the whole thing a pass. But this is about a broad regional benefit. The deal will create a seamless rail route around Chicago, reducing freight traffic delays. It will promote more use of rail freight, reducing the number of trucks hauling cargo. It will spur rail infrastructure improvements. It will create new opportunities for commuter lines. The respected, nonpartisan Chicago Metropolis 2020 reported that this deal would create more than 600 jobs and boost the regional economy by $60 million a year.

Source: ChicagoTribune.com, January 6, 2009.

4.5.2   Safety and Security

Safety and security are particularly important and emotional social concerns. Public reactions to projects seldom derive from a calm, rational assessment of the costs and benefits. Sometimes the public response is driven by fears and emotions, whether the fears relate to the potential for disaster or for national security. Proponents are likely to downplay the potential problems, while opponents are likely to stir up people's emotions. The classic case is nuclear power. While very few serious accidents involving nuclear power plants have occurred in the United States, the public has been fearful of such accidents and leery of proposals for sequestering nuclear waste. If nuclear power plants are built to modern safety standards, and if radioactive waste is properly sequestered, then nuclear power would seem to provide an efficient, clean alternative to the use of fossil fuels for generating electricity. However, public fears have forced the taking of extraordinary measures to limit the risks of such plants, and some countries have banned the plants altogether.

In the United States, a rather inconsequential incident at the Three Mile Island Power Plant in 1979 was "the most serious in U.S. commercial nuclear power plant operating history, even though it led to no deaths or injuries to plant workers or members of the nearby community."[6] The accident led to very little off-site release of radiation. The average dose (1 millirem) of radiation to the population of 2 million closest to the site was only one-sixth of the dose received during a full set of chest X-rays. In addition, the Nuclear Regulatory Commission (NRC) reports that studies have determined that the environment sustained no more than negligible effects. However, the accident at Three Mile Island led to sweeping changes in the regulation of the nuclear industry, including the addition of many costly safety procedures. The negative media attention that it received also created a terrible public image for the nuclear power industry as a whole, and no new plants were built in the United States for 30 years. While public opposition to the nuclear power industry was strong enough to cripple the nuclear power industry in the United States and in some other countries, nuclear power became the dominant source of electrical power in France (see Box 4.3).

Box 4.3   Projects in the News

Nuclear Power—What's the Future?

Nuclear power has the potential for vastly reducing our reliance on fossil fuels, but cost, safety and security and security concerns have hindered its use. The United States has been slow to develop nuclear power, and some countries have begun to cut back existing facilities:

Germany disconnected the first of its 19 nuclear power stations yesterday, beginning an unprecedented phase-out that underscores the differences between some European nations and the United States on securing future energy supplies. . . . Germany is the first major industrialized nation to renounce the technology. Under a deal negotiated after years of wrangling between the government and power company bosses, all Germany's nuclear reactors are to close by 2020. . . .

Not all European countries have joined the trend, though. France relies heavily on nuclear energy—its 58 nuclear reactors provide more than three quarters of the country's electricity.

Source: Stephen Graham, "Germany Begins Nuclear Phaseout," *Boston Globe*, 15 November 2003.

---

[6] *Fact Sheet on the Three Mile Island Accident*, 1 March 2004. U.S. Nuclear Regulatory Commission, 10 November 2004. www.nrc.gov/reading-rm/doc-collections/fact-sheets/3mile-isle.html.

4.6   SUMMARY

Differences in Public and Private Perspectives

The public sector's perspective regarding projects differs in several important ways from that of the private sector. Entrepreneurs view projects as a means of making their fortune; investors view projects as a way to earn a return on their investment; public officials and charitable organizations view projects as a way to achieve goals related to society's needs or desires. Many of the differences can be captured by the distinctions between financial analysis and economic analysis. Financial analysis addresses cash flows, whereas economic analysis also considers safety and security, growth in the economy, and provision of education, infrastructure, and other public services.

Another difference between the public and private perspective is that environmental and social concerns are more likely to motivate public projects, whereas such concerns are apt to be constraints on private projects. This is not an absolute difference. In particular, many infrastructure projects may have very negative consequences on the environment or upon people who live near the projects or who must move to accommodate them. Construction of roads through cities, creation of dams that displace thousands of people, irrigation projects that require flooding of fertile valleys, and use of public land for forestry and mining are examples of public efforts that can lead to serious controversy among agencies and among those who favor or oppose the projects. When controversies arise about public projects, a change in government, legal battles, or the passage of new legislation may be needed to resolve the issues. When controversies arise with respect to the social or environmental impacts of private projects, there could be legal and legislative battles to determine what kinds of activities are allowed and what kinds are prohibited.

Benefit-Cost Analysis

A principle underlying public projects is that the benefits should exceed the costs. The project itself does not have to produce the cash necessary to cover the investment or even the operating costs, since the government has the power to raise taxes to pay for projects. However, the project should provide measurable economic benefits that are at least as great as all of the economic costs of the project. Just as the benefits can be more than the cash flow directly related to the projects, the costs can include social, economic, and environmental impacts that are directly or indirectly related to the project. The logic underlying benefit-cost analysis can easily be misinterpreted. The proper interpretation of this kind of analysis is that the government should not pursue projects if the expected benefits are less than the expected costs. This does not mean that projects where benefits exceed costs should be approved; it only means that they deserve further consideration. Many controversies about the need for projects boil down to controversies as to what counts as a benefit, what counts as a cost, and how values should be put on factors such as safety, air quality, or job creation.

Economic Impacts

The economic impacts of a project will include

- Construction jobs and income
- The multiplier effect of construction
- Jobs and income related to the eventual operation of the new project
- The multiplier effect of operation of the new project
- Continuing productivity benefits resulting to citizens, users, industries, or public agencies as a result of the project
- Growth in the economy related to the productivity benefits provided by the project

Environmental Impacts

The environment impacts of a project will include changes related to

- Ecosystems
- Pollution
- Wetlands, aquifers, and drainage
- Wildlife habitat
- Renewable versus nonrenewable resources
- Climate change

The goal of a project may be to improve the environment, as in the creation of parks or wildlife refuges. More often, the location and design of a project and the choice of construction methods and materials will be decided in part because of concerns about environmental impacts.

In many countries, an environmental impact assessment must be submitted to and accepted by a government agency before work can begin on the project. A draft impact assessment is made available to the public, and a period is allowed for public comment on the project. The agency may conclude that the project has no significant impact on the environment, so that it may proceed as planned; or the agency may require mitigation of certain environmental impacts. Mitigation may include a change in size, design, location, methods, or materials; it may also require additional investments, such as the construction of new wetlands to compensate for the destruction of wetlands by the proposed project. In extreme cases, the public agency may be authorized to halt the project. The environmental review process ensures that the public is made aware of the environmental impacts of proposed projects and has a chance to comment on those projects. Having an open process ensures that the public has a chance to review the options considered for a project, the types of environmental impacts expected, and the opportunities for mitigation. However, the environmental review process may take years for a major project, and these delays can prove costly for the project. By thinking about environmental issues from the outset, developers and entrepreneurs can anticipate and deal with environmental issues more effectively than if they wait until the end of the planning process to tack on a short statement of environmental impacts.

Social Impacts

Social impacts of a project may be either positive or negative. If positive, they make the project more acceptable to the public; if negative, they can lead to public resistance to a project and perhaps cancellation of a project. Anticipating and mitigating negative social impacts is thus an important part of project evaluation.

Negative social impacts of a project could include the following:

- Relocation of people
- Deaths and injury during construction or operation of a project or resulting from infrastructure failure
- Disruption of neighborhoods and local cultures
- Disruption of the local economy as a result of environmental degradation or introduction of new competitors
- Reduced quality of life, including loss of privacy, degraded aesthetics, and declining safety or security

Social impacts like these must generally be addressed in the environmental impact assessment required for a project. Social scientists have developed methods for measuring these impacts, and they should be involved in the evaluation process for major projects that are likely to have (or are perceived by the public to have) serious social consequences. Table 4.7 provides a framework that can be used to structure social impact assessment; this framework considers demographic, political, cultural, and family impacts during the different stages of a project's life-cycle.

Three broad strategies are available for mitigating negative social impacts:

1. Adjust the design so as to avoid or reduce the social impacts.
2. Require mitigation as a condition for approval of the project.
3. Compensate those who are hurt by the project.

Whether any or all of these strategies are necessary is ultimately decided by the governments, developers, other stakeholders, and the courts. Over time, as the magnitude of social impacts becomes clearer, it may be desirable or necessary to require improvements to existing infrastructure and more stringent requirements for locating, designing, and constructing new projects.

FINAL COMMENT: SUSTAINABLE INFRASTRUCTURE

This chapter has addressed the key factors that must be considered in enhancing the sustainability of infrastructure systems: financial feasibility, economic impact, environmental impact, and social impact. Finances are important, because cash is needed to construct, maintain, and operate infrastructure. Economic impacts are important because they include many types of short- and long-term impacts that can help justify a good project or prevent a bad project. Environmental impacts are important because an infrastructure system that is too destructive to the environment or that requires excessive use of limited natural resources will not last long. Social impacts are important because ultimately, society decides whether to proceed with infrastructure projects and bears the brunt of failures to consider hidden costs of projects.

THE NEXT STEPS

What are the next steps? It is clear that many diverse factors influence what needs are addressed, what projects are considered, and how projects are evaluated. There are no easy methods for determining the best projects, and no simple ways to gain public support for a particular project. The next chapter introduces various methods and concepts that can be used in reaching a consensus about what is needed and what should be done.

ESSAY AND DISCUSSION QUESTIONS

4.1 What is the difference between financial and economic analysis? Why is the private sector more concerned with financial analysis while the public sector is more concerned with economic analysis?

4.2 If urban sprawl is so bad, why do we have so much of it? When you fly across the country, why is so much of it empty? What is the marginal value of farmland, why are farmers going broke, and what happens to all the chemicals in that fertilizer? Is suburbia worse than a forest or another cornfield? Comment on the nature of the development of an urban area and the surrounding suburbs and the types of projects that might produce better places for us to live—with each other and with nature.

4.3 The twenty-first century is viewed by some as the century when we begin to fight wars over access to water. Hundreds of millions of people worldwide lack easy access to clean water. L. A. and many other large cities seek farther and farther for water, and the consequences for the environment are severe. The Three Gorges Dam in China is billed by many as an environmental disaster that will displace a million people. Boston completed a multibillion-dollar cleanup of Boston Harbor that quintupled water and sewer bills in the 1990s, causing great public consternation. Even with this increase, however, a typical family's water bill in Boston averaged only about 25 cents per day per person. Comment on the nature of the crisis and the types of projects that might be most relevant.

4.4 In the United States, most power plants use fossil fuels to generate electricity, despite the availability of nuclear power plants that are much cleaner and require no fossil fuel. Most intercity transport involves automobiles and airplanes, despite the availability of more energy-efficient rail and bus services. Nuclear power plants, which have experienced no fatal accidents in the United States, are feared as too risky whereas automobiles, which in fact cause 30,000 or more fatalities per year, are viewed as essential. Discuss some of the ways that public policy, technological development, and public perceptions of risk have influenced or could influence the extent to which the United States depends upon imported fossil fuels for generating electricity and for transportation. What kinds of projects would you advocate for reducing our dependence upon imported fossil fuels?

4.5 In preindustrial times, many people lived in small villages whose houses were located around a common meadow used for grazing by cattle or sheep. Each family was allowed to let its cattle or sheep use the common, and each family helped in the upkeep of fences. As long as the number of animals was within the limits of what the common could support, all was well. However, at some point—perhaps in a year that was too cold or too dry or too wet—the animals were more than the common could support. They either ate everything, roots and all, or trampled everything underfoot—and everyone suffered. The so-called **tragedy of the commons** applies to other situations, including fishing grounds where there is a possibility of overfishing, beautiful rural lakes whose tranquility could be ruined by failure to limit development along the shoreline, and the more general problems of air and water quality. Select one of these situations, or another that could suffer from the tragedy of the commons, and discuss how the commons could be better managed for sustainability.

4.6 Technological change can lead to social change—desirable perhaps, but perhaps not. Identify some of the positive and negative social changes brought about by recent rapid advances in information technology and communications. How have these changes affected your interactions with friends and family? How have these changes influenced the way you view the world as opposed to the ways that previous generations viewed the world?

PROBLEMS

4.1 Identify the major economic, environmental, and social impacts likely to be associated with one the following projects:

a. Construction of a superhighway through an urban area

b. Construction of a major dam that is intended to control flooding, generate electricity, and provide irrigation to the surrounding region

c. Construction of a nuclear power plant near the coast in a region that depends heavily on tourism and ocean-based recreation (Once the nuclear plant is operating, a nearby coal-fired power plant will be closed.)

d. Construction of a wind farm along a ridge in the rural part of a region that is noted for two things: its beauty and its poverty

4.2 Choose one type of project considered in the previous question. Search online to find an actual, recently completed project that would be expected to have similar environmental and social impacts. Continue searching online to find the environmental impact statement for this project, along with commentary in newspapers or blogs or elsewhere about the project. What appear to be the major social and environmental issues that were raised? What mitigation was required? What was the public response to the project after it was completed and began operations?

4.3 Look online to find the budget for a city, town, or state that you are familiar with. How much of the budget is allocated for transportation, recreation, housing, and recreation? How much of the budget is allocated to paying interest on loans or on bonds? What major projects are mentioned in the budget? What funding is available for continuing programs (such as school construction or highway maintenance)?

CASE STUDY: GREEN INFRASTRUCTURE

The text described why it is important to maintain a "green infrastructure" that is adequate to support the plants and animals native to a region. The green infrastructure includes four elements: (1) a few wild areas that are large enough to provide room and habitat for the animals requiring the largest territories, (2) a few small- to medium-sized areas that protect rare or endangered habitats, (3) many small- to medium-sized areas that are distributed throughout the region and include all of the varied habitats to be found in the region, and (4) corridors that connect the various open spaces, so that animals do not become isolated.

To understand green infrastructure, it is best to spend some time looking at maps and aerial photographs, which can be found online. As a region develops, it will be important to recognize what could be important parts of the green infrastructure and take steps to protect open spaces and connecting corridors. The various state and local governments in a region can work together in developing a long-term plan to ensure there will always be adequate green infrastructure. The first step, which is what you will do in this case study, is to become familiar with what exists and to identify the major elements of the green infrastructure and the corridors that might be protected.

Questions

a. In a state or region you are familiar with, identify the existing parks, wildlife refuges, and other conservation areas that are the major open spaces anchoring the region's green infrastructure. This information should be available on a good map, and you can probably find something online.

b. You should be able to get a bird's-eye view of the region using online aerial photographs. Describe the number and distribution of small- to medium-sized open areas that form another key part of the green infrastructure. Are there plenty of such spaces, or are there too few? Is the open space nearly all agricultural, or are there numerous diverse, wild ecosystems as well?

c. How well are these open spaces linked by corridors wide enough to allow free movement of wildlife? (The corridors would have to be at least 50–100 yards wide, and they could include agricultural land, wetlands, and undeveloped land.) How much of this land is publicly owned? (This may be evident by looking at a good map of the region, and you may be able to obtain online information from local environmental or natural resources agencies.) What would be necessary to protect wildlife corridors between the existing large, open areas identified in part a?

Chapter 5

Comparing Strategies for Improving System Performance

The basic idea is to define alternatives to just sufficient a level of detail to allow different stakeholders to at least rank order them in terms of desirability according to each identified criterion.

Theodor J. Stewart, "Thirsting for Consensus: Multicriteria Decision Analysis Helps Clarify Water Resources Planning in South Africa," *OR/MS Today* (April 2003): 30–34

CHAPTER CONCEPTS

Section 5.1 Introduction

Section 5.2 Presenting All Results in Monetary Terms
Advantages of using a monetary metric for multiple criteria
Difficulty of expressing all measures in monetary terms

Section 5.3 Net Present Value: Comparing Current and Future Costs and Benefits
Discounting of future costs and benefits
Net present value of costs and benefits

Section 5.4 Measuring Cost Effectiveness
Definition of cost effectiveness
Usefulness in comparing any quantifiable options

Section 5.5 Using Weighting Schemes in Multi-Criteria Decision Making
Usefulness of a weighting scheme
Political nature of any weighting scheme
Using public input to clarify objectives, identify alternatives, and establish priorities

Section 5.6 Seeking Public Input
Public input helps identify and rank problems in each region
Public input helps rank alternatives for managing development
Identifying possible scenarios as a basis for public input
Using workshops to achieve consensus regarding development strategy

5.1 INTRODUCTION

This chapter presents methodologies that can be helpful in developing and comparing strategies for improving system performance when there are multiple objectives and various different measures of performance. Various methodologies are presented:

- *Presenting all results in monetary terms:* Use the best estimates available to determine the value of the changes in each aspect of performance.
- *Discounting and net present value:* Compare future costs and benefits with current costs and benefits.
- *Cost effectiveness:* Determine the cost per unit of improvement in any quantifiable but non-monetary objective.
- *Weighting schemes:* Provide weights in order to compare different kinds of measures.
- *Public input:* Use surveys or workshops involving interested parties or experts to determine the relative importance of different kinds of measures.

To demonstrate the weaknesses inherent in using weighting schemes to rank alternatives, a hypothetical example of evaluating ways to improve the performance of a bus system is provided. To illustrate how public input can be used to reach consensus in defining a problem and structuring a solution, this chapter includes two case studies involving development priorities in Africa.

5.2 PRESENTING ALL RESULTS IN MONETARY TERMS

As discussed in the previous chapter, it is may be possible to express different aspects of performance in monetary terms. Sometimes the translation into monetary terms may be straightforward, as in measuring changes in congestion by using the average values of time for the automobiles and trucks that are stuck in traffic. Other factors, such as safety, may require potentially controversial assumptions (e.g., the value to society of avoiding or incurring injuries or fatalities). Still others, such as aesthetics, can only with great difficulty be given a monetary value.

Economists and other researchers have been very creative in developing methods of estimating the value of changes in measures that at first glance appear to be purely qualitative. For example, it is possible to link air quality to health and also to more mundane matters such as the need for window washing and periodic cleaning of buildings.

The benefit of stating all costs and benefits in monetary terms is obvious: the methods used to compare financial costs and benefits can then be used to compare all of the economic costs and benefits. The danger of trying to state all costs and benefits in monetary terms is that the evaluation may simply disregard important measures that cannot (or have not yet) been converted into monetary measures. The fact that it is difficult to put a price tag on beauty or equity is no reason to ignore aesthetics or forget about being fair.

5.3 NET PRESENT VALUE: COMPARING CURRENT AND FUTURE COSTS AND BENEFITS

For any infrastructure project, costs and benefits are likely to be spread across of period of many years or decades. Usually a large investment is made over a period of a year or several years that creates or improves a system; the basic question is whether the expected future benefits will be sufficient to justify the initial cost. To answer this question, it is necessary to compare current and future costs and benefits. For any cost or benefit that can be monetized, a process known as **discounting** can be used. Discounting provides a means of reducing future costs or benefits so that they can be expressed as equivalent **present values** that will be directly comparable to current costs and benefits. Discounting is used in both the public and the private sectors; the public sector considers broad categories of costs and benefits and the private sector concentrates on cash flows. Discounting and related methodologies are examined in greater detail in Part II of this text.

Discounting is necessary because money in the future is worth less than it is today; in other words, money today can be invested and grow into a larger sum in the future. The basic relationship between present and future value is a function of the discount rate (r) and the time period (t):

$$\text{Present Value} = \text{Future Value}/(1 + r)^t \qquad (Eq.\ 5.1)$$

By use of this equation, future costs or benefits can be discounted to be equivalent to a current cost or benefit. For example, if an expected benefit of $1 million in year 4 were discounted at a rate of 5% per year, it would be worth only $822 today:

$$\text{Present value} = \$1\ \text{million}/(1.05)^4 = \$822\ \text{thousand}$$

The higher the discount rate, the lower the present value of a future sum of money. For example, the same $1 million in year 4 would be worth less than $700 thousand today if it were discounted at 10%:

$$\text{Present Value} = \$1\ \text{million}/(1.1)^4 = \$683\ \text{thousand}$$

The **net present value (NPV)** of a project is the sum of the discounted values of all the benefits and costs associated with the project. If the NPV is positive, then the benefits outweigh the costs; if the NPV is negative, then the costs outweigh the benefits. Using a spreadsheet, it is possible to discount any arbitrary stream of future cash flows to get the NPV, as discussed in detail in Part II of this textbook.

Maximizing NPV may be viewed as the primary objective of a project, particularly in the private sector. In the public sector, this objective corresponds to maximizing the overall net benefits to society, taking into account any and all costs and benefits that can be expressed in terms of dollars and cents.

If only financial costs are considered, as would commonly be the case in the private sector, then this objective would be stated as "maximize the net present value of cash flows." When cash flows are being considered, NPV can be related to financial measures such as return on investment (ROI). As discussed in Chapter 2, the return on investment in year t is the annual profit (P_t) divided by the total investment (I):

$$ROI_t = P_t/I \qquad (Eq.\ 5.2)$$

The ROI varies from year to year, in part because profits vary and in part because additional investments may be made. For now, it is useful to consider the special case where there is an investment I at time zero and a constant annual profit P over an infinite time period. The annual return on this investment will be P/I every year, and the average ROI for the project will be equal to P/I.

If we manipulate this formula, we get the following relationship:

$$I = P/ROI \qquad (Eq.\ 5.3)$$

Now think about a related question. If we have an annual profit of P over an infinite time horizon, and if we use a discount rate of r, then how much of an investment can we justify? This question is readily answered if our discount rate is interpreted as our **minimum acceptable rate of return (MARR)**. If so, then the question can be answered using Eq. 5.3: we can justify investing an amount M such that:

$$M = P/r \qquad (Eq.\ 5.4)$$

M is therefore the present value of this infinite stream of profits. The NPV of this project is thus the difference between what we would be willing to invest (M) and the original investment (I):

$$NPV = M - I = P/r - P/ROI \qquad (Eq.\ 5.5)$$

This value is positive if the ROI on the project is greater than r, and it is negative if the ROI is less than R. This is a very useful result since it allows us to make some simple calculations for finding NPVs.

This section has just introduced the basic concepts of discounting and NPV, but even these basic concepts are powerful. If we are given costs and benefits that occur over any stretch of time, and if we have a discount rate, we now know how to calculate the NPV of these benefits. We also have a simple formula (Eq. 5.4) that can be used to calculate the NPV of an infinite stream of constant benefits.

EXAMPLE 5.1   NPV of an Infinite Stream of Constant Benefits

An investment of $200 million in a project to create a busway is expected to have annual net benefits of $15 million in terms of reduced congestion and faster travel times for commuters and better utilization of buses for the transit agency. What is the net present value of the benefits? Of the project? Is the project worth pursuing? Assume that the government uses a discount rate of 7%.

Using Eq. 5.4, the NPV of the benefits will be $15/0.07 = $214.3 million, which is greater than the investment of $200 million. Since the NPV is positive, the project is worth pursuing.

The choice of a discount rate is an important factor in project evaluation. A very high discount rate emphasizes initial costs and makes it more difficult to justify projects that require large investments; a very low discount rate favors projects with net benefits that continue over a very long time horizon. Private companies typically use a discount rate of 10% or greater in evaluating their projects; higher rates are used if projects are perceived to be riskier.

Public agencies, which do not have to pay taxes, typically use a lower discount rate. The rationale for choosing a discount rate is discussed at length in Chapter 8; for now, let's consider some of the options that may be suggested for use by a public agency for a particular project:

- Use a zero interest rate since the costs or benefits to future generations are just as important as they are to current generations and so should not be discounted.
- Use the interest rate on bonds that were specifically authorized for this project.
- Use the interest rate on general-purpose bonds that are issued by the agency.
- Use a higher discount rate to reflect the opportunity cost of raising money by taxing people who have other investment alternatives.

Use of a zero discount rate may sound good when discussing how important it is to consider global warming or the eventual collapse of an ecosystem due to continued development. A zero discount rate makes it impossible to ignore future costs like these, so some economists and engineers have favored its use. However, a zero discount rate means that benefits or costs to future generations are just as important as benefits or costs today. If that is true, then it is possible to justify very costly or very repressive projects that are devastating to people right now because "the end justifies the means." A zero discount rate ignores the reality of the time value of money, and it is not a good way to ensure that future catastrophic possibilities are considered.

If special bonds have been authorized to promote a specific type of project, then it may be acceptable to use their interest rate as the discount rate. For example, a state may guarantee the interest payments on bonds that are sold to raise money for low-income housing. The motivation for providing housing is to improve the health and general quality of life of people who would otherwise be homeless or living in crowded or substandard housing. The projects may depend upon rents to pay off the bonds, but the whole purpose of the projects is to lower the costs of housing for people with low incomes. The minimum acceptable rate of return therefore could be the cost of capital for the housing program, which would be the interest rate on the special bonds issued to pay for the housing.

Some agencies are created with fairly wide authority to issue bonds for construction, maintenance, and operation of public infrastructure. The interest rates on their bonds will be based upon the authority's credit rating, not the state's. Since such authorities cannot raise money by direct taxation of the general public, they have to convince investors that the operations of their authority will provide the income necessary to pay the interest on the bonds. Thus these bonds would be viewed differently than bonds issued by the state or bonds for which the state guaranteed interest payments, as in the housing authority example.

For example, an airport authority may be authorized to sell bonds that are backed by the fees received from airport users. The authority may be encouraged to expand the airport in order to promote the economic health of the region, and it may be authorized to invest in airport facilities so long as it can cover all of its investment and operating costs without further funds from state or local governments. In such a situation, the minimum acceptable rate of return for the airport authority must exceed its costs of capital, which is the interest rate on its bonds. If the airport enjoys continued growth, and if air travelers and airlines are able to afford fees that can be increased over time, then the airport authority could raise a great deal of money to invest in its terminal facilities. Note that expansion projects funded by airport authorities and other special-purpose authorities (e.g., turnpike authorities or port authorities) do not have to compete with other public projects for funds. So long as they are able to cover their interest payments, such authorities are able to expand or even gold-plate their facilities.

If a proposed project seeks to use funds raised by general taxes on the public, then that project must compete with education, health, water, and all other public investment possibilities. Moreover, it must be remembered that an opportunity cost exists for the money raised from taxpayers, who certainly could find many other uses for the money that they pay in taxes. The U.S. Government Accountability Office (GAO) periodically defines a discount rate (typically 7–8%) to be used by federal agencies in evaluating projects.

5.4   MEASURING COST EFFECTIVENESS

It is always possible to calculate the cost effectiveness of any investment with respect to any nonfinancial, but quantifiable objective. The cost effectiveness is related to the ratio of the investment cost to the improvement in the measurement of interest (i.e., the $ per unit improvement). If different options are compared, the most costeffective option has the lowest cost per unit of improvement.

Consider the case shown in Table 5.1, which presents three alternatives for reducing the impact of highway noise on the residents of two new apartment buildings that will be built adjacent to a heavily traveled interstate highway. The first option is to erect a quarter-mile-long sound barrier along the edge of the property that abuts the highway. The second option is to require special soundproofing for all of the windows and walls that face the highway. The third option is to design floor plans to minimize the windows on the walls facing the highway (e.g., by placing bathrooms, closets, and stairways along that wall). The effectiveness of each measure is estimated in terms of the average reduction in noise levels that would be experienced by residents living in the structure. In this hypothetical example, the three strategies result in a similar reduction in noise levels, and redesigning the floor space is the most cost-effective way to gain peace and quiet.

Table 5.1   Cost Effectiveness of Three Options for Reducing Noise Levels

Option	Cost	Reduction in Noise Levels	Cost Effectiveness ($/% improvement)
Noise barrier	$500,000	20%	$ 25 thousand
Soundproofing	$200,000	22%	$ 9.1 thousand
Redesign floor space	$100,000	21%	$ 4.8 thousand

5.5   USING WEIGHTING SCHEMES IN MULTI-CRITERIA DECISION MAKING

Any major project, but especially public projects, tries to satisfy multiple objectives, many of them nonfinancial. There are likely to be many competing designs, some with markedly different approaches. Each of these projects can be evaluated in terms of each of the objectives, producing a (very large) matrix showing the predicted impacts of each option on each of the objectives. It is unlikely for any option to be rated best on all measures; instead, one design will be the cheapest, another will have the highest capacity, another will have the least impact on the environment, and another will be the easiest to construct. Which alternative is the best depends upon how much weight is placed on each objective.

Choosing the best project thus ultimately requires making judgments about the relative importance of the various objectives and the validity and uncertainty of the evaluation process. Unless a single individual has authority to make all design decisions, selecting the best project is inherently a political process: people with different perspectives and agendas must work out a process to determine the best way to proceed.

Many methods can be used to help structure the political process. Weighting schemes can be developed and applied to each of the criteria. However, as shown in the hypothetical example in the next section, weighting schemes may seem more objective than they really are. This is not an argument against using weights, and it is certainly not an argument against having multiple criteria. However, it is a caution to avoid thinking that it will be easy to agree upon criteria or weights or to think that the public will agree with whatever criteria or weights are proposed by any of the parties.

Choosing weights is simply another way of making judgments. In situations where there is general agreement concerning (1) the options to be considered, (2) the relevant criteria, and (3) the relative importance of the criteria, a structured weighting scheme can be helpful in ranking the options. However, weighting schemes are not of much use if there is strong disagreement about which alternatives should be considered, which criteria are most important, and how impacts should be measured. Presenting a weighting scheme as a way to obtain an "objective ranking of the options" is impossible in such cases because the different groups will simply push for weights that favor their own preferred options. Difficulties in reaching consensus are exacerbated if many of the criteria are highly qualitative.

The role of the analyst is clear: provide the best information possible within the available time and budget; identify what is certain, what is likely, and what is possible. Let people know how much faith you put in each part of the analysis. Explain the assumptions, and indicate whether you or your expert colleagues believe such assumptions to be reasonable. Where you had to make guesses, say where those guesses are most uncertain. You also want to make sure that the range of options is wide enough to cover the major strategies that might be considered. Finally, in public meetings or in private meetings with stakeholders, you can try to ensure that the discussion deals with the actual data and credible options, and you can try to provide insight into the cost effectiveness of various proposals with respect to various criteria.

EXAMPLE 5.2   Weighting Schemes for Evaluating Options for Increasing Bus Capacity

Table 5.2 shows how four hypothetical options for improving the performance of a bus system might affect various aspects of performance. The system currently has 300 buses, some of which are old and inefficient. Strategies for improving performance of the system include buying new buses, creating a busway within the downtown area to allow buses to avoid congestion, and developing a control system that would improve performance by enabling dispatchers to monitor the location and number of people on every bus. Buying new buses would have two effects: increasing the number of scheduled operations, thereby improving service, and allowing the system to handle more passengers. The new buses could replace the oldest 25 buses, and the average fuel efficiency of the fleet would increase while the average emissions would decline. If new buses are purchased, ordering cleaner—but more expensive—hybrid buses would reduce emissions. If more buses are acquired, the agency plans to create a new servicing facility for the buses on land that the agency owns; several businesses that currently rent space from the transit agency would have to be moved from this site. If the city builds a busway and creates some bus-only

Table 5.2   Predicted Cost and Performance for Expanding Capacity of a Bus System

	Cost	Improvement in Travel Times	Increase in Ridership	Reduction in Emissions per Bus-Mile	Families and Businesses to Relocate
Buy 100 new buses	$ 50 million	5%	20%	10%	10
Buy 75 new hybrid buses	$ 60 million	4%	15%	40%	10
Create a busway	$200 million	20%	30%	8%	30
Install a control system	$ 20 million	10%	5%	5%	0

lanes, it can provide much faster service and run additional trips without increasing the size of the fleet. Finally, if the city installs a state-of-the-art control system, it will gain some improvement in travel times, ridership, and emissions for a much lower cost than any of the other options.

None of the proposals dominates all of the others, and three of the proposals look best based on at least one of the criteria. Buying hybrids provides the greatest reduction in emissions, creating a busway would lead to the greatest increase in ridership, and installing a control system would be the least costly. Table 5.3 ranks the proposals based on each criterion. These are called **ordinal rankings**: first, second, third, or fourth. The "Total" column simply adds the five numbers, so that it is a measure that weights each of the criteria equally. If the best option is the one with the lowest total, then the best option would be to buy 100 new buses.

Table 5.3   Summing Ordinal Rankings for Each Criterion to Obtain an Overall Ranking

	Cost	Improvement in Travel Times	Increase in Ridership	Reduction in Emissions per Bus-Mile	Families and Businesses to Relocate	Total
Buy 100 new buses	2	3	2	2	2	11
Buy 75 new hybrid buses	3	4	3	1	2	13
Create a busway	4	1	1	3	3	12
Install a control system	1	2	4	4	1	13

It is unlikely that any group of planners or financial managers or government officials or public interest groups would simply accept this result. Those who really want the busway, such as the transit agency's Strategic Planning Group, might argue that capacity and ridership are the main goals. They might propose weighting travel time and ridership three times as heavily as the other criteria. As shown in Table 5.4, creating a busway now looks best.

Table 5.4   Weighting Scheme Proposed by the Transit Agency Strategic Planning Group

	Cost	Improvement in Travel Times	Increase in Ridership	Reduction in Emissions per Bus-Mile	Families and Businesses to Relocate	Total
Weight:	1	3	3	1	1	
Buy 100 new buses	2	3×3	2×3	2	2	21
Buy 75 new hybrid buses	3	4×3	3×3	1	2	27
Create a busway	4	1×3	1×3	3	3	16
Install a control system	1	2×3	4×3	4	1	24

The operators and the bus passengers association, who really want some immediate relief from overcrowded, unreliable buses, support the concept of bus lanes and busways; but what they want most is new buses. They point out that the busway will take three years to complete, and they also wonder why it should be possible to get hundreds of millions for capital improvements when budgets have been so tight that it has been necessary to freeze salaries for managers and raise fares. The general manager of Bus Operations argues for high weighting only for cost and ridership, which he views as the key factors, and he claims that relocating small businesses that rented space is not an issue, since those businesses knew very well that their building would eventually be needed for the bus servicing facility. He therefore argues for a revised set of weights as shown in Table 5.5.

Table 5.5   Weighting Scheme Proposed by the General Manager of Bus Operations

	Cost	Improvement in Travel Times	Increase in Ridership	Reduction in Emissions per Bus-Mile	Families and Businesses to Relocate	Total
Weight:	2	1	2	1	0	
Buy 100 new buses	2×2	3	2×2	2	2×0	13
Buy 75 new hybrid buses	3×2	4	3×2	1	2×0	17
Create a busway	4×2	1	1×2	3	3×0	14
Install a control system	1×2	2	4×2	4	1×0	16

Table 5.6   Weighting Scheme Proposed by Environmental Groups

	Cost	Improvement in Travel Times	Increase in Ridership	Reduction in Emissions per Bus-Mile	Total
Weight:	2	1	2	1	
Buy 100 new buses	50/20 = 2.5	20%/5% = 4	30%/20% = 1.5	40%/10% = 4	12
Buy 75 new hybrid buses	60/20 = 3	20%/4% = 5	30%/15% = 2	40%/40% = 1	11
Create a busway	200/20 = 10	20%/20% = 1	30%/30% = 1	40%/8% = 5	17
Install a control system	20/20 = 1	20%/10% = 2	30%/5% = 6	40%/5% = 8	17

The local environmental groups, who are a major political force in the city, push very hard for investing in clean buses that would dramatically improve the city's air quality. Furthermore, they argue, the city should set a high standard when it comes to cleaning up the environment. They say the city should buy as many hybrid buses as it can afford, to establish a long-term commitment to improving the environment. They also dislike the ranking scheme as structured in all of the above charts, since some of the differences among options are small while others are very high. They recommend normalizing each of the measures by dividing by the measure for the best option for each criterion where a lower number is better and by using the inverse when a higher number is better. Hence, the hybrid bus cost of $60 million would be divided by $20 million, the cost of the lowest cost option, to get a value of 3 for the cost criterion. The hybrid bus value of 40% reduction in emissions is in fact the best, so dividing 40% by 40% would give a value of 1 for the hybrid buses reduction in emissions. They also agree with the notion that no real relocations would be caused by creating the bus servicing facility, so they simply dropped that criterion. Finally, they interpreted the use of weights as merely an exercise in promoting special interests; once the relevant criteria have been identified, they weight everything equally. The ranking proposed by the local environmental groups is shown in Table 5.6.

Note that the participants in this example were not debating the information that they were given. They did not dispute the costs of the proposed systems, the ability of each system to improve service or capacity, or the effect of improved service and higher capacity on ridership or emissions. Anyone who has been involved in evaluating such competing projects knows that extended debates over any or all of these matters could well take place. Still, even though all of the participants accepted the predictions of cost and impacts, it was possible to devise a scheme to make any one of the choices look the best.

LESSON

It is at best difficult—and more likely impossible—to define a "correct" weighting scheme when there are competing options, multiple objectives, and differences in priorities among those who participate in making the decision. The best that can be done is to use some sort of participatory process to reach a consensus on the weights that are used and the rankings that result. Extreme weights and contorted measurement schemes will be apparent to most people and to diverse groups, so if the measurement schemes and weighting options are presented fairly and subjected to general discussion, there is some hope for reaching consensus.

5.6   SEEKING PUBLIC INPUT

Input from the public and key stakeholders can be helpful in identifying how to look at a complex problem. Input is most helpful—and the exercise is likely to be most productive—when there is general awareness that something needs to be done about a problem, but no one knows (or thinks they know) what is best to do. The main reasons for seeking input from the public and from stakeholders are to clarify the nature of the problem, to identify potential alternatives or strategies for dealing with the problem, and to discuss the relevant criteria for selecting and evaluating specific alternatives. Preliminary discussions can also be very useful in identifying where there is consensus about needs and opportunities, where additional information is needed, and where potential controversy is most likely. This feedback is helpful in allocating research and planning resources and in determining how best to structure the process.

The next two sections present two case studies illustrating ways that public input can be obtained and used to help prioritize needs and evaluate development strategies. The first study (Section 5.6.1) sought to improve life in the coastal zone of Ghana and required effective balancing of economic, social, and environmental issues. The study used input from citizens to help prioritize the problems in four regions and then identify ways to address these problems. Using quite simple measures of effectiveness, the study was able to identify the most cost-effective ways to improve conditions in the coastal zone of this country.

The second case study (Section 5.6.2) began when a forest products company sought approval to expand its forestry operations. The plan would increase jobs and economic activity, but it also would threaten wildlife and virgin grasslands. Consultants were hired to conduct a series of workshops that used possible development scenarios as a way to solicit responses concerning what was most important to the region. The consultants identified a process that could be used more generally to find consensus about how best to address potentially controversial development proposals.

5.6.1 Case Study: Identifying the Most Effective Strategies for Balancing Economic, Social, and Environmental Issues—"Integrated Coastal Zone Management Strategy for Ghana"

A study by the World Bank sought "to identify economically, socially and environmentally appropriate interventions and projects in the coastal zone that improve the prospects for human development."[1] Ghana's coastal zone accounts for just 6.5% of the country's land area, but 25% of its population. The people in the region suffer from poor health and poverty while continuing development leads to environmental degradation; this completes a vicious cycle by causing greater poverty and more health problems. The study built upon regional development plans and 25-year forecasts for macroeconomic factors and demographics.

The study sought to understand the interrelationships among the coastal population and natural resources in order to recommend approaches for successful, sustainable development. In cooperation with the Environmental Protection Agency of Ghana, the study gathered input from interested parties at the local, regional, and national level. Based upon this participatory process, seven major environmental problems were identified and prioritized for four regions of the coastal zone that were found to have the highest rates of environmental degradation (Table 5.7). The highest priorities were contamination of the water supply near the city of Accra and within the Volta Delta and degradation of the fisheries near Accra, since these problems directly affected health and local employment. Other problems, such as the loss of wetlands, erosion, and forest degradation, reflected the impact of rapid development on the region.

The study then identified four major categories of intervention that could be applied to these environmental problems:

1. Direct investments in new construction or in upgrading technology
2. Economic or regulatory incentives, including taxes, subsidies, and licenses

Table 5.7   Summary of Regional Environmental Priorities within Ghana's Coastal Zone

Environmental Problem	Western Region	Central Region	Greater Accra Urbanized Area	Volta Delta
Domestic sanitation	Moderate	High	High	High
Fisheries degradation	Low	Low	High	Moderate
Wetland and mangrove degradation	Moderate	Low	Moderate	Moderate
Industrial pollution	Moderate	Low	Moderate	Low
Erosion	Low	Low	Low	Moderate
Forest degradation	Moderate	Moderate	Low	Moderate
Aquatic weed development	Low	Low	Low	Low

Source: World Bank, *Findings 113* (June 1998).

[1] World Bank for Reconstruction and Development, "Integrated Coastal Zone Management Strategy for Ghana," *Findings 113* (June 1998).

3. Education initiatives to promote better management of coastal resources by increasing knowledge and awareness of the problems and opportunities for improvement

4. Institutional or policy reforms

Within each category, various intervention strategies were identified and evaluated. The duration and net present value were estimated for each strategy, along with anticipated benefits. Six criteria were used to weight the benefits of each strategy:

1. Reduction of the extent of the problem
2. Improved ecosystem health
3. Improved human health
4. Reduced risk of extinctions
5. Reduced incidence of toxic and hazardous substances
6. Improved economic efficiency

Because they lacked the data to calibrate models of environmental degradation, the World Bank researchers did not conduct in-depth studies concerning the impact of strategies on these criteria. Instead, they categorized the impacts qualitatively as high, medium, or low for each of the six evaluation criteria and calculated an average score. They created a cost-effectiveness index by normalizing the cost of the intervention and its impacts for a standard population of 100,000: cost effectiveness equals the normalized cost per person divided by the average impact.[2] Then they converted the ratio of the normalized cost to the normalized impacts to a simple rating of high, medium, or low. This approach, while quite qualitative and dependent upon expert judgment, enabled the researchers to identify the strategies with the greatest potential for alleviating the environmental problems and improving conditions for people in the coastal region. Table 5.8 shows the results for four of the highest-priority problems.

5.6.2 Case Study: Scenario-Based Planning as a Means of Achieving Consensus Regarding Controversial Development Proposals

In South Africa, attempts to enhance the local forestry industry appeared to be in direct conflict with the need to conserve water and also the desire to maintain virgin grasslands.[3] North East Cape Forests (NECF), a private company, had planted 38,000 hectares of forests but believed that its operation needed to expand in order to be economically sustainable. The company was in a position to plant another 37,000 hectares of forests, but planting any new forest required a permit from the National Department of Water Affairs and Forestry (DWAF). Conservationists were concerned about the loss of virgin grasslands and the consequences for wildlife. Controversy over the proposed expansion of the forestry operations had already delayed granting NECF the necessary permits and threatened any expansion of the project.

Upset with the delays in securing a permit, NECF hired consultants to help bring together the key stakeholders so they could reach some consensus on what to do. The consultants facilitated workshops attended by representatives of NECF, DWAF, national and provincial conservation groups, local government officials, and hydrologists. Instead of trying to look at each discrete plot proposed for forestation, the consultants convinced the workshop participants to consider the overall question of how to develop an overall forestry plan for the region.

---

[2] For example, assume that the cost per person would be $10 and that the six criteria were rated on a scale of 1 to 10, where 10 indicates the most favorable impact. If the average impact were judged to be 5, then the cost effectiveness would be $2 per unit of improvement. Another option might have a cost per person of $2 and an average impact of 4, so that this option would have a cost effectiveness of $0.5. The lower the cost effectiveness, the better; the study rated strategies with CE < 1 as "high," those with CE >2.5 as "low," and those in between as "moderate."

[3] This case is based on Theodor J. Stewart, "Thirsting for Consensus: Multicriteria Decision Analysis Helps Clarify Water Resources Planning in South Africa," *OR/MS Today* (April 2003): 30–34.

Table 5.8 Cost Effectiveness Summary of Potential Interventions

Intervention Category	Domestic Sanitation	Fisheries Degradation	Wetland and Mangrove Degradation	Industrial Water Pollution
Direct investment	Small-scale waste collection H Constructed wetlands M Centralized waste treatment L Large-scale treatment L	Fish landing sites H Aquaculture M	Mangrove planting H Sensitivity mapping H Protected area acquisition H	Process modifications H Secure landfills M Centralized treatment L
Incentives	Regulatory enforcement H Fines and user charges H Payments/subsidies H	Nearshore licensing H Regulatory reforms M	Regulatory reforms H Biodiversity strategy H Management plans H	Incentives H Regulatory reforms M Pollution Fines M
Education initiatives	Health awareness H Recycling/reuse H Specialized training H Curriculum H	Appropriate fishing techniques H	Government awareness H Curriculum H	Industry awareness H Industry standards H Training for nongovernmental organizations H
Institutional reforms	Small-scale local management H Land use planning H Privatization options study H CZM coordination H	Small-scale local management H Marine use zoning H CZM Coordination H	Diversity research H Biodiversity research unit H NGO support for monitoring H Traditional regulation H	EPA strengthening H Land use planning H District strengthening H Privatization options study H

Source: World Bank, *Findings 113* (June 1998).

Table 5.9 Steps in Scenario-Based Planning

1	**Initial problem structuring**: At a workshop involving the major stakeholders, identify criteria of concern and also the primary "policy elements" that should be considered; construct a (possibly very large) set of possible policy scenarios based upon combinations of the policy elements.
2	**Screening analysis**: Use available models to quantify consequences related to each criterion of concern for the various policy elements or for some or all of the possible policy scenarios; select a small number of scenarios (i.e., five to nine) for consideration by the workshop.
3	**Assessment and ranking of scenarios**: At a workshop, have participants discuss and rank the alternatives according to each of the criteria—by whatever means they are comfortable with (i.e., the facilitator may suggest methods, but would not require any particular method); identify where there is still no consensus.
4	**Refinement of the scenarios or of the problem definition**: Refine the scenarios and seek additional information in order to provide better grounds for achieving consensus.
5	**Ranking of options using each group's weighting preferences**: Let each group assess and weigh the consequences for each scenario, then compare the ratings to clarify where views are similar and what particular scenarios or options must be eliminated to achieve consensus.
6	**Iterations**: Two or three rounds of workshops are likely to be required to identify two or three options with broad acceptability.

Source: Theodor J. Stewart, "Thirsting for Consensus: Multicriteria Decision Analysis Helps Clarify Water Resources Planning in South Africa," *OR/MS Today* (April 2003): 30–34.

The first workshop developed scenarios that would then be analyzed in further detail. The two key elements of the plan were (1) how much of the land to use for commercial forestry (i.e., how much of the additional 37,000 hectares to develop); and (2) whether to process logs in local sawmills, a centralized pulp mill for the region, or in mills in other regions. The group defined six scenarios based upon these two elements, plus the status quo (no expansion of commercial forestry operations).

At a subsequent meeting, two of the six scenarios were eliminated by consensus. The group also decided that more detail was needed about the location as well as the extent of the forested areas, because some locations were much more sensitive in terms of the impacts on water or on virgin grasslands. This led the group to adjust the scenarios so that development of any area could be permitted, with restrictions to limit the impact on the most environmentally sensitive areas, based upon analysis by a team of environmental experts.

The refinement of the scenarios and the addition of safeguards for the environmental issues enabled the group to reach a consensus on the most desirable option: develop the entire 37,000 hectares, subject to conservation restrictions, and process the logs at a single pulp mill within the region. The consultants then analyzed this broad consensus in more detail to identify the implied trade-offs between conservation and profitability of the forestry enterprise. DWAF eventually issued permits that were close to what was recommended by the workshop participants.

Table 5.9 summarizes the steps in scenario-based planning as used by these consultants. This example has not delved into the details of the scenarios, because the intent is to demonstrate one approach for achieving consensus about a controversial development proposal. Chapter 11 provides a more detailed discussion of scenarios and a more detailed example of how they can be used to help a company prioritize its investments in infrastructure.

5.7 SUMMARY

Large projects are always trying to meet multiple, sometimes conflicting objectives. Sometimes there are obvious ways to define economic or financial criteria that capture the essence of a project. More often, it is necessary to make assumptions about the monetary value that might be associated with some of the impacts of a project. And sometimes the problems and potential solutions are primarily concerned with factors such as aesthetics or equity, which are inherently difficult to quantify.

Figure 5.1 San Antonio's River Walk. Cleaning up this stretch of the river transformed an eyesore into a dramatic setting for restaurants, pubs, cruises, and walking. Benefits include reduced risks to public health, increases in land values along the river, and enhanced attractiveness of San Antonio for conventions and tourism.

Expressing Costs and Benefits in Monetary Terms

If the most important objectives can be expressed in economic terms, then it is possible to create a single monetary estimate of the costs and benefits associated with a project. Many private sector projects are primarily concerned with financial matters: if this project proceeds, will the revenues be sufficient to cover the investment and operating costs? Many public projects, as discussed in the previous chapter, are concerned with broader economic benefits, such as economic growth, job creation, and average income. While these economic benefits are not necessarily readily tied to the project or to specific individuals, they can certainly be expressed in monetary terms.

Comparing Current and Future Costs and Benefits

When evaluating costs and benefits over the life of a project, it is necessary to discount future costs and benefits for comparison with present costs and benefits. By using a discount rate, any future cost or benefit can be reduced to an equivalent present value. By summing all of the discounted costs and benefits, it is possible to obtain the net present value (NPV) of a project. If the NPV is positive, then the project provides net benefits and may be worth pursuing. Discounting, net present value, and related issues are covered in more detail in Chapter 7.

Discounting can be used in financial or economic analysis. Discount rates are determined by market forces and the opportunities available to entrepreneurs and investors. Discount rates for public projects are lower than the rates used by private companies and investors, because the public agencies do not have to pay taxes. Public projects may also be viewed as less risky, since governments have the power to raise taxes if necessary to cover unexpected costs of a project. If tax money is used to fund a project, then the proper discount rate should reflect the opportunity cost to the taxpayers; if taxes were lowered, then taxpayers would be able to use more of their income for their own purposes. In the United States, a discount rate of about 7-8% has commonly been used for public projects that depend upon taxes.[4] If a special agency is created to deal with a particular problem, whether transportation or housing for the elderly or parks and recreation, then that agency may be able to raise funds by selling bonds with a low interest rate. Such an agency could use the interest rate on its bonds as its discount rate in evaluating projects. The choice of a discount rate is examined further in Chapter 8.

Cost Effectiveness

Cost effectiveness is a useful concept when dealing with criteria that are quantifiable but difficult to monetize. Cost effectiveness is the ratio of the cost of a proposed project to the change in

---

[4] The Office of Management and Budget (OMB) provides federal guidance for the use of discount rates and other aspects of financial analysis for public projects (see OMB Circular A-94).

performance. So long as it is possible to quantify a key aspect of performance, such as risk or capacity or service quality, it should be possible to evaluate alternatives by considering their cost effectiveness in improving this aspect of performance. Cost effectiveness is less relevant in projects that have multiple objectives, so it is impossible to focus on just one critical aspect of performance.

Weighting Schemes for Projects with Multiple Criteria

When there are multiple objectives, some of which cannot be monetized, then the evaluation cannot focus on a single metric. Instead, some kind of weighting scheme—whether objective or subjective—is required to compare alternatives. If a decision is made by an individual or by the vote of a committee or by a referendum, then each person involved can make his or her own subjective judgment in determining a preferred option. Often a structured process for making the decision is followed, thus requiring participants to consider multiple options, to consider impacts upon various objectives, and to follow a specific procedure for ranking the options. If so, then some sort of a weighting scheme may be helpful. With a weighting scheme, any set of multiple measures can be collapsed into a single measure of effectiveness.

It is critical to remember that there is no objective way to determine the "correct" weighting scheme or even the selection of the "proper" criteria. The choices of what criteria to consider and how much weight to give to each one could cause intense debate among those trying to address a problem or those trying to promote a particular project.

Using Public Input in Identifying and Evaluating Projects

Input from the public and/or stakeholders can be helpful in identifying and evaluating projects. Public input can help in clarifying the nature of the needs or problem, as well as in identifying measures or criteria that can be used to define the problem and evaluate potential solutions. In some cases, very rough measures of potential impacts may be useful in reaching consensus about what to do, how to do it, and when to begin.

ESSAY AND DISCUSSION QUESTIONS

5.1 Some people have argued strongly that the traditional approach to project evaluation, which they often call cost-benefit analysis, fails to deal with many important issues because of its emphasis on financial matters and its use of discounting. Others call for new approaches, such as sustainability assessment, to ensure better consideration of environmental matters. Discuss how cost-benefit analysis, environmental impact analysis, and sustainability assessment relate to each other and to overall project evaluation. You may want to consider the extent to which these are complementary methods or incompatible approaches.

5.2 The bus example demonstrated some of the difficulties in establishing a weighting scheme and showed how supporters of different projects could come up with a weighting scheme that made their project look best. Assume that you are about to coordinate a meeting of the major stakeholders to discuss how to interpret and evaluate the results of the study. Assume that each of the stakeholders is coming to the meeting prepared to present their analysis.

a. What additional weighting schemes might have been used to compare these alternatives?
b. What additional criteria could have been considered?
c. What additional alternatives could have been considered?
d. How would you structure the meeting so as to promote discussion and consensus rather than angry debate and conflict?
e. What additional information would you seek to clarify the key issues?

5.3 *Integrated Coastal Zone Management Strategy for Ghana* (Section 5.6.1): What actions do you recommend for addressing environmental issues in the coastal zone of Ghana? Which types of strategies do you believe will be most effective?

5.4 *North East Cape Forestry* (Section 5.6.2): This example suggests that workshops involving stakeholders, public officials, and the public can be helpful in achieving consensus about how best to proceed with controversial projects or programs. Compare and contrast this approach with the approach used to address a controversial project or program in your city or state.

PROBLEMS

5.1 A company can raise money for a project by selling bonds. A typical bond has a face value of $1,000 and a 30-year life. The company will pay interest on the bond at the end of each year, and at the end of 30 years it will also redeem the bond (i.e., pay back the original $1,000). The interest rate is determined by market forces. What is the NPV of a 30-year bond paying 6% interest to someone with a discount rate of 6%? To someone with a discount rate of 5%? To someone with a discount rate of 7%?

NOTE: This question can be answered by creating a spreadsheet and showing the interest of $60 received at the end of years 1 to 30 and the return of the $1,000 at the end of year 30. Discount all of these payments using the investor's discount rate, and sum to get the NPV of the bond to each investor.

5.2 A zero-coupon bond has no interest payments. Investors buy the bonds for less than face value, and they receive the face value of the bond at the end of the bond's life. A 30-year, zero-coupon bond that has a face value of $1,000 therefore is a promise to pay the investor $1,000 at the end of 30 years. If investors have a discount rate of 6%, what would they be willing to pay for this bond? What would they be willing to pay if their discount rate was 5% or 7%?

5.3 Which of the projects in the following table are worth pursuing if the public agency is required to use a discount rate of 7%? The other benefits include growth in the regional economy, reduction in congestion, improved safety, and improvements in health. These are all economic benefits to the region, although they produce no revenue that can be used to help cover the investment or operating costs.

Problem 5.3: Possible Projects for a Public Agency

Investment	Annual Revenue	Annual Cost	Net Value of Other Benefits
$ 10 million	$ 6.5 million	0	0
$ 10 million	$ 8 million	0	0
$ 10 million	$ 1 million	$0.2 million	$ 0.2 million
$ 10 million	$ 0.2 million	$0.2 million	$ 1 million
$ 50 million	$ 2 million	$3 million	$ 3 million
$100 million	$10 million	$1 million	($ 3 million)
$100 million	$20 million	$1 million	($10 million)

5.4 A project is expected to produce benefits of $10 million in year 10 and $100 million in year 30. What is the net present value of these benefits, assuming a discount rate of 7%?

5.5 A proposed transit system is expected to require annual maintenance costs of $1 million, and it will require major rehabilitation in year 10 and year 20. What is the net present value of the maintenance and rehabilitation costs?

5.6 Some members of the city council would like the city to construct facilities that would provide housing for people with low income. Federal funds are available to assist, but the city would have to match the federal funding. If the city uses general revenues, then the project would have to use the city's standard discount rate of 7% when evaluating the project. James Bondworthy, a member of the council, recommends creating a Housing Authority that would be authorized to sell bonds to raise the cash needed to match the federal funds. According to Bondworthy, the interest rate on the bonds would be no more than 4.5%, and the proposed rents would be sufficient to pay the interest on the bonds. Other members immediately objected, claiming that Bondworthy was simply trying to create a housing bureaucracy that he could dominate. According to them, the merits of the project would not change at all by creating a new agency. In defense of Bondworthy, what advantage would the creation of a Housing Authority have for the economic assessment of the proposed housing project?

5.7 Consider a proposed project in which an initial investment of $10 million would produce an annual financial benefit of $3 million. The project is intended to improve water quality and increase the efficiency of the water delivery system in the region. The financial benefits relate to two factors:

- There would be less leakage in the pipes used to transport water to the city from the reservoir.
- The proposed treatment program would be more efficient.

The life of the project is expected to be 100 years. Two-thirds of investment is related to replacing the pipes to eliminate the leakage, while the rest is related to upgrading the treatment facility. The project would be financed either by selling bonds or borrowing money from a local bank; either way, the annual interest rate would be about 4–5%. A hearing has been scheduled for the project, and several speakers have been asked to present their viewpoints. What is the main point that you expect each of the following people to make at the hearing?

a. A banker whose bank has made many loans to the city to finance a variety of projects. (The banker is known to prefer projects whose financial benefits are sufficient to cover the financial costs of the investment.)

b. An environmental advocate who recently wrote an article entitled "Discounting Is Unethical—We Must Not Overlook Benefits to Future Generations."

c. A doctor who believes that the new water treatment system will lead to a measurable improvement in health for the region.

d. A farmer whose crops depend upon a reliable supply of water

e. A politician who is in the midst of an election campaign; his slogan is "It's high time for low taxes."

f. His opponent, whose slogan is "Build now for our children and for their children."

g. A civil engineer at the local university who recently completed a study of the water supply system. (Among the results of the study was a finding that two major leaks caused 80% of the loss of water; these leaks could be fixed for less than $5 million, and the rest of the system should continue in reasonable condition for another 20–30 years.)

Problems 5.8–11: Options for Improving Transit Performance

Option Weights:	Budget Increase 40%	Service (on-time performance) 20%	Capacity (ratio of new to current capacity) 20%	Improvement in Safety and Security 20%
A	$20 million	80%	110	10%
B	$25 million	85%	110	10%
C	$30 million	90%	120	20%
D	$35 million	85%	140	10%

h. The chair of a citizen's advisory committee that advises the mayor on all major infrastructure projects that involve use of taxes to cover operations or investment. (This group used the federally recommended discount rate of 7% to determine that the benefits of $3 million per year over the life of the project would have a present value of more than $40 million.)

i. Finally, since you are related to the mayor and you are known to be an excellent student with an interest in infrastructure, you have been asked to listen to all of the previous speakers and summarize what you feel should be done.

5.8 A major city is considering upgrades to its transit system that would improve the quality of service (measured as on-time performance), increase capacity, and improve safety and security. The system currently operates with an annual budget of $500 million, 70% on-time performance, and 200 incidents per year that involve personal injury, assault, or theft. Using the criteria, the estimated changes in performance, and the weighting scheme shown in the following table, rank the alternatives from best to worst.

5.9 Use the information given in the previous table to determine the ordinal rankings for each alternative for each criterion. Using the same weights shown in the table, rank the alternatives based on their ordinal rankings.

5.10 Your boss has complained that the measures used for the various criteria are inconsistent: cost and service are shown as the actual numbers, while capacity is shown as a percentage of current capacity, and safety and security is measured as the percentage improvement in performance. This time, use a more consistent set of measurements and determine the ranking of the alternatives.

5.11 After trying various schemes to rank the options, an analyst suggests that it would be better to consider the cost effectiveness of the various options considered in the previous three questions. Using a consistent scheme for estimating the service, capacity, and safety benefits (e.g., as used in the previous question), what is the cost effectiveness of each strategy?

5.12 There are four options for building a new road that would shorten the distance between two cities, reduce congestion on existing roads, and improve safety by diverting traffic from a narrow road with many dangerous curves. A committee has been formed to select the best option, and they are reviewing the results of analyses as presented in the following table. Explain how you would create a single measure that encompasses all of these criteria.

5.13 There are four options for building a new sports stadium. The owners and the city are considering the relative advantages and disadvantages of each option, and they agree that the main factors to consider are seating capacity, accessibility by transit or car, and cost. Option A is best if transit accessibility is the only criterion, and the mayor favors this option; option D is best if automobile access and cost are the dominant criteria, and the owners favor this option. The owners hope to obtain considerable financing from the city, whichever option is chosen, as they believe that the stadium and the team create a great deal of favorable economic activity for the entire region. A public hearing has been scheduled to solicit input regarding the selection of a site for the stadium.

a. You are a major landowner near site B, and the value of your property would increase sharply if site B were chosen. However, you really don't want the city, the team, or the public to know about your plans. Create a plausible weighting scheme

Problem 5.12: Options for Building a New Road

Option	Investment Cost	Maintenance Cost/Year	Miles Saved Relative to Existing Route	Expected Reduction in Fatalities/Year
A	$100 million	$10 million	1	1
B	$120 million	$10 million	1	1.5
C	$140 million	$ 8 million	2	2
D	$160 million	$ 7 million	3	2.5

Problem 5.13: Options for Building a New Stadium

Location	Seating	Accessibility by Transit	Parking	Cost
A (Nearest to city center)	60,000	Excellent	4000 spaces	$140 million
B	60,000	Good	8000 spaces	$100 million
C	80,000	Good	20,000 spaces	$130 million
D (farthest from city center)	80,000	None	Unlimited	$ 80 million

such that option B is the best option, so that one of your associates can make the case for site B in the public hearing.

b. You own a hotel and two restaurants near site C, and you believe that the stadium will be very beneficial for your business. Create a plausible weighting scheme such that you can explain why option C is the best option for the facility when you speak at the public hearing.

c. You are part of a team organized by a nonprofit organization whose mission is to promote rational development that is in the overall best interests of the region. Your role at the hearing is to discuss the main issues that will influence the decision and to identify further information that would be needed to determine which site is in the public interest. What will you say?

5.14 The following table shows the costs of various options for protecting wetlands along with comments concerning the political viability of each option.

a. Which are the most cost-effective strategies for protecting wetlands?

b. Given the political comments, what strategy would you pursue as an advocate of preserving wetlands?

Problem 5.14: Strategies for Protecting Wetlands

Strategy	Potential Cost to the State	Acres of Wetland That Can Potentially Be Protected	Comment
Create wetlands in highway interchanges and along highways.	$20 million to convert state-owned land	300	This amount is small compared to costs of constructing highways.
Acquire wetlands adjacent to state parks and state-owned land.	$20 million to purchase land	1000	Annual budgets for acquiring conservation projects are less than $5 million per year.
Purchase conservation easements from owners of large tracts of wetlands.[a]	$10 million to purchase easements	10,000	There is currently no program allowing the state to purchase easements.
Pass legislation that would provide more stringent limits on the development of wetlands.	$1 million for education and enforcement	50,000	Previous attempts to pass such legislation were easily defeated.
Pass legislation creating a "wetlands bank."[b]	$2 million for administration	2000	Several developers have advocated such a plan.

[a]A conservation easement can prohibit development of land while continuing to allow normal use of the land for agriculture, forestry, fishing, hunting, etc. An easement is a legally binding document that becomes part of the deed to a property.
[b]A wetlands bank would receive payments from developers or cities that destroy wetlands as part of their projects; the payments would then be used to protect wetlands (by purchase or through buying an easement) or to create new wetlands.

CASE STUDY: WE'LL CROSS THAT BRIDGE WHEN WE COME TO IT (PART I)

Cammitibridge, a prosperous city of a million residents, is located on the north side of the Cammiti River. It is adjacent to, and to some extent hemmed in by, a state forest to the northeast and a series of steep hills to the northwest (see map at end). The town is named for its historic bridge, first built in 1720 and most recently rebuilt in 1850. The bridge is a narrow, two-lane bridge that spans the river so as to connect Cammitibridge with the main road to the capital, Boslondale, some 100 miles away. The bridge connects to a highly scenic road that winds for several miles toward the west below sandstone cliffs along the edge of the river. Everyone in the city agrees that a new bridge is needed, for various reasons:

- Development in Cammitibridge is increasingly limited by a lack of vacant land close to the city center. On the outskirts of town, people are starting to move up the slopes of the hills, but there just isn't much room for expansion downtown or along the major roads serving the city.
- There is considerable traffic between Cammitibridge and Boslondale, so that the old bridge is often congested, irritating those who must use it and limiting opportunities for development on the south side of the river.
- The current bridge provides little access to the land available for development south of the river. A new bridge in a new location would open up this land for development.
- The local construction industry supports the construction of a new bridge, because of the opportunity for new activity and profits.
- A consortium of private citizens, led by Canwy Bildem, has even proposed that they will replace and expand the existing bridge at their own expense if they are allowed to charge a toll of no more than $2; after a period of 30 years, they would turn the bridge over to the city.

There are four competing options for the bridge:

- *Bridge 1:* Expand the existing bridge at a cost of $20 million. The plan would basically build a second two-lane bridge next to the existing bridge. Once the new bridge is operating, the old bridge would be rehabilitated to handle heavier trucks and then reopened.

- *Bridge 2:* Replace the existing bridge with a new four-lane bridge at a cost of $50 million. The old bridge would be torn down upon completion of the new bridge. An additional $5 million would be required to modify the roads to match up with the new bridge.
- *Bridge 3:* Build a new four-lane bridge that connects to the developable land to the southeast. Since the river is wider at this location, the cost would be greater. Initial estimates are that the bridge would cost $75 million if built at the narrowest location while access roads would cost $20 million.
- *Bridge 4:* A variation on Bridge 3, this option would change the bridge location slightly to improve access to the city while requiring a longer bridge to get over the swamps; the costs would be $90 for the bridge plus $10 million for access.

The bridge could be financed in several ways:

- The city could sell revenue bonds to the public and pay the interest on the bonds out of general tax revenue.
- The city could sell revenue bonds to the public and charge tolls on the bridge sufficiently high to pay the interest on the bonds.
- The city could authorize Mr. Bildem to proceed with his plan to build a toll bridge without using public money.
- The city could probably fund the required connections (but not the bridge) from its ongoing budget for road construction and maintenance.

The economic benefits from construction of the bridge fall into several categories:

- Reductions in travel time for people who currently use the bridge
- Increased opportunities for development south of the river, leading to higher land values, new jobs, and greater real estate taxes for Cammitibridge

Preliminary analysis suggests that

- Expanding or replacing the current bridge will have minor effects on traffic volume or development, since very little open land is suitable for development near the existing routes.
- Building a bridge at the east end of town will provide a spark to development of that region; total traffic across the river is expected to grow quickly if a new bridge is built in that location. Also, some traffic will divert to the new bridge, reducing congestion in the city.

Questions

a. What major economic benefits could be obtained by expanding the existing bridge or building a new bridge? How can the city estimate the magnitude of these benefits?
b. What major environmental concerns are related to expanding the existing bridge or building a new bridge?
c. What are the major social issues?
d. What additional information is needed to make a decision about which bridge to build?
e. What main criteria would you use in comparing the bridges? What kind of weighting scheme would you recommend for comparing the options?
f. From the public perspective, what are the advantages and disadvantages of Mr. Bildem's proposal?

NOTE: this case study is continued at the end of Chapter 11.

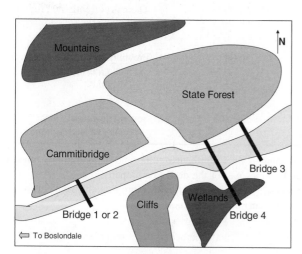

Schematic Diagram of Cammitibridge and Possible Bridge Locations

Chapter **6**

The Panama Canal

To Europeans, the benefits of and advantages of the proposed canal are great; to Americans they are incalculable.

Ulysses S. Grant, President of the United States of America

6.1 INTRODUCTION

This chapter concludes Part I of this text by presenting a case study of the Panama Canal, one of the great infrastructure projects of the last 200 years and one of the most important links in the global transportation system. This case study captures some of the excitement and vision that motivate people to pursue vast projects, and it also illustrates the many pitfalls and dangers that can ruin those who attempt to do too much, too soon, without the necessary resources. As a case study, this chapter differs markedly from the others in the text, because it is intended to provide enough contextual detail to spark discussion and encourage debates concerning the various concepts that have already been introduced. By moving away from concepts and methodologies, it offers insights into the overall theme of Part I— "Building Infrastructure to Serve the Needs of Society."

The construction of the Panama Canal was one of the great engineering feats of the twentieth century, culminating centuries of efforts to create more efficient transportation routes across this rain-soaked isthmus between the Atlantic and the Pacific. This case describes the kinds of motivations, challenges, setbacks, and achievements of those who attempt to build upon a grand scale. Cost, service, capacity, safety and security, and other aspects of system performance introduced in Chapter 2 have all been matters of concern in Panama. The methods used to analyze small, nineteenth-century canals in New England can be applied to the Panama Canal as well. The importance of financial feasibility, the competition among modes of transportation, the potential of opening up new global markets, and trade-offs among cost, capacity, and service were all factors in the final design of the Panama Canal. Objectives ultimately included economic and strategic goals, as well as financial success, as more of the efforts were undertaken or supervised by public agencies. The relative importance of the different objectives clearly changed over time as different problems were encountered and better technologies became available.

The historical context is always relevant to large-scale projects, and personalities, geography, unforeseen difficulties, and technological innovations can all be major factors leading to success or failure. The case therefore begins with the sixteenth-century search for routes from Europe to the Pacific, and it describes the development of roads and railroads as well as the eventual construction of the Panama Canal. The case also identifies the challenges facing the canal at the beginning of the twenty-first century, as the need for expanding capacity conflicts with heightened concerns for the environment and for social responsibility. This case therefore illustrates how an infrastructure-based system can evolve over a very long time—a topic that is revisited in Chapter 15.

As you read this case, consider the following questions as they relate to each of the projects discussed:

1. Who was in charge of the project?
2. What were their objectives?
3. What were the major problems or issues?
4. What were the major costs and benefits, and how were they evaluated?
5. What were the results of the project? How successful were they?
6. What else might have been done?

6.2   EARLY ROUTES ACROSS THE ISTHMUS

Columbus, sailing west in hopes of opening a trade route between Europe and Asia, instead ran into the Americas. While promising in terms of future development and precious metals, the Americas were not accepted as the limits of Spanish interest. The ultimate goal remained opening up a trade route to the civilizations of Asia.

Balboa, in 1516, was the first to establish a freight route across the Isthmus of Panama. Having "discovered" the Pacific Ocean, with help from the local Indians, he of course wanted to continue his explorations. To do this, he needed to build ships on the Pacific side of the isthmus, and the trees required for the task grew only on the Atlantic side.

The terribly onerous labor of collecting the material and carrying it on their backs to its destination was imposed upon the Indians, of whom thousands were gathered together for the purpose, and impelled to the unaccustomed work by the merciless severity of their taskmasters. Many months were consumed in this grim struggle for a passage of the Isthmus, which, in many respects, foreshadowed the endeavors of the modern successors of these hardy pioneers. Hundreds of the wretched aborigines, Las Casas says their number fell little short of two thousand, lost their lives in the undertaking, but it succeeded, and four brigantines were carried piecemeal from sea to sea and put together on the Pacific coast.[1]

Balboa's efforts were, for him at least, for naught. Before he could depart on his expedition, he was tried, convicted, and executed on trumped-up charges by the jealous Spanish governor of the region. While exploration by sea was put on hold, the need for a "permanent highway to take the place of the Indian trails which were poorly adapted to the traffic which had now begun to move over them became apparent."[2] With great difficulty, a paved road wide enough for two carts was constructed in 1521, linking Old Panama on the Pacific with Nombre de Dios on the Atlantic. After about 10 years, the use of the route was modified to an intermodal route: light sailing vessels left from Nombre de Dios and sailed up the Chagres to meet up with the road. By the end of the sixteenth century, the Atlantic terminus of the cross-isthmus road was shifted to Porto Bello. By then, this road was the "richest highway in the world," since it was the critical link between the Atlantic and Old Panama (the "most important Spanish City in the New World"), the mines of Peru, and the major regional fairs in Cartagena and Porto Bello.[3]

The Spanish continued to look for an all-water route to the orient. In 1519, Magellan found the southern route to the Pacific, passing through the Straits of Tierra del Fuego. In 1522, using Balboa's vessels, Gil Gonzales sailed north, looking for a waterway, and eventually found Lake Nicaragua; in 1529, Diego Machuca explored the lake and, with difficulty, navigated the San Juan River from the lake to the Atlantic. Over the next century, this became an important commercial route for "vessels making ports in Spain, the West Indies and South America . . . for more than a hundred years, a constant stream of gold, pearls, and other products of Spain's island possessions flowed across the

---

[1] Marshall, Logan, *The Story of the Panama Canal* (L.T. Meyers, 1913), 19.

[2] Ibid., 19–20.

[3] Ibid., 20–21.

Isthmus."[4] At about this time, Cortes established a transcontinental trade route across Mexico, from the mouth of the Coatzacoalcos River to the port of Tehuantepec on the Pacific, which was used as a trade route between Spain and the Americas as well as a link to the Philippines.

Military considerations put plans for a water route on hold. Philip the Second feared that a water route across the isthmus would simply give enemies easy access to Spain's new possessions. This policy lasted for two centuries. In the late 1700s, the possibility of a canal along the Cortes route was investigated by Manuel Galisteo. Although his conclusion was unfavorable, British engineers accompanying Galisteo felt the project was feasible. When war broke out with Spain, the British sent Captain Horatio Nelson to the region, who viewed his mission as follows:

> *In order to give facility to the great object of the government I intend to possess the Lake of Nicaragua, which for the present, may be looked upon as the inland Gibraltar of Spanish America. As it commands the only water pass between the oceans, its situation must ever render it a principal post to insure passage to the Southern Ocean, and by our possession of it, Spanish America is divided in two.*[5]

Nelson and his men indeed grabbed control of Lake Nicaragua, but climate and illness forced them out, and Spain retained control of the "canal region" at the beginning of the nineteenth century. New investigations by Humboldt generated interest in a canal, and in 1814 Spain passed legislation authorizing the construction of a canal. Before any work could begin, revolutions in South and Central America overthrew the Spanish dominance and

> *opened up new possibilities in connection with the much-mooted question of a waterway and claimed the attention of capitalists and statesmen of all the commercial nations. From this time the matter is taken up with definiteness of purpose and never allowed to rest.*[6]

The United States officially became interested in 1825, when Secretary of State Henry Clay entered into negotiations with the Republic of Central America for building a canal across Nicaragua "the execution of which . . . will form a great epoch in the commercial affairs of the whole world."[7]

In 1827, Colombia commissioned J. A. Lloyd to study possible rail and water routes across the Isthmus of Panama. Lloyd considered plans for a canal to be premature and instead recommended an intermodal route combining water and rail to take the place of inadequate roads.

In 1838, a French company obtained a concession from what was now New Granada to construct highways, canals, or railroads from Panama to the Atlantic, by any feasible route. The pressure and interest were growing; but the difficulties were not yet well understood, and the costs of construction were greatly underestimated by all parties involved.

From the perspective of the United States, the strategic importance of the isthmus changed immensely when California was acquired in 1848 as a result of the war with Mexico:

> *The requirements of travel and commerce demanded better methods of transportation between the Eastern States and the Pacific coast, but there were other reasons of a more public character for bringing these sections into closer communication. The establishment and maintenance of army posts and naval stations in the newly acquired and settled regions in the Far West, the extension of mail facilities to the inhabitants, and the discharge of other governmental functions, all required a connection in the shortest time and at the least distance that was possible and practicable. The importance of this connection was so manifest that the*

---

[4] Marshall, *Story of the Panama Canal*, 19.

[5] Ibid., 25.

[6] Marshall, *Story of the Panama Canal*, 27.

[7] Ibid., 28.

Government was aroused to actions before all the enumerated causes had come into operation, and negotiations were entered into with the Republic of New Granada to secure a right of transit across the Isthmus of Panama.[8]

A treaty was ratified between the United States and New Granada in 1848 for the Panama route in 1848. In 1849, a treaty was ratified with Nicaragua to allow construction of railroads, highways, or canals across Nicaragua. For many years thereafter, it appeared most likely that the United States, if it did build a canal, would choose the Nicaraguan route.

The discovery of gold in California vastly increased the demand for transportation across the isthmus, as thousands of "forty-niners" flocked to the gold fields. This led to the construction of the Panama Railroad, as discussed in the next section.

6.3 THE PANAMA RAILROAD

The idea of a railroad across Panama remained just that until the gold rush raised the possibilities of profits to much higher levels. After earlier concessions expired without producing any construction, New Granada granted the railway concession to the Panama Railroad Company, which was incorporated in 1849 with strong financial backing from Wall Street. The concession gave the company a railroad monopoly across the isthmus and allowed it to sell its assets at a fair price to any company that was authorized to build a canal (since a canal, once built, would likely destroy the railroad both financially and literally).

The benefits of a railroad were clear. For passengers, the railroad, once built, would cut the transit time across the isthmus from a hazardous 5 to 10 days to a relatively luxurious couple of hours. For freight, the time savings were even greater; months could be saved by not going around Cape Horn. For a trip from New York to San Francisco, the all-water route was 13,000 miles, whereas the intermodal route via the Panama Railroad would be only 5000 miles—a savings of 8000 miles.

The engineering work for the railroad began in 1849, and construction was estimated to require 2 years and a cost of less than $2 million.[9] In fact, the first train operated across the 48-mile broad gauge[10] line in 1855, and the construction cost was more than $8 million—six times the initial estimates. In addition, the railroad acknowledged more than 1200 fatalities in the workforce, which averaged 5000 men over the 5 years of construction. More likely, there were more than 6000 fatalities from disease.[11]

The railroad was a financial success, showing profits of 12 to 22% per year (i.e., $1 to $2 million annually). The railroad also reshaped the economic geography of Panama: Colon, at the terminus of the railroad, replaced Porto Bello as the major Atlantic port. The financial success was to be expected given the potential for the railroad to capture the lion's share of the benefits of not having to ship around South America or trek through the jungles of Panama. Passengers were quite happy to pay the fare of $25, and the railroad had 40,000 passengers per year from 1856 to 1966.[12] Shippers were also willing to pay a bill that might be nearly as much per ton:

With the opening of the railroad, a large traffic across the isthmus sprang into existence and grew rapidly with the advance of time. The products of Asia and the countries upon the Pacific coast were carried [on the railroad] from Panama to Colon, there to be distributed amongst

---

[8] *Report of the Isthmian Canal Commission.* Washington, 1899–1901; cited in Marshall 1913, 34.

[9] McCullough, David, *The Path Between the Seas: The Creation of the Panama Canal, 1870–1914* (New York: Simon & Schuster, 1977), 34.

[10] Gauge is the distance between the rails. Broad gauge, at 5 feet, is slightly wider than the standard rail gauge of 4 feet 8.5 inches, which reputedly can be traced back to the gauge of wagons used in Roman times. The persistence of the standard gauge from ancient times reflected the need for wagons to fit the ruts found in dirt and gravel roads.

[11] McCullough, *Path Between the Seas*, 37.

[12] Ibid., 35–36.

steamships making the ports of Europe, Canada, the United States and the West Indies. Moving in the reverse direction, goods from these countries reached, by the same trans-isthmian route, South and Central American and San Francisco. From the last named port, reshipment was made to the Pacific islands and points on the Asian mainland. A number of steamship lines made regular calls at the terminal ports of the railroad. The line occupied a commanding position as the essential link in this chain of traffic, and took full advantage of the fact. Its charges were exorbitant and its profits enormous for many years. Its rates were based on, in general, fifty per cent of the through tariff. For instance, of the total cost of shipping goods from New York to Valparaiso [in Chile], one half represented the charge of the railroad company for its share of the carriage. For many years the road carried enormous quantities of coffee to Europe. The through rate was about thirty dollars per ton. The railroad company received fifteen dollars and the two steamship companies that handled the goods divided a similar sum.[13]

In the first 6 years of operation, the Panama Railroad had cumulative profits of more than $7 million, nearly recouping the entire construction cost. At one point, the stock price of the Panama Railroad Company reached $295 per share, the highest on the New York Stock Exchange for a market valuation of $21 million and an indication to potential investors of the financial possibilities of a transcontinental connection.[14]

In 1879, the railroad was offered for sale to the (French) Panama Canal Company for $14 million; 6/7 of the stock of the company was eventually sold to the Panama Canal Company for $18.6 million.[15]

When the United States took over construction of the Canal from the French in 1904, it acquired the railroad as well. Shippers took the change in ownership as an opportunity to challenge the monopolistic pricing practices of the railroad. For example, the railroad had a contract that gave the Pacific Mail Steamship Company the exclusive right to issue through bills of lading from San Francisco to New York; all other steamship lines would have to pay full fare for the local rail move. When the United States took over, the monopolistic rates were replaced with rates designed only to provide a fair rate of return over costs.[16] This change in pricing policy ended the period of independent prosperity for the railroad.

However, the Panama Railroad's finest hour was yet to come. Constructing the canal required extensive use of the railroad; but the railroad first had to be moved since much of it would otherwise be flooded by the creation of Lake Gatun, a critical step in the construction. The cost of the new, double-track line was $8.9 million (i.e., roughly the same as the cost of the original line). During the height of canal construction, the line handled 700 to 800 dirt trains daily, each consisting of a locomotive and 18 flatcars with a total load of 500 tons. The peak year was 1910, when the line moved approximately 300 million tons of freight—a truly phenomenal amount that is roughly 20% of the record levels of tonnage handled by the entire U.S. rail system (during World War II and again in the late 1990s). Despite tremendous technological advances in both track and equipment, the highest-density lines in the United States (and in the world) carried less than half that amount of tonnage at the end of the twentieth century.[17]

6.4   THE FRENCH EFFORT

The first serious effort to construct a trans-isthmian canal was initiated by the Panama Canal Company, a French company. The firm was headed by Ferdinand de Lesseps, who had conceived, organized, and completed the construction of the Suez Canal. A number of engineering conferences had been held to debate the route, the nature of the canal, and the resources required. Panama was the route favored by de

---

[13] Marshall, *Story of the Panama Canal*, 52–53.

[14] McCullough, *Path Between the Seas*, 35–36.

[15] Marshall, *Story of the Panama Canal*, 109.

[16] Ibid., 55–56.

[17] Association of American Railroads, *Railroad Facts*, various editions.

Lesseps, who also insisted on a sea-level canal. The construction time for and the cost of constructing the canal were variously estimated at 2 years and $100 million (by a contractor eager to do the job), at 8 years and $168 million by a national technical commission, and at 12 years and $214 million by the Paris Congress. De Lesseps, in promoting the project, chose a figure of $131 million. The canal was clearly going to be much more expensive than the railroad, since all of these estimates were on the order of 100 times the initial estimates of constructing the railroad that was eventually built for $8 million.

De Lesseps estimated the first year's traffic as 6 million tons, which would assure revenue of $18 million (at $3 per ton). Since a sea-level canal would have low operating costs, most of the toll revenue would be expected to be profit. Hence, according to the company's statements, the project would return approximately 10–15% once it opened for business.

Writing in 1913, just before the canal finally opened, author Logan Marshall (and the whole world) knew that the cost estimates were way too low. He also stated that the revenue estimate "was claimed to be a very conservative assumption, whereas, it was in reality almost beyond the possibility of realization."[18] Given that the railroad freight charge was as much as $15 per ton, a price of $3 per ton might well have seemed reasonable at the time. However, by 1913, with the United States running the Panama Railroad, the rates were no longer so extravagant and, with larger ships operating, the prices of ocean transport had also fallen. Marshall may also have been concerned with the projected volume of traffic. (In actual fact, the canal handled 4.9 million long tons of cargo and earned revenues of $4.4 million in its first full year of operation in 1915; by 1923, with cargo of nearly 20 million tons, revenues reached $17.5 million.[19])

The canal was viewed as a tremendous financial opportunity, and the effort was therefore undertaken by the private sector. The Panama Canal Company was formed in 1879 with an initial capitalization goal of $80 million and, given the general excitement, double that amount was quickly raised.

Construction began in 1883, and troubles were almost immediately encountered. The amount of excavation required was more than expected, the soil conditions were much softer and less stable than anticipated, and the 20-foot difference in tides between the Pacific and the Atlantic was recognized as a major problem. Yellow fever and malaria took a fearful toll among the workers and their families, and the project took on the aspect of a military campaign.

Nevertheless, de Lesseps still expected the canal to be completed by 1888. His engineers were more realistic, and they recommended that a lock canal be built to reduce the need for excavation. De Lesseps, however, refused to break his promise for a sea-level canal. Marshall emphasizes that this particular dispute concerned matters of financial rather than engineering feasibility:

> *The point of their decision was whether a sea level canal could be constructed at a cost and in such time as to make its after operation a profitable business for the shareholders. Time, of course, is a great factor in the cost of an operation involving hundreds of millions. Interest increases at an enormous rate during the later years. Therefore, considerations which would preclude the pursuit of a project solely contemplating commercial results might not be of sufficient weight to deter a government from following the same lines. . . . Even though the operation of the canal should fail to return any interest upon the money invested, the Government might well consider itself fully compensated for the outlay by the political advantages secured, the great savings in the movements of warships, and other desiderata.*[20]

The labor force was nearly 10,000 men by 1887; the standard wage was $1.50 per day for about 20 days per month. The payroll for laborers was therefore on the order of $5 million per year. Another 1500 or so company employees added several more million to the payroll, and costs and transportation

---

[18] Marshall, *Story of the Panama Canal*, 98.

[19] Office of Executive Planning, Panama Canal Commission, Historical Reports—*Panama Canal Traffic: Oceangoing Commerce.*

[20] Marshall, *Story of the Panama Canal*, 104.

of machinery averaged several million per year. The big problem, however, was that the costs were rising and the possibility of ever making a profit was disappearing. By 1889, the company had raised (and spent) $265 million and was looking for more than $100 million more. Interest charges were already $16 million per year, and it was apparent that they would rise to more than $30 million by the time the canal was opened (assuming that it could be completed). Since revenues were projected to be only $18 million, the prospects were nil, and this was evident to everyone with any money to invest. A final effort to raise $160 million in "lottery bonds" that would provide 4% interest plus participation in semimonthly drawings for cash prizes was approved by the government, but only about $60 million was raised. This was the last hope for the French effort, and the Panama Canal Company went into bankruptcy in 1889.

At this point, if the canal project were to be terminated, the company and its shareholders would lose everything. Since the work completed at that time was valued at about $100 million, a major effort was made to reorganize the effort to at least salvage this value. Colombia, which also stood to gain from construction of the canal, was also anxious for work to proceed. Consequently a new company, the New Panama Canal Company, emerged from the chaos of the old and was given an extension until 1904 to complete the canal.

At this time extensive engineering surveys were conducted by a commission established to review the status of the canal. The commission believed that "a lock canal might be completed in eight years at a further cost of $100 million." The New Panama Canal Company therefore studied how such a canal might be built.

The company also saw another way to escape from its problems: sell its concession, equipment, and the completed portions of the canal to the United States, which at that time was pursuing the possibility of digging a canal in Nicaragua. After several years of negotiations and study, the New Panama Canal Company offered to sell everything to the United States for $109 million. The Isthmian Canal Commission, in its report to President Roosevelt, set the value of the property at only $40 million and concluded that the Nicaragua route would actually be the "most practical and feasible route." Since there was only one possible buyer for the property, the New Panama Canal Company, in a panic to salvage something, quickly reduced its asking price to $40 million. In turn, the Isthmian Canal Commission revised its opinion concerning the route on the grounds that Panama was preferable at the lower price.

This deal cleared the way for the United States to take over construction of the canal, although not before a good deal of political theater. Suffice it to say that Colombia attempted to raise the annual fee that it would collect; that the United States balked; and that Panamanian citizens, fearful of losing the canal to Nicaragua, declared the independence of Panama. Historian David McCullough describes the intrigue in absorbing detail; author Joseph Conrad describes the emotions of the times in his great novel *Nostromo* (1904), which was modeled on these events.

6.5   THE U.S. EFFORT

When the United States took over, it had to deal with several major design issues related to the cost, capacity, and performance of the canal:

- Sea-level canal versus a canal with locks
- Number and height of the locks
- Length and the width of the locks
- Height of the canal above sea level and the size of the lake

The United States also had to deal with the tropical illnesses. Fortunately, mosquitoes had been identified as the transmitters of malaria and yellow fever, so it was possible to formulate and implement a strategy for eliminating mosquitoes as a way of controlling disease. That fascinating story is covered by McCullough; suffice it to say that first priority was given to eradicating the mosquito within the Canal Zone, and the diseases were successfully eliminated.

6.5.1 A Sea-Level Canal versus a Canal with Locks

A sea-level canal would have the advantage of lower lock cost and easier operations, but it would require more excavation. Since the tides on the Pacific vary by 20 feet from high to low, a tidal lock would be needed on that end of the canal, even for the sea-level route (otherwise tidal currents would be too strong to safely move large ships through the canal). A lock canal would reduce the excavation costs and reduce the time required to open the canal to operations. Like De Lesseps, most people wanted a sea-level canal, if it were reasonably possible to construct one. President Roosevelt stated the case elegantly, in language that could readily be adapted to the evaluation of the alternatives being considered for any mega-project (see feature box).

I hope that ultimately it will prove feasible to build a sea-level canal. Such a canal would undoubtedly be best in the end, if feasible, and I feel that one of the chief advantages of the Panama Route is that ultimately a sea-level canal will be a possibility. But, while paying heed to the ideal perfectibility of the scheme from an engineer's standpoint, remember the need of having a plan which shall provide for the immediate building of the canal on the safest terms and in the shortest possible time.

If to build a sea-level canal will but slightly increase the risk, then of course, it is preferable. But if to adopt a plan of a sea-level canal means to incur hazard, and to insure indefinite delay, then it is not preferable. If the advantages and disadvantages are closely balanced I expect you to say so.

. . . Two of the prime considerations to be kept steadily in mind are: 1. The utmost practicable speed of construction. 2. Practical certainty that the plan proposed will be feasible; that it can be carried out with the minimum risk.''

President Theodore Roosevelt's instructions to the International Board of Consulting Engineers that was assembled to consider the principal problems in construction of a canal, September 1905 (Marshall 1913, 134).

6.5.2 The Number and Height of Locks

The commission recommended a sea-level canal, but the American members filed a minority report recommending a lock canal that would reach 85 feet above sea level. Congress accepted the minority report, and that was the basis for what was built.

The number and height of locks represents a balance among lock technology, operating costs, and construction costs. The height of the lock chamber and the mitre gates must be several feet higher than the draft of the largest ships (close to 40 feet) plus the height of the lift plus several feet of water under the ship plus a foot or two above water level when full. For an 85-foot lift, this would require two locks with a lift of 42.5 feet each, or three locks with a lift of 27.4 feet each; even the smaller lift would be higher than any other locks yet constructed, and the required chambers and gates would be about 80 feet high. A three-lock system was selected for each end of the canal (Figure 6.1).

6.5.3 The Length and Width of the Locks

The goal for the canal was to handle the largest ships being planned at that time; these were in fact battleships. The original dimensions of the lock chambers were to be 900 feet long by 95 feet wide (by 81 feet deep), but the Navy requested an increase to 1000 feet long by 110 feet to allow for larger ships in the future. A debate ensued about whether to go to the larger size. Colonel Goethals, in charge of canal construction, advanced the case for staying with a 100-foot beam:

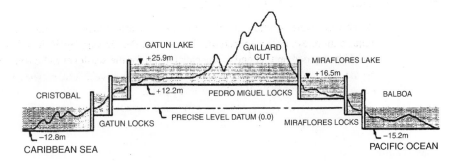

Figure 6.1 A Cross Section of the Panama Canal. Three sets of locks on each side bring ships up to the level of Lake Gatun.

Source: Commission for the Study of Alternatives to the Panama Canal, "Operating Characteristics and Capacity Evaluation Study," May 1993.

The present lock designs provide intermediate gates dividing the locks into lengths of 600 and 400 feet. About 98 per cent of all ships, including the largest battleships now building, can be passed through the 600-foot lengths, and the total lock length will accommodate the largest commercial vessels now building, which, I believe are 1,000 feet long and 88 feet beam. It is true that ships may increase in size so as to make the present locks obsolete, but the largest ships now afloat cannot navigate the Suez Canal, nor the proposed sea level canal at Panama. It must also be remembered that the commerce of the work is carried by the medium-sized vessels, the length of only one of the many ships using the Suez Canal being greater than 600 feet.[21]

Marshall was happy to report that Goethals was overruled and the lock dimensions were set at 1000 by 110 feet:

Our new battleships have a beam of 97 feet and upwards, which will leave a clearance in the lock chambers of less than 13 feet in all, or about 6 feet on either side. Commercial vessels now built, and others whose keels have been laid, have a beam of 96 feet, so that it is quite possible that the locks may prove to be too narrow before they are found too short.

6.5.4 Water Requirements

The height of the canal above sea level and the size of the locks represent a balance between the availability of water and the size of ships that can be handled. Larger locks require more water, but they can handle larger ships. If the canal is higher and the chambers are deeper, then more water is also required.

A canal requires a source of water, such as a large lake, that will be sufficient to operate the locks year-round. If there is a distinct dry season, then the lake ideally would have enough reserve capacity to operate throughout the dry season without affecting operations. Given the size of the lock chambers, a great deal of water is used for each lockage. The basic requirements are (1) the total water lost through lockages must be less than the average annual inflow to the lake, and (2) the surplus water in the system must be sufficient at the beginning of the dry season to last until the rains return.

A height of 85 feet above sea level was selected for Lake Gatun, the 164-square-mile lake created by damming the Chagres River (Figure 6.2). This lake impounded water from a basin of 1320 square miles that received extremely heavy tropical rains from early May through the beginning of December.

---

[21] Quoted in Marshall, *Story of the Panama Canal*, 186.

Figure 6.2 Lake Gatun. A vast supply of water must be available for operating the locks. Lake Gatun was formed by damming the Chagres River, using material excavated from the cut to create the dam. Stumps from the forest that was submerged by the lake are still visible, and divers visit the remains of the old Panama Railroad.

The amount of water was quite considerable. In the 22 years before the opening of the canal, the outflow of the river ranged from a low of 132 billion cubic feet in 1912 to a high of 360 billion cubic feet in 1910; this amount was nearly double the 183 million cubic feet contained by the lake when full. Marshall used the 1912 season to illustrate the adequacy of the lake for supporting the canal operations. If Lake Gatun entered the rainy season on December 1 with an elevation of 87 feet and operated with 48 lockages per day, then the lake level would decline to 79.5 feet by May 7, when the rains returned. At this water level, the cut would have 39 feet of water, which would provide sufficient depth for navigation (for the ships of that time). Allowing for evaporation and seepage, there would still be enough water for 41 lockages a day, which was more than could actually be done because of the time required per vessel. The width of the cut was originally set at 200 feet, but this was not wide enough for two large ships to pass. The width was therefore increased to 300 feet. A wider cut required considerably more excavation, but it also increased canal capacity.

6.5.5   Construction Cost and Pricing Policy

The Panama Canal was the largest project undertaken by the United States until that time. After the United States had been at work for 3 years, it was estimated that the cost would be $375 million, including original payments of $10 million to Panama and $40 million to the French company.[22] This proved to be accurate, since the actual construction cost was $352 million for the U.S. portion of the work (and a total of $639 million for the French and American efforts). The loss of life associated with the construction was staggering. Disease and accidents claimed 5609 lives between 1904 and 1914, but this was far better than the French experience; they lost approximately 20,000 people. If the French companies and their prospective stockholders had understood the financial and human costs, they would never have been able to raise the money to begin. If the U.S. Congress had realized the magnitude of the effort in 1904, it might also have balked.

It is an irony of history that the first vessel went through the canal on August 3, 1914, the day that World War I began. Though the canal was completed 6 months ahead of schedule, it would be 10 years before traffic would grow to the expected levels of 5000 ships per year. Toll revenues reached $27 million in 1929 and 1930, but they did not reach this level again until 1953 due to the effects of economic depression and another world war (Table 6.1).

The U.S. Congress required the Panama Canal to operate on a break-even basis; that is, it had to cover both operating and capital costs from tolls. This prevented the Panama Canal Commission from incurring debt or from achieving exorbitant profits. It also meant that there was no attempt to recover the capital costs of constructing the canal. If the canal had been financed through private sources at 5%, the interest costs during the 10-year construction period would have added well over $100 million to

---

[22] McCullough, *Path Between the Seas*, 610.

Table 6.1   Oceangoing Traffic through the Panama Canal, 1915–1996

Fiscal Year	Transits/Year	Transits/Day	Cargo (Long Tons)	Toll Revenue ($ millions/yr)	Revenue/Ton
1915	1058	2.9	4.9	$ 4	$0.90
1920	2393	6.6	9.4	$ 9	$0.91
1925	4592	12.6	24.0	$ 21	$0.89
1930	6027	16.5	30.0	$ 27	$0.90
1935	5180	14.2	25.3	$ 23	$0.92
1940	5370	14.7	27.3	$ 21	$0.77
1945	1939	5.3	8.6	$ 7	$0.84
1950	5448	14.9	28.9	$ 24	$0.85
1955	7997	21.9	40.7	$ 34	$0.83
1960	10795	29.6	59.3	$ 51	$0.86
1965	11834	32.4	78.6	$ 65	$0.83
1970	13658	37.4	114.3	$ 95	$0.83
1975	13609	37.3	140.1	$142	$1.01
1980	13507	37.0	167.2	$292	$1.75
1985	11515	31.5	138.6	$299	$2.15
1990	11941	32.7	157.1	$354	$2.25
1994	12337	33.8	170.5	$417	$2.44
1995	13459	36.9	190.3	$460	$2.42
1996	13536	37.1	198.1	$483	$2.44

Source: Data prepared by Office of Executive Planning, Panama Canal Commission, May 8, 1997.

the initial cost, and the carrying charges would have been on the order of $25 million annually thereafter. Given the tremendous savings in distance, it is quite possible that tolls could have been raised to cover this cost during good times; but it is also quite likely that the canal would have gone bankrupt during the wars or the depression.

During World War I, the Panama Canal played no strategic military role because the first flotilla of warships to transit the canal was composed of ships returning home after the war. In World War II, the canal played a major role; it allowed rapid deployment of ships from the Atlantic to the Pacific theater of operations.

6.6   THE PANAMA CANAL IN THE TWENTY-FIRST CENTURY

In 1996, the Panama Canal handled 13,536 oceangoing commercial vessels carrying 198 million long tons of cargo and earned revenues of $483 million from tolls. While the average toll per ton remained less than the $3 projected by De Lesseps, the total tonnage and total revenue greatly exceeded his projections of $6 million and $18 million respectively. The cumulative toll revenue from the opening of the canal reached $9 billion by 1997.

6.6.1   Transfer of the Canal to Panama

Operation of the canal was transferred to Panama on December 31, 1999, culminating a 20-year transition period in which responsibility was shifted from the United States to Panama. In anticipation of the transfer, the Panama Canal Commission and the U.S. Army Corps of Engineers conducted a thorough inspection of the canal and locks. In general, the locks and the canal were believed to be in excellent condition, and programs were in place for maintenance and rehabilitation of the major components of the canal.

A greater concern was the capacity of the canal, both in terms of the size of ships that can fit through the locks and the number of ships that can be handled on a sustainable basis. In 1996, the canal handled a record-breaking 37.5 oceangoing ships per day, which caused the Canal Waters Time (the time from arrival at one end of the canal until departure from the other end) to rise from the target level of 24 hours to more than 30 hours. This increase in delay signaled potentially serious capacity problems for the canal.

6.6.2 Post-Panamax Ships

Aircraft carriers and oil tankers were the first ships that exceeded the dimensions of the locks. In the 1980s, a new class of container ships (Post-Panamax) was designed for use in transpacific operations; to reduce the cost per container, these ships were built wider than the 110 feet that the canal could accommodate. Even larger ships were being planned for the future. Since container shipping was one of the fastest-growing areas of commerce, the existence of a large number of Post-Panamax ships was a strategic concern for the canal.

6.6.3 Increasing the Operating Capacity of the Canal

More Resources for Lock Operations

The capacity of the locks is limited by the average time required to move a large ship through one chamber. This time is about an hour, thus suggesting a maximum service rate of two vessels per hour (since there are two parallel channels in each set of locks). Additional time is required to position ships as they arrive at the locks; and since the locks must be closed periodically for routine inspections and maintenance, the sustainable capacity drops to 37–38 vessels per day.

Some efficiency can be gained in lockages by increasing the number and reliability of the specialized railroad locomotives that are used to guide ships through the locks. Several minutes can be lost in repositioning locomotives when several large ships are going through simultaneously. The Panama Canal Commission therefore authorized $90 million to increase the fleet from 82 to 110 locomotives.[23]

Widening the Gaillard Cut

The Gaillard Cut was originally a minimum of 91.5 m wide for its entire 12 km length. Widening of the cut to 152 m, begun in the 1930s and completed by the early 1970s, allowed unrestricted two-way traffic for almost all ships operating at that time. However, by the 1980s, a substantial and growing

Figure 6.3 Pedro Miguel Lock. Note the locomotives that are attached by cables to the container ships. The ship moves through the lock under its own power while the ship's pilot directs the locomotive engineers to tighten or loosen the cables to keep the ship properly aligned. Also notice how little room there is between the ships and the sides of the locks.

---

[23] Spillway Newsletter, "Canal Accelerates Modernization Plan, Improvement Work," Panama Canal Commission, September 20, 1996.

Table 6.2   Panama Canal Traffic (Long Tons), by Commodity Group, 1994–1996

Commodity Group	1996	1995	1994
Grains	42.34	44.07	34.07
Petroleum and petroleum products	32.77	27.48	26.96
Containerized cargo	25.62	24.91	22.44
Nitrates, phosphates, and potash	15.94	15.91	15.44
Coal and coke (excluding petroleum coke)	11.38	11.32	9.34
Ores and metals	11.52	10.76	10.10
Lumber and products, including pulp wood	11.03	10.71	9.47
Chemicals and petroleum chemicals	11.37	10.11	9.71
Manufactures of iron and steel	8.35	9.17	7.85
Canned and refrigerated foods	6.95	6.86	7.00
Minerals, miscellaneous	6.87	5.43	5.79
Other agricultural commodities	5.16	4.92	4.54
Machinery and equipment	1.93	2.14	2.05
All other	6.62	6.52	6.61
Total	198.07	190.30	170.54

Source: Office of Executive Planning, Panama Canal Commission (Report TRA 1-3, November 18, 1996).

number of vessels using the canal were Panamax ships. They were too large and unwieldy, and too valuable, to risk passing in the Gaillard Cut or operating in the cut after dark (Table 6.2). Fleets of these large ships were forced to move single file through this 9-mile stretch during the day. This requirement complicates scheduling and restricts capacity of the canal.

A widening program begun during the mid-1990s was initially scheduled to be completed by 2005. At a cost of $200 million, this program increased the cut to 192 m in straight sections and up to 222 m in curves in order to allow bidirectional operation of Panamax vessels, which would increase the capacity to approximately 42 ships per day. The program was spread out over so many years in order to allow the work to be done largely with the existing workforce and equipment. When the number of Panamax vessels grew rapidly during the mid-1990s, the capacity problem became more critical, and the program was accelerated.[24]

System Control

A $20 million effort to develop a computerized scheduling system enabled higher utilization of the Gaillard Cut by adjusting the sequence of large ships through the system. These management efficiencies increased overall capacity by one ship per day.

Figure 6.4 Widening the Gaillard Cut, 1996. This process required dredging to widen the canal, cutting back the hillsides along the cut, and providing sluices to channel water from heavy rains into the canal. The process could be accelerated by devoting more resources to the task.

---

[24] Panama Canal Commission, *Gaillard Cut Widening Program,* 1996.

Table 6.3 Conceptual Alternatives to the Existing Canal

	Existing Canal (in millions)	High-Rise Locks[a] (in millions)	Low-Rise Locks[b] (in millions)	Sea-Level Locks[c] (in millions)
Vessel size (dead wt. tons)	65	250	250	300
Rise (feet above sea level)	85	90	55	0
Number of lifts	3	2	1	0
Number of lanes	2	2	2	Half with 1; half with 2

[a] This option uses taller gates so that only two lifts would be required; a somewhat deeper lake would necessitate a higher rise. (One variation of this option is to keep the existing locks and add a third set that can handle larger ships, thereby increasing capacity 70%.)

[b] This option requires a new location for the locks, and it needs only a single lift.

[c] This option requires an entirely new sea-level canal (see one possible route in Figure 6.5) that still needs locks because the great difference in tides would result in unacceptably rapid currents through the canal.

Source: Commission for the Study of Alternatives to the Panama Canal, "Operating Characteristics and Capacity Evaluation Study," May 1993.

6.6.4 Expanding the Canal

By the mid-1990s, it was evident that the Panama Canal would have to be expanded to handle the demand projected for the first half of the next century. Even the pessimistic scenarios for growth foresee traffic growing to more than 50 vessels per day by 2050, which is at least 10-20% above what the previous improvements could allow on a sustainable basis.

Several major options were considered, including adding a third set of locks, replacing the existing locks with larger locks, and constructing a new sea-level canal (Table 6.3). A sea-level canal was estimated to cost on the order of $12 billion to construct, and it would require large ships to move single file through the canal. A sea-level canal would require carving out an entirely new route (Figure 6.5). Moreover, parts of the route would be only one lane wide, so that the largest ships would be unable to pass. Hence, the capacity of the new canal would be only half the capacity of the existing canal, and building such a canal would provide only a 50% increase in overall capacity. Adding a new set of locks, even a

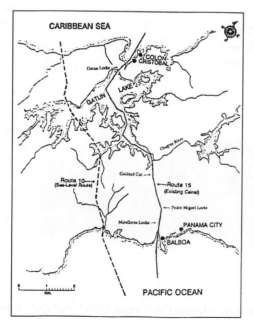

Figure 6.5 Map Showing Existing Canal and Possible Sea-Level Route

Source: Commission for the Study of Alternatives to the Panama Canal, "Operating Characteristics and Capacity Evaluation Study," May 1993.

much larger set of locks, would be less costly and would provide a greater boost in capacity. With the ability to handle larger ships, canal capacity would be increased by 70% at a cost estimated at about $2 billion by the U.S. Army Corps of Engineers in 1995. The third set of locks was therefore selected as the most cost-effective way to increase capacity, and work began on this project after the canal was transferred to Panama.

6.6.5   Water Supply

In 1997–98, the canal experienced the lowest rainfall in its history. With rainfall a third less than normal, it was necessary to restrict operations to ships with a draft of 37 feet compared to the normal restriction of 39.5 feet. This was the first such restriction in 16 years. The water problem is viewed quite seriously because of the prospects for development that will either drain off water for other uses or eliminate wetlands that currently are able to store water during the rainy season.[25]

It appears that the calculations cited above by Marshall were pretty much on the mark. He felt there would be enough water to allow for 41 lockages a day and still maintain 39 feet of water in the cut for operations. Today, with perhaps 37 lockages, a problem has emerged; the solution has been to reduce draft to something closer to what Marshall anticipated (a ship with 39.5 feet draft will need perhaps 44 feet of water in the cut; the estimates cited by Marshall assumed that 39 feet would be sufficient—with larger ships, the limits are reached sooner).

The 1997–98 drought also cast doubts on the feasibility of adding a third set of locks, unless some system is put in place to reuse the water. If the third set of locks is longer, wider, and deeper, as recommended, then the water requirements will be much greater and restrictions during dry periods are much more likely.

ESSAY AND DISCUSSION QUESTIONS

6.1 Discuss what motivated Spanish, French, Panamanian, and U.S. efforts to build roads, railroads, or canals across the isthmus.

6.2 Why did the French effort to construct the Panama Canal fail?

6.3 Discuss the differences in the pricing policies of the Panama Railroad in the nineteenth century and the Panama Canal in the twentieth century.

6.4 Discuss how and why the financial, economic, social, or environmental issues changed over time. How did they affect the design, construction, and operation of transportation between the oceans?

6.5 What are the major issues related to expanding the capacity of the Panama Canal?

6.6 What do you think would happen today if a major infrastructure project were to result in thousands of fatalities in the first year of a planned multi-year construction effort?

6.7 What is the current construction status of the third set of locks for the Panama Canal? What issues does Panama currently face in managing the canal?

---

[25] Thomas T. Vogel Jr., "Drought Leaves Panama Canal in Dire Straits," *Wall Street Journal*, April 21, 1998.

Part II

Comparing Economic and Financial Impacts Over the Life of Proposed Projects

P art I of the text is an overview of many of the processes needed to assess the performance of infrastructure projects and to evaluate alternatives for improving performance. Inevitably, many aspects of performance will be considered, and many possible impacts on society or the environment must be minimized or mitigated. Part I also emphasizes the difference between financial analysis, which is concerned with the cash flows directly related to a project, and economic analysis, which also includes the impacts of a project on the overall economy. Both financial and economic impacts can be measured in monetary terms; which types of impacts are considered will depend upon who is doing the analysis. Owners, developers and users are largely interested in financial matters; public agencies that must approve projects are concerned with broader economic matters, such as job creation and regional prosperity.

Part II shows how engineering economics provides the methodologies needed to compare projects in terms of financial and economic impacts, including the effects of taxes and depreciation. Part III then goes into greater detail about various topics that are relevant to projects and programs aimed at improving infrastructure performance.

STRUCTURE OF PART II

As discussed in Chapter 5, it is possible to discount future financial or economic costs and benefits so that they can be compared to current costs or benefits. Chapter 7 develops the basic relationships that can be used to transform an arbitrary stream of financial or economic benefits into an equivalent present value, an equivalent value at some future time, or an equivalent annuity. These relationships make it possible to compare the financial and economic impacts of multiple alternatives. The alternative with the highest net present value (NPV) will also have the highest future value and produce the largest equivalent annuity. Thus, from a financial or economic perspective, it makes sense to choose the alternative that maximizes net present value.

The equivalence relationships all depend upon the choice of a discount rate, which is far from a simple, objective task. Chapter 8 goes into considerable detail about the factors affecting the choice of a discount rate. It emphasizes the idea that different actors involved in implementing, using, or investing in a project may have different perceptions of the project and therefore may use different discount rates when evaluating a proposal.

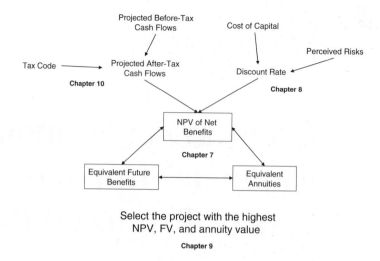

Once the NPV of financial or economic benefits have been calculated, it is straightforward to select the alternative with the highest positive NPV. However, complications may arise if a different measure is used. Companies commonly use the internal rate of return (IRR) when evaluating projects. The IRR is the discount rate that makes the NPV equal zero. If the IRR is greater than a company's discount rate, then that is a valid project. However, if there are mutually exclusive projects, then a small project with a high IRR might appear to be better than a larger project that has a lower IRR, but actually has a higher NPV. Chapter 9 presents methods that show how to deal with this issue. So long as the IRR analysis is applied properly, it will provide the correct ranking of mutually exclusive projects.

Public policy can affect the cash flows of a project in many ways. Zoning restrictions may limit what can and cannot be done on a site. Regulations may determine what kinds of materials or construction techniques can be used. The intricacies of the tax code can be manipulated by lawmakers to promote or hinder certain types of development or certain types of investments. In particular, major investment expenses usually cannot be immediately deducted from taxable income; instead, capital investments are depreciated over many years. Exactly how depreciation is treated in the tax code will determine when this expense is incurred. Since expenses affect profits, and profits result in income tax payments, it is necessary to consider depreciation in order to obtain a valid after-tax analysis of a project. Chapter 10 provides many examples of how depreciation and other "rules of the game" affect project evaluation.

Chapter 7

Equivalence of Cash Flows

What it comes down to is pieces of paper, numbers, internal rate of return, the net present value, discounted cash flows—that's what it's all about. . . . Sure, we want to build quality and we want to build something that is going to be a statement, but if you can't do that and still have it financed and make a return, then why are we doing it?"

Terry Soderberg[1]

CHAPTER CONCEPTS

Section 7.1 Introduction

Section 7.2 Time Value of Money

Reasons for discounting future cash flows: potential for growth, inflation, and risk

Section 7.3 Equivalence Relationships

Present value of an arbitrary stream of cash flows

Future values and annuities that are equivalent to a present value

Notation for describing equivalence relationships

Using tables that show these equivalence relationships

Section 7.4 Continuous Compounding: Nominal versus Effective Interest Rates

Compound interest: nominal and effective interest rates

Continuous equivalence relationships

Section 7.5 Financing Mechanisms

Mortgages and mortgage payments

Bonds: interest rate, life, redemption

How changes in interest rates affect the value of a bond

Section 7.6 The Use of Equivalence in Costing

Converting life-cycle expenses into an annuity (equivalent uniform annual cost) that can be included in creating cost functions and estimating unit costs

Section 7.7 Case Study: Building an Office Tower in Manhattan

How discounting and equivalence relationships enter into financial decisions concerning a major project

Discount rates used to evaluate a project are not the same as interest rates on bonds or mortgages

Different parties to a project may use different discount rates because they have different perceptions of potential risks and returns

---

[1] Terry Soderberg was in charge of leasing a 50-story office tower for the Worldwide Plaza; he is quoted by Karl Sabbagh in *Skyscraper: The Making of a Building* (New York: Penguin Books, 1991), 377.

7.1 INTRODUCTION

This chapter provides the basic methodology needed to compare the economic and financial impacts over the life of proposed projects, which is the theme for Part II of this text. Economic and financial impacts can both be expressed in monetary terms. Financial impacts concern the cash flows that are directly related to the project; these cash flows will determine the profitability of the project and the returns to those who invest in the project. If there is not enough cash to support construction, then a project will not be completed; if it appears that revenues will be insufficient to provide a desirable rate of return for investors, then they will not invest in the project. In Panama, the French effort to build the canal suffered from the extraordinary loss of life; but the project failed because the interest charges on the construction debt rose higher than predicted revenues, and the canal was nowhere near being finished. As discussed in Chapter 3, economic impacts are broader than financial impacts because they may include costs and benefits related to such things as consumer surplus, multiplier effects of construction, or safety—all of which are important factors to the public, even though they may not be show up as revenues or expenses related to the investment in or operation of the project. Financial and economic impacts together are likely to motivate many projects, but social and environmental objectives and constraints also must be considered in proposing and evaluating projects that are aimed at enhancing the sustainability of infrastructure-based systems. The Panama Canal was eventually completed only when the U.S. government, which had strategic interests in a shorter all-water link between its east and west coasts, was able to invest what was needed to complete a less ambitious project—namely, a canal with locks rather than a sea-level canal.

In Part II, the focus is on economic and financial aspects of projects, not because these are the most important measures of success, but because certain well-defined techniques are commonly used to evaluate financial feasibility and economic desirability of a project. The concept of **net present value (NPV)**, which was introduced in Chapter 5, provides a very useful method for determining whether the predicted future benefits of projects justify the investment. NPV also provides a useful way for comparing different alternatives that may be proposed for a project.

When seeking ways to improve infrastructure performance, there will always be many alternatives to consider. As seen in the bus system and coastal management examples in Chapter 5, it may be possible to improve performance by investing in markedly different types of infrastructure, by regulating land use or development, or by subsidizing certain types of activities. To determine which is best, from an economic or financial perspective, the costs and benefits must be compared over a long time period. Calculating the NPV for each option provides a convenient way to make such a comparison.

To simplify the presentation, let's focus on financial matters.[2] For each major alternative, cash flows need to be predicted over a long time horizon, taking into account the costs of construction, the continuing costs of maintenance, and the costs and revenues related to operations. A typical proposal will have cash flows similar to those shown in Figure 7.1, which shows the net annual cash flows over the life of a hypothetical project. Net cash flows are the sum of revenues, subsidies, and any other source of income minus investment, operating, maintenance, and any other type of expense. In a typical proposal, cash flows are negative at the outset because of the expenses related to planning, site acquisition, and construction. Once the project is completed, revenues start to offset continuing costs of operation and maintenance. Eventually, as the structures age or as competitors capture more of the market, net cash flows begin to decline. At the end of the life of a project, there may be expenses related to tearing down the structure.

The alternatives that must be investigated may have sizable differences in terms of investments, construction costs, performance capabilities, and projected operating costs and revenue potential.

---

[2] Discounting and NPV analysis can be applied to any stream of costs and benefits that can be expressed in monetary terms. It is convenient to begin by focusing on the cash flows that are directly related to the project since the cash flows are well defined and easily understood. Moreover, for most investors, private companies, and entrepreneurs, the analysis of cash flows dominates their concerns.

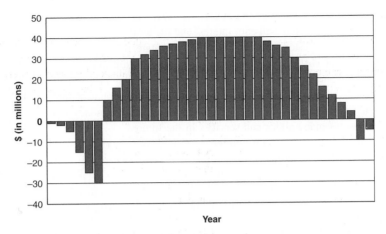

Figure 7.1 Cash Flows for a Typical Infrastructure Project

Comparisons among options with markedly different cash flows will be difficult. For example, how should a low-cost option be compared with an option that requires much higher investment but offers a chance to earn more money over a longer time frame? It is therefore desirable to convert these complex cash flows into something equivalent that is more easily understood and that will make it possible to rank the alternatives in terms of financial performance.

This chapter introduces the key concept of **equivalence**. If someone—an individual, a public agency, or a company—is indifferent between two projected streams of cash flows, then those cash flows can be viewed as equivalent for that person, agency, or company. It is particularly useful to be able to take the complex flows of a typical project and compare them to something that is equivalent, but easier to understand. One obvious possibility is to determine the amount of cash that would be equivalent to each projected stream of cash flows. The projected cash flows for each alternative can then be reduced to something as simple as a deposit to or a withdrawal from a bank account, either today or at some future time. Comparing alternatives is then trivial, at least in terms of financial matters: the bigger the deposit the better; any deposit is better than any withdrawal; and if the best option is equivalent to a withdrawal, then clearly there had better be some nonfinancial objective for pursuing the project! This is why net present value is such a widely used measure of financial performance.

Section 7.2 goes into more detail concerning the time value of money, the need to discount future cash flows, the concept of a discount rate, and the concept of equivalence. Equivalence relationships can most easily be understood in the context of fixed interest payments, where there is a well-defined relationship between money invested today and the interest to be earned over time. Section 7.3 shows how present or future value can also be equivalent to annuities (i.e., to a series of equal payments received at regular intervals for a specified period of time). Interest can be credited at discrete intervals or continuously, and equivalence relationships can be calculated for continuous compounding of interest, as shown in Section 7.4. Section 7.5 provides some examples of equivalence that are commonly encountered in project evaluation, including financing mechanisms such as bonds and mortgages. Section 7.6 shows how equivalence relationships can be used to develop cost functions in situations where investment costs must considered along with operating costs. Section 7.7 illustrates how equivalence relationships affected the evaluation and financing of the Worldwide Plaza, a major urban redevelopment project completed in Manhattan in 1989.

7.2 TIME VALUE OF MONEY

Cash today is worth more than a promise that you will receive the same amount of cash in the future. This statement is true for several reasons, including the opportunities for investing the cash today, the likelihood of inflation, and the risk that the promised cash will not materialize. Let's consider each of these in turn.

If the money is invested in a low-risk investment, such as savings bonds or a savings account at a bank, then the money will earn interest and the total amount available will be greater in the future. If the money is invested in stocks, bonds, or real estate, the returns could be even greater. Thus, there is an **opportunity cost** if money is available at some future time rather than being available today.

The second major reason for preferring money today rather than in the future is that **inflation** will generally reduce what can be bought with a given amount of money. Having the same amount of money in the future will not be as good as having the money today, because that money will not purchase as many goods and services in the future.

The third major reason for preferring money today is **risk**: something could go wrong, causing the future payment to be smaller or later than expected. If the money is coming from the anticipated sale of property, the return could be less than expected if the housing market declines. If the money is coming from the repayment of a loan, perhaps the borrower will be unable to make the payment. If the money is linked to some sort of international deal, perhaps a change in government will reduce the revenue from the deal.

These and other factors all affect the time value of money. Since people have different needs and expectations about the future, they have varying perceptions of investment opportunities, inflation, and risk; and different people and different organizations put different relative values on current and future sums of money. For this chapter, suffice it to say that there are several major reasons why money in the future is less valuable than money at hand in the present. Therefore, it is necessary to **discount** future cash flows, as was discussed earlier in Section 5.3. *Discount* comes from a Latin term meaning "count for less," so discounting future cash flows means that they will count for less when evaluating a project. The **discount rate** is defined as the annual percentage by which future cash flows (a **future value**) must be reduced (discounted) to a **present value** for comparison with current cash.

The simplest way to understand discounting is to consider the benefits of investing in something safe that earns a respectable, steady, i% interest per year. After one year, the money will have increased by a factor of $(1 + i\%)$. If the same interest rate is maintained for t years, the money will have grown by a factor of $(1 + i\%)^t$. In other words, assuming an interest rate of i%, M dollars today is equivalent to $M(1 + i\%)^t$ dollars in the future. To look at this same situation from the perspective of the future, M dollars in the future would be equivalent to $M/(1 + i\%)^t$ dollars today if those M dollars were invested and earning i% per year. Examples 7.1 and 7.2 show how these calculations would relate to money invested in a savings account in order to have funds available for a future purchase.

EXAMPLE 7.1 Future Values of Money Deposited in a Savings Account

Suppose you deposit $1,000 in a bank that pays 4% interest at the end of each year. How much will you be able to withdraw at the end of 5 years? To determine the future value of your deposit, it is necessary to begin by calculating the annual interest that will be received at the end of the first year and adding this interest to your account. If the interest rate is 4%, then the value of the account will increase by 4% at the end of the year. The same procedure can be repeated for four more years. The results will be as shown in Table 7.1; the value at the end of 1 year equals the value at the beginning of the next year, and the value at the end of 5 years will be $1,216.65. This result could also be obtained directly as $1,000 * $(1.04)^5$ = $1,216.65.

Table 7.1 Future Value of Money Deposited in a Bank Account

Year	Value at Beginning of Year	Interest Rate	Interest Received at End of Year	Value at End of Year
1	$1,000	4%	$40	$1,040.00
2	$1,040	4%	$41.60	$1,081.60
3	$1,081.60	4%	$43.26	$1,124.86
4	$1,124.86	4%	$44.99	$1,169.86
5	$1,169.86	4%	$46.79	$1,216.65

Figure 7.2 shows the cash flows associated with this example. In this diagram, it is assumed that cash flows occur at the beginning of the period. From your perspective, there are just two cash transactions: a deposit of $1,000 is made today, and a withdrawal of $1,216.65 will be made 5 years later. The annual interest payments will be added to your account and will not be taken as cash; they therefore are not included in the cash flows shown on this chart. The chart (a bar chart prepared in Excel) shows the deposit at the beginning of month 1 and the withdrawal 5 years later at the beginning of month 61.

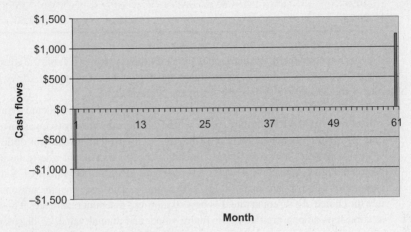

Figure 7.2 Cash Flows for Table 7.1

Spreadsheets are very useful in charting cash flows because it is possible to create bar charts, like the one in Figure 7.2, that are linked to cash flows that are calculated and presented in tables. The charts can be updated automatically by the program whenever the values in the tables are changed.

EXAMPLE 7.2   Deposit Required Today in Order to Have a Desired Future Value

Suppose that you hope to have $5,000 available by the end of 5 years so you can buy a used car when you finish grad school. How much would you have to deposit today in order to have $5,000 in the account at the end of 5 years? To answer this, you have to work backward from the end. You will need to have an amount deposited at the beginning of year 5 that will earn interest to bring the total to $5,000 by the end of year 5. Since the interest rate is 4%, the amount needed at the beginning of the year will be $5,000/1.04 = $4,807.69. This calculation can be repeated to determine what amount is needed at the beginning of year 4: $4,807.69/1.04 = $4,622.78. Continuing through years 3, 2, and 1 indicates that you should deposit $4,109.63 at the beginning of year 1. You can obtain the same result by calculating $5,000/(1.04)^5 = $5,000/1.21665 = $4,109.65 (the tiny two-cent discrepancy is caused by rounding errors in the calculations). The cash flow diagram would be similar to Figure 7.2 because there would be a deposit in month 1 and a withdrawal in month 61.

Table 7.2   Amount to Deposit in Order to Meet Future Goal

Year	Value at End of Year	Discount Factor Rate	Value at Beginning of Year	Interest Received at End of Year
5	$5,000	1.04	$4,807.69	$192.31
4	$4,807.69	1.04	$4,622.78	$184.91
3	$4,622.78	1.04	$4,444.98	$177.80
2	$4,444.98	1.04	$4,274.02	$170.96
1	$4,274.02	1.04	$4,109.63	$164.39

The choice of a discount rate is a crucial issue in project evaluation, where large investments in the near future must be justified by benefits that are achieved over a long time period. The more those

Table 7.3   Present Value of $100 Received at Time t

Discount Rate	5 Years	10 Years	20 Years	50 Years
1%	$95	$91	$82	$61
5%	78	61	38	8.80
10%	62	38	15	0.90
20%	40	16	2.60	0.11

benefits are discounted, the harder it is to justify the investment. Table 7.3 illustrates the present value of $100 received 5–100 years in the future and discounted at a rate of 1–20%. The higher the discount rate or the longer the period, the lower the present value.

The discount rate for an entrepreneur or an investor may be considered their **minimum acceptable rate of return (MARR)**. If they are presented with an opportunity to invest in a project, they will do so only if that project appears attractive relative to their other investment options.

Public agencies and companies commonly specify a discount rate to be used in evaluating their projects. Public agencies typically use discount rates in the range of 5–10%, while companies commonly use discount rates of 15–20% or more. The next chapter presents the factors that enter into the choice of a discount rate. For now, assuming a discount rate makes it possible to develop precise relationships among present value, future value, and annual value, as discussed in the next section.

7.3   EQUIVALENCE RELATIONSHIPS

Equivalence is a key concept in project evaluation and project finance. Two projected streams of cash flows are equivalent for someone if they are equally acceptable (i.e., if the individual is indifferent to receiving one or the other). As noted in the introduction, our goal is to transform an arbitrary stream of cash flows into an equivalent cash flow that is easily understood, such as net present value.

7.3.1   Cash Flow Diagrams

The first step in many financial analyses is to prepare a **cash flow diagram** that depicts the net cash flow for each period by a bar or an arrow at the end of the period. Any expenditures made at the outset of the project are therefore shown as occurring at time 0, which can be interpreted as being the end of period 0 and the beginning of period 1. Even though expenses may be incurred and income received throughout the period, the net cash flows are conventionally shown at the end of the period. If this assumption appears to be unrealistic, then a shorter time period can be used. For example, if construction and maintenance expenses are much greater in the summer than in the winter, then it would be better to use monthly rather than annual time periods in structuring the analysis. Figure 7.2 above is a very simple cash flow diagram. Example 7.3 shows how to create a more complex cash flow diagram.

EXAMPLE 7.3   Cash Flow Diagram for Construction of an Apartment Building

Create a preliminary cash flow diagram for the following situation: a real estate company plans to construct a new apartment complex on land that it has just acquired next to a new transit station. The company plans to complete the first phase of construction within a year or so, which means that it will begin to receive rents in year two. The project is planned in two phases, and the first phase is expected to be fully rented by year 5. At that point, the company expects to begin planning for additional units, which will be constructed during year 9. The plan is to sell the entire apartment complex in year 12.

A preliminary cash flow diagram is shown in Figure 7.3. Arrows that point up indicate income; arrows that point down indicate expenses. The land purchase, major expense at the outset, is shown as a downward-pointing arrow in year 0. Construction generates major expenses in year 1; but after that, rents begin to come in and cash flows will be positive.

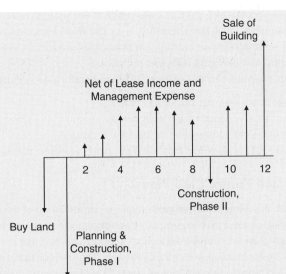

Figure 7.3 Cash Flow Diagram for the Two-Phase Construction of an Apartment Complex

In years 8 and 9, expenses related to Phase II construction offset rental incomes, but net incomes recover once Phase II is completed. Finally, the sale in year 12 provides significant positive cash flow.

Even without specifying the exact amounts of money involved, this diagram illustrates the general nature of the cash flows. As the company proceeds with its planning, it will be able to estimate reasonable values for these cash flows. Once that is done, it will be able to calculate the net present value of the project.

Given a cash flow diagram, it is possible to find the net present value of a project as well as other equivalent measures, including the future value and the equivalent uniform annual value over a specified period of time (Figure 7.4). Present value, future value, and equivalent uniform annual value may also be referred to as present worth, future worth, or annuity worth.[3] Given a discount rate, it is possible to calculate these three measures so that each is equivalent to a given stream of cash flows:

1. *Present value or present worth (P)*: The equivalent present value of the cash flows (what is the projected stream of cash flows worth today?)
2. *Future value or future worth (F)*: The equivalent future value of the cash flows (what is it worth at a specified time in the future?)
3. *Annual value or annuity worth (A)*: The equivalent annuity amount (what is it worth in terms of receiving a uniform cash flow of A at the end of each period for a specified number of periods?)

P, the present value of the cash flows, is clearly the easiest to understand. If we have a choice among various alternatives, each of which is equivalent to receiving a lump sum of money today, then the larger the present value the better (assuming for now that money is our chief object in life and that the money related to the various projects is indeed legally acquired!).

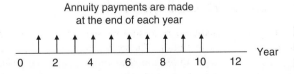

Figure 7.4 Example of an Annuity. Equal annual payments are made at the end of the year, every year for 10 years.

---

[3] *Present value, future value,* and *equivalent uniform annual value* are the terms most commonly used in general business. Present worth (PW), future worth (FW), and annuity worth (AW) are encountered in many engineering economics textbooks. This text uses the two sets interchangeably; the very common use of net present value may make the use of present and future value more desirable.

The other two measures are also easy to understand. If our time of reference is some time in the future, we will presumably want the alternative with the maximum future value. If we are more comfortable dealing with monthly or annual cash flows, then we can express options in terms of an annuity and choose the one that pays the most per period.

The basic tasks in financial analysis of a project can therefore be summarized as follows:

1. Predict the cash flows over the life of the project.
2. Estimate the net present value of the cash flows.
3. Calculate the equivalent future value or annuity, if desired.
4. Rank projects by P, F, or A (since they are equivalent measures, the ranking will be the same).

7.3.2 Equivalence between Present and Future Cash Flows

To convert future cash flows into a present value requires consideration of the time value of money. To begin, let's assume that we are a private company considering various projects that we can finance by borrowing from a bank at an interest rate of i%. For simplicity, assume that we can also put money into a savings account at the bank and also receive i% interest. Our choices therefore are either to invest in more projects or to put money into our bank account. If a project earns more than i%, then we will invest; if it earns less than i%, we'd be better off putting money in the bank.

Let's begin with the simplest possible cash flows: the project requires an investment of P at time 0 and generates a lump-sum return F at the end of period N. Is it better to invest in this project or to simply put the money in the bank? If the money is deposited in the bank, it will increase by a factor of $(1 + i)$ per period, as illustrated above in Examples 7.1 and 7.2. By the end of period N, our bank account will have grown to the following amount:

$$\text{Bank Account} = P(1+i)^N \qquad \text{(Eq. 7.1)}$$

Thus, we will prefer to invest if the expected future amount F is greater than this, and we will prefer to put the money in the bank if F is projected to be less than this. To compare values for time 0 and for time N, we need to use the **discount factor** $(1 + i)^N$ to discount the future value. The process of discounting, therefore, is the process of reducing the current value of future cash flows to take into account the time value of money and the uncertainties that may exist between today and tomorrow.

It is a small step from discounting one future payment to discounting all anticipated cash flows over a project's time horizon. The present value of the entire stream of cash flows is the sum of the individual discounted cash flows. It is a major accomplishment to be able to convert any future stream of cash flows to a single present value.

It is useful to introduce some notation for the discount factors. [F/P,i,N] can be used to denote the factor that calculates the future value F as a fraction of the present value P, assuming a discount rate of i% for N periods. [P/F,i,N] denotes the factor that is used to determine the present value P given the future value F, again assuming a discount rate of i% for N periods. The two factors are used as follows to compare future and present values:

$$F(N) = P * [F/P, i, N] = P * (1+i)^N \qquad \text{(Eq. 7.2)}$$

$$P = F * [P/F, i, N] = F/(1+i)^N \qquad \text{(Eq. 7.3)}$$

EXAMPLE 7.4 Notation for Discounting

Example 7.1 showed that $1,000 invested at 4% would grow to $1,216.65 by the end of 5 years, because $1,000 * (1.04)^5 = $1,000 * 1.21665 = $1,216.65. In this case, the factor $(1.04)^5$ can be symbolized as [F/P,4%,5]. This is the factor that will calculate the future value F, given the present value P, the discount rate of 4%, and the time period of 5 years. The inverse of this factor is $1/(1.04)^5$, and this value can be symbolized as [P/F,4%,5].

If you have a spreadsheet, you can use these equations repeatedly to convert an arbitrary cash flow into either a present or a future value. As we have seen previously, the term *net present value* is commonly used to denote the present value of a stream of costs and benefits over a designated time horizon. Given:

C (t) = Costs during period t
B (t) = Benefits during period t
Discount rate = i

Then it is straightforward to calculate the net benefits during any period, the present value of those benefits, and the net present value of the entire project:

$$\text{Net benefits during period } t = B(t) - C(t) \qquad \text{(Eq. 7.4)}$$

$$\text{Present value of net benefits during period } t = NPV(t) = (B(t) - C(t))/(1+i)^t \qquad \text{(Eq. 7.5)}$$

$$NPV(\text{project}) = \Sigma((B(t) - C(t))/(1+i)^t) \text{ for the life of the project} \qquad \text{(Eq. 7.6)}$$

Sometimes it is desirable to consider an annuity rather than a present or future value. An annuity can be compared to other measures reported annually, such as revenue or profitability. There are multiple ways to find the equivalent annuity. Since an annuity of A per period is certainly one possible cash flow, you can find the annuity that is equivalent to either a present or a future value using the above equations. If you make interest rate, annuity amount, and the number of periods a variable, you can easily find the annuity amount that is equivalent to any present or future value. However, it can be more elegant (and less time-consuming) to use algebraic expressions to convert P or F into annuities (or to convert annuities into P or F), as shown in the following section.

7.3.3 Calculating the Future Value of an Annuity [F/A,i,N]

The equivalent future value F of an annuity A is the amount that would be accumulated by the end of the last payment, assuming that all payments were invested so as to grow at a rate equal to the discount rate i. As was the case in calculating the relationships between P and F, there will be an equivalence factor that can be used to find F as a function of A. This factor depends upon the discount rate per period and the number of periods. It is denoted [F/A,i,N], and it is called the **uniform series compound amount factor**. It is assumed that a payment of A is made at the *end* of each period and that each payment is invested at i% per period for the remaining periods. Given i and N, the future value will be proportional to A, and [F/A,i,N] will be the proportionality factor:

$$F(N) = A * [F/A, i, N] \qquad \text{(Eq. 7.7)}$$

There is a simple algebraic expression for [F/A,i,N]:

$$[F/A, i, N] = [(1+i)^N - 1]/i \qquad \text{(Eq. 7.8)}$$

Values for the expression can be found in tables, and spreadsheets have functions that will calculate the expression for you.

Derivation of the Uniform Series Compound Amount Factor [F/A,i,N]

The derivation of this expression for [F/A,i,N] is based upon well-known relationships for geometric sequences. First, we can calculate F just by converting each payment A(t) to a value at the end of the Nth period. The payment at the end of period t will earn interest for another N – t periods, so its future value will be

$$A(t) = A(1+i)^{N-t} \qquad \text{(Eq. 7.9)}$$

Summing over the entire N periods:

$$F(N) = A((1+i)^{N-1} + (1+i)^{N-2} + \cdots + (1+i)^{N-t} + \cdots + (1+i)^0) \qquad \text{(Eq. 7.10)}$$

If we let $b = 1 + i$, this is equivalent to a simple geometric sequence:

$$F(N) = A(b^{N-1} + b^{N-2} + \cdots + b^{N-t} + \cdots + b^0) \qquad \text{(Eq. 7.11)}$$

If we rearrange the terms, then

$$F(N) = A(1 + b + \cdots + b^{N-t} + \cdots + b^{N-1}) = A\,[F/A, i, N] \qquad \text{(Eq. 7.12)}$$

and

$$[F/A, i, N] = (1 + b + \cdots + b^{N-t} + \cdots + b^{N-1}) \qquad \text{(Eq. 7.13)}$$

Now use a mathematical trick: multiply this by one, expressed as $(1 - b)/(1 - b)$, to get a more elegant result:

$$[F/A, i, N] = [1/(1-b)][(1 + b + \cdots + b^{N-t} + \cdots + b^{N-1}) \qquad \text{(Eq. 7.14)}$$
$$- (b + b^2 + \cdots + b^{N-t+1} + \cdots + b^N)]$$

$$[F/A, I, N] = [1/(1-b)][1 - b^N] = [1 - b^N]/(1 - b) \qquad \text{(Eq. 7.15)}$$

Substitute $(1 + i) = b$ and rearrange terms to get the uniform series compound amount factor:

$$[F/A, i, N] = [(1+i)^N - 1]/i \qquad \text{(Eq. 7.16)}$$

7.3.4   Expressions for the Other Equivalence Factors

Expressions for the other factors can readily be found. Since $F(N) = A * [F/A,i,N]$, we can invert Eq. 7.16 to get $[A/F,i,N]$, which is known as the **sinking fund factor**:

$$[A/F, i, N] = i/[(1+i)^N - 1] \qquad \text{(Eq. 7.17)}$$

A sinking fund may be established by a local government or a company as a means of paying off a future debt. An amount is paid each year into the sinking fund, which could be a savings account or another safe investment. Each year, the sinking fund would earn interest, and at the end of the time period, enough would have accumulated to pay off the debt (or fix the roof on the town hall or deal with whatever problem the sinking fund was established to solve). The relevant question is the size of the annuity.

A related question involves the present worth of an annuity. This amount can be calculated using what is known as the **uniform series present worth factor** and symbolized as [P/A,i,N]. To calculate this factor, recall that the equivalent future value of the annuity is represented by [F/A,i,N] (see Eq. 7.16). Taking the present value of that future value will produce the desired uniform series present worth factor:

$$[P/A, i, N] = [F/A, i, N]/(1+i)^N \qquad \text{(Eq. 7.18)}$$

The equation for the uniform series present worth factor can now be obtained by substituting for [F/A,i,N] using Eq. 7.16:

$$[P/A, i, N] = [(1+i)^N - 1]/[i(1+i)^N] \qquad \text{(Eq. 7.19)}$$

The inverse of this expression will give the **capital recovery factor**, which can be used to determine the size of an annuity that is required to recover an initial capital investment:

$$[A/P, i, N] = [i(1+i)^N]/[(1+i)^N - 1] \qquad \text{(Eq. 7.20)}$$

These expressions may not seem too elegant, nor are they easy to remember. However, note that when N gets large, they become very clear and simple:

$$[P/A, i, N] = 1/i \qquad \text{(Eq. 7.21)}$$

$$[A/P, i, N] = i \qquad \text{(Eq. 7.22)}$$

These relationships are useful to keep in mind because they allow easy, approximate conversions between annuities and present value.

EXAMPLE 7.5   Estimating Relationships between Present Values and Annuities

The approximate expressions for the sinking fund and the capital recovery factors can be very useful in estimating the present value of a long-term annuity or in estimating the present value of a long-term annuity. For example, suppose a toll road generates \$1 million per year in profit. What would that be worth to a potential purchaser with a discount rate of 10%? We could approach this by assuming a life of 30, 40, or 100 years and looking up the values for [P/A,10%,30], [P/A,10%,40] and [P/A,10%,100]. If we did this, we would find the following:

$$(\$1 \text{ million}) [P/A, 10\%, 30] = \$1 \text{ million} (9.4269) = \$9.4 \text{ million}$$
$$(\$1 \text{ million}) [P/A, 10\%, 40] = \$1 \text{ million} (9.7791) = \$9.8 \text{ million}$$
$$(\$1 \text{ million}) [P/A, 10\%, 100] = \$1 \text{ million} (9.9993) = \$10 \text{ million}$$

If we had just used the approximation, we would immediately have said that

$$\$1 \text{ million} (1/0.1) = \$10 \text{ million}$$

This result is very close for 40 years and almost exact for 100 years. In many analyses, particularly preliminary analyses where most of the numbers are imprecise, the approximation will be quite adequate.

In going from present value to annuities, we find a similar result. In this case, the question concerns the annual profit that would be required to justify an investment of \$10 million in a turnpike. The approximation says that the long-term annuity would be approximately \$1 million multiplied by the discount rate of 10% or \$1 million per year. The more precise calculations would call for somewhat higher returns, but nothing markedly greater than the quick estimate of \$1 million:

$$\$10 \text{ million} * [A/P, 10\%, 30] = \$10 \text{ million} (0.1061) = \$1.06 \text{ million}$$
$$\$10 \text{ million} * [A/P, 10\%, 40] = \$10 \text{ million} (0.1023) = \$1.02 \text{ million}$$

Using these approximations is sometimes called the **capital worth method**.

Figure 7.5 summarizes the concept of equivalence. The chart at the upper left shows a typical stream of cash flows for a project. There are substantial investments during the first 4 years, profitable operations beginning in year 5, a dip in earnings midway through the life of the project reflecting the need to expand or rehabilitate the project, and ultimately a decline in profitability and a decommissioning expense. It is impossible to tell for sure how good this project is simply by looking at it. The other three charts show the equivalent present value, annuity, and future value, each of which is easy to understand. If one of these values is positive, all of them will be positive. If the present value is positive, then investing in this project is better than investing in something that earns interest equal to the discount rate. If the present value is negative, then the project is not as good as investing in something that earns interest equal to the discount rate.

Table 7.4 summarizes the six factors derived above. F refers to the future value, P to the present value, and A to the equivalent annuity amount. The equations are all functions of the discount rate i and the number of periods N. The final two rows of this table highlight two easily remembered factors that can be used for situations where N is large. Sometimes referred to as the capital worth method, these factors provide an easy way to get a quick estimate of the value of an annuity (A/i) or the annuity that is equivalent to any present amount (Pi), as was illustrated in Example 7.4.

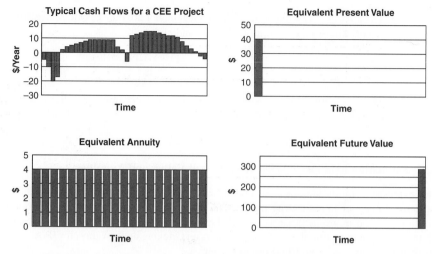

Figure 7.5 Converting Complex Cash Flows into Equivalent, More Readily Understood Cash Flows. For a given discount rate, any arbitrary stream of cash flows will be equivalent to a present value, a future value, or an annuity.

The next two figures illustrate the relationship between annuities and present or future worth. Figure 7.6 shows the uniform series, compound amount factor (the amount by which the future value exceeds the amount of an annuity depending upon the length of time and the earnings rate). This factor is useful in determining how much should be invested each year toward retirement. Figure 7.7 shows the annuity that is equivalent to a present value for various time periods and interest rates. This factor is called the uniform series, capital recovery factor, and an example would be the amount that is paid each year on a mortgage.

Table 7.4   Summary of Equivalence Factors, Discrete Compounding

Symbol	Name	Comment	Value
[F/P,I,N]	Future value given present value	How much growth can be expected	$(1 + i)^N$
[P/F,I,N]	Present value given future value	Discounted value of a future amount	$1/(1 + i)^N$
[F/A,I,N]	Uniform series compound amount factor	If I save some each period, how much will I accumulate?	$[(1 + i)^N - 1]/i$
[A/F,I,N]	Sinking fund payment	How much must I save each period to meet my retirement goals?	$i/[(1 + i)^N - 1]$
[A/P,I,N]	Capital recovery factor	What will my mortgage payment be?	$[i(1 + i)^N]/[(1 + i)^N - 1]$
[P/A,I,N]	Uniform series present worth factor	If I can pay A per month, how large a mortgage can I afford?	$[(1 + i)^N - 1]/[i(1 + i)^N]$
[A/P,I,infinity]	Capital recovery factor for very long time periods	Capital worth method	i
[P/A,I,infinity]	Uniform series present worth factor for very long time periods	Capital worth method	$1/i$

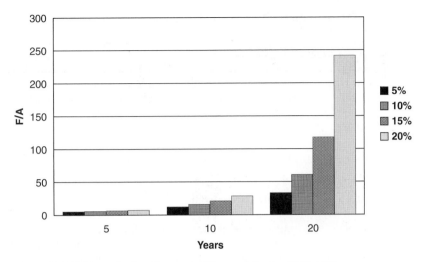

Figure 7.6 Uniform Series, Compound Amount Factor [F/A,i,N]

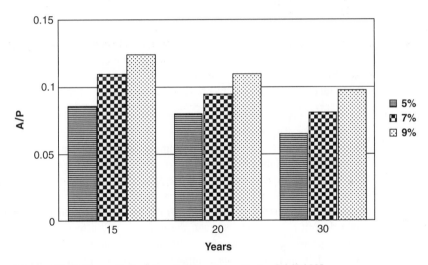

Figure 7.7 Uniform Series, Capital Recovery Factor [A/P,i,N]

7.4 CONTINUOUS COMPOUNDING: NOMINAL VERSUS EFFECTIVE INTEREST RATES

Interest rates are normally expressed in terms of annual returns, and the **nominal rate of interest** is the interest rate you would receive if interest were compounded annually. If interest is compounded more frequently, then there will be a higher **effective rate of interest**. With a 12% nominal rate, an investment of $1,000 on January 1 would earn $120 interest on December 31. However, if you compound interest semiannually, then you would get more interest. First, after 6 months, you would earn 6% or $60 on your initial investment. Then, at the end of the year, you would earn another $60 on the initial investment plus $3.60 on the interest you received at the end of June. For the year, you therefore earned an effective rate of 12.36% interest on your investment, even though the nominal rate was 12%.

More rapid compounding gives a bit higher effective rate, but there are diminishing returns:

- Quarterly rate is 12.55%.
- Bimonthly rate is 12.62%.

Table 7.5   Summary of Equivalence Factors, Continuous Compounding

Symbol	Name	Comment	Value
[F/P,I,N]	Future value given present value	How much growth can be expected	e^{rN}
[P/F,I,N]	Present value given future value	Discounted value of a future amount	$1/e^{rN}$
[F/A,I,N]	Uniform series compound amount factor	If I save some each period, how much will I accumulate?	$[e^{rN} - 1]/r$
[A/F,I,N]	Sinking fund payment	How much must I save each period to meet my retirement goals?	$r/[e^{rN} - 1]$
[A/P,I,N]	Capital recovery factor	What will my mortgage payment be?	$[r(e^{rN})]/[e^{rN} - 1]$
[P/A,I,N]	Uniform series present worth factor	If I can pay A per month, how large a mortgage can I afford?	$[e^{rN} - 1]/[r(e^{rN})]$
[A/P,I,infinity]	Capital recovery factor for very long time periods	Capital worth method	r
[P/A,I,infinity]	Uniform series present worth factor for very long time periods	Capital worth method	$1/r$

- Monthly rate is 12.68%.
- Daily rate is 12.75%.

Clearly, this effect is approaching a limit as interest is compounded more frequently. It turns out that all of the formulas for discrete cash flows can be revised into expressions for continuous compounding by substituting e^{rN} for $(1 + i)^N$, where r is the nominal rate of interest and i is the effective rate (Table 7.5). For example, if the nominal rate is 12%, as in the example we have been discussing, then $e^{rN} \sim 2.718^{0.12} = 1.1275$, which equals 1 plus the effective daily rate of 12.75% that we just calculated.

The nuances of nominal versus effective rates are important in certain areas, notably in banking and finance, where contracts will specify interest rates and compounding periods. If you are buying a car or making a deposit to a savings account, these details will affect the size of your payments or your earnings. In project evaluation—especially during the early stages of the process—the differences between discrete and continuous compounding will be minor compared to the uncertainties in estimating costs, revenues, time periods, and other factors that affect the outcome of the analysis.

Some Useful Approximations

The continuous compounding formulations for [F/P,r,N] and [P/F,r,N] are very useful because they are easy to remember and can readily be used for quick approximations. Since [F/P,r,N] $= e^{rN}$ and [P/F,r,N] $= 1/e^{rN}$, it can be convenient to remember a few useful results:

- If $rN = 1$, $e^{rN} = 2.718\ldots$
- If $rN = 0.7$, $e^{rN} = 2.013\ldots$ (approximately 2)
- If $rN = 1.1$, $e^{rN} = 3.004\ldots$ (approximately 3)
- If $rN = 1.4$, $e^{rN} = 4.055\ldots$ (approximately 4)

You can use these relationships to figure out how long it will take to double, triple, or quadruple your money. If $rN = 2$, then money invested at r% for N years will double in value. If the nominal interest rate is 10% per year, then it will take 7 years to double your money; if the nominal rate is 5%, then it will take 14 years to double your money. Likewise, if $rN = 1.1$, then $e^{rN} \sim 3$ and the future value will be about three times the present value. Thus, if you earn 10% per year for 11 years, you can triple your money.

You can also use the inverse relationship to calculate present values. For example, if the nominal interest rate is 7%, then the present value of something received in 10 years will be half its future value.

In short, it is possible to make mental estimates of quite complex functions made up of incomprehensible expressions such as $(1 + i)^n$. Mental math has been an increasingly undervalued skill since the invention of the electronic calculator, but you will certainly find the above relationships useful if you are ever involved in face-to-face discussions, negotiations, or debates related to project evaluation when it would be inconvenient or inappropriate to use your calculator or computer. You *can* do present value analysis in your head—and that *will* give you an advantage in negotiation!

Proof that $e^{rN} = (1 + i)^N$

To prove this relationship, we can begin by writing an algebraic expression for what we are doing as we compound more frequently. If r is the nominal interest rate, but we compound M times per year, then the effective rate will be

$$i = [\, 1 + (r/M)\,]^M - 1 \qquad \text{(Eq. 7.23)}$$

and the factor $[\,F/P,r\%,1\,]$ will be

$$[\,F/P,r\%,1\,] = [\, 1 + (r/M)\,]^M \qquad \text{(Eq. 7.24)}$$

Note that the term r% represents the nominal interest rate r and the use of continuous compounding.

If we let p = M/r and rewrite this equation, we get

$$[\,F/P,r\%,1\,] = (1 + 1/p)^{rp} = ((1 + 1/p)^p)^r \qquad \text{(Eq. 7.25)}$$

This is a classic relationship because the limit of $(1 + 1/p)^p$ as p approaches infinity is e = 2.7128 . . . ! Thus, we have

$$[\,F/p,r\%,1\,] = e^r \qquad \text{(Eq. 7.26)}$$

and therefore

$$[\,F/P,r\%,N\,] = e^{rN} \qquad \text{(Eq. 7.27)}$$

which somewhat unexpectedly gives us a very nice relationship:

$$[\,F/P,r\%,N\,] = e^{rN} = (1 + i)^N \qquad \text{(Eq. 7.28)}$$

where the exponential expression assumes continuous compounding using the nominal rate, and the other expression uses the effective rate. Hence, we can revise all of the formulas for discrete cash flows into expressions for continuous compounding by substituting $e^{rN} = (1 + i)^N$.

7.5   FINANCING MECHANISMS

Equivalence relationships enable financing of large projects. Those investors or banks with money to invest are willing to make cash available for implementing a project in return for future interest payments, mortgage payments, or dividends. This section describes the financing mechanisms used for infrastructure projects.

7.5.1   Mortgages

Calculating the Mortgage Payment

A mortgage is a loan that is backed by property. If the borrower defaults on a payment, then the lender can seize the property and either use it or sell it to recover their costs. From the lender's perspective, a

Table 7.6 Capital Recovery Factor [A/P,i%,N] for Selected Interest Rates i% and Years N

Years	3%	4%	5%	6%	7%	8%	9%	10%
5	0.2184	0.2246	0.2310	0.2374	0.2439	0.2505	0.2571	0.2638
10	0.1172	0.1233	0.1295	0.1359	0.1424	0.1490	0.1558	0.1627
15	0.0838	0.0899	0.0963	0.1030	0.1098	0.1168	0.1241	0.1315
20	0.0672	0.0736	0.0802	0.0872	0.0944	0.1019	0.1095	0.1175
25	0.0574	0.0640	0.0710	0.0782	0.0858	0.0937	0.1018	0.1102
30	0.0510	0.0578	0.0651	0.0726	0.0806	**0.0888**	0.0973	0.1061

mortgage is less risky than a long-term, unsecured loan and therefore merits a lower interest rate. A mortgage typically is limited to about 80% of the assessed value of the property in order to reduce the risk to the lender in case the owner defaults. If the mortgage is less than the value of the property, then the bank will be able to sell the property and regain its investment.

The monthly payments on a mortgage depend on the amount of the loan (the principal amount), the interest rate, and the period of the mortgage. If the mortgage principal is PRICE, the annual interest rate is i%, and the term is N years, then the monthly payment will be

$$M = (\text{PRICE})[\,A/P, i\%, N\,] \qquad\qquad \text{(Eq. 7.29)}$$

where $[\,A/P, i\%, N\,] = [\,i(1+i)^N\,]/[\,(1+i)^N - 1\,]$ as shown in Section 7.3. This factor can be found in several ways. It can of course be calculated using the formula, but it can also be obtained from a table (see Table 7.6), and it can be obtained by using a function available on many spreadsheets. In Excel, the function PMT(i%,N,P) will give the desired answer.[4]

EXAMPLE 7.6 Calculating the Amount of a Mortgage Payment

What are the annual and monthly payments on a 30-year mortgage for $1,000,000 at 8% interest? The mortgage payment will be the annuity that—for the bank—is equivalent to the principal amount of the loan. To obtain the annual payment, the amount of the loan needs to be multiplied by the capital recovery factor [A/P,i%,N]. This factor could be retrieved from a table such as Table 7.6, which indicates that this factor would be 0.0888 for a 30-year mortgage with an 8% interest rate. The annual payment would therefore be approximately $88,800:

$$\text{PMT} = (\text{PRICE})[A/P, 8\%, 30] = \$1,000,000\,(0.0888) = \$88,800$$

If payments were to be made monthly, then the PMT would be approximately one-twelfth of the annual amount, or $7,333.33. These amounts are approximate because Table 7.5 shows the capital recovery factor to only four significant digits. More accurate answers for annual or monthly mortgage amounts can be obtained by using the formula for the capital recovery factor, or more readily by using the PMT function in Excel:

$$\text{Annual Payment} = \text{PMT}\,(8\%, 30, \$1,000,000) = \$88,827.43)$$

To obtain the monthly payment, it is necessary to use the monthly interest rate and 360 monthly payments. Using the PMT function, the result is:

$$\text{Monthly Payment} = \text{PMT}\,(8\%/12, 360, \$1,000,000) = \$7,337.65$$

These results are slightly higher than the results obtained using the factor from Table 7.6. The bank would be sure to use the actual number, but the difference is far too small to make any difference in project evaluation, which naturally must deal with many ill-defined numbers when comparing options. Neither the amount of the loan nor the interest rate would be known with certainty until a particular option has been chosen and the project is almost completed.

---

[4] The Excel function is PMT (interest rate, number of periods, present value). Table 7.6 and the more complete tables in the appendix can easily be created in any spreadsheet.

Figures 7.8 and 7.9 show the cash flows for the bank and for the borrower (known as the mortgagor and the mortgagee). If the cash flows in these two figures were combined, the net would be zero.

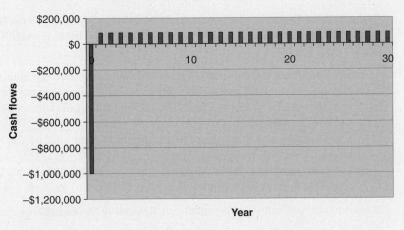

Figure 7.8 Cash Flows from the Perspective of the Mortgagor

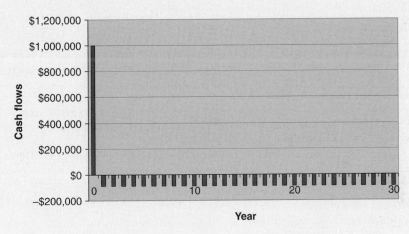

Figure 7.9 Cash Flows from the Perspective of the Mortgagee

Refinancing a Mortgage

An owner might have several reasons for wanting to refinance a mortgage. The most likely reasons for refinancing are to reduce monthly payments or to obtain cash:

- Interest rates may be lower, so that the mortgage payment could be reduced.
- The owner would like to extend the mortgage in order to reduce payments.
- The owner may be able to increase the size of the mortgage and extend the term of the mortgage without increasing the size of the monthly payment.
- The value of the property may have increased, and the owner may need additional money for another project.

Refinancing involves three steps:

1. Figure out the remaining balance on the initial loan.
2. Negotiate terms for the new loan.

3. Close on the new mortgage:
   a. Use some of the proceeds to pay off the original loan.
   b. Give a check to the mortgagee for any additional funds that are borrowed.

The remaining balance on the new loan can be calculated as of a point in time immediately after a regular mortgage payment has been made. The remaining balance could be calculated in multiple ways:

1. Read the statement; the monthly or annual statements show the amount outstanding as of the previous payment, the amount due for the current payment, and the portion of the payment that goes toward interest and principal. This information is what homeowners look at when they are considering refinancing.
2. Calculate the value of all the payments that have been made so far, and subtract it from the original amount of the loan.
3. Calculate the value of the remaining payments, using the interest rate of the loan as a discount rate. This result is equivalent to what must be repaid to the bank.

These calculations will involve multiple applications of the equivalence factors summarized above in Table 7.7.

EXAMPLE 7.7   Refinancing a Mortgage

Suppose that payments had been made once a year for 15 years on the 30-year, 8% mortgage for $1 million that was described in the previous section. How much of the mortgage has been paid off? If the borrower has a chance to refinance with a 20-year mortgage carrying a 6% interest rate, what would the new payments be?

The first step is to calculate the amount remaining on the mortgage. Assume that the borrower just made the 15th payment, so that 15 more payments are due on the original 30-year mortgage. Although half the payments have been made, much less than half the loan is paid off because most of the payments have gone toward interest. The amount remaining can be calculated in various ways.

One way would be to subtract the amounts paid off each year from the initial purchase price. The first payment of $88,827 included $80,000 interest (8% of $1 million) and therefore a principal payment of $8,827. With a spreadsheet, it is possible to continue the analysis, reducing the remaining principal by the proper amount after each payment, as shown in Table 7.7. The amount remaining after the 15th year is the amount shown at the beginning of year 16, namely $760,317.

Table 7.7   Payments of Principal and Interest and Remaining Balance Over the Life of a 30-Year Mortgage for $1 Million at 8% Interest

Year	Mortgage balance	Payment	Interest	Principal
0	$1,000,000	$88,827	$0	$0
1	$1,000,000	$88,827	$80,000	$8,827
2	$991,173	$88,827	$79,294	$9,534
3	$981,639	$88,827	$78,531	$10,296
4	$971,343	$88,827	$77,707	$11,120
5	$960,223	$88,827	$76,818	$12,010
6	$948,213	$88,827	$75,857	$12,970
7	$935,243	$88,827	$74,819	$14,008
8	$921,235	$88,827	$73,699	$15,129
9	$906,106	$88,827	$72,488	$16,339
10	$889,767	$88,827	$71,181	$17,646
11	$872,121	$88,827	$69,770	$19,058
12	$853,063	$88,827	$68,245	$20,582
13	$832,481	$88,827	$66,598	$22,229

14	$810,252	$88,827	$64,820	$24,007
15	$786,244	$88,827	$62,900	$25,928
16	$760,317	$88,827	$60,825	$28,002
17	$732,314	$88,827	$58,585	$30,242
18	$702,072	$88,827	$56,166	$32,662
19	$669,410	$88,827	$53,553	$35,275
20	$634,136	$88,827	$50,731	$38,097
21	$596,039	$88,827	$47,683	$41,144
22	$554,895	$88,827	$44,392	$44,436
23	$510,459	$88,827	$40,837	$47,991
24	$462,468	$88,827	$36,997	$51,830
25	$410,639	$88,827	$32,851	$55,976
26	$354,662	$88,827	$28,373	$60,454
27	$294,208	$88,827	$23,537	$65,291
28	$228,917	$88,827	$18,313	$70,514
29	$158,403	$88,827	$12,672	$76,155
30	$82,248	$88,827	$6,580	$82,248
31	$0			

Another approach would be to start with the purchase price and subtract the value of the 15 payments of $88,827 that have already been made, first calculating everything as of the time the mortgage agreement was signed:

$$\text{Outstanding amount (as of time 0)} = \text{PRICE} - ((\text{PRICE})[A/P, 8\%, 30]))[P/A, 8\%, 15]$$
$$= \$1,000,000 - \$88,827\,[8.5595]$$
$$= \$1,000,000 - 760,314 = \$239,685$$

The second term of this equation should be examined carefully since it illustrates how multiple factors can be used to obtain the desired amount. The first part of the term is simply the amount of the annual mortgage payment, which is calculated as the PRICE multiplied by the factor [A/P,8%,30], which converts a present value into a 30-year annuity assuming a discount rate of 8%. The final factor [P/A,8%,15] then converts the 15 mortgage payments into an equivalent value at time 0, using the 8% interest rate on the mortgage.

However, this amount of less than a quarter of a million dollars would have been the value at time 0 of the sum of payments 16 through 30. Now it is 15 years later, so the value has increased considerably:

$$\text{Outstanding amount (as of time 15)} = \$239,916 * [F/P, 8\%, 15]$$
$$= \$239,916 * 3.1722$$
$$= \$760,329$$

This process was a bit convoluted. Another, much more elegant way is to think about the value of the remaining payments rather than worrying about the contributions so far to principal and interest. From the lender's perspective, they are receiving 15 more payments at the original 8% interest rate. The value of this annuity, discounted at 8%, will also give the amount remaining on the mortgage:

$$\text{Outstanding amount} = \$88,000 * [P/A, i\%, 15] = \$88,800 * 8.5595 = \$760,084$$

The differences among these three answers are due to rounding errors from using the factors with only 4 or 5 significant digits. With any of these answers, we can express the outstanding amount as $761,000 and calculate the new mortgage payment:

$$\text{New payment} = \$761,000 * [A/P, 6\%, 20] = 761,000 * 0.0872 = \$66,359$$

Thus, by refinancing, the borrower can reduce the annual mortgage payment by $22,000. The annual payments are lower because of the lower interest rate and also because the repayment period has been extended by 5 years (15 remaining years on the original mortgage versus 20 years on the new mortgage). Figure 7.10 shows the reduction in mortgage payments over the extended life.

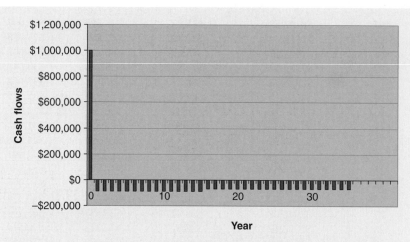

Figure 7.10 Cash Flows Assuming the Mortgage Is Refinanced after 15 Years

This example illustrates how multiple paths may be used to reach the same answer. The equivalence relationships can be used repeatedly, in different ways; and so long as the logic is correct along each path, each path will reach the correct destination.

7.5.2   Bonds

Bonds provide a way for companies or agencies to raise money. A bond is offered with a face value V, a life of N years, and annual interest of interest%. Figure 7.11 shows the cash flows associated with a 12-year bond that is purchased at the beginning of year 0. Interest payments are made at the end of each year for 12 years. At the end of the 12th year, the owner receives the final interest payment and the bond is redeemed for its original face value. Bonds are commonly sold in denominations of $1,000 with a 30-year term, but other options are available. Bonds can be bought and sold over their lifetime, so it is possible to buy a 30-year bond that will become due in less than 30 years.

Bonds can even be offered as **zero-coupon bonds**, in which the seller pays no interest but promises to pay the face value at the end of the term; for these bonds, the purchase price is much less than the

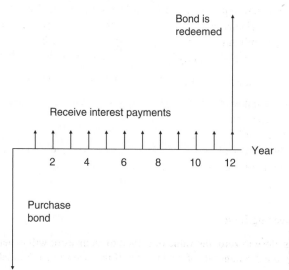

Figure 7.11 Cash Flow Diagram for Someone Who Purchases a 12-Year Bond at the Beginning of Year 1

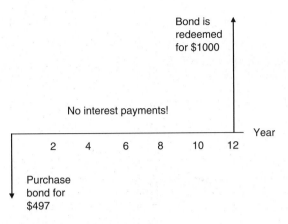

Figure 7.12 Cash Flow Diagram for Someone Who Purchases a 12-Year Zero-Coupon Bond That Pays 6% Interest

face value. Figure 7.12 shows the cash flow diagram for a zero-coupon bond that will be redeemed in 12 years for $1,000. The purchase price will be much less than the face value of the bond because there will be no interest payments.

Bonds are supported by the credit of the issuing agency or company; if the company fails to pay interest when it is due, then the bondholders can force the company into bankruptcy. If the company declares bankruptcy, then the assets of the company are divided among the creditors, including the bondholders. Various credit agencies rate the quality of bonds, so that investors have a reasonable idea of the risks involved in buying the bonds. The higher the perceived risks of the issuing agency or company, the lower the credit rating on the bonds, and the higher the interest rates that must be offered to attract investors.

Three interesting questions are (1) the value of the bond to an investor, (2) the change in the value of the bond as interest rates change, and (3) the change in the value of the bond as perceived risks associated with the issuing agency or company change. The value of the bond to an investor depends upon the investor's perception of the risks involved, the investor's discount rate for bonds with such risks, and the interest rates offered for the bond. If the investor's discount rate is lower than the interest rate offered, the investor will consider buying the bond. Market forces related to the supply and demand for fixed interest securities will determine what interest rates are actually required to sell bonds.

Assume that the bond is sold at time 0. The seller agrees to pay interest of i% per year for N years and, at the end of N years, to pay back the face value of the bond. The seller offers the bonds to the marketplace, and potential purchasers decide whether they want to buy the bonds. The purchasers may plan to hold the bonds until maturity, or they may merely view the bonds as a short- or medium-term investment. The value of a high-quality bond to an investor can be calculated as follows:

$$\text{Value} = \text{V}(\text{interest}\%)[\,P/A, i\%, N\,] + \text{V}[\,P/F, i\%, N\,] \qquad \text{(Eq. 7.30)}$$

The first term is the present worth of the N annual interest payments, and the second term is the present worth of the final redemption of the bond at its original face value. It is crucial to recognize that the investor's discount rate of i% can be higher or lower than the interest rate on the bond. If there is a possibility that the bond will default, both terms could be reduced by a factor representing the probability that interest or the final redemption would not be made.

EXAMPLE 7.8   Value of an Interest-Bearing Bond

Assuming that the probability of default is close to zero, the value of a bond to an investor will depend upon the face value, the years to maturity, the interest rates, and the discount rate of the investor. If the face value is $1,000 and the interest rate is

6% for a 30-year bond, then the value to a potential purchaser who also has a 6% discount rate will be exactly $1,000:

$$\text{Value} = (\$1,000)\,(0.06)\,[P/A, 6\%, 30] + \$1,000\,[P/F, 6\%, 30]$$
$$= \$60\,(13.7648) + \$1,000\,(0.1741)$$
$$= \$826 + \$174 = \$1,000$$

For someone with a discount rate of 5%, the bond will be worth much more:

$$\text{Value} = \$60\,(15.3725) + \$1,000\,(0.2314)$$
$$= \$922 + \$231 = \$1,153$$

Likewise, someone with a discount rate of greater than 6% would value the bond at less than $1,000. In each case the equivalence factors were obtained from the tables in the appendix.

   The same type of calculations can be used to show how the value of a bond could change if interest rates change. For example, suppose that 20 years have gone by and interest rates on similar bonds have fallen to 5%. The 6% bond is therefore paying more interest than a bond with similar risk would have to pay today; hence, investors would find that bond more appealing. Someone with a 5% discount rate would be willing to pay $1,000 for the bond paying 5%, but would pay more for the 6% bond with 10 years until maturity:

$$\text{Value} = \$60\,[P/A, 5\%, 10] + \$1000\,[P/F, 5\%, 10]$$
$$= \$60\,(7.7217) + \$1,000\,(0.6139)$$
$$= \$463 + \$614 = \$1,077$$

Thus, when interest rates fall, bond values rise. In this instance, note that the redemption value is now much greater than the value of the interest payments. As the bond approaches maturity, the redemption value dominates.

EXAMPLE 7.9   Value of a Zero-Coupon Bond

For zero-coupon bonds, the calculations are very simple because the value of the bond is simply the present value of the eventual redemption of the face value of the bond. For example, consider the 12-year, zero-coupon bond illustrated by the cash flow diagram in Figure 7.10 above. A potential purchaser using a discount rate of 6% would value the bond at $1,000/(1.06)^{12}$ or just under $500. A potential purchaser with a discount rate of 5% would value the bond as $1,000/(1.05)^{12}$ or just over $550. The actual price of the bond would depend upon the market for bonds: if there is sufficient demand among investors who would value the bond using a discount rate of 5% or less, then the bonds could be sold for $550 or more. If not, then the purchase price would fall. If there are enough potential investors who would discount the bond at no more than 6%, then the market price of the bond might be about $500. Zero-coupon bonds may have a significant tax advantage for wealthy investors since all of the gains would eventually be taxed at a capital gains rate that is lower than the tax rate they would pay annually on interest-bearing bonds. The effect of tax policy on financial decisions is covered in Chapter 10.

7.5.3   Sinking Funds

A sinking fund can be established to meet expected future capital needs, such as paying the principal on bonds when they become due or conducting a major rehabilitation of a factory at some distant time in the future. Companies or agencies can pay a constant amount into a fund that is maintained solely to cover this future capital need. As shown above, the sinking fund factor [A/F,i%,N] can be used to determine the annual amount A to invest at i% to reach the goal of having an amount F at the end of N periods.

   The calculations will require another step if the future need is itself an annuity. For example, planning for retirement involves consideration of three questions: First, how much will be needed in annual income after retirement? Second, how much will be needed in savings to produce this level of retirement income? Third, how much will need to be saved each year (e.g., in a sinking fund) in order to have accumulated the money that will be used to purchase the retirement annuity? The first question is a matter of personal needs and desires; in effect, it is necessary to determine how large an annuity will be needed to support your desired lifestyle in retirement. The second question is a matter of equivalence: how much will be needed to purchase the retirement annuity? This is the retirement goal. The third question is a different matter of equivalence: how much do you have to save each year to reach your retirement goal? The relevant equations are as follows:

$$\text{Retirement Goal} = \text{Retirement Annuity} \left[P/A, i\%, N \right] \qquad \text{(Eq. 7.31)}$$

$$\text{Savings Goal} = \text{Retirement Goal} \left[A/F, g\%, M \right] \qquad \text{(Eq. 7.32)}$$

The key unknowns are (1) how much will really be needed, (2) what interest rate ($i\%$) can you expect for the annuity, and (3) what annual return on investment ($g\%$) can you expect on your savings. Perhaps the most interesting variable for many people will be M, the years until retirement. The more that is saved, the higher the return on investment; and the higher the interest rate on the retirement annuity, the sooner you can retire. The longer you work, the more time you will have to reach your retirement goal and the less you will need to save each year to reach that goal.

EXAMPLE 7.10   How Much to Save for Retirement?

A couple in their thirties is planning for retirement. They would like to save a constant amount per year in a tax-deferred investment account. After working for another 30 years, they hope to have accumulated enough funds to provide an annuity that will last them until they are 100 years old or older. If they can achieve earnings of 8% per year on their investments, how much should they save each year in order to be able to have $80,000 per year if they retire at age 60 and live forever?

  To answer this, assume a discount rate of 8% for all of the calculations. While it is unlikely that they will live forever, that hope at least makes the analysis a bit easier: the anticipated 8% return on their accumulated retirement investments will be $80,000, so they will need to accumulate $80,000/0.08 = $1 million by age 60 (note that this approach uses the simple capital worth method to convert the desired annuity into a required sum). The question therefore is, what level of annual investment is needed to accumulate a million dollars in 30 years if returns are 8%?

$$\text{Annual Investment} = \$1{,}000{,}000 * \left[A/F, 8\%, 30 \right] = \$1{,}000{,}000 * (0.0088) = \$8{,}800$$

This amount may seem steep. If the couple extends their planned retirement 10 years to age 70, then they will need to invest only $3,900 per year:

$$\text{Annual Investment} = \$1{,}000{,}000 * \left[A/F, 8\%, 40 \right] = \$1{,}000{,}000 * (0.0039) = \$3{,}900$$

7.5.4   Toll-Based Financing

For highways and bridges that carry a lot of traffic, it may be possible to use the projected toll revenues to justify issuing bonds that are sufficient to cover the cost of construction. Some quick estimates may indicate whether toll-based financing will work. The interest on the bonds must be compared to the net revenue from the tolls, which is what remains after paying for the annual operating costs of the bridge. If the net toll revenue is well above the anticipated interest payments, then this would be a good candidate for toll-based financing.

EXAMPLE 7.11   Will Toll Financing Be Feasible for a Proposed Bridge?

A proposed bridge is expected to cost $50 million to construct, and its annual operating and maintenance costs will total $3 million. The bridge is expected to serve 5 to 10 million vehicles per year, and the governor believes that the public would view a toll of $1 to $2 as reasonable. Will the tolls cover the cost of bonds that could be sold to pay the construction costs of the bridge? Interest rates on the bonds are expected to be 4%.

Given the projected traffic volume, the bridge would earn $5 to $20 million per year with a toll of $1 or $2. After deducting $3 million per year for operating expenses, the net revenue would $2 to $17 million. Given the interest rate of 4%, then annual interest on the $50 million investment would be $2 million. Thus even the lowest estimate of net revenue would cover the annual interest payments on bonds. With a $2 toll, funds would be available for related projects, such as improving access roads or providing support to other transportation projects.

7.6   THE USE OF EQUIVALENCE IN COSTING

Earlier chapters have emphasized the importance of long-run average costs and long-run marginal costs. The difference between fixed and variable costs was shown to be critical in determining the break-even volume at which a company would begin to make some profit and the break-even volume at which a more expensive but also more efficient technology becomes cost competitive with cheaper technologies. In those discussions, capital costs were at most assumed to be given, and they were lumped together with other fixed costs. However, it is clear from earlier sections of this chapter that the capital costs will vary with the interest rates and other terms of the loans that are obtained or the bonds that are sold. It is therefore necessary to consider the actual financing mechanisms associated with a project and to use the proper equivalent annual cost when estimating average costs. Example 7.12 shows how to do this.

EXAMPLE 7.12   Converting Investment Costs into Annual Fixed Costs

You have two options for constructing a transportation network, as shown in Table 7.8. Develop an equation for the long-run average cost curve that includes all three categories of cost, and determine the range of volumes for which each technology is the best choice. Assume that the investment will last indefinitely, and assume a discount rate of 10%.

Table 7.8   Two Competing Technologies

Technology	Investment	Annual Fixed Costs	Variable Cost/Vehicle	Capacity Vehicles/Year
Low capacity	$20 million	$3 million	$6	700,000
High capacity	$30 million	$4 million	$2	1,200,000

The basic problem in estimating the long-run cost curve is that the investment is a one-time expense that cannot be directly compared to the annual fixed costs and the variable costs. This problem is one of the fundamental problems motivating this text: how can we compare initial investment costs with future net benefits? To deal with this problem, we must use the concept of equivalence. We need to make the initial investment "equivalent" to an annual expense that can be compared with the annual fixed and variable costs. We therefore need to multiply the investment by the factor [A/P,10%, life of project] to get the equivalent annuity. You could assume a project life and obtain [A/P,10%,life] using a spreadsheet function or equations from the text. An easier way, however, is strongly suggested by the statement that "the investment will last indefinitely." If so, then [A/P,10%,life] = 10% using the capital worth method, and the equivalent annual costs for the investments are very easily estimated as $2 million for the low-capacity option and $3 million for the high-capacity option. Note that when you are comparing future options, you cannot treat the proposed future investment as a sunk cost; the nature and extent of the future investment is precisely what is at issue, so it cannot be ignored.

The long-run average cost curve is one of the fundamental economic concepts presented in Chapter 3 and part of the framework for the Panama Canal case study. The long-run average cost is the cost per unit for the best of the available technologies for a particular volume of activity. In this question, there are just two technologies:

$$\text{Average Cost} = (\text{Annual Investment Cost} + \text{Annual Fixed Cost} + (\text{Var.cost/veh.})V)/V,$$
where V = vehicles per year   (Eq. 7.33)

$$\text{Low} - \text{Capacity Average Cost} = (\$2 \text{ million/year} + \$3 \text{ million/year} + \$6 \, V)/V \qquad \text{(Eq. 7.34)}$$

$$\text{High} - \text{Capacity Average Cost} = (\$3 \text{ million/year} + \$4 \text{ million/year} + \$2 \, V)/V \qquad \text{(Eq. 7.35)}$$

The question of which technology is best can be solved graphically or algebraically. At the break-even point, the average costs will be equal. Therefore:

$$2 + 3 + 6 \, V = 3 + 4 + 2 \, V \qquad \text{(Eq. 7.36)}$$

Solving this equation shows that $4V = 2$, so the break-even volume is 0.5 million vehicles per day. For greater volumes than this, the high-capacity option will have a lower average cost. If the long-run cost curve did not include investment costs, the break-even point would be lower.

7.7 CASE STUDY: BUILDING AN OFFICE TOWER IN MANHATTAN

A real estate developer is looking for opportunities to create value by constructing buildings. If the estimated value of the newly created space is worth more than the expected development costs, then there is a development opportunity. For an office building, the value will be based upon the leases that can be obtained for the office space. The development costs will include the cost of the land, preparation of the site, design and engineering, construction, and possibly various costs related to the approval process (e.g., in return for building a new entrance to a subway station, the developer may be allowed to build more intensively).

In Manhattan and other urban centers, land becomes a very expensive resource that causes strong economic pressures for intensive development. In very general terms, the value of a building will be proportional to the usable space it contains (i.e., the space that can be leased to clients). Doubling the size of the building therefore roughly doubles the usable space and therefore doubles the value of the building. On the other hand, development costs are not at all proportional to the size of the building. The price of the land depends upon the local real estate market, not the value of what you intend to build; whether you build a single-story warehouse or a 50-story office building, the cost of the land will be the same. Moreover, whether that office building is 20, 50, or 80 stories tall, it will require access to local streets, a lobby, and a roof. For a large building, adding stories simply means replicating the designs and materials used for one story over and over again. While certain structural components will need to be stronger for a taller building, the added costs are rather minor for a steel-framed structure.

Since there are economies of scale in building, the incremental cost of adding another story will be well below the average cost, while the incremental value of another floor of leasable space will not diminish (assuming the space can be leased!). Hence, adding more stories and maximizing the usable space on each story will increase the value of the project while reducing the average cost per square foot of the project. The developer therefore has a strong incentive to build the largest possible building.

The building that can or will be built is subject to various size constraints:

- Zoning regulations may limit the portion of the site that can be developed or the total floor area ratio (FAR, the ratio of floor space to the area of the site).
- Technological capabilities may limit the height (although the limit is obviously more than 100 stories and seldom if ever a real limit today).
- Market considerations may limit the amount of space that the developer wishes to make available today.

Karl Sabbagh, in a highly readable book called *Skyscraper* (1991), described the redevelopment of an entire block in Manhattan during the mid-1980s. The site, formerly occupied by Madison Square Garden, was between 49th and 50th Streets and 8th and 9th Avenues—a rather rundown area somewhat west of the prime office locations in Manhattan. Developing the site as an upscale office building was somewhat risky—not because the rents would be lower than in the best locations, but because the space might be impossible to rent at any price. Developer Bill Zeckendorf bought the land

Table 7.9   Projected Costs of the 50-Story Office Tower

Cost Element	Estimated Cost
Land acquisition (office tower portion of the site)	$ 58 million
Preparation of case for development (architects and lawyers)	$ 5 million
Architects, engineers, and borrowing costs	$145 million
Construction cost	$145 million
Project management	$ 17 million
Total	$370 million

for $100 million, but only when he was reasonably sure that the site would be able to attract tenants to what he called the "Worldwide Plaza."

The zoning regulations called for a FAR of 12, which was increased by the city to 14 as a bonus for Zeckendorf's agreeing to make some improvements to the subway station on the site and to provide an acre of open space as part of the project. This provided an opportunity for 1.9 million sq ft of usable space, of which 1.5 million was in a 50-story office tower. Zeckendorf expected to be able to lease the space in the office tower at rates of $20–$30/sq ft/year, with possible increases to $40 in the future. These estimated lease rates were discounted by about $5/sq ft from the rents achievable a few blocks toward the other side of Manhattan. At $20/sq ft, the annual rent would be $32 million for the office tower; at $30/sq ft, the annual rent would be $48 million.

The estimated costs for the entire project were expected to exceed $500 million, and the costs for the office tower were estimated to be $370 million (Table 7.9). The basic plan was to use a construction loan to cover the construction costs and to refinance to a 30-year mortgage at a lower interest rate once the building opened. The construction loan would have a high interest rate due to the risks of delays and overruns in construction and the possibility of being unable to lease all of the space. If all went well, the building space would be leased at favorable rates to long-term tenants, and the lease payments would be more than enough to justify a mortgage sufficient to repay the construction loan.

Interest rates and lease rates were the keys to the success of the project. Interest rates were likely to be on the order of 10% or more for the construction loan and on the order of 8% for the mortgage. The costs of the construction loan were included in the estimated cost of the building, but delays and unexpected expenses could lead to higher interest payments. With an interest rate of 10% on the construction loan, the monthly interest on a balance of $370 million would be about $3 million (10% per year/12 months/year)($370 million).

If the construction costs were indeed on the order of $370 million, and if Zeckendorf could obtain an 8% mortgage, then the annual mortgage payments would be approximately $33 million:

$$\text{Annual Payment} = \$370 \text{ million } [A/P, 8\%, 30] = \$370 \text{ million } (0.0888) = \$32.9 \text{ million}$$

(Eq. 7.37)

If the building could indeed be rented at $30/sq ft, then the $48 million annual revenue would seem to provide enough cash to cover this mortgage payment plus some operating expenses for managing the property. However, if the average lease rate were only $20/sq ft, then the cash flow would be about the same as the mortgage payment, with no reserve for managing the building. Thus, for the project to succeed, it would be essential to complete the project on time and on budget, to secure long-term leases with favorable rates, and to secure long-term financing sufficient to cover the costs of construction.

> *The whole magic of our industry is twofold. One is to build a beautiful building but, more important, it's got to be successful. The only way it becomes successful is if you start collecting rent. The sooner you start collecting rent, the sooner the building becomes more successful.*

> *The minute you start collecting rent, all the sins of the father are forgiven. Everything that we've done wrong, they forget—we're all friends again.*
>
> Marvin Mass, HVAC Contractor,
> Worldwide Plaza (quoted in *Skyscraper*, p. 306)

The building was actually constructed for about $380 million due to minor overruns in several areas related to construction or material problems. It was rented at rates of $26 to $32/sq ft; the lowest rate went to a major tenant who became a part owner of the building and committed to leasing 600,000 square feet of space at the outset of the project. The next-largest tenant obtained a rate of $29/sq ft, which was lower than the owners wanted, but it was accepted in the uncertain aftermath of the stock market crash of October 1987, just before the building was ready for occupancy. Smaller tenants paid rates of about $32/sq ft. By the end of the project, monthly interest costs were close to $3 million and deferring rentals was costing close to $4 million per month.

The building was constructed on a fast-track basis to minimize borrowing costs during the construction period and to begin lease payments as soon as possible (Table 7.10). The projected lease payments were sufficient to justify a permanent mortgage that enabled Zeckendorf to repay the construction loan. Despite the unexpected downturn in the Manhattan real estate market, the building project was successful.

The overall viability of the building depended upon being able to complete construction at close to budget without major delays and being able to rent the building at something close to the expected rates. Both of these requirements were met. However, the objectives of the different actors were not based upon the overall perspective:

> *In this particular building you have a pretty characteristic group. You have the architect, who has the design as his main consideration. He wants to put up a monumental building, something everybody is going to see and say, "Hey, wow! That's great!" It's his entry into posterity. The construction manager, HRH, they're interested in having a building up that they don't get sued over, that's going to stay in place. Each of the individual trades have the same interest as the construction company. The only difference is, each of the trades says, "I'm only going to do so much. The rest is someone else's responsibility." So then you have to argue out who's actually doing what part of the interface between the various trades. The consultant is working to represent the interest of the owner. Again, he's after a viable building, something the guy can make money with. He's not investing money to lose it. He also wants to make sure it's sound. I tell you, he has about the same interest as the construction manager.[5]*

Table 7.10   Project Timetable for the Worldwide Plaza

Event	Date
Site acquisition	1985
Secure major tenant as co-owner	1985
Groundbreaking ceremony	November 12, 1986
Initial target for making space available to tenants	November 25, 1988
All tenants in the building, working on finishing their space	March 1989
Tenants start moving in	May 15, 1989
Permanent mortgage obtained for the entire project	May 31, 1989

---

[5] Sabbagh, *Skyscraper*, 199.

Other participants would include the banks that provide the construction loan and the permanent mortgage. For them, the project involves providing a large sum of money up front in the hopes of receiving a larger sum a few years in the future (when the construction loan is repaid) or receiving an annuity that will provide a guaranteed return over a much longer period (via mortgage payments over a 30-year period). The banks should take care to ensure that the size of the mortgage is limited by the ability of the owners to make mortgage payments; the banks will offer lower interest rates if they are more certain that the project will be successful. The "Skyscraper" assignment at the end of Chapter 8 is loosely based upon the construction of the Worldwide Plaza. That assignment explores the project from the perspective of the owner and the banks, and it also considers the prospects for selling the office tower.

7.8   SUMMARY

When evaluating projects, it will be necessary to compare costs and benefits that are incurred over a period of many years. The project costs are generally concentrated at the outset of the project because benefits do not begin until construction is at least partially completed. One basic question for any proposed project is determining whether the eventual benefits will be sufficient to justify an initial investment in a project. In most situations, multiple alternatives must be considered, each with its own investment requirements, timetable, and projected stream of future costs and benefits. A second basic question is to determine which project—which projected set of cash flows—is most desirable. These are not the only questions that must be answered to determine whether a project can be justified, but they are questions that will certainly be considered by investors, bankers, and entrepreneurs who are considering participating in such projects. It is therefore essential to understand how such financial comparisons are made.

Time Value of Money

The first critical concept concerns the time value of money. For several reasons, money in the future is worth less than the same amount of money available in the present. First, if money is available today, there is an opportunity cost: the money could be put into a savings account or invested so that it will be worth more in the future. Second, inflation is likely to reduce the purchasing power of money, so that today's money will buy more goods and services than the same amount of money would be expected to purchase in the future. Third, there is a risk that the actual money that becomes available in the future will be less than was predicted. It is therefore necessary to discount future cash flows in order to determine their present value. If F is the predicted future cash flow in year N, and the discount rate is i% per year, then the present value P will be given by the following expression:

$$P = F/(1+i)^N \qquad \text{(Eq. 7.38)}$$

The higher the discount rate that is used, the lower the present value. The greater the potential for growth, the higher the expected rate of inflation, and the greater the risk associated with the proposed investment, the higher the discount rate that will be used. An investor or entrepreneur has a minimum acceptable rate of return (MARR) that they are likely to use in discounting future cash flows. Nuances of discounting are discussed at length in Chapter 8.

A cash flow diagram illustrates the net cash flows over the life of a project. By discounting all of these cash flows, it is possible to determine the net present value (NPV) of the project. If the NPV is positive, then the project is worth more than doing nothing; if the NPV is negative, then—at least from a financial perspective—the project is not worth pursuing.

By estimating the NPV for a set of alternatives, it is possible to determine which alternative is worth the most financially. Maximizing the net present value of cash flows is a common financial objective for the private sector, although public projects are likely to have more complex objectives.

Equivalence Relationships

Discounting provides a means of establishing equivalence between two sets of cash flows. If the cash flows are equivalent in a financial sense for a company or an individual, then that company or individual is indifferent between them. The conceptual power of discounting is that it provides a straightforward methodology for establishing an NPV that is equivalent to any arbitrary set of projected cash flows.

Using the same discount rate, it is possible to establish equivalence between a present value P and a future value F or an annuity A.[6] An annuity is a stream of identical cash flows that are received at the end of each of a series of N periods. Another term for an annuity is the equivalent uniform annual cost (EUAC). Factors that relate P, F, and A are functions of the discount rate and time, as derived in Section 7.3 and summarized in Table 7.4. These equivalence factors make it possible to compare cash flows in various ways, depending upon what is most useful for a particular analysis. For example, investment opportunities commonly are compared using the NPV; a mortgage translates an amount of money borrowed to pay for a house or a building into equivalent monthly payments; a retirement plan translates a long series of monthly contributions into a future value that will be sufficient to purchase an annuity that will support you in your retirement.

Continuous Compounding

Section 7.4 shows how interest payments and discounting can be done on a continuous as well as a discrete basis. If interest payments are made more frequently, then the annual interest will be slightly higher even if the nominal interest rate remains the same, because the interest earned during each interval will itself earn interest in subsequent intervals. For example, if the nominal rate is 12% per year, compounding quarterly will actually pay an effective rate of 12.55% per year, and compounding monthly will pay an effective rate of 12.68% per year. In the limit, as the compounding interval becomes ever shorter, interest can be compounded continuously, which is captured in a very interesting and useful relationship:

$$[\,F/P, \underline{r}\%, N\,] = e^{rN} = (1 + i)^{N} \qquad \text{(Eq. 7.39)}$$

where the exponential expression assumes continuous compounding using the nominal rate, and the other expression uses the effective rate. Hence, we can revise all of the formulas for discrete cash flows into expressions for continuous compounding by substituting $e^{rN} = (1 + i)^{N}$. The differences between the nominal and effective interest rates are important in determining the monthly payments for a mortgage or a car, so it is important to read the small print when taking out such loans. In evaluating major projects, monthly or annual cash flows typically are used, so it is not necessary to use the exponential formulations for equivalence.

Nevertheless, the relationship between the discrete and continuous formulations is well worth remembering because it forms the basis for some quick approximations. All you need to remember are some useful values of rN:

- If rN = 1, e^{rN} = e = 2.718 . . . .
- If rN = 0.7, e^{rN} is approximately 2.
- If rN = 1.1, e^{rN} is approximately 3.
- If rN = 1.4, e^{rN} is approximately 4.

These relationships provide a quick and easy way to estimate compound growth factors or to discount cash flows. For example, an investment that returns 10% for 7 years will approximately double over that period, while a projected cash flow of $1 million that would be received in 14 years would have an NPV of just a quarter million dollars at a discount rate of 10%.

---

[6] These three factors may also be referred to as present worth (PW), future worth (FW), and annuity worth (AW).

Financing Mechanisms

Section 7.5 shows how equivalence relationships are used to determine mortgage payments, the value of bonds, and other financing mechanisms. A mortgage is a financing mechanism by which a bank or other financial company provides a large payment in return for a series of monthly payments over a period of many years. If the mortgage payments are not made, then the mortgage is in default, and the bank can foreclose on the property. A company or agency can sell bonds to raise money for a project; the bonds are purchased by investors who receive interest payments over the life of the bond and also receive the full value of the bond at the end of life of the bond. If the company or agency fails to make the interest payments or is unable to redeem the bonds as they become due, then the company or agency may be forced into bankruptcy.

The Use of Equivalence in Costing

Section 7.6 shows how equivalence can be used in cost models, where it is often necessary to consider both investment costs and annual operating costs. The investment cost must be converted to an equivalent annual cost in order to be added to operating cost and used to determine such things as the cost per unit of capacity or the cost per user.

Case Study: Building an Office Tower in Manhattan

Section 7.7 shows how various equivalence relationships are relevant to decisions about constructing, renting, or purchasing office buildings. The value of the project depends upon many factors, including the discount rates of the owner, the interest rates available on mortgages, and the expected lease rates and occupancy rates. The discount rates used by the owner, the bank, and potential purchasers of the building are all different, in part because they have a different share of the risks and in part because they have different perceptions of the risk.

THE NEXT STEP

The next chapter examines the choice of a discount rate from the perspective of an owner, an entrepreneur, investors, and public agencies. Different perceptions of risks and opportunity costs become key considerations in evaluating projects. Many projects create value by reducing risks, thereby increasing the discounted value of future cash flows. Chapter 8 includes a case study based upon the office tower in the Worldwide Plaza, which was described in Section 7.8. Successfully completing that case study will demonstrate a thorough understanding of the equivalence relationships presented in this chapter and of the factors affecting discount rates, as presented in Chapter 8. That case also provides insight into how a well-financed construction project can create value.

ESSAY AND DISCUSSION QUESTIONS

7.1 Discuss the differences in financing a project by using internal funds, obtaining a mortgage, and selling bonds.

7.2 There clearly is a difference between nominal and effective interest rates, so it is necessary to pay close attention to the terms of any loan agreement. How important is it to worry about this difference when discounting cash flows to obtain a net present value for a project?

7.3 A consultant with long experience in project evaluation once remarked, "All I ever learned about equivalence relationships—and all I ever needed to know—was contained in the formula Present Value $= 1/(1 + i)^n$. I never learned the equations for sinking funds, capital recovery factors or anything else, nor did I ever use continuous discounting. I was always able to convert present values to annuities using the simple spreadsheet function for calculating mortgage payments." Does this statement seem credible to you?

7.4 Section 7.7 presented the history of the development of the Worldwide Plaza in New York City. This project created office and residential space that became available to rent just as the real estate market collapsed in 1987. How did the developer manage to survive that recession? What were the major factors that contributed to the success of that project?

7.5 What is the current status of the Worldwide Plaza, following the dramatic real estate collapse of 2008?

PROBLEMS

7.1 Create a spreadsheet that you can use to estimate equivalent cash flows for an arbitrary sequence of cash flows over 50 periods. You want to be able to use this spreadsheet to convert an arbitrary sequence of cash flows into a present worth PW, a future worth FW at any time t, or an annual worth AW over N periods. You want to be able to do this using both discrete and continuous compounding factors, and you want to be able to compute effective interest rates. Take some care in designing your spreadsheet so you can easily do sensitivity analysis on interest rates and N, and so you can easily print out a compact and attractive report showing results. Test your spreadsheet:

a. *Salvage value*: You are involved with a project that is expected to last 50 years and have a salvage value of $10 million. A consultant has advised your company that an expenditure of only $100,000 at the end of every 5 years will double the salvage value. Your company generally uses a MARR of 15%—do you buy the consultant's recommendation?

b. *Bonds*: A bond has an initial purchase price of $1,000 and an interest rate of 6% paid at the end of each year for 30 years. At the end of year 30, the bond is redeemed and the owner is repaid the initial payment of $1,000.

  i. What is the value of the bond at the beginning of year 6 to someone with a MARR of 7%? At the beginning of year 29?

  ii. What is the value of the bond at the beginning of year 6 to someone with a MARR of 5%? At the beginning of year 29?

7.2 A new rail car costs $100,000 and will last for 20 years, assuming that $20,000 is spent on a major overhaul at the end of year 10. Routine servicing and maintenance are expected to cost $2,000 per year. The car is expected to be used in revenue service for 300 days per year. What is the equivalent cost per day-in-use over the 20-year life of the car, assuming a discount rate of 6%, 8%, and 10%?

7.3 A new Toyota Camry costs $20,000 and will last for 5 years in rental service, according to "Toyotas-R-Us Car Rental." At the end of the 5 years, the car will be sold for an expected price of $5,000. Routine servicing and maintenance are expected to cost $3,000 per year, and the car is expected to be rented for 250 days per year. Your MARR is 10%. Under these conditions, what is the minimum rental rate that will cover the cost of the car?

7.4 Very simple exponential equations can be used to estimate the equivalence between present and future values when interest is compounded daily. Use mental arithmetic (as described in Section 7.4) to answer the following:

a. What is the future value of $1,000 invested for 7 years at 10%, then reinvested for 10 years at 7%?

b. What is the present worth of $100,000 due in 14 years, assuming that your MARR is 10%?

c. Which is worth more: $4,000 in 14 years or $6,500 in 22 years, assuming MARR is 5%?

d. Which is worth more: $10,000 per year forever, or $90,000 now, assuming MARR is 10%?

e. What is the future value of $2,000 invested for 11 years at 10%, then reinvested for 7 years at 20%?

f. What is the present worth of $600,000 due in 11 years, assuming that your MARR is 10%?

g. Which is worth more today: $10,000 that you would receive in 22 years or $6,500 that you would receive in 14 years, assuming your MARR is 5%?

7.5 You have just won the lottery. The prize is $1.2 million, payable in 20 equal annual installments. You received the first installment of $60,000 today, and you are so happy to be a millionaire. However, your friend claims you really aren't a millionaire! What is the present worth of your winnings, assuming your MARR is 8%?

7.6 Congratulations! You have just won the Mass Millions Lottery with a jackpot of $15 million! The prize is payable in 30 equal annual payments of $500,000, beginning right now. If your opportunity cost is 12%, how much is this lottery really worth?

7.7 Having won the lottery (as described in Question 6), your only (financial) problem now is that you are just 20 years old and expect to live (in grand style) for well over 30 years. How much of your first year's installment would you have to invest in order to ensure that you could earn $500,000 per year for another 30 years after the Mass Millions bonanza comes to an end? Assume that you can invest in stock mutual funds and earn 15% per year until it is time to start the annuity. Assume that the annuity will be calculated based upon 30 end-of-the-year payments and an interest rate of 8%.

7.8 A student takes an education loan from a bank. He expects to borrow $40,000 a year for the next 4 years (the first amount is drawn immediately, on 2/28/2010). The annual interest rate being charged is 8%; for the first several years, the interest is not paid, but it is added to the loan at the end of each year. He is required to repay the loan in equal annual payments starting 7 years from the date of the loan (i.e., the first payment will be made on 2/28/2017), with one payment per year for 15 years.

a. What will be the amount of each loan payment?

b. Right after making the fifth annual payment, he decides to repay his entire loan. How much will he have to pay?

c. How much interest will he end up paying?

7.9 Assume that you can borrow money from your credit union to finance a new car at 4% interest for 5 years. Also assume that your

MARR is actually 6%. You go to buy a car, and you are offered either 0% financing for a 3-year auto loan or $1,000 off the sticker price of $11,000 if you pay cash (and borrow the money from the credit union). What is the equivalent annual worth of each option over a projected 10-year life of the car?

7.10 You are trying to decide whether to bid on a construction contract for a new bridge. You think that it will take 30 months to build and that construction costs will be $2 million per month. You expect tolls to be $10 million per year once the bridge opens, which will be offset by toll collection and maintenance costs of $2 million per year. Your MARR is 15% per year. To bid on the project, you specify the price you are willing to pay to the state (in cash, at time 0) for the right to build the bridge and operate it for a period of 30 years. At the end of 30 years, the ownership and operation of the bridge revert to the state.

a. You can obtain a construction line of credit at 10% per year that can be used to cover all the construction expenses plus all of the accrued interest. What will be the outstanding balance when the bridge is completed?

b. Once the bridge is open, you will have a steady stream of income, so that you can refinance the construction loan at a lower interest rate, say 8% per year, and pay off the loan in 30 years. What will the annual payments be on this loan?

c. What is the cash flow (toll revenue minus payments on your loan) from operating the bridge worth to you at the end of month 30 when the bridge opens?

d. What are you willing to bid for the bridge?

7.11 You have inherited an apartment building in Cambridge that cost $3 million to construct in 1964. Annual rents total $200,000 and annual operating expenses total $100,000. The building is in reasonable condition, and your real estate agent expects that similar profits could be earned for another 20 years.

a. You receive an offer from MIT to buy the building for $2.5 million; MIT wants to convert it to dormitory space. From a purely financial perspective, explain why you should or should not accept the offer.

b. Your uncle, a building contractor, recommends that you upgrade the building by adding a pool, renovating the lobby, and redoing all of the kitchens. This would cost $2 million, but rents could be increased by 25%. Explain why you should or should not renovate the building.

c. Your best friend advises you to drop out of MIT, live off the profits for 20 years, and then sell the building to cover your retirement. What do you think you could sell the building for in 20 years, assuming MIT is no longer bidding and that the asking price will be what entrepreneurs will be willing to pay for the cash flows? Would you follow your friend's recommendation?

7.12 The estimated construction costs for a new apartment building are given in the table below. Expenses can all be assumed to be paid at the end of each month. The developer has obtained a line of credit for $200 million that will be used to cover the costs of constructing the building. The nominal interest rate on the line of credit is 10% per year; interest payments are not made in cash,

but are added to the outstanding balance at the end of each month. Thus, no interest is due at the end of month 1, and the balance at the end of month 1 will equal the expenses incurred in month 1.

Construct a spreadsheet that can be used to calculate the costs of construction, including interest costs on the construction loan. Structure the spreadsheet so that it shows the various cost categories across the top and the months down the side. Have columns for total monthly cost in each of the various cost categories, a column for the cumulative cost as of the beginning of the month, a column for the monthly interest that will be added to the debt, and another column for the cumulative costs as of the end of the month (which equals the beginning of the next month). With this information, you can calculate the monthly interest charged on the cumulative costs at the end of each month (remember to use the monthly, not the annual rate of interest!). Use the spreadsheet to answer the following questions:

a. What will the total balance be at the end of month 24 when the building is completed?

b. What will the monthly interest be at the end of month 24?

c. If the developer's discount rate is 12%, then what is the present worth of the sum of money that would be required to pay off the construction loan?

Problem 7.12 Estimated Construction Costs for an Apartment Building

Expense	Amount	Months Incurred
Engineering design	$20 million	1–6
Construction	$100 million	7–20
Landscaping	$20 million	21–24
Project management	$1 million per month	1–24

7.13 The general contractor for the building described in the previous problem has identified a way to save $10 million in construction costs. However, savings would come from using a smaller workforce and extending the construction period by 2 months. Completing the project therefore would take 26 months, not 24 months. What would the balance be on the line of credit when the project is completed at the end of month 26? Should the owner allow the contractor to pursue the slower, but less costly construction plan?

7.14 The developer for the building described in the previous two problems actually has the cash on hand to pay for the construction costs as they are incurred. Using the developer's discount rate of 12%, what is the present value of the construction costs? As in the previous questions, assume that all expenses are incurred at the end of the month. Remember, since the developer is not borrowing any money, there is no interest expense. Compare the present worth of construction costs as calculated in this question with the present worth of the sum required to pay off the construction loan in part c of the initial question. List the advantages and disadvantages of using a construction loan.

7.15 A developer has just completed construction of an office building. The construction cost of $300 million (including interest) was covered by a line of credit with 10% interest. The building currently has long-term leases that provide expected net income (lease income minus operating expense) of $20 million per year.

a. The developer would like to obtain a 30-year, 8% mortgage on the new building. What would the monthly payments of principal and interest be for this $300 million mortgage?

b. The bank is willing to provide an 8% mortgage to the developer, but it has imposed a restriction on the maximum amount it is willing to loan. The annual mortgage payments must be no more than 80% of the expected net income from the building. What is the maximum amount the bank would be willing to lend?

7.16 The developer of an office building expects total construction costs to be equivalent to $100 million as of the completion of construction. The developer anticipates that it would be possible to obtain long-term leases from several major tenants that would provide annual net income of $10 million per year for 30 years. The developer is wondering whether to pursue the project.

a. *Using internal financing*: The developer is able to use existing funds to pay for the building construction. If the developer's discount rate is 12% and the project life is 30 years, what is the NPV of the leases as of the completion of construction? Is the expected income from the leases enough to justify the $100 million project? (In other words, calculate and comment on the NPV of the entire project as of the completion of construction.)

b. *Using a long-term mortgage*: Suppose the developer could obtain a 30-year, 7% mortgage that would cover the $100 million construction costs. What would be the NPV of the entire project as of the completion of construction?

c. Comment on the differences between the results of the two prior questions.

7.17 The owner of an office building is interested in selling the building to raise capital for development of a large shopping mall.

The building has a 30-year, 7% mortgage with 20 years of remaining payments; the original mortgage principal was $200 million. The building is fully occupied by tenants who have long-term leases of at least 20 years. The owner enjoys net income of $1 million per month after paying all operating expenses and the mortgage payment. The new owner would be able to take over the existing mortgage.

a. What is the minimum offer that the owner would accept, assuming that the owner's discount rate is 20% (reflecting the owner's opportunities to pursue lucrative new developments)?

b. What is the maximum offer that a purchaser would make, assuming that the new purchaser (1) plans to take over the existing mortgage and (2) has a discount rate of 12% (reflecting potential purchasers who are interested in obtaining relatively safe, long-term investments)?

c. What is the maximum offer that a purchaser would make, assuming that the new purchaser (1) is able to refinance the remaining amount on the mortgage as a 30-year, 7% loan and (2) has a discount rate of 10% (reflecting the lower risks associated with securing the new financing).

d. The owner expects that there could be potential purchasers such as those described in parts b and c. What sale price would you recommend? Why might you recommend a sale price that is substantially higher than the minimum that the owner might accept?

7.18 Section 7.7 describes some of the costs and benefits associated with the construction of the Worldwide Plaza in New York City. The success of that project depended upon many factors, including the interest rates on loans, the rents that were charged, construction costs, and the time required to construct the building. Estimate the impact of the following potential problems on the financial success of that building:

a. The interest rate on the construction loan increased by 1%.

b. The interest rate on the permanent mortgage increased by 1%.

c. The construction took an additional 6 months.

d. The average rents declined by 10%.

Chapter 8

Choosing A Discount Rate

The world's largest iron-ore producer [sold] $1 billion in investment grade bonds due in 2016 priced to yield 6.254%. Cia. Vale do Rio Doce, which last year became the first Brazilian corporation to win an investment grade rating, issued the debt to help fund its repurchase of $300 million of its 9% bonds due in 2013, thereby cutting the company's borrowing costs.
"CVRD Issues Record Bond," *LatinFinance* (February 2006), 4

CHAPTER CONCEPTS

Section 8.1 Introduction
Section 8.2 Related Financial Concepts
    Profits and rate of return versus net present value of cash flows
    Leveraging
Section 8.3 Factors Affecting the Discount Rate
    Minimum acceptable rate of return (MARR)
    MARR as a function of opportunity cost, inflation, and risk
    MARR from the perspective of investors, entrepreneurs, companies, and public agencies
Section 8.4 Choosing a Discount Rate: Examples
    How different parties might choose a discount rate for evaluating a project
Section 8.5 Dividing Up the Cash Flows of a Major Project
    Risks of investment depend upon who gets paid first
    Perceived risks of a project change as it proceeds from design to implementation
    If risks are reduced, the value of a project increases

8.1 INTRODUCTION

The previous chapter introduced the important concept of equivalence. Given a discount rate, it is possible to relate any arbitrary sequence of cash flows to an equivalent present worth or future worth or to an annuity that continues indefinitely or for a fixed number of periods. The concept of equivalence is essential to evaluating infrastructure projects that require substantial investments before any benefits are obtained. Given the projected costs and benefits, it is possible to calculate the net present value (NPV) of a project, which can then easily be compared to the NPV for other projects. However, this extremely useful concept depends upon having a discount rate, and the selection of a discount rate is a complicated and ultimately quite subjective matter. The discount rate cannot be established by fiat, nor is there a methodology for determining the exact discount rate that someone should use in evaluating a project. Moreover, the various people and organizations considering a project are likely to use quite different discount rates in evaluating the same project. Since the discount rate determines the importance of future financial costs and benefits, it is necessary to give some thought to the choice of a discount rate.

The discount rate is similar, but not identical to the rate of return. The **discount rate** is a conceptual figure that is useful in establishing equivalence of cash flows. As described in Section 8.2, the **rate of return** is an accounting term that is used in describing past or predicted financial performance of companies. Historically, rates of return have been higher for riskier investments, because investors discount future earnings more heavily for such investments. The greater the perceived risks associated with stocks or bonds, the lower the prices they will command—and the higher the costs for the company or agency trying to raise capital for a project. Section 8.2 also introduces the process of leveraging, which entails borrowing money for investments. Leveraging makes it possible to undertake very large-scale projects, and it may make it possible to achieve a higher rate of return, but it also increases the risks associated with the venture. Section 8.3 discusses the factors associated with determining a discount rate. A key concept was introduced in the previous chapter; namely, the **minimum acceptable rate of return (MARR)**. The discount rate used to evaluate projects should be at least as high as the rate of return that could be obtained via other investments with similar risks. Another key concept is the **weighted average cost of capital**, which is the cost of money to the project. The minimum acceptable rate of return for a company will be at least equal to its cost of capital—and perhaps much higher.

The concept of a discount rate may seem to be rather arcane, but it has clear and important consequences for a company or a government agency. The quote at the beginning of this chapter describes how CVRD was able to reduce its interest costs from 9% to 6.254% because the financial community upgraded the company's credit rating. With lower interest on $300 million worth of bonds, CVRD would save approximately $8 million per year in interest from 2006 through 2013. How the financial community views the risks associated with a company can be extremely important to that company's profitability and even its survival.

8.2   RELATED FINANCIAL CONCEPTS

8.2.1   Profits and Rate of Return versus Net Present Value

Companies and investors often think in terms of profits and return on investment as well as or instead of present worth or future worth. Profit and return on investment are both accounting terms. Profit is the difference between revenue and expense, while return on investment (ROI) is the ratio of profit to the total amount invested. In the simplest case, an investment of I at time 0 results in annual profits of A/year over a very long time horizon. In this simple case, the annual return on investment is always

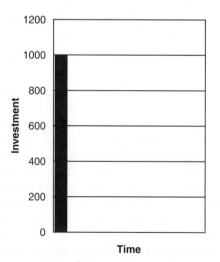 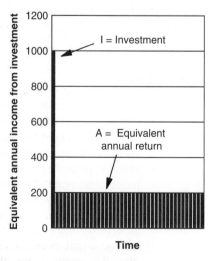

Figure 8.1 Rate of Return. If A is the annual income from an investment I, then the return on investment is A/I.

A/I. While this simple logic is fine for, say, buying bonds, it is insufficient for investments in infrastructure. The first problem is that the investment does not occur at time 0, but may in fact require years of effort. The second problem is that the revenues and expenses associated with the project are likely to vary over the life of the project. The third problem is that tax laws and accounting conventions determine what is called an expense and what is called an investment; the way that legislators and accountants consider financial matters may be quite different from the ways that entrepreneurs, companies, and investors do their analysis.

The first two problems can be addressed by using the concept of equivalence, assuming that an appropriate discount rate is given. The actual investments that take place over a period of months or years can be related to an equivalent total investment either at time 0 or at the time the project is completed and the operation begins. Likewise, the net benefits can be converted to an equivalent long-term annuity that begins either at time 0 or at the time that operation begins. The return on investment would then be the ratio of the equivalent annuity to the equivalent investment.

EXAMPLE 8.1   Estimating Return on Investment

Suppose that a project has the cash flows shown in Table 8.1. Using the investor's discount rate of 10% and assuming that all cash flows occur at the end of the period, the present worth of the investment will be $1,736, while the present worth of the revenue stream will be $1,924. The net present value of the project is therefore $1,924 − $1,736 = $188, and the project is worthwhile (at least from a financial perspective).

Table 8.1   Calculating the Present Worth of Cash Flows Associated with a Project

Year	1	2	3	4	5	6	7	8	9	10	Total
Investment	$1,000	$1,000									
PW of Investment	$909	$826	$0	$0	$0	$0	$0	$0	$0	$0	$1,736
Revenue	0	100	200	400	600	600	600	500	300	100	
PW of Revenues	$0	$83	$150	$273	$373	$339	$308	$233	$127	$39	$1,924

For this investor, the present worth of the revenue stream would be equivalent to an annuity of $192.40 per year, continuing indefinitely. From the perspective of this investor, the project is therefore equivalent to a project that required an immediate investment of $1,736 and produced an infinite revenue stream of $192.40/year. The ROI for such an investment would be $192.40/$1,736 = 11.1%.

Another way to look at this would be to determine the equivalent future value of the revenues at the end of the project's 10-year life. Then we could determine the growth rate that would be required for the present worth of the investment to reach the future value of the revenues. At the investor's discount rate of 10%, the revenue stream that has a present worth of $1,924 would have a future worth of $1,924 (1.1)^{10} = $4,991 at the end of 10 years. The present worth of the investment is $1,736. By trial and error, we find that $1,736 (1 + ROI)^{10} = $4,991 when ROI = 11.1%, which is the same result previously obtained. Thus we conclude that this project would provide an 11.1% return on investment for an investor with a discount rate of 10%.

The problems related to accounting and taxes are trickier. Profit and return on investment are defined by legislation and accounting rules, and it is not possible to adjust the definitions of these terms to be consistent with some other view of the world (e.g., the emphasis on net present value of cash flows that is presented in this text and other texts on engineering economy, management science, and project evaluation). The generally accepted belief is that managers and investors will do better by focusing on net present value (or its equivalents) rather than focusing on profits and return on investment.

In Chapter 10, we discuss how taxes and accounting rules affect cash flows. We also consider how changes in tax laws and accounting can be used to promote different types of projects. The basic message in that chapter is that it is important to consider taxes and accounting because accounting rules affect profits, profits affect taxes, and taxes affect cash flows. For the purposes of this chapter, however, we are keeping our focus on cash flows.

8.2.2   Leveraging

Borrowing the funds needed for an investment will reduce the initial investment required from the owner and potentially increase the expected return on the investment, but it will increase the risk of the project. Consider the case from the previous subsection in which a project's cash flow has been transformed into the equivalent cash flows represented by an investment of I at time 0 and annual profits A that are received at the end of every year thereafter. If all the investment funds are provided by the owner of the project, then the owner's return on investment will be A/I. If the owner borrows a portion of the investment, then two things happen: the owner's investment declines by the amount borrowed, and the annual profit declines by the amount paid in interest. The owner's return on investment is now

$$\text{Owner's ROI} = (A - \text{loan interest})/(I - \text{loan principal}) \qquad \text{(Eq. 8.1)}$$

Figure 8.2 illustrates what would happen in a situation where the initial annual return A is $20 million on an investment I of $100 million. The y-axis on this chart should be interpreted as millions of dollars for income and debt payments, but for ROI it should be read as the annual %. The chart on the left shows the initial situation in which all of the investment is provided by the owner; the annual income is $20 million, there is no interest payment on the debt, and the ROI is 20% (A/I = $20 million/ $100 million).

The chart on the right shows what happens if the owner provides only half of the investment and borrows the other $50 million at an interest rate of 10%. In this case, the annual income remains unchanged at $20 million, but there is an interest payment of $5 million. The owner's ROI jumps to 30% because the net income to the owner after paying the interest is $15 million, which is 30% of the owner's investment.

Borrowing money will increase the owner's ROI so long as the interest rate is lower than the return on the original project. The process of borrowing money in order to increase the amount that can be invested is called **leveraging**. Leveraging may increase or decrease the profitability of the project. Since infrastructure projects generally require substantial investments, most such projects are highly leveraged. Minimizing interest costs becomes a major concern in such projects.

Whether leveraging is undertaken to increase ROI or simply to enable the project to be constructed, it will increase risks due to the possibility of a default on interest payments. Figure 8.3 shows why this is so. As suggested by the chart on the left, cash flows may vary from year to year. In this situation, annual profits range between $5 million and $70 million. In the unleveraged situation, smaller annual

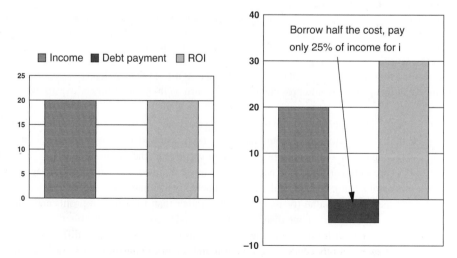

Figure 8.2 Benefits of Leveraging. Debt financing increases expected returns if the interest rate is lower than the unleveraged ROI.

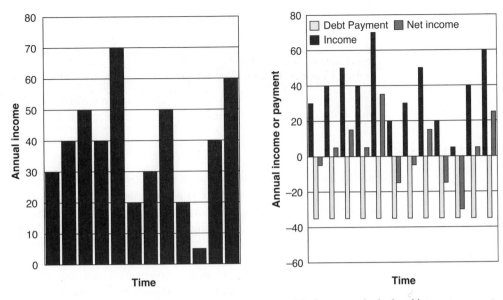

Figure 8.3 Risks of Debt Financing. Leveraging increases risks because principal and interest payments must be paid when due.

profits simply reduce the profitability for that year, and the owner still enjoys positive cash flow. In the chart on the right, it is assumed that the project is highly leveraged, requiring interest payments of $35 million per year. As a result, the cash flows in some years are insufficient to cover the interest payments, and the owner must use other sources of funds to make the interest payments and avoid bankruptcy.

The greater the uncertainty in predicting future cash flows, the greater the risks from leveraging. Leveraging works best in situations where there is a stable source of revenue, so that there is little risk of interest expenses exceeding actual cash flows.

8.3 FACTORS AFFECTING THE DISCOUNT RATE

8.3.1 Minimum Acceptable Rate of Return (MARR)

As described previously in Chapters 5 and 7, the discount rate reflects the time value of money to a particular individual or organization. Future cash flows must be discounted because of at least three factors.

1. **Opportunity cost**: The opportunity cost represents the potential financial benefits that must be forgone once a decision is made to invest in a particular project. Instead of investing in a particular project, individuals and companies could invest in other projects, in stocks and bonds, in real estate, or in other ventures. If they have borrowed money, they could pay off their debt. If they have sold stocks to the public, they could buy back some of the stocks.
2. **Inflation**: Inflation is likely to erode the purchasing power of money received in the future. As a result, the same amount of money would purchase less in the future than it would today.
3. **Risk**: The money that is anticipated to be received sometime in the future may or may not materialize.

These factors overlap to some extent since the opportunity cost depends to some extent on expectations concerning inflation and risk. Also, each of these factors will be treated differently by those promoting a project and those investing in a project. Nevertheless, both promoters and investors are likely to have a **minimum acceptable rate of return (MARR)** as they contemplate undertaking or investing in a project. The MARR represents the rate of return that promoters or investors believe they could achieve via other investments with similar perceived risks. Discounting cash flows using the MARR as the discount rate will indicate whether the proposed project is as attractive as pursuing the other available options. Promoters and potential investors have different investment options, and they have different perceptions of the risks associated with a particular project. Therefore the MARR used by those proposing a project is likely to be different from the MARR used by those who are asked to invest in the project. The differences are worth discussing in further detail.

8.3.2   MARR for Investors

Potential investors have quite a different perspective from those proposing a project. They are not necessarily concerned very much, or at all, with the project objectives; instead, they are primarily interested in maximizing their financial returns. They probably do not understand nearly as much about the technologies, locations, opportunities, or possibilities as do the promoters of a project—and the investors are apt to be leery of promoters who perhaps have reasons to oversell their project proposals.

Investors have numerous investment opportunities, and they have access to many qualified investment analysts who rate the financial attractiveness of many of those investment opportunities. They are likely to have a very good understanding of the financial markets, and they have their own track record with respect to investing in different sectors and in different types of securities.

In general, investors can seek higher returns by investing in riskier endeavors, as illustrated in Figure 8.4. Savings accounts insured by the U.S. government and bonds issued by the U.S. government may be viewed as a risk-free investment, and they may offer an interest rate of 5% or less. Riskier bonds will require higher interest rates in order to attract investors. The risk that a bond will default can be estimated by analyzing the finances of the company or government agency issuing the bonds. Companies such as Moody's rate bonds to indicate their creditworthiness; the best bonds are rated as AAA, then AA, B, and so on. Bonds with lower ratings require higher interest rates because the probability of default is greater; bonds with very low ratings may be deemed unsuitable as investments for some very conservative pension funds or mutual funds. In Chapter 7, Section 7.5.2 showed how bond prices reflect the discounted value of the stream of interest payments plus the discounted value of the ultimate redemption of the face value of the bond. The lower the credit rating, the more the interest

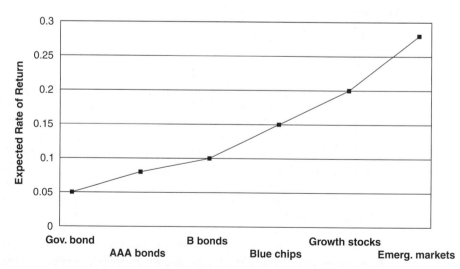

Figure 8.4 Risk versus Expected Return. The higher the perceived risk, the higher the return that is necessary to attract investors.

and the ultimate redemption of the bond will be discounted. Since bonds commonly are sold in standard denominations (e.g., $1,000), the interest rate must be increased for riskier bonds to justify the sales price of $1,000.

Figure 8.4 shows stocks as having a higher expected return than B-rated bonds. This makes sense, because the interest must be paid on the bonds before any cash is available to the stockholders. "Blue chips" are stocks in the largest, most well-known, and very financially stable and attractive companies. Blue chips have an excellent history of good financial performance, and so they are viewed as having relatively low risk. The value of blue-chip stocks is largely dependent upon a continuation of past performance rather than anticipation of future improvement. Growth stocks are issued by companies that anticipate rapid growth in earnings; the value of the stocks is based more on hope for the future rather than continuation of the past. Since the earnings are expected, not proven, investors discount future cash flows more heavily than they discount blue-chip cash flows. Even higher discounts are applied to cash flows projected by companies involved with new technologies or companies that do substantial business in regions of the world where financial markets are relatively undeveloped. The opportunities may well be much greater in such markets, but the risks are also much higher.

The risk-return curve plotted in Figure 8.4 is determined by market forces and economic conditions. The shape of the curve is affected by the state of the economy, government policy concerning interest rates, government debt levels, the number of new offerings of stocks and bonds, political uncertainty in various regions around the world, public perceptions concerning new technologies, wars, and other factors. The discount rates implied by the risk-return curve can be interpreted as resulting from the sum of three factors:

1. The real return available on risk-free investments
2. The expected rate of inflation
3. A risk premium that reflects investors' collective views concerning the riskiness of the company, including judgments concerning the company's financial situation, the outlook for the entire industry, and the social and political conditions in the country or countries where the company is located or where the company's products are sold.

Each of these factors is used to "discount" or diminish the current value of future earnings. For example, given the risk-free return (rf), the annual risk premium, and the inflation rate (inf), the present worth of the cash flow (CF) in month N would be

$$\text{Present Worth of CF(N)} = CF(N)/((1 + rf)^N (1 + risk)^N (1 + inf)^N) \qquad \text{(Eq. 8.2)}$$

If each of the three factors is small, then

$$\text{Present Worth of CF(N)} \sim CF(N)/(1 + rf + risk + inf)^N \qquad \text{(Eq. 8.3)}$$

The discount rate, in this case, would be (rf + risk + inf).

It is important to understand what the chart in Figure 8.4 shows and what it does not show. It *does* show that investors will discount future cash flows much more severely if they believe that a lot of risk is associated with a project. It does *not* show that investments in riskier projects will make more money. It *does* show that reducing risks allows a developer or a company to raise more money *today* based upon its projected cash flows for *tomorrow*.

It is even more important to understand that the above chart represents the market's evaluation of risks and investment opportunities. This is not the same as the owner's or developer's assessment of opportunities. Each individual and each company has their own minimum acceptable rate of return (MARR) and their own preferences for risk and return. Their MARR is based upon their opportunities, perceptions of their risks, and their preferences for risk and return. They will discount projections of future cash flows in their projects using their MARR for projects they deem to have similar risks. Because of their own experience or knowledge or ignorance, they may well believe that a project can be completed as planned and that it will indeed achieve the anticipated cash flows. Hence, they may see projects as being less risky (and therefore a better alternative) than the same projects would be viewed by the market. They will therefore go to rather great lengths to convince potential investors that the project is feasible and that the risks are not as great as the investors might fear.

One way to gain insight into the risk-return curve is to look at the historical returns of mutual funds. A mutual fund is managed to meet a specific investment objective, which could be one of the types of securities listed in Figure 8.4, a mixture of such securities (e.g., 30% bonds and 70% blue chips), or a more restricted type of securities (e.g., stocks of companies involved in biotechnology or electronics). The manager of the mutual funds buys stocks and bonds, and the value of the fund is based upon the value of the individual securities that it owns. The performance of a mutual fund is monitored, and quarterly and annual reports are issued to shareholders. The annual returns from mutual funds over a 5- to 10-year period therefore provide a good indication of the risk-return curve that is available to investors. Figure 8.5 shows returns in 2004 for selected mutual funds offered by Fidelity Investments; this was a year in which interest rates were low; bank accounts and money market mutual funds paid only about 1% interest. The Standard & Poor's 500, an index based upon the average returns for the 500 largest U.S. companies, returned 11%. Biotechnology stocks and small companies did better (*small cap* means "small capitalization," i.e., companies with a lower total value of stocks and bonds than the "large cap" companies), while international and emerging markets did much better. The best sector that year was construction and housing; stocks for such companies rose by nearly 30%.

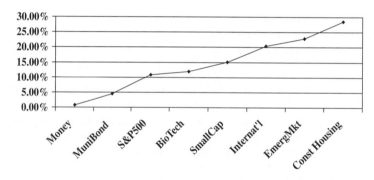

Figure 8.5 Equity Returns, Selected Securities, for the Year 2004.

Source: "Fidelity Mutual Fund Performance," *Fidelity* (February 2005), 20–27.

The returns for longer periods are more indicative of the market's view of the sectors. Figure 8.6 shows the 5- and 10-year returns for the same sectors as well as the 1-year returns.

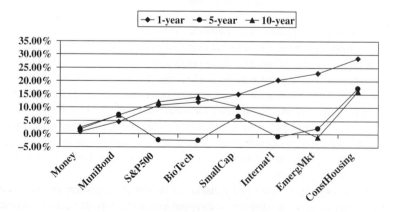

Figure 8.6 Equity Returns, Selected Securities, for the Years 1995–2004.

The risks associated with similar financial securities can be very different in different regions of the world. Figure 8.7 illustrates the different rates on various bonds issued between 2000 and 2006 by the U.S. government, a mining company in Brazil, the government of Brazil, and the government of Panama. The lowest interest rates were on the U.S. government bonds, and the highest were on the bonds issued by Panama in 2000 (P00-20).

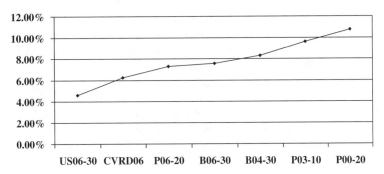

Figure 8.7 Interest Rates for Selected Bonds Issued in the United States and Latin America between 2000 and 2006. U.S. 30-yr Treasuries; 10-yr bonds for CVRD, a large mining company in Brazil; Brazil 30-yr bonds; Panama 10- and 20-yr bonds.

Source: "Deals," *LatinFinance* (February 2006), 4.

8.3.3   MARR for Owners and Promoters

Project owners and promoters have a much different perspective than investors. The promoters may be a developer, a company, or an agency promoting a particular project; there may be a "champion" who has long been advocating the project and seeking political and financial support from many agencies, organizations, and companies. The promoters are seeking to achieve various benefits from the project; some of them may be financial, some may be economic and quantifiable as monetary benefits, and some may be social or environmental or other benefits that are difficult to quantify. They understand that they need to finance the project (i.e., come up with the cash necessary to pay for the investment and the continuing expenses of the project). Presumably they believe that the project is worth doing, that it is a good way to achieve the benefits that are sought, and that it has a reasonable chance of success. They will have considered similar projects, and perhaps they have already completed similar projects. They likely have consulted experts regarding the feasibility of construction, the demand for the project, and the risks associated with the proposed project. They also have a good idea of their own opportunity costs, based upon their experience and their investment options. They presumably do not expect their next project to be their best project ever, but they do expect to achieve returns similar to what they have achieved in the past and to what they believe they could achieve in other projects.

If the promoters have sufficient funds to construct and operate the project, then they can do pretty much what they want to do. If they need to raise money for the project, then they have additional financial concerns. If they borrow money, they will worry about the interest rates they will have to pay. If they need to sell stock, they will worry about what price their stock will command in the market and what proportion of the ownership must be transferred to the new stockholders. Their MARR must be greater than their cost of capital, and it must be greater than what they earn in other endeavors.

The cost of capital is an important factor for companies. Many companies raise money by selling bonds or selling stock. The cost of capital is lower for bonds than for stocks, as described in the previous section. For bonds, the cost of capital is the interest rate that is paid. For stocks, the cost of capital can be estimated as the historical returns to the owners of the stock. The annual return for a stock can be calculated as follows:

$$\text{Annual return} = (\text{Final Value} - \text{Initial Value} + \text{Dividends})/\text{Initial Value} \qquad \text{(Eq. 8.4)}$$

For example, consider a stock that was valued at $100/share on January 1, paid a $2 dividend on December 31, and had a value of $110 on January 1 of the following year. The stock increased in value by 10%, and the dividend was 2% of the initial value, so the total return was 12%. The average historical returns for a stock can be used as an estimate of the cost of capital for that company.

A share of stock represents ownership in the company, which means that stockholders have "equity" in the company. The **weighted average cost of capital (WACC)** depends on the relative amounts of debt and equity in the company and can be calculated as follows.

$$\text{WACC} = \%D(\text{average interest on debt}) + \%E(\text{average return on equity}) \qquad \text{(Eq. 8.5)}$$

where

$\%D$ = debt as a percentage of total value of debt plus stock
$\%E$ = equity as a percentage of total value of debt plus stock

---

EXAMPLE 8.2   Weighted Average Cost of Capital

What is the weighted average cost of capital (WACC) for a company whose ratio of debt to equity is 3 to 2, whose average interest rate on debt is 8%, and whose historical returns to equity have been 16%? Using the above formulation (Eq. 8.5), the company's average cost of capital is 11.2%:

$$\text{WACC} = 60\% \, \text{debt}(8\% \, \text{interest on debt}) + 40\% \, \text{equity}(16\% \, \text{return on equity})$$
$$= 0.6(0.08) + 0.4(0.16) = 0.48 + 0.64 = 11.2\%$$

As discussed above, a company may reduce its credit rating and increase the risk of bankruptcy if it increases its proportion of debt. If a company increases the ratio of debt to equity, it will eventually encounter higher interest rates and difficulties in selling bonds. To maintain their credit ratings, many companies endeavor to maintain a constant ratio of debt to equity. If this ratio is constant, then the WACC represents what it costs (or has cost) the company to raise capital for new projects. Over time, as the debt-equity ratio and market conditions change, the cost of capital will also change.

Analysis of the cost of capital can be made much more complicated. Instead of the weighted average cost of capital over the past year or past 5 years, a company is likely interested in its current cost of capital. In regulatory affairs and legal cases, lawyers and experts debate projected cash flows and discount rates. In Wall Street or other financial centers, analysts are trying to figure out the true values of stock prices, which presumably reflect the projected cash flows for the company, the discount rate applied to those cash flows, and the number of shares outstanding. Company financial officers present their projections of cash flows to financial analysts, and the financial analysts make recommendations to their clients as to whether the stock should be bought, sold, or held. The financial analysts may make their own judgments regarding future cash flows of the company, and they may select a discount rate based upon their perceptions of risk. Their judgments are critical in determining share prices as well as the interest rates the company will have to pay on future sales of bonds.

The benefit to a company in issuing stock instead of bonds is that there is no requirement to issue dividends, which is very helpful during unprofitable periods. If a company defaults on interest payments on bonds, it can be forced into bankruptcy; but if it decides not to issue dividends, there is no such risk. The negative aspect of selling stock is that the sales price may reflect investors' application of very high discount rates to the company's projection of future cash flows. As a result, the current owners may give up more of the company (and its future earnings) than they want to if they try to sell more shares of stock.

EXAMPLE 8.3   Perceptions of Risk May Make it Impossible to Finance a Good Project

Suppose a group of engineers and entrepreneurs have created a company that is trying to raise $100 million for constructing what they believe to be a lucrative project in a developing country. The company has done extensive research concerning construction costs, operating cost, and potential revenues. While they recognize that there are some risks related to regional economic conditions, they are confident that their proposal can be constructed on time and on budget and that it will attract high demand almost immediately. They have prepared a prospectus in which they describe the project in great detail, and they anticipate annual profits to reach $20 million per year for at least 30 years. Using a discount rate of 10%, they estimated the net present value of the revenue stream to be $200 million as of the time the project begins operations, which is far greater than the construction cost. They will have no income for several years while the project is being constructed, so they hope to raise the $100 million from a combination of low-interest loans and sale of stock to get started. Almost immediately, they discover that they would have to pay 12% interest for a construction loan of no more than $20 million; by the time construction was completed, they estimated they would have to pay $5 million in interest. Then they approached several potential investors to see if it would be feasible to sell stock in their company to raise the funds necessary to cover the remaining $80 million of construction costs plus the $5 million in interest. They indeed found investors interested in infrastructure projects in this country—but those investors were discounting cash flows by 25% or more when they evaluated such projects. Hence, for those investors, the cash flows of $20 million per year were worth only $80 million (estimated as the equivalent present value of an annuity for an indefinite period = $20 million/year [P/A, 25%, infinity] = $20 million/(0.25) = $80 million). Thus, to these investors, the returns from the proposed project were much less than their MARR, so they would not be willing to pay enough for the stock for the developers to finance their project even if they purchased 100% of the company. The project therefore collapsed because it was impossible to raise sufficient funds to get started.

8.3.4   MARR for Large Companies

A large company's minimum acceptable rate of return (MARR) will never be less than its weighted average cost of capital (WACC), because it can always buy back its stock and bonds if it has an excess of cash and no suitable investments. Since the WACC reflects market valuations of its stock and bonds, the WACC and therefore the MARR will be affected by macroeconomic factors such as inflation and recessions as well as by the market's perception of risks associated with the company and the industry.

The company's MARR must also take into consideration the company's opportunities for investments. Investments in the company may be desirable to expand capacity, increase efficiency, support moves into new types of business, or improve safety. Company officials will have many competing internal requests for capital, and they will have (or should have) an excellent perspective concerning the potential benefits for these competing projects. A company may also consider broader investments. It can invest in the financial markets, just like any investor, or it could attempt to buy or merge with other companies.

In a completely predictable world, a company perhaps could invest in any and all projects with a return greater than the WACC. After all, if a company can raise money by selling stocks and bonds at an average cost of, say 11%, then it can increase its profits by investing in all projects with an expected return greater than 11%. If so, then the MARR for this company would be the WACC.

In actual practice, however, the company faces many uncertainties. The projections of cash flows are based upon numerous assumptions, and there is in fact a potential distribution of cash flows associated with each project. The expected return may be greater than 11%, but there may be a good chance of earning less or even substantially less than 11%. The more highly leveraged the company becomes, the greater the potential that it will be unable to cover its interest payments. The company may not be willing to risk losing money and therefore may not wish to invest in projects unless it is very sure that the projects will have a return greater, and perhaps substantially greater, than its WACC. This kind of financial discipline will be imposed at three levels:

1. The chief financial officer will scrutinize proposals and impose criteria for ranking and selecting projects, taking into account the need to maintain a debt-equity ratio that is acceptable to investors.

2. The board of directors may limit the total capital budget for the company in order to limit the overall risks associated with the company.

3. The capital markets may decide that a company is a credit risk, which means that someone will issue a report or a study or a credit ranking that reflects poorly on the company. Companies, countries, or agencies with a poor credit rating will have to pay higher interest on any future bonds that they issue, and they will probably see a reduction in their stock prices or more difficulty in raising money for their projects.

Thus, in practice, a company typically is unable to invest in all of the potentially profitable projects. Instead, it is likely to invest in the most profitable projects, which means that it will set its MARR (often called a hurdle rate) well above its average cost of capital.

8.3.5   MARR for State, Local, and Federal Governments

Governments are often able to sell bonds at very low interest rates, so it may appear that the cost of capital for a public agency is very low. However, the low rates reflect in part the ability of governments to raise money by taxation. The actual cost of capital reflects not only the interest costs on public bonds but also the opportunity cost associated with taxation. What could the public have otherwise done with the money they paid in taxes? When economists consider this question, they come up with a higher cost of capital and therefore a higher discount rate or MARR for public projects.

Consider a taxpayer who has a minimum acceptable rate of return of 12%. This individual or business would be unwilling to invest in a scheme that offered less than 12% profit. The taxpayer would not keep all of the profits, because any applicable federal, state, or local taxes would have to be paid. If those taxes amounted to a third of the profits, then the minimum acceptable after-tax profit for the taxpayer would be 8% of the investment. A taxpayer concerned with efficiency in government might feel that government-funded projects should be subjected to the same kind of test: the government should not invest in projects unless they also provide an after-tax benefit of 8%. Since governments do not pay taxes, this taxpayer might conclude that the government should use a MARR of 8% for projects funded by general tax revenues. The taxpayer might prefer projects to have a financial return of 8% but perhaps would be willing to include clearly documented economic benefits received by society.

There is certainly some credibility to the argument that governments should not use tax revenues for projects with only modest benefits if the same money could have been better used by the taxpayers themselves. Deciding what rate of return is required is a matter of policy more than logic. In the United States, the General Accountability Office from time to time issues guidelines for the discount rates that should be used for public projects. These rates have recently been 7–8%, which is consistent with the logic expressed above. Given the many difficulties in measuring and monetizing the public costs and benefits of a project, the choice of a discount rate is just one of many analytical assumptions that must be considered in evaluating a public project.

8.3.6   MARR for Special Government Agencies

Governments sometimes create special independent agencies that are authorized to raise money by selling bonds. These bonds are backed not by taxation, but by the financial credit of the agency. For these agencies, the cost of capital really is the interest rates that they must pay on their bonds. These agencies are therefore able to invest in projects with low financial returns—presumably justified by the broader socioeconomic benefits provided by the agency. Examples include port authorities, turnpike authorities, and housing authorities.

8.4   CHOOSING A DISCOUNT RATE: EXAMPLES

This section presents several examples illustrating the logic behind the choice of a discount rate. Note that the choice is heavily dependent upon the perspective of the individual or organization that is making the choice. In some situations, there may be no clear answer, and there is usually no precise answer.

EXAMPLE 8.4   Determining the Interest Rate for Corporate Bonds

As the VP of Finance for Acme Construction, you have been asked to estimate the interest rates on bonds that your company plans to sell in order to finance the construction of a new toll bridge. Long-term U.S. Treasury bonds currently pay just under 5% interest, and you know that investors usually consider these to be nearly risk-free investments. You expect that the rate of inflation (currently 2%) will increase to 3% by the time your company is ready to sell the bonds. You expect that your company's bonds will continue to have a risk premium of 2% relative to U.S. Treasury bonds. What interest rate should you plan to pay on these bonds?

Answer: This question addresses the interest rates that investors will require to invest in your project. From a financial analyst's perspective, the interest rate will be determined by the market. As seen in Section 8.3, the market requires higher returns for riskier securities (see Figure 8.4, which illustrates a risk-return curve). Discount rates for each type of security will equal the sum of three factors:

1. The interest rate on a risk-free, inflation-free investment (i.e., the basic time value of money, a couple of %)
2. The inflation rate
3. A risk premium

Here, we have 5% interest on U.S. Treasury bonds, which includes inflation but is presumed to be risk free. We expect inflation to increase by 1%. And the risk premium will be 2%. Therefore we expect the rate to be 5% + 1% + 2% = 8%.

EXAMPLE 8.5   Choosing a Discount Rate Based upon a Firm's Cost of Capital

Brothers K, a construction firm specializing in prison security, borrows money at 6%; its stock is priced by the market to provide a 12% return. The company's debt-equity ratio is currently 2. What discount rate should the company use to evaluate new projects so as to at least cover its weighted average cost of capital?

Answer: We know that the company's MARR should be at least as great as the weighted average cost of capital. We are given no further information about the company, except that it is an established firm with prior projects and an ability to raise money by borrowing or selling stock. The debt-equity ratio of 2 means that loans account for two-thirds of the market valuation of the company, while stocks account for one-third of the market valuation. The weighted average cost of capital will be 2/3(6%) plus 1/3(12%) = 8%. It would be possible to argue for a higher rate based upon the likelihood that the MARR will be above, or even significantly above, the WACC.

EXAMPLE 8.6   Choosing a Discount Rate for a City

Cities must justify investment in infrastructure by comparing the present worth of net benefits over the life of the project to the present worth of the construction costs. Cities raise money through income tax, property tax, or the sale of bonds. Interest paid on municipal bonds is about 4%, which is lower than the interest rates that private companies pay on their bonds (about 6%) and much lower than the returns required to sell stock (10% or more). Private companies often use a discount rate of 12% or more, because they have many attractive investment opportunities. What discount rate should the city use in evaluating the present worth of its investments?

Answer: A city's discount rate must be greater than the interest it pays on bonds (4%) because it raises money from taxpayers—individuals and companies—who have other options. The cost of capital for taxpayers will reflect some mixture of debt and equity financing, that is, about 8% if the Brothers K firm in the previous example is typical. As noted above, the General Accountability Office requires the U.S. federal government to use something like 7–8% for discounting related to public investments.

EXAMPLE 8.7   Choosing a Discount Rate Based upon a Firm's Investment Options

Earp Enterprises is a developer in Arizona that specializes in building relatively inexpensive, but functional corrals that they call OK Corrals. Because the company uses advanced planning techniques and standardized components, it has consistently been able to make a return of 14–18% on its projects. The company's weighted average cost of capital is 10%, and it is a highly profitable company. A recent PhD from MIT, Flora Holliday, would like Mr. Earp to invest in a second home development focused on a new marina at Lake Powell. She has prepared a business plan showing the cash flows that she expects from this venture, which she calls Holliday Docks. What discount rate should Earp Enterprises use to determine whether to shoot down this proposal?

Answer: Earp's MARR must be greater than or equal to his cost of capital, which is 10%. If capital is unlimited, then any project that earns the cost of capital will be acceptable. On the other hand, capital probably is not unlimited and the question certainly implies that Earp has done very well on his prior projects. Since Earp has won many financial victories at his OK Corrals, he seems to have opportunities to make 14–18% "consistently." Hence, his MARR should reflect the lucrative potential for continued investment in OK Corrals and be higher than the WACC, perhaps 14–15%.

8.5   DIVIDING UP THE CASH FLOWS OF A MAJOR PROJECT

It is important to understand that the various players involved in a major project will have markedly different cash flows, different risks to worry about, and different MARRs. Consider a case where a developer has secured a line of credit from Bank One for constructing a building. The bank will pay all the construction costs and charge the owner interest; no payments will be made on the loan until the project is completed, at which time loan payments will begin. The construction loan is likely to have a high interest rate, since there are risks related to the feasibility, time, and cost of construction. The owner plans to refinance the loan with Bank Two when the building is completed at the end of year 3. Refinancing should provide a lower interest rate because the building will in fact be completed and will (hopefully) have tenants who are paying on long-term leases. If all goes well, the new loan will cover the construction loan, and the monthly loan payments plus operating costs will be less than the revenue from the tenants.

The cash flows for the three major players will be as follows:

- Bank One—pays all construction costs as they are incurred over the 3 years of construction; receives reimbursement plus interest when the loan is refinanced at the end of year 3.
- Bank Two—gives the owner an amount large enough to pay off the construction loan at the end of year 3; receives monthly payments of principal and interest for the life of the mortgage.
- Owner—pays nothing during construction period, since all of those costs are covered by Bank One; receives a large amount from Bank Two at the end of 3 years but immediately uses that money to pay off the loan from Bank One; collects lease payments, pays for operations, and makes loan payments to Bank Two over the life of the loan.

Bank One has completed its role by the end of year 3. Bank Two just starts its role at that time. The owner, if all goes well, doesn't have to put up his own money to construct the building, and then has sufficient cash flow to cover the mortgage payments. Bank One prefers to earn high interest rather than hold onto its cash; the owner prefers to have the cash as needed. Bank Two is willing to accept lower interest, but is also creating a long-term annuity for itself. The owner would rather pay the interest on a long-term loan than pay for the building when it is constructed.

We could go into more detail and consider such things as the cash flow for the construction firms and suppliers or the possibility of selling the building upon completion. Each actor will have different perspectives on whether this is a good project, and each will have a different level of exposure to the risks that might be associated with the project. (To explore these issues in more detail, complete the "Skyscraper" assignment at the end of this chapter.)

8.6   SUMMARY

Discounting is a mechanism for converting an arbitrary stream of cash flows into a present value, a future value, or an annuity. For example, using a discount rate of i%, the present value of a sum of money M received in N years is worth $M/(1 + i)^N$ today. Three main factors must be considered in choosing a reasonable discount rate for such a calculation:

1. *Investment opportunities:* What alternative opportunities are available for investment?
2. *Risk:* Is the proposed project more or less risky than the other options?
3. *Inflation:* How much will inflation reduce the future purchasing power of our money?

Discount rates and the notion of a minimum acceptable rate of return (MARR) are very important and potentially very confusing topics. The confusion is a result of the differences in perspective among the various actors involved in designing, building, and financing a project.

Developers and entrepreneurs are generally in the position of raising money for constructing projects that they believe to be justifiable in terms of their future benefits. Since they lack the funds to build these projects, they must convince others to invest in them—possibly by citing the importance of the project, but more likely by demonstrating the potential profitability of the investment. Developers have various strategies for raising funds for their projects, including borrowing money from a bank, selling bonds, and selling stock:

- *Loans*: A loan may be secured by a mortgage on the property, or it may be unsecured. A mortgage typically calls for the loan to be repaid over a fixed period of years, with interest rates that may be fixed or that may vary with market rates. If the mortgage payments are not made, then the mortgage holder may take ownership of the property. Since the mortgage holder has a claim on a physical asset if payments aren't made, the interest rates on mortgages will be lower than the interest rates on long-term, unsecured loans.
- *Bonds*: A bond is a financial instrument whereby the seller agrees to pay a specified amount of interest over the life of the bond and eventually repay the original face value of the bond. The buyer does not own any of the company, but the seller must pay the interest to avoid bankruptcy.
- *Stock*: When a company sells stock, it does not have a commitment to make mortgage or interest payments, although it may plan to pay dividends at regular intervals. Hence, when a company sells stock, it is able to raise cash without incurring a fixed cost of paying interest. However, the stockholders are actually purchasing part ownership of the company, sharing both the risks and the potential for future profit.

Notice the marked difference in perspectives between developers and investors. Developers are thinking of receiving rents or tolls or operating profits, which they hope will be enough to cover the mortgage or interest payments or to justify a high price for their stock. For them, the cost of money is similar to the cost of energy or labor, and their main concern is about the long-term success of their projects. For the bankers and other investors, the nature of the project is much less relevant than the prospect of making money from mortgage payments, loan payments, bond interest, or rising values for the stock they have purchased.

Financial markets exist for stocks, bonds, mortgages, mutual funds, and other types of financial assets. The price of these financial assets depends upon the market—that is, upon the price that a willing seller will accept from a buyer. The value of these assets (to an investor or a securities analyst) is based upon a projection of cash flows, an estimation of the risks associated with these cash flows, and the availability and price of other assets with similar levels of risks. Different potential investors may view the cash flows, the risks, and the alternative opportunities quite differently, which is part of the reason that securities are continuously bought and sold.

The different perspectives of financial analysts, governments, independent government agencies, and private companies result in different ways of determining their MARRs:

1. Financial analysts—in discounting securities, the discount rate is equal to the sum of three components:
   a. The real (before inflation) returns available from a risk-free investment
   b. The rate of inflation
   c. A risk premium based upon the company issuing the security
      Financial analysts are not particularly concerned about the merits of a company or a project; their job is to determine the risks associated with stocks, bonds, or loans associated with a company or a project. A company with good credit can easily raise capital to invest in bad projects; a company with no past history may be unable to raise funds even for projects that may appear to the public to be highly desirable.
2. Local, state or federal governments—in discounting costs and benefits of proposed projects, the discount rate should reflect the average returns available to taxpayers, not the low interest rates on public bonds.
3. Special public agencies—in discounting cash flows, the MARR should be at least equal to the agency's cost of capital. An independent agency that is authorized to issue bonds may use the interest rate on those bonds as its MARR.
4. Private companies—the MARR should be at least as great as the weighted average cost of capital, and it should be at least as high as the rate of return achievable on alternative projects or on investments in the financial markets. In practice, the financial risks associated with leveraging will generally result in a MARR well above the weighted average cost of capital.
5. In any situation, the discount rate is a rather fuzzy number, so it will be wise to consider a range of discount rates when evaluating a project.

In major projects, it is usually necessary to raise funds for construction from banks (loans) or financial markets (stocks and bonds). An entrepreneur or a company presents estimates of costs and benefits to banks and investors, who then evaluate the risks and choose discount rates consistent with their own MARRs. If a commitment is made to pay the banks before paying interest on bonds or dividends on stocks, then the banks' investments in the project are less risky than the investments made by those buying stocks and bonds. If the project is being undertaken by a government agency or a large company, it may be possible to get a low interest rate for loans based upon the agency's or company's credit rating rather than a rate based upon the riskiness of the project.

If you can reduce the perceived risks of your project, you can raise more money because investors will apply a lower discount rate to the same future benefits. In particular, a project that has been completed or that has very clear commitments for cash flows (leases for a building; approved tolls for a highway; approved public subsidies) will have lower risks than a new project with uncertain time to completion and no guaranteed source of income.

Leveraging is a term used when money is borrowed for your project and a pledge is made to (1) repay the loan or (2) turn the project over to the lender if loan payments are not made. Borrowing reduces the amount of their own money that the owners must put into the project and allows a chance for a greater return on their investment, but it also creates greater financial risks.

ESSAY AND DISCUSSION QUESTIONS

8.1 MARR: Discuss the factors that influence the minimum acceptable rate of return. Why might entrepreneurs, investors, banks, and public agencies use different discount rates for evaluating the same project?

8.2 Build for the Future: A left-wing government assumes power in a small country and declares that it will build for the future, so that the people will not always languish in poverty. The government plans to embark on monumental infrastructure projects whose enormous costs are being justified by the benefits to be enjoyed by many future generations. Is this program superior to a less capital-intensive program that would provide more immediate benefits to people suffering from a lack of clean water, inadequate roads, and substandard housing?

8.3 Zero Discount Rates: The "Clean & Green Party" has taken over the government in a small country in Europe. Citing the need to clean up the environment for future generations, the government announces that it will use a zero discount rate when evaluating its major environmental projects. That way the estimated costs of future pollution will be better balanced against the current costs of construction, forcing greater consideration of pollution control. Discuss why the government should or should not let the discount rate equal zero for environmental projects. (Assume that the Clean & Green Party and the "Business-as-Usual Party" both agree on the costs of pollution.)

8.4 **Discounting Shortchanges the Future:** Discounting certainly means that whatever happens in the future is discounted or counted for less in any analysis that we undertake. On the other hand, there really are markets for money, and people (and more important, banks and investors) do value today's money more than tomorrow's. Is discounting simply a convenience for doing financial analyses? Is it a way to ensure the dominance of the capitalist barons? Is it a way to ensure that financial resources will be available when needed to deal with future problems? Is discounting part of the problem or part of the solution with respect to achieving more sustainable development? Will careful attention to financial matters help or hurt efforts to improve generational equity (us vs. our grandchildren) and social justice (rich vs. poor)?

PROBLEMS

8.1 Suppose that you are evaluating various projects, each of which has an expected net cash flow of $100,000 per year for 20 years. Choose the discount rate that you would recommend in each of the following situations, state why that is the best choice, and indicate how much you would invest based upon your choice of a discount rate.

a. You work for Enormous Paper Company (EPC), which owns vast forests in northern New England. The $100,000 will come from sales of paper products from a new mill that EPC is planning to construct. EPC has built several similar mills throughout the region that have provided returns on investment of 10% to 20%. What discount rate do you use as you are trying to decide whether to build another mill: 5%, 10%, 15%, 20%, or 25%?

b. You are a banker, and a very large, financially stable company is willing to pay $100,000 a year for 20 years toward principal and interest on a loan. The prime rate for corporate customers is currently 7%. What rate do you offer this very attractive customer: 5.5%, 6.5%, 7.5%, or 8.5%?

c. Your friend has a great scheme for a website and wants you to cash in your trust fund in order to finance his new Internet business. He promises to pay you $100,000 per year for 20 years. What discount rate do you use in deciding the upper limit of what you might invest: 5%, 10%, 15%, 20%, or 25%?

8.2 A "Special Commission for Beaches (SCB)" is formed to design, build, and operate a series of beach resorts along the Charles River, which is expected to get cleaner year after year. The SCB is allowed to raise money by selling tax-free bonds (currently selling at about 4% for similar agencies). Revenues will come from user fees and, if necessary, from Cambridge, Boston, and other cities and towns along the river. What discount rate do you use in deciding how fancy to make the bathhouses, the restaurants, and the boating facilities?

a. 4%
b. 7%
c. 10%
d. 13%

8.3 As a financial analyst, you often advise companies what interest rates they can expect to pay on a new bond issue. You have been asked to estimate the interest rates on bonds that a company plans to sell in order to finance the construction of a new toll bridge. Long-term U.S. Treasury bonds currently pay just under 4% interest, and you know that investors usually consider these to be nearly risk-free investments. You expect the rate of inflation (which was just 1% last year) to rise to 2% by the time the company is ready to the sell bonds. You consider private toll bridges to be more risky than airport projects, but less risky than port projects. Interest rates on bonds issued by airports averaged 5.25% last year, while interest rates on port projects averaged 6.75% last year. What interest rate should you plan to pay on these bonds?

a. 5%
b. 6%
c. 7%
d. 8%
e. 9%

8.4 Anna K, a Russian firm specializing in railway safety systems, borrows money at 8% and its stock is priced by the market to provide a 20% return. The company's debt-equity ratio is currently 3. What discount rate should the company use to evaluate new projects so as to at least cover its weighted average cost of capital?

a. 8%
b. 9.5%
c. 11%
d. 14%
e. 20%

8.5 A company has a debt-equity ratio of 0.5. It pays 9% interest on its bonds, and its stock has averaged 15% returns for the past several years. What is its weighted average cost of capital? Explain why this is or is not its MARR.

8.6 A city can sell general revenue bonds that pay 5% interest. Explain why the city can or cannot use this as its MARR.

8.7 Draw cash flow diagrams for Bank One, Bank Two, and the developer based upon the information in the example in Section 8.5. Assume that the project costs $65 million to construct plus $10 million in interest on the construction loan; the long-term mortgage will be for $70 million, with annual payments of

$6 million; the net rents will be $8 million per year (after paying operating expenses but before paying the mortgage).

> *NOTE:* The next two problems are unfortunately close to reality. In the early twenty-first century, housing markets boomed around the world, allowing developers and home owners to become wealthy, especially if they kept buying and selling real estate. This question follows the sad trail of a couple who give up secure jobs and a tiny mortgage to pursue high-paying jobs and their dream house. They moved to California just before the housing bubble burst, plunging the country into a deep recession. Problem 8 looks at the questionable decisions that seem to have been made by bankers who failed to consider the possibility that housing values could ever seriously decline. The issues addressed in these two problems fooled most of the financial experts in the world, so don't worry if some of this seems complicated (especially Problem 9)—feel free to make whatever simplifications or approximations are needed to continue with the analysis. The ultimate problems that occur when the real estate bubble is pricked will not depend upon nuances of the analysis, but upon the fallacy of believing that real estate values can only go up.

8.8 Mortgage Crisis—the Home Buyer: A bank has a policy that it will approve a mortgage for someone only if they have a steady income that is sufficient to cover the costs of home ownership. In addition to monthly payments of interest and principal, the bank requires the homeowner to make monthly payments into an escrow account that the bank will used to pay property taxes, property insurance, and mortgage insurance. The bank requires this arrangement so that it can be certain that the property taxes are in fact paid on time, that the property is insured in the case of fire or other disaster, and that the mortgage payments will be made even if the homeowner dies. The bank uses two common rules of thumb in reviewing mortgage applications. First, the buyer should make a down payment of at least 20% of the cost of house (and homeowners are not allowed to borrow this money from someone else—they must be able to produce the cash from their own resources). Second, the sum of the mortgage payment plus the escrow payment should be no more than 30% of the homeowner's income.

a. Consider a family with annual income of $60,000. Will they be able to obtain a 6%, 30-year mortgage for a house that costs $100,000, has property taxes of $3,000 per year, and will cost $1,200 per year for property insurance and mortgage insurance?

b. The couple decides they would like a bigger house. What is the largest amount that the bank would be willing to lend this family for purchasing a house with taxes and insurance that total $4,200 per year?

c. The family goes ahead and buys a bigger house for $300,000, and they take out a $200,000 mortgage at 6% for 30 years. After 10 years, they decide to move to California in order to get high-paying, high-tech IT jobs. If property values have risen 11% per year, how much will they be able to gain from the sale of their house? Assume that closing costs such as the real estate agent's fee will be 5% of the sale price. Also assume that a portion of the proceeds from the sale will be used to pay off a total of $300,000, consisting of the remaining portion of their original mortgage plus the outstanding balance on their home equity loan.

d. When they reach California, the family finds that the housing market is much more expensive than they expected, but they have all that money from the sale of their previous house, and they now make $300,000 per year. They conclude that they really are rich, so they should buy the best house they can afford, maybe something with a view of the mountains. If they are willing to put all of their profits from the sale of their previous house into their new house, how much will they be able to afford to pay per year for mortgage and escrow payments?

e. Unfortunately, what the bank offers is not enough for them to purchase the $4 million house of their dreams, which has $30,000 per year for taxes and insurance. So they decide to go to a mortgage company. What they would like is a mortgage with only a 10% down payment, so that they could afford their dream house (and, it must be added, they decided to include their expected bonuses of $41,000 and $72,000) in their application, stating their income as $413,000 per year, even though they knew they were unlikely to get such good bonuses again.) They were not concerned about stretching to make ends meet, and the agent noted that the old 30% limit had been increased to 35%, reflecting the fact that housing prices were rising faster in California than anywhere else in the country. The agent and the couple agreed that even if things were tight for a couple of years, housing prices would rise, their salaries would rise, and they would be able to refinance their mortgage for an even larger amount. The agent for the mortgage company agreed to provide the mortgage under the suggested terms, and they bought their dream house. And, it turns out, the agent quietly rounded off their income to $450,000 per year when he forwarded their mortgage application to his home office for approval. Assuming the company required a 10% down payment and limited mortgage plus escrow to 35% of what was reported to be their income, how much would they be able to borrow?

f. Unfortunately for our intrepid IT experts, they happened to buy their dream house at the top of the market. Three years later, housing prices had dropped by 25%. And then their company went through a massive restructuring, and both of them lost their jobs, along with a great many others in the region. Their only option was to sell the house at a loss, move back to the east, and take up their old jobs at a huge cut in pay. Approximately how much will they lose?

> *NOTE*: Problem 9 provides some insight into what motivated banks to create mortgage-backed securities, why these securities were riskier than anticipated, and how failure of these

securities could lead to a larger financial crisis. Don't get lost in the details, and make reasonable simplifications and

essentially 0% for the other mortgages. Assume that a default rate of 5% per year means that mortgage payments for the riskier mortgages are actually only 95% of what should have been paid.
~~ for every
~~ve-

contract is signed, Local Bank may simply settle down to the routine process of sending out monthly bills to the homeowner, receiving checks from the homeowner, and paying the taxes and insurances. Local Bank could also decide to sell the mortgage. Suppose that Local Bank has just closed on a 30-year, 6% mortgage for $100,000 on a house valued at $250,000 and owned by a couple who earn $200,000 per year. This is clearly a very safe mortgage; the mortgage principal would be only 40% of the value of the house, and the couple could afford to pay a much higher mortgage. In fact, another bank may be interested in buying such mortgages and would in fact be willing to pay more than $100,000 to acquire it. Why? Perhaps because that bank's discount rate for investments with that low level of risk is less than 5.5%.

a. Assuming that Big International Bank (BIB) values such mortgages using a 5.5% discount rate, what would they be willing to pay to acquire it?

b. Let's say that BIB actually acquires 100 similar mortgages at a total cost of $10 million. The total expected income from all of these mortgage payments, after BIB's expenses, is expected to be $5.5 million. BIB now devises a scheme to use the interest payments on the mortgages to cover the interest on bonds that it hopes to sell in the international financial markets. In publicity to potential investors, BIB emphasizes that all of its mortgages are of the highest quality and that rising home prices make the investments especially safe. Based upon its studies of international markets, BIB believes that overseas investors and investment bankers in the United States would be willing to buy such bonds if the bonds paid 5% per year. How much would they raise by selling such bonds?

c. A group of bright young MBAs at BIB believed they could greatly expand this business of mortgage-backed securities. In effect, they believed that aggregating hundreds of mortgages into a single bundle would result in securities that had less risk than the original mortgages. After all, some families might have health problems, and other homeowners might lose their jobs, but past statistics showed that only a small percentage of homeowners actually defaulted on their mortgages. Spreading the risk of default over hundreds of mortgages would surely make it easier to predict the expected number of defaults, and any losses would be spread out over hundreds of mortgages and thus have a minor impact on cash flow. These MBAs convinced BIB to purchase some riskier mortgages (e.g., mortgages with only a 10% down payment or ones that allowed homeowner to spend 35% of their income on their payments). Assume that 10% of the mortgages purchased by BIB were riskier mortgages carrying higher interest rates (7% instead of 6%), but had a default rate of 5% per year rather than a default rate of

ii. BIB's bonds would have to p~~
interest if they were backed in part by riskie.
Instead of 5%, what interest rate would be needed to a
investors?

d. BIB's board of directors is impressed by the profits made from the mortgage-backed securities. One of their brightest members, however, notes that some risk could be involved if the economy turned bad and defaults increased, because BIB would no longer be able to make the interest payments on the bonds. He suggested that BIB simply act as a middleman, buying mortgages from small banks like Local Bank and then reselling them to investment bankers who could create the mortgage-backed securities and arrange for someone else to manage the actual mortgage payments. BIB could make its profits by charging a small fee, say 1% of the mortgage value, in return for acquiring and making it possible to aggregate the securities.

  i. Under this scheme, if BIB processed $100 million in mortgages, what fee would it earn?

  ii. If 30% of the risky mortgages were defaulted within the first 3 years, what would it cost BIB?

  iii. What would happen to the market value of the mortgage-backed securities in which 50% of the mortgages were safe and 50% were risky (still assuming that 30% of the risky mortgages had defaulted by the end of year 3)?

  iv. How might a collapse in real estate values lead to widespread financial panic?

8.10 A pair of entrepreneurs, Bonnie Ash and Clyde Woods, have a great plan for extracting lumber from tropical forests in Bolivia. However, they themselves would have difficulty raising the $10 million needed to purchase the land, create access roads, and build housing and other amenities for their employees. Bonnie suggests, hopefully in jest, that they simply rob a bank; but Clyde instead suggests bringing their financial plan to the bank and asking for a loan. After all, they anticipated making a profit of $2 million per year on the deal. The bank officer, Frannie Maple, is actually quite interested in the scheme but is willing to put up only half of the money, as a 20-year mortgage with an interest rate of 10%; the high rate of interest was necessary, she says, because of potential risks to the scheme and the entrepreneurs' lack of experience in South America. She suggests they seek the remaining funds from a private equity company that specializes in the region. The partners in that firm are indeed interested in the scheme, and they are willing to invest an amount based upon their assessment of the cash flows and their MARR of 25% for forestry projects in South America. Will this be enough for Ash & Woods to finance their project?

CASE STUDY: TIME IS MONEY: FINANCING A SKYSCRAPER

To work on this assignment, you must thoroughly understand the differences between interest rates and discount rates, the differences in risks as a project progresses from concept through implementation, and the differences in perspectives of entrepreneurs, owners, and investors. If you can complete this assignment, then you have an excellent grasp of the concepts and applications of equivalence, and you will gain some clear insight into the financial mechanisms underlying a large infrastructure project.

Financing a Skyscraper

A simple cost model for the construction of a skyscraper could be based upon the following elements:

a. The land area (acres) and the price of the land ($0.1 to $10 million/acre)

b. The costs of clearing the land ($10 to $50,000/acre)

c. The foundation ($2–4 million/acre)

d. The weight-bearing structure

e. The exterior of the building, including windows, exterior materials, waterproofing, and insulation

f. The elevators

g. Heating, air conditioning, ventilation

h. Creation of a dramatic entry area and mini-parks

i. Interior walls

j. Interior finishing

For any infrastructure development, we are generally considering whether to begin a particular type of project. We probably have several basic design and location options plus numerous minor variations on these options (e.g., variations in size or quality of components). From the underlying logic of the project, we can develop a function that relates the project costs to the design and location possibilities. Separate functions can be used to estimate the initial investment costs as well as the fixed and variable operating costs. As engineers, it is our job to understand what options are available, including new designs or construction techniques as well as tried-and-true approaches.

Revenues

We also would like to estimate the demand for the services provided by our project, taking into account the quality of the service provided, the capacity of our project and competing facilities, and the price that we and our competitors charge. For a skyscraper, the revenues will be based upon the rental rates, commonly expressed as the annual rent per square foot. A skyscraper makes financial sense if the projected rents (net of any continuing expenses) will be sufficient to justify a mortgage sufficiently large to cover all of the construction costs, including interest on a construction loan.

Timing and Cost Assumptions

This assignment is based upon the construction of the Worldwide Plaza, which was presented as a case study in Section 7.7 of the previous chapter. Karl Sabbagh's very interesting book describes

the people and the processes involved in constructing a 50-story office building plus some smaller buildings in Manhattan. As is the case with most engineering books, the focus is on the technical rather than the financial matters. Hence, in addition to the information presented previously (which is about all there is in the book), some additional assumptions are needed to evaluate various options for the project:

a. The time between land acquisition and beginning of construction is exactly 18 months. All of the engineering and architectural work is completed during this period.

b. The time from groundbreaking until the time that tenants can move in is exactly 24 months (i.e., months 19 to 42 of the project if it is completed on time).

c. The permanent mortgage is obtained 3 months after the tenants move in (at the end of month 45 if the project is completed on time).

d. The tenants have 4 months (to fix up the interior) before they make their first rent payment (at the end of month 46 if the project is completed on time).

e. The fixed costs include $58 million for land acquisition, $5 million for preparation of the case for development, the architects' and engineers' fees (assumed to be $90 million of the $145 million for fees and borrowing costs cited in the note), $45 million of the construction costs (for the foundation, landscaping, entry, lobby, and roof), and $5 million of the project management costs.

f. The $90 million architect/engineering fees are paid uniformly over the first 18 months of the project; the $145 million construction costs are paid uniformly over the period of construction.

g. The construction could be stretched out over 3 years with a savings in construction of $5 million and essentially no chance of an overrun.

h. The project costs are covered by a construction loan with interest of 10% per year charged from the point that costs were incurred. (This type of loan works like a checkbook—the owner writes checks for all payments, and the amount of the check is added to the outstanding amount of the loan. Interest is charged and added to the outstanding balance at the end of each month. Note that the owner doesn't actually pay any cash until the construction loan is repaid.)

i. When the project is completed, the construction loan can be refinanced as a 30-year mortgage at a rate of 8%.)

j. All invoice and loan payments are made on the last day of the month.

k. Additional space, if available, could likely be rented at $30/sq ft per year or more.

Questions

1. *Project cost:* Construct a spreadsheet that can be used to calculate the costs of construction, including interest costs on the construction loan. Structure the spreadsheet to show the various cost categories across the top and the months (1 to 48) down the side. Have columns for total monthly cost in each of the various cost categories, a column for the total, and another column for the cumulative costs as of the end of the month. With this information, you can calculate the monthly interest charged on the cumulative costs at the end of each month (remember to use the monthly, not the annual rate of interest!). You can also calculate the present worth (PW) of the total project costs (use discreet discounting, compounded monthly). Have an area at the top of the spreadsheet where you can enter key parameters (including interest rate on the construction loan, length of the construction period, and the owner's discount rate) when doing sensitivity analysis. Use the spreadsheet to answer the following questions:

a. Estimate the total project cost as of the end of month 45 when—if all goes well—the permanent mortgage is secured. (The total cost is the sum of interest cost plus construction cost, which will equal the total amount of the construction loan as of the end of month 45. This amount should be roughly comparable to the $370 million cited in the case study.)

b. What is the present worth of the total project cost as of the beginning of month 1, assuming a discount rate of 15%?

c. How would the total project cost (as calculated in part a) and the PW (as calculated in part b) vary under the following circumstances?
   ◦ If the interest rate for the construction loan were 8% or 12% rather than 10%?
   ◦ If the time required for construction were 30 or 36 months rather than 24 months?
   ◦ If the owner's discount rate were 12% or 18%?

2. *Project revenue:* Assume that the project has one major tenant who pays $26/sq ft for 600,000 sq ft and another who pays $29/sq ft for 200,000 sq ft, while all other space is rented at $30/sq ft (these are annual rates per square foot for long-term leases).
   a. Calculate the monthly revenue assuming that the building is fully leased from the outset.

b. Calculate the PW of the revenue as of the beginning of the first month in which rents are received (use the owner's discount rate of 15%).

c. Calculate the PW of the revenue as of the beginning of month 1 (using the owner's discount rate of 15%).

d. From the owner's perspective, is this a worthwhile project?

3. *Refinancing:* Once the building is up and rented, the owner can refinance the building at a lower interest rate because the bankers now can see the completed building and the paying tenants. Assume that the bank will provide a 30-year mortgage with the maximum amount no greater than the minimum of (a) project cost or (b) 80% of the net present value of the rents (calculated with a discount rate equal to the interest rate on the loan). What would the maximum amount of the mortgage be? What would the monthly payments be?

4. *Selling the building:* Once the building is refinanced, the owners may get bored simply collecting rents and making mortgage payments.
   a. What is the minimum sale price they should accept assuming that their discount rate remains 15%? (Assume that the new owners will take over the mortgage, and be sure to include the balance on the mortgage in the sale price!)
   b. Suppose they see great opportunities in South American gold mines, where they expect to make a 25% return—how low would they go?
   c. What would you recommend as a reasonable sale price for the building? Why?

5. *Redesigning the building:* Suppose that, early in the planning stage, the city planning department offers you a chance to increase the floor area ratio to 15. (This would allow you to add three stories and 100,000 sq ft of rentable space). All you have to do is provide enclosed, all-weather walkways to neighboring buildings as part of a "downtown mall" concept. Assume that it would take 2 months longer to construct the somewhat larger and more complex project. How much additional cost would you be willing to pay to construct the all-weather walkways in order to gain the benefits of the added rentable space? (Be careful to consider all of the relevant costs and revenues.)

Chapter 9

Financial Assessment

I will gladly pay you Tuesday for a hamburger today.

J. Wellington Wimpy (Popeye's friend)

CHAPTER CONCEPTS

9.1 INTRODUCTION

No project is undertaken solely to make money, but money is a consideration in every project. Entrepreneurs are in the game primarily to make a lot of money, even though the projects they contemplate are at some level aimed at satisfying someone's needs or desires. Even if a project is contemplated solely for some marvelous cultural or aesthetic benefit, it is still necessary to pay the carpenters and buy the lumber. We therefore must be prepared to deal with money, to understand why someone would be willing to invest in a project, and to understand how entrepreneurs and investors think about money.

In this chapter we continue to deal with the typical infrastructure project that requires a significant initial investment in order to address societal needs for a long period of time. The chapter focuses on the cash flows: where does the money come from to build the project, how much money is the project expected to generate once it is completed, and how and when do the investors achieve a return on their investment?

To address these questions, we rely primarily upon the principles of equivalence and discounting that were introduced in the previous two chapters. Given a discount rate, it is possible to convert any arbitrary sequence of cash flows into an equivalent amount of money that could be received today, tomorrow, or at regular intervals for any period of time. There are two significant components to this assertion:

1. *Equivalence relationships*: Equivalent amounts of money can be found by mathematical means, as presented in Table 7.4 in Chapter 7.
2. *Discount rate*: The math works only if the discount rates are known—and these rates will vary with perceived risks associated with the project and with the investment opportunities available to the various participants (owners, developers, investors, bankers, users), as discussed in Chapter 8.

To create their projects, owners and entrepreneurs need funds up front; they expect future profits to be sufficient to provide an attractive return on their investment. If they try to raise money from a bank or from investors, they must prepare a financial plan showing how the project will generate sufficient cash flows to pay off the interest on the loans or bonds while increasing the value of the company for stockholders. When trying to raise money, what the owners and entrepreneurs think the project is worth does not necessarily matter very much. What matters is what the bankers and other potential investors think the project is worth. The investors will use their own discount rates and their own perceptions of the risks associated with the project to evaluate the project. If they perceive the project to be very risky, they will use a higher discount rate. If their portion of the project can be made less risky, then they will use a lower discount rate and be willing to invest more in the project. It is conceivable that projects that appear very profitable to the proponents may appear to be too risky to investors, who will therefore be unwilling to provide the funds needed.

In general, a project must satisfy three criteria to be worth pursuing:

1. The benefits expected from the project must be greater than its costs.
2. The project must be viewed as a good way to achieve these benefits—there may be engineering or institutional alternatives that are as good or better.
3. There are no better ways to use the resources to be devoted to this project—for example, maybe it would be better to invest in housing than in transportation.

This chapter shows how to assess competing projects based on analysis of the projected cash flows and economic impacts associated with each project. Investors and entrepreneurs are primarily concerned with the cash flows, but the public and public agencies will want to consider the broader economic impacts of the project. The methods presented in this and the previous chapters can be used for any types of costs and benefits, so long as they can be expressed in monetary terms.

9.2 MAXIMIZING NET PRESENT VALUE

The net present value (NPV) of a project is the difference between the present value of the net benefits over the life of the project and the present value of the investment. The NPV of a project depends upon the costs and benefits that are considered, the project life, and the discount rate. In general, the objective is to maximize the net present value when evaluating alternative projects.

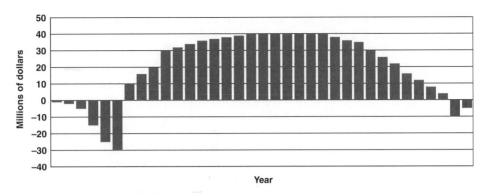

Figure 9.1 Typical Cash Flows for an Infrastructure Project. Cash flows expected from any project can be converted to an equivalent NPV or an equivalent uniform annual net benefit.

Figure 9.1 shows the typical pattern of cash flows for an infrastructure project. From Chapter 7, we know that we can convert this or any other set of cash flows to an equivalent net present value. The cash flows could also be converted into an equivalent worth for any time in the future or to an annuity that begins now or sometime in the future.

If the NPV is positive, then any equivalent annuity and any equivalent future worth will also be positive. If the NPV of one option is better than the NPV of another project, then any equivalent annuity or future worth will also be better for this option. Whichever measure is used, the ranking of any options will be the same. Depending upon the situation, it may make sense to focus on future worth or annuities rather than NPV.

EXAMPLE 9.1   Which Measure Should Be Used: NPV, Annuities, or Future Values?

In project evaluation, it does not matter which of these measures is used to compare different alternatives or different types of projects. If NPV is greater than zero, then the other measures will also be greater than zero. If NPV is better for option 1 than for option 2, then the future value and any equivalent annuity will be better for option 1 than for option 2. The choice of a measure is therefore based on the needs of the analysis and what will allow the best presentation of results. Here are some situations where something other than NPV would be most useful:

- *Planning for a major future event, such as purchasing a house or retiring*: The basic question is how much will be needed sometime in the future, so it makes sense to think about what the situation will be at that time. For example, perhaps an individual believes he could retire at age 60 if he had $2 million in his retirement accounts. How much would he have to save, and what annual returns would he need, to accumulate $2 million by the time he reaches age 60? This person is in effect creating a sinking fund that he will use to purchase an annuity that will be sufficient for his retirement (this type of problem was solved in Example 7.10).
- *Investments aimed at improving the environment, where the benefits may be measurable but not in monetary terms*: If the benefits cannot be monetized, then it is impossible or meaningless to talk about their NPV. Instead, convert the investment cost into an equivalent annuity over the life of the project. That way, the annual cost of a project is compared with its measured benefit, and each project option can be compared based on its cost effectiveness.
- *Construction of an office building*: The critical time is likely to be the completion of construction, so it may be useful to consider the future value of construction costs as of that time and to discount rentals to that time as well. It will probably be useful to convert the construction cost into an annuity that can be used in conjunction with annual rent payments and maintenance costs.
- *Incorporating equipment costs in operating budgets*: Operating budgets can easily include weekly or monthly expenses. Converting the purchase price into an equivalent weekly or monthly cost is therefore a convenient way to allocate costs of equipment.

If the NPV is greater than zero, then the project may be worth pursuing. It is not automatically worth pursuing, for reasons cited above: nonmonetary costs may offset the financial benefits of this project, the benefits of this project may be achieved in better ways, and entirely different kinds of benefits may be more desirable to pursue.

For now, just consider the financial benefits from the owner's perspective. The owner's goal is to maximize the NPV of cash flows, using the owner's discount rate for discounting any future cash flows. It is worth recalling why focusing on the NPV (or the equivalent future or annuity values), rather than on the much more detailed and informative cash flow diagram, is so useful. The main reason is that comparing projects based on their cash flows is extremely difficult, while comparing different projects based on their NPV, future value, or annuity value is extremely simple. The higher the value, the better. That is why the principle of equivalence is so powerful. It translates the cash flows of complex projects into equivalent amounts that are very easy to understand and compare.

Of course, the principle of equivalence is valid only if a reasonable discount rate is used. That is why Chapter 8 was devoted to the meaning of the discount rate, emphasizing the fact that different parties involved in a project may have different discount rates.

The choice of a discount rate is extremely important in determining what kinds of projects are most appealing. If a very high discount rate is used, then the NPV will be based primarily upon what happens in the first 5–20 years of a project. Small projects with immediate benefits will look better with a higher discount rate, whereas large projects with benefits that extend far into the future may fare poorly. If a very low discount rate is used, the opposite is true: future costs and benefits will be much more heavily weighted.

In planning for public projects, use of a low discount rate may promote undertaking large-scale projects while ignoring important current needs. On the other hand, use of a high discount rate may prevent a company or a country from ever undertaking large-scale projects.

9.3   IMPORTANCE OF THE PROJECT LIFE

Projects need to be evaluated over a reasonable project life. Several factors enter into the choice of a "reasonable" life:

- The economic life of the project—the period of time for which the project is expected to be in use.
- The period of time for which discounted cash flows are relevant to the analysis.
- Knowledge concerning any dramatic costs or benefits that might be expected in the distant future.

The economic life of a project can be much less than the physical life of the structures that are constructed. If a railroad is built to a mine, the railroad might be expected to last indefinitely so long as it is maintained properly. The facilities at the mine may also be constructed to standards that would ensure a life of 30–50 years. However, the ore may be gone after just 20 years, so that the economic life of both the railroad and the mine would be 20 years.

Unless very low discount rates are used, a 20- to 50-year life is usually sufficient for analysis. Because of discounting, the costs and benefits from more distant years do not add much to the NPV, so it is not necessary to include them in the calculations.

Ignoring the out-years is viewed by some as a very bad practice since it means that the analysis would be ignoring the impacts of current decisions on unborn generations. Some have called for the use of a zero discount rate so that the needs of future generations are considered properly. In a financial analysis, however, it is foolish to talk of a zero discount rate since in fact investors and the financial markets that provide the funding for projects do discount cash flows—and the amount of money that can be raised for projects depends upon their discount rates. Potential benefits that occur in the far distant future will not attract additional funding from the markets.

Of course there may be some merit to the argument that current projects may be damaging the environment or creating hazards or promoting serious financial problems that will not be apparent for 20 or more years. If a 20-year life is used, then such problems may conveniently be overlooked. This problem can be dealt with by requiring additional considerations in the choice of the time period:

- Benefits are expected to continue to exceed costs for an indefinitely long period.
- The project could be decommissioned at the end of its useful life, and the cost of that decommissioning is included at the end of the assumed project life.
- No known catastrophic consequences are associated with the project beyond what is considered within the chosen period for the analysis.
- No known extraordinary costs or benefits are associated with the project during the out-years.

EXAMPLE 9.2   How Project Life Can Affect Evaluation

Consider the example in Figure 9.2, which includes four charts. The first chart (A) shows cash flows for a project over a 10-year life. There is an initial investment of $5 million in year 1 plus $10 million in year 2, followed by net revenue of $5 million per year for 8 years. The chance to earn $40 million in a short period of time based on a $15 million investment looks pretty good, although we would need to get the NPV to be sure. Perhaps the use of a 10-year life is too short for this project. If the cash flows are expected to continue for at least 25 years, then we would have the cash flow such as that in Chart B. Now the project looks much more attractive, with an excellent return on investment for a very long time. On the other hand, maybe someone had used a 10-year life to avoid the unfortunate problem that increasing competition and maintenance problems are actually expected to lead to declining revenues and heavy losses. The cash flows shown in Chart C are unlikely to attract anyone's interest, since the long-run trend is decidedly negative. Nevertheless, this may be part of a long-run development strategy that is expected eventually to reap major benefits because the initial project established an operation in what everyone expects to become a prime location; the negative cash flows during years 10–20 may simply reflect ongoing investments in the site that are expected to produce large benefits in the future, as suggested by Chart D. The logic of the proposal should determine the choice of the time period for the analysis. Blindly using a specified time frame can lead to blunders in project evaluation.

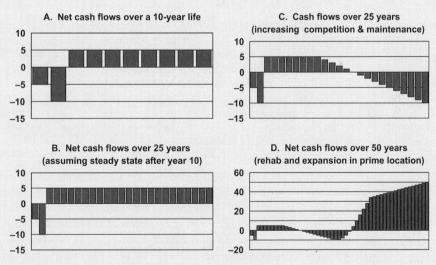

Figure 9.2 Choice of the Project Life Can Be Critical.

In summary, a 20–50 year project life is intended to be long enough to capture the relevant costs and benefits associated with a project. For discount rates of 5% or more, the out-years contribute very little to the analysis, and it is rather meaningless to make projections further into the future. If the economic life of the project is less than 20 years, then a shorter life should be used. If there is reason to expect extraordinary costs or benefits that would be apparent only after a period of 20–50 years, then the project life should of course be extended. In normal circumstances, using a discount rate and limiting the life of the project should not be seen as somehow damaging to future generations—it is simply reflecting the reality of money and the principle of equivalence.

9.4 DOES DISCOUNTING IGNORE FUTURE CATASTROPHES?

To answer this question, we need to define the term *catastrophe* and consider what the costs of a catastrophe might be. For perspective, we can look at the more dismal side of history. In numerous instances, natural disasters—earthquakes, hurricanes, or tsunamis—have killed tens of thousands of people, and outbreaks of disease have killed that many people in a single year in many different cities.[1] Millions of people died during the Great Influenza of the early 1920s, and the Black Plague reduced the population of Europe by a third during the fourteenth century. Diseases introduced by Europeans wiped out an even larger proportion of the native populations of North and South America during the sixteenth century. Wars with tens of thousands of casualties are commonplace in history, and the world wars of the twentieth century killed tens of millions. Presumably, natural disasters, disease, and warfare will continue to haunt humanity with catastrophic consequences far into the future. So we might want to begin by considering the present worth of such catastrophes or the benefits to be gained by limiting the frequency or the consequences of such catastrophes.

EXAMPLE 9.3   Potential Benefits of Averting a Catastrophe

Consider the possibility of an epidemic that could break out in 50 years, taking 1 million lives. Suppose that steps could be taken today to reduce the expected fatalities by 90% or defer the epidemic for another 50 years. What would the benefits be, assuming a discount rate of 8%?

First, we need some way to quantify the magnitude of the disaster. As discussed in Chapter 2, various countries have incorporated a factor in their planning and regulatory processes that indicates the value to be placed on a reduction in expected fatalities. In the United States and in Europe, a value in excess of $2.5 million has been used in transportation planning. This means that government safety regulations can be justified if the costs of improving safety are less than $2.5 million per expected life saved.

If we use this number, then an event that led to 1 million deaths would have a cost to society of 1 million deaths multiplied by $2.5 million per fatality, for a total of $2.5 trillion (i.e., 2.5×10^{12}). This is an extremely large number. Even if this event occurs 50 years in the future, the NPV of such a disaster is very large. With a discount rate of 8%, the NPV would be $2.5 trillion $(1/1.08)^{50} = $2.5 trillion $(0.02132) = $53 billion, which is a substantial sum of money! Thus efforts that could reduce the expected fatalities by 90% would be worth nearly $50 billion today.

If the disaster occurred 100 years in the future rather than 50, then the $2.5 trillion would be further discounted by another factor of 0.02131, and the NPV would be reduced from $53 to a bit more than $1 billion. Thus either reducing the magnitude of this catastrophe by 90% or deferring the epidemic for another 50 years would each have an NPV of approximately $50 billion. Discounting does not allow us to ignore future catastrophes.

What would projects look like that had the effect of reducing the frequency or consequences of future disasters? For reducing the probability and severity of a pandemic, doctors can work to develop better drugs, public health officials can work to eliminate unhealthy slums and improve water supplies, and governments can stockpile emergency supplies of medicine and other supplies. To reduce the consequences of earthquakes and other natural disasters, governments can impose building codes that limit or require sturdier construction in dangerous areas, they can provide better communications and warning systems, and they can prepare for rapid response to natural disasters. A lot can be done in each of these areas for $50 billion!

---

[1] This section was written before the horrendously devastating earthquake in Haiti that destroyed much of Port Au Prince and other cities, killed on the order of 100,000 people, and left more hundreds of thousands injured or homeless. In a few minutes, this earthquake caused double the amount of casualties suffered by U.S. troops in all of this country's wars from Vietnam to Afghanistan, and it caused 30 times the loss of human life suffered on 9/11. Let us hope that the scale of this disaster will at least lead to much more careful consideration of ways to reduce the all too real risks from natural disasters that endanger hundreds of millions of people throughout the world.

9.5   INTERNAL RATE OF RETURN

9.5.1   Calculating ROI Given a Discount Rate

The annual return on investment (ROI) is calculated as the annual profit from the investment divided by the amount of the investment. In some cases, it is easy to determine the ROI of an investment. For example, if you buy a bond for $1,000 and receive $60 per year in interest, then your ROI is $60/$1,000 = 6% per year.

In the case of a project with rather arbitrary cash flows, it is not so straightforward to calculate the ROI. After all, the initial investment may be spread out over several years, and the anticipated cash flows vary from year to year. As discussed in Chapter 8, it is possible to convert the initial investment into an equivalent investment (I) at time 0 and to convert the net revenues into an equivalent annuity (A) that will continue forever. Once this is done, the return on investment is readily seen to be A/I. This is a useful concept; but the result depends upon the discount rate used, as demonstrated in Example 9.4.

Since the choice of a discount rate depends upon the user's perspective, it may be disconcerting to find that calculation of the ROI depends upon the choice of a discount rate. It would be much more convenient to have a measure that does not depend upon subjective input.

EXAMPLE 9.4   ROI Depends upon the Discount Rate

Table 9.1 shows the same cash flows used in Chapter 8 (see Example 8.1). In this example, the present worth (PW) of the investment and the project revenues were calculated using a discount rate of 10%. The present worth of the revenues was $1,924, which is equivalent to an annuity paying $192.40 per year forever. The return on investment was 11.1%, which was calculated as $192.40 divided by the present worth of the investment ($192.40/$1,736 = 0.111).

Table 9.1   Calculating NPV of Cash Flows Associated with a Project Using a Discount Rate of 10%

Year	1	2	3	4	5	6	7	8	9	10	Total
Investment	$1,000	$1,000									
PW of investment	$ 909	$ 826	$ 0	$ 0	$ 0	$ 0	$ 0	$ 0	$ 0	$ 0	$1,736
Revenue	0	100	200	400	600	600	600	500	300	100	
PW of revenues	$ 0	$ 83	$150	$273	$373	$339	$308	$233	$127	$39	$1,924

Table 9.2 shows the same cash flows discounted at 8%. With the lower discount rate, the present worth of the investment increases to $1,783 and the present worth of the revenues increases to $2,142, which is equivalent to an infinite annuity of $2,142 (0.08) = $171.40. The ROI using an 8% discount rate is $171.40/$1,783 = 9.6%. Using this methodology to compute the rate of return clearly produces a different answer if a different discount rate is used.

Table 9.2   Calculating NPV of Cash Flows Associated with a Project Using a Discount Rate of 8%

Year	1	2	3	4	5	6	7	8	9	10	Total
Investment	$1,000	$1,000									
PW of investment	$ 926	$ 857	$ 0	$ 0	$ 0	$ 0	$ 0	$ 0	$ 0	$ 0	$1,783
Revenue	0	100	200	400	600	600	600	500	300	100	
PW of revenues	$ 0	$ 86	$159	$294	$408	$378	$350	$270	$150	$46	$2,142

9.5.2   Calculating the Internal Rate of Return

If we take another look at the situation described in Example 9.3, we can see a way to get a more objective measure, one that does not depend upon a predetermined discount rate. In Example 9.3, the return on investment varied with and differed from the discount rate that was used in the analysis. However, perhaps there might be a discount rate for which the present worth of the investment equals the present worth of the revenues, in which case the ROI would equal the discount rate. In fact this may be true, as shown in Example 9.5.

EXAMPLE 9.5   Choosing a Discount Rate Such that the Discounted Investment Equals the Discounted Revenues

This example continues analyzing the cash flows of the previous example. By trying several discount rates, it was possible to determine that a rate of 12.7% resulted in the present worth of the investment equaling the present worth of the revenues, as is shown in the "Totals" column of Table 9.3. The equivalent infinite annuity of revenues would be $1,674 (12.7%) = $212.56, and the ROI would be $212.56/$1,675 = 12.7%.

Table 9.3   Calculating the NPV of Cash Flows Associated with a Project Using a Discount Rate of 12.7%

Year	1	2	3	4	5	6	7	8	9	10	Total
Investment	$1,000	$1,000									
PW of investment	$ 887	$ 787	$ 0	$ 0	$ 0	$ 0	$ 0	$ 0	$ 0	$ 0	$1,675
Revenue	0	100	200	400	600	600	600	500	300	100	
PW of revenues	$ 0	$ 79	$140	$248	$330	$293	$260	$192	$102	$30	$1,674

In short, the practice of discounting cash flows suggests a rather nice way to calculate the return on investment without first defining a discount rate. Simply find the discount rate that will make the NPV equal zero, in which case the ROI will equal the discount rate. This rate is known as the **internal rate of return** (**IRR**). Given cash flows such as those given in Figure 9.1 or 9.2, the IRR can be found by trial and error as illustrated in Examples 9.3 and 9.4. The higher the IRR, the better; and companies in the private sector commonly use this method to characterize the profitability of proposed projects.

9.5.3   Ranking Projects with the Internal Rate of Return

It is possible to estimate the internal rate of return for a set of competing projects. One might think that the best projects are those with the highest internal rate of return, but that is not the case. The best projects from a financial or economic perspective are those with the highest net present value, and these projects do not necessarily have the highest internal rate of return. The problem in ranking projects arises when projects are mutually exclusive. A smaller project may have a higher internal rate of return, but a larger project may have greater total returns, as illustrated in Example 9.6.

EXAMPLE 9.6   The Best Project Might Not Have the Highest IRR

The investment and the equivalent uniform annual net benefits are given in Table 9.4 for four projects. You are trying to decide if the annual net benefits are large enough to justify any of the investments. The minimum acceptable rate of return for your firm is 10%.

Table 9.4   Investment and Equivalent Annual Net Benefits for Four Projects

Project	Investment (NPV as of time zero)	Equivalent Annual Net Benefits
A	$1 million	$ 90,000
B	$2 million	$440,000
C	$3 million	$600,000
D	$4 million	$480,000

If we assume that the benefits will continue for a very long time, then we can estimate the NPV of the benefits by dividing the annual benefits by the discount rate. The annual net benefits of $90,000 for project A are therefore worth $90,000/10% = $900,000. Since this is less than the investment cost, project A has a negative NPV and should not be pursued. For the other three projects, the present value of the benefits exceeds the investment cost, so that the NPV for each of these projects is greater than zero. Projects B, C, and D can therefore all be justified financially. If only one of the projects can be undertaken, then C is the best since it has the highest NPV (Table 9.5).

Table 9.5   Investment and Benefit Data for Four Projects

Project	Investment (NPV as of time zero)	Equivalent Annual Net Benefits	Present Value of Benefits (using capital worth method)	Net Present Value of Project
A	$1 million	$ 90,000	$0.9 million	($0.1 million)
B	$2 million	$440,000	$4.4 million	$2.4 million
C	$3 million	$600,000	$6.0 million	$3 million
D	$4 million	$480,000	$4.8 million	$0.8 million

The internal rate of return can easily be calculated for these projects by dividing the equivalent annual net benefits by the investment, with the results shown in the "IRR" column of Table 9.6. Since the IRR is greater than the firm's minimum acceptable rate of return (MARR) of 10% for projects B, C, and D, these projects are all acceptable; but project A with its return of only 9% is unacceptable. If only one of these projects can be undertaken, then selection based on IRR would choose project B—but didn't we just figure out that project C was the best? What's going on? Why doesn't the IRR method result in the same choice as the NPV method? This is certainly a problem, but it is a problem that can be fixed. With some adjustment, the IRR approach will indeed give the same rankings as the NPV approach.

Table 9.6   Investment and Benefit Data for Four Projects

Project	Investment (NPV as of time zero)	Equivalent Annual Net Benefits	Present Value of Benefits (using capital worth method)	IRR
A	$1 million	$ 90,000	$0.9 million	9%
B	$2 million	$440,000	$4.4 million	22%
C	$3 million	$600,000	$6.0 million	20%
D	$4 million	$480,000	$4.8 million	12%

The fact that projects cannot simply be ranked by their IRR is just one problem with this measure. In fact, there are three problems with the IRR. First, the process for estimating the IRR could produce multiple answers. Second, the IRR methodology assumes that any cash received during the course of the project can be reinvested at the IRR while future costs can be discounted using the IRR. For projects with a very high IRR, both assumptions could be unrealistic. Third, as illustrated above, the project with the best IRR might not be the best project, which is a serious problem when selecting among mutually exclusive projects. We will consider each of these problems in turn.

The next example describes a project that has multiple internal rates of return. This example is structured as the kind of situation that a student studying project evaluation might encounter in a subsequent internship with a construction company or some other large business. Section 9.6 describes the external rate of return, a method similar to the IRR that avoids the problems of multiple solutions and unrealistic assumptions. Sections 9.7 through 9.9 show how to use both the NPV and the IRR so that they provide the same, correct ranking of projects.

EXAMPLE 9.7   Multiple Internal Rates of Return for a Stadium Project

You have accepted an internship position that gives you a chance to do what you always wanted to do—work in the front office of a major league team. With your background in project evaluation, you have been asked to calculate the financial potential of the proposed new stadium. At a meeting with the president, the general manager, the vice president of finance, and several consultants, you learn the general plan. The stadium will be constructed over 2 years at a total cost of $250 million. Constructing the stadium will lead to two very large immediate payoffs from selling 20-year leases on corporate boxes and selling real estate around the stadium to various hotel and restaurant chains. Revenue will be about $8 million per year in the first 10 years, after which a major rehabilitation and expansion is planned at a cost of $50 million. With a somewhat larger stadium, revenues are expected to rise. Management plans to use all of that revenue plus some additional funds for further development of the area; thus, cash flows are expected to be negative for several years before another big payout occurs in year 20, when the organization is able to complete and sell more of the real estate. Table 9.7 summarizes the expected cash flows.

Table 9.7   Expected Cash Flows for Stadium Project

Year	Investment	Revenue	Cost	Cash Flow
0	100	0		−100
1	150	10		−140
2		150	2	148
3		150	2	148
4		10	2	8
5		10	2	8
6		10	2	8
7		10	2	8
8		10	3	7
9		10	3	7
10		0	50	−50
11		30	20	10
12		30	25	5
13		30	30	0
14		30	35	−5
15		30	40	−10
16		30	50	−20
17		30	60	−30
18		30	70	−40
19		30	90	−60
20		100	10	90

You set up a spreadsheet with all of this information and set out to determine the internal rate of return. First you try 8%, but that gives a positive NPV of $2.8 million; then you try 12%, but that gives a negative NPV of $4.8 million. You figure the NPV will be zero somewhere in the middle of these two, and you eventually determine that the NPV is zero with a discount rate of

Figure 9.3 Lucas Oil Stadium, Home of the Indianapolis Colts. (Under construction, January 15, 2008)

9.85%. You therefore prepare a presentation that concludes that the stadium project has an internal rate of return of "just under 10%." You even have prepared a portfolio filled with pictures of similar projects, such as the recently completed football stadium in Indianapolis (Figure 9.3). You're quite proud of your work and rush in to show your boss your result (Table 9.8).

Table 9.8   IRR for Stadium Project Is Nearly 10%.

Year	Cash Flow	Discounted Cash Flows		
		8%	12%	9.85%
0	−$100.0	−$100.0	−$100.0	−$100.0
1	−$140.0	−$129.6	−$125.0	−$127.4
2	$148.0	$126.9	$118.0	$122.6
3	$148.0	$117.5	$105.3	$111.7
4	$ 8.0	$ 5.9	$ 5.1	$ 5.5
5	$ 8.0	$ 5.4	$ 4.5	$ 5.0
6	$ 8.0	$ 5.0	$ 4.1	$ 4.6
7	$ 8.0	$ 4.7	$ 3.6	$ 4.1
8	$ 7.0	$ 3.8	$ 2.8	$ 3.3
9	$ 7.0	$ 3.5	$ 2.5	$ 3.0
10	−$ 50.0	−$ 23.2	−$ 16.1	−$ 19.5
11	$ 10.0	$ 4.3	$ 2.9	$ 3.6
12	$ 5.0	$ 2.0	$ 1.3	$ 1.6
13	$ 0.0	$ 0.0	$ 0.0	$ 0.0
14	−$ 5.0	−$ 1.7	−$ 1.0	−$ 1.3
15	−$ 10.0	−$ 3.2	−$ 1.8	−$ 2.4
16	−$ 20.0	−$ 5.8	−$ 3.3	−$ 4.4
17	−$ 30.0	−$ 8.1	−$ 4.4	−$ 6.1
18	−$ 40.0	−$ 10.0	−$ 5.2	−$ 7.4
19	−$ 60.0	−$ 13.9	−$ 7.0	−$ 10.1
20	$ 90.0	$ 19.3	$ 9.3	$ 13.7
Total		**$ 2.8**	**−$ 4.3**	**$ 0.0**

Your boss is rather more subdued than you are, because he understands that 10% is no great rate of return for the crafty men and women who run the major league team. He also notes that the cash flows are weird—they show several shifts from positive to negative (Figure 9.4). He fears that there might be a problem with the IRR you calculated. He therefore runs the cash flows through his program and quickly obtains the result shown in Table 9.9.

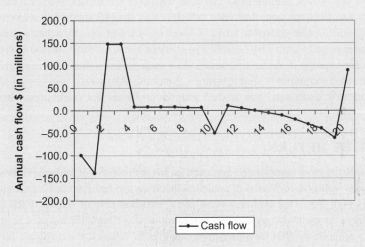

Figure 9.4 Expected Cash Flows for the Stadium Project.

Table 9.9   NPV of Stadium Cash Flows, Showing Sensitivity of Discounted Cash Flows to Discount Rate

Year	Cash Flow	Discounted Cash Flows for the Given Discount Rate								
		1%	2%	4%	6%	8%	10%	12%	14%	16%
0	−100.0	−100.0	−100.0	−100.0	−100.0	−100.0	−100.0	−100.0	−100.0	−100.0
1	−140.0	−138.6	−137.3	−134.6	−132.1	−129.6	−127.3	−125.0	−122.8	−120.7
2	148.0	145.1	142.4	136.8	131.7	126.9	122.3	118.0	113.9	110.0
3	148.0	143.6	139.7	131.6	124.3	117.5	111.2	105.3	99.9	94.8
4	8.0	7.7	7.4	6.8	6.3	5.9	5.5	5.1	4.7	4.4
5	8.0	7.6	7.3	6.6	6.0	5.4	5.0	4.5	4.2	3.8
6	8.0	7.5	7.1	6.3	5.6	5.0	4.5	4.1	3.6	3.3
7	8.0	7.5	7.0	6.1	5.3	4.7	4.1	3.6	3.2	2.8
8	7.0	6.5	6.0	5.1	4.4	3.8	3.3	2.8	2.5	2.1
9	7.0	6.4	5.9	4.9	4.1	3.5	3.0	2.5	2.2	1.8
10	−50.0	−45.3	−41.2	−33.8	−27.9	−23.2	−19.3	−16.1	−13.5	−11.3
11	10.0	9.0	8.1	6.5	5.3	4.3	3.5	2.9	2.4	2.0
12	5.0	4.4	4.0	3.1	2.5	2.0	1.6	1.3	1.0	0.8
13	0.0	0.0	0.0	0.0	0.0	0.0	0.0	0.0	0.0	0.0
14	−5.0	−4.3	−3.8	−2.9	−2.2	−1.7	−1.3	−1.0	−0.8	−0.6
15	−10.0	−8.6	−7.5	−5.6	−4.2	−3.2	−2.4	−1.8	−1.4	−1.1
16	−20.0	−17.1	−14.7	−10.7	−7.9	−5.8	−4.4	−3.3	−2.5	−1.9
17	−30.0	−25.3	−21.6	−15.4	−11.1	−8.1	−5.9	−4.4	−3.2	−2.4
18	−40.0	−33.4	−28.3	−19.7	−14.0	−10.0	−7.2	−5.2	−3.8	−2.8
19	−60.0	−49.7	−41.6	−28.5	−19.8	−13.9	−9.8	−7.0	−5.0	−3.6
20	90.0	73.8	61.2	41.1	28.1	19.3	13.4	9.3	6.5	4.6
Total	**−$8.0**	**−$3.3**	**$0.0**	**$3.8**	**$4.4**	**$2.8**	**−$0.3**	**−$4.3**	**−$8.9**	**−$13.8**

A quick look at the bottom row of this table indicates that the NPV is zero when the discount rate is 2%, suggesting that the IRR is a dismal 2%. On the other hand, the table also supports your calculation—the NPV is also zero for something a little less than 10%.

Now, you and your boss can't go to the CEO and the stadium committee and say that the project is perhaps OK with an IRR of nearly 10%, except that it may be dismal with an IRR of only 2%. You need to fix this problem—and you need to fix it fast! Read on . . .

The problem of dueling IRRs can arise whenever the stream of annual cash flows switches from positive to negative more than once, which is why the IRR method is seldom praised by academics. Most projects presented to the board of directors, however, specify a pretty clear initial investment that produces positive annual net benefits that continue for an indefinite period with at most a minor cost for decommissioning in the distant future. With such projects, the result is unambiguous, which is why this method is commonly used in business. Since the IRR method is so commonly used in business, it is important to understand how to deal with the problems that might arise when it is used.

9.6   EXTERNAL RATE OF RETURN

A somewhat more complicated approach avoids the problem of multiple values for the internal rate of return as well as the difficulty of assuming that costs and benefits can be discounted with what could be a very high IRR. This approach is called the **external rate of return** (**ERR**)—and it is in fact very similar to the methodology that was first introduced in Chapter 8 and used in the beginning of Section 9.5 to estimate the ROI for a particular example. In Tables 9.1 and 9.2, all the investment costs were discounted to time 0 while all of the revenues were converted to an equivalent annuity or future value.

Calculating the external rate of return also uses equivalence relationships to create an easily understandable comparison between costs and benefits.

1. First, divide all of the periods considered in the analysis into periods where the cash flow is negative and periods where the cash flow is positive. There is no need to distinguish between investment costs, rehabilitation costs, or operating losses.

2. Next consider the periods with negative cash flow. For each such period, we could establish a fund that is expected to grow over time so that it can be used when needed to cover the negative cash flow. The size of the fund can be determined by using a discount rate that is consistent with the company's expected overall return on investment during the intervening years. This discount rate can perhaps be the company's minimum acceptable rate of return or the company's average rate of return. All of the negative cash flows can be converted to an equivalent present value using this same discount rate.

3. Next, using the same rate of return, convert all of the positive cash flows into a future value. The logic in extrapolating these funds to the future is that any extra cash generated by a specific project will be used to promote the overall activities of the company. For example, if the company has historically enjoyed a rate of return of about 10%, and if conditions in the future are expected to be no different, then the company can expect that the earnings from any new project in year t can be reinvested and earn 10% per year from year t until the end of the analysis period.

Figure 9.5 summarizes the process of calculating the ERR.

The rate of return used in these calculations is called the *external* rate of return to indicate that the rate of return is based on factors unrelated to the specific project being investigated. The same external rate of return is used for evaluating any project; it is not something that has to be defined for each specific project.

Given an external rate of return, the following comparison between the future value of the positive cash flows and the present value of the negative cash flows can then be used to determine the return on investment (ROI) for this project:

$$\text{FV positive cash flows} = (1 + \text{ROI})^n (\text{PV costs}) \qquad \text{(Eq. 9.1)}$$

In this equation, the external rate of return (also known as "e") is used to calculate both the present value of costs and the future value of benefits. The ROI that satisfies the equation can readily be obtained by trial and error using a spreadsheet. The ROI could by coincidence equal "e," but it most

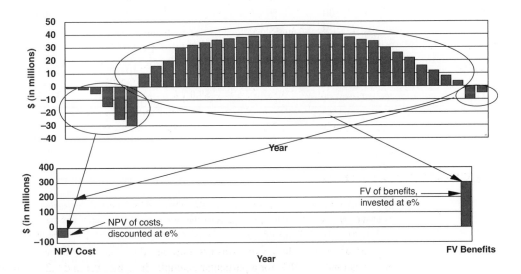

Figure 9.5 Calculating the External Rate of Return.

likely will be higher or lower.[2] The next example shows how this approach provides a reasonable way to calculate the ROI for the stadium project.

EXAMPLE 9.8   Calculating the External Rate of Return for the Stadium Project

The external rate of return for the stadium project can now be calculated. The costs are all discounted to the present using a rate of 8%, which is the MARR for the stadium owners. The future value of annual revenues is calculated as of year 20 (using the same rate of 8% and [F/V, 8%, 20 years]). The NPV of the costs is $399 million, while the FV of the benefits is $1,873 million, as shown in Table 9.10. The ERR is the annual rate of return at which $3,999 would grow to $1,873 million in 20 years. The answer can be gained by solving the following equation for i%.

$$FV = NPV\,(1 + i\%)^{20}$$

Trial and error on a spreadsheet shows the answer to be 8.04%, so the report to the CEO and the committee could indicate that the project has an expected ROI of 8%.

Table 9.10   Calculating the External Rate of Return for the Stadium Project

Year	Investment	Revenue	Cost	Cash Flow	Cost	NPV of Costs	FV of Benefits
0	100	0		−100.0	−100.0	−100.0	0.0
1	150	10		−140.0	−150.0	−138.9	43.2
2		150	2	148.0	−2.0	−1.7	599.4
3		150	2	148.0	−2.0	−1.6	555.0
4		10	2	8.0	−2.0	−1.5	34.3
5		10	2	8.0	−2.0	−1.4	31.7
6		10	2	8.0	−2.0	−1.3	29.4
7		10	2	8.0	−2.0	−1.2	27.2
8		10	3	7.0	−3.0	−1.6	25.2
9		10	3	7.0	−3.0	−1.5	23.3
10		0	50	−50.0	−50.0	−23.2	0.0
11		30	20	10.0	−20.0	−8.6	60.0
12		30	25	5.0	−25.0	−9.9	55.5
13		30	30	0.0	−30.0	−11.0	51.4
14		30	35	−5.0	−35.0	−11.9	47.6
15		30	40	−10.0	−40.0	−12.6	44.1
16		30	50	−20.0	−50.0	−14.6	40.8
17		30	60	−30.0	−60.0	−16.2	37.8
18		30	70	−40.0	−70.0	−17.5	35.0
19		30	90	−60.0	−90.0	−20.9	32.4
20		100	10	90.0	−10.0	−2.1	100.0
Total				**−$8.0**		**−$399.1**	**$1,873.2**

---

[2] This section uses the term *external rate of return*, which is commonly seen in engineering economic textbooks. ERR is presented as a measure that is similar to, and better than, the internal rate of return; but it is not clear what is really meant by "e," the external rate of return. In Sections 8.2.1 and 9.5, a method was presented whereby a company or individual would use their own discount rate to determine the present worth of the costs and the future worth of the benefits. *Return on investment* for a particular project would then be defined as the annual growth needed for the present worth of project costs to grow into the future value of project benefits at the end of the project life. With this approach, there is no need to introduce "e" as something new, because the usual discount rate would be used in the calculation. The ROI for the project would then be seen as a clearly defined measure that is naturally dependent upon the use of the proper discount rate (just as NPV is dependent upon the use of the proper discount rate).

Academics favor the ERR approach because it avoids the necessity of implying unreasonable returns for reinvesting profits, and it provides a reasonable means of dealing with future periods with negative cash flow. However, this approach is unlikely to be encountered outside of textbooks. Public agencies are apt to consider benefit-cost ratios rather than ROI, as discussed in Chapter 4, whereas private companies use the internal rate of return as an easier and apparently more objective result.

9.7   CONSTANT DOLLAR VERSUS CURRENT DOLLAR ANALYSIS

In Chapter 8, we saw that the discount rate observed in the financial markets reflects three factors: the return available on risk free investments, a risk premium, and inflation. Any analysis that uses historical or projected interest rates is using data that reflects past and future expectations concerning inflation. Inflation expectations are also among the many factors that affect the price of stocks and real estate. Inflation is also a factor in MARRs of individuals, companies, and government agencies.

It is important that inflation be treated consistently when evaluating projects. In estimating costs and revenues, it is often convenient to ignore inflation. So long as the major elements of cost and the major sources of revenue all increase at about the same rate, a constant dollar analysis will result in a reasonable projection of cash flows. If some components are expected to perform much differently, then adjustments would have to be made in projecting cash flows. For example, the costs of computers and communications have declined for many years, so it is reasonable to assume that these costs will continue to decline relative to other costs. In recent years, energy costs have risen sharply relative to other costs, and it is reasonable to assume that this trend will continue. Any project that has significant costs related to communications, computers, or energy therefore might require adjustments in projections of constant dollar costs and revenues.

Given projections of cash flows, it is necessary to ensure that the discount rates are consistent with the assumptions about inflation. Two sets of assumptions are reasonable:

1. *Constant dollar*: Neither cash flows nor discount rates consider inflation.
2. *Current dollar*: Cash flows and discount rates both reflect inflation.

If cash flows are provided in constant dollars, but are discounted with real discount rates, then future cash flows will be discounted too much. If cash flows are provided in current (i.e., inflated) dollars, but are discounted with discount rates that do not consider inflation, then future cash flows will be insufficiently discounted.

Inflation in even the most stable economies during the most stable economic conditions is usually at least 1–2% a year; in other circumstances, inflation can easily be 3–4% per year in the most stable economies and much higher elsewhere. This is not a factor to be overlooked, because the mistakes could be considerable.

9.8   CHOOSING AMONG INDEPENDENT INVESTMENT OPTIONS

Consider a company that has many independent investment opportunities. These opportunities are independent in the sense that choosing any one of them does not require or preclude any of the others. The company could decide to choose none, any, or all of the options. In theory, the company could decide to invest in any project with positive NPV. If the NPV is positive, then the project will produce cash flows that will, when discounted at the company's MARR, be equivalent to having more money today. The company's MARR will be at least as great as its weighted average cost of capital, and the cost of capital conceivably could rise if the company attempted to raise excessive amounts. The company's executives and board of directors would also have some concerns about the quality of the analysis and the possibility that some projects might prove less successful than they had hoped. As a result, the funds available for projects would likely be limited, and only the best projects would be chosen. The objective would therefore be to maximize NPV subject to a capital budget constraint, which is equivalent to maximizing the return on investment for the capital that is budgeted.

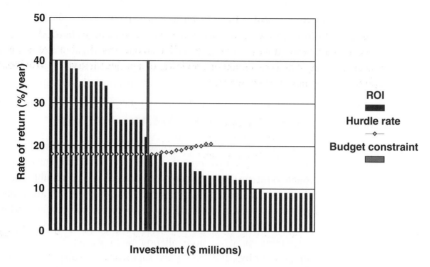

Figure 9.6 Selecting Projects Based on a Hurdle Rate of Return.

In practice, many companies use the IRR in evaluating independent projects. The IRR can easily be used to rank all projects according to a common metric. The IRR is deemed acceptable so long as it is higher than the company's MARR (usually viewed as the company's weighted average cost of capital). The decision process is straightforward (Figure 9.6), at least related to strictly financial matters: choose the projects with the highest IRR, so long as the IRR exceeds the hurdle rate and the total investment is within the budgeted amount.

This process of ranking projects by their IRR assumes that the risks associated with the project are similar, so that each project can be compared to the same hurdle rate. A large company with many diverse opportunities for cutting costs or expanding markets in fact has many investment options with similar risks: they know what to expect if they decide to make the investment. If the company is moving into a new type of business, or if an investment is believed to have unusual risks, then a higher hurdle rate could be used. If an investment is deemed essential to the company's safety or to its continued operation, then the investment will be made even with an IRR lower than the hurdle rate.

The same approach could be taken by using the external rate of return as described above. For the typical projects that involve clear investments and produce continuing annual benefits, there is unlikely to be much difference in the projects that are selected by the ERR as opposed to the IRR method. However, for projects like the stadium project investigated previously in this chapter, there could be significant differences in which projects are selected. In the stadium example, those favoring the project would use the "nearly 10%" figure that the "brilliant new intern" had estimated. Meanwhile, those trying to bury the project would highlight the 2% return that shows up in the company's standard financial analysis, and a consultant brought in to provide a neutral opinion might recommend using the 8% return that was calculated using the ERR method.

As with so many elegant frameworks, this clear and logical process for selecting projects may not work so clearly or logically in practice. While a company indeed should know its MARR, that may be a subject of debate; or it may not ever be explicitly defined. The elegant model indicates that all projects whose expected return exceeds the MARR should be approved, ignoring the reality that capital expenditures always have some kind of limit. The limit is undoubtedly flexible, but that means marginally acceptable projects may be approved only if they are supported by people with the power or persuasiveness necessary to convince the board of directors. The decision model depicted above also assumes that there is an ordered list of all the feasible projects, none of them mutually exclusive. No one who has ever seriously considered design assumes that they can ever know all of the alternatives, many of which certainly are mutually exclusive. In a large organization, whether public or private, leaders from each department are promoting their own projects; those who are more diligent, more eloquent, or closer to the senior officials may get their projects approved.

There may well be many better projects that no one thought of or that no one wanted to champion. If you are an analyst or a consultant or a reviewer of a project, it is your job to look for some of those other options. Some possibilities include

- Use of better materials and techniques to build the same facility
- A better structural design to serve the same purpose
- A different location for a similar project
- A different scale—many smaller projects or fewer larger projects

In general, no one can prove that their design or proposed project is the best. They can only defend, refine, or abandon their proposal in response to whatever feedback and opposition they receive.

9.8.1 Independent Projects Can Be Ranked Using Present Value, Future Value, Annual Value, or IRR

If making money is your objective, then ranking projects by present, future, or annual value certainly seems to be the correct approach. The rankings obtained by using any of these three approaches would be the same since the differences among them depend upon factors that vary only with the discount rate and the period of the investments. If the projects are independent and budgets are unlimited, then any project with positive NPV is worth pursuing. If the NPV is greater than zero, then the internal rate of return will be greater than the discount rate, so that the IRR will also identify which projects should be pursued. However, as shown in the next subsection, the projects are not always independent of each other.

9.9 CHOOSING AMONG MUTUALLY EXCLUSIVE PROJECTS

Sometimes competing projects cannot all be pursued. They may both use the same land (should we build a hotel or an office building on this site?), they may offer different solutions to the same problem (should we build a bridge or a tunnel?), or they may be related to competing strategies of production and distribution (should we build small retail outlets in every neighborhood or large box stores to serve the entire region?). Projects will also vary in their design and scale of effort: should the sports stadium seat 30, 50, or 75 thousand spectators? Should the bridge have four or six lanes? Should apartments have four rooms or five rooms? In cases like this, once a particular design is selected, the others are no longer available; the choices are mutually exclusive.

When selecting from a group of mutually exclusive projects, it does make sense (from a financial perspective) to maximize the net present value of cash flow. The best project will indeed be the one that is equivalent to the largest amount of money today.

However, if a company evaluates projects by choosing those with the highest IRR, problems are likely to arise. Example 9.6 has already shown that the project with the highest IRR may not be the best project. Example 9.9 shows how to use IRR to determine the best project. The key in properly using IRR is to consider the rate of return on each increment of investment. If an incremental investment exceeds the MARR, then that increment can be justified even if it lowers the IRR for the project.

EXAMPLE 9.9   Examining the Incremental IRR

Table 9.11 summarizes the investment and expected annual net income for four options for developing a site: build a parking lot or construct a building with 1, 2, or 3 stories. If we assume that the same net income will continue indefinitely, then the annual rate of return would be the net income divided by the investment. For example, the rate of return for the parking lot would be $22,000/$200,000 = 11\%$. We can also estimate the present worth of the income using the capital worth method: present worth equals annual income divided by our discount rate, and the NPV of the project would be the present worth of the

Table 9.11   Mutually Exclusive Options for Developing a Site (amounts in $1,000s)

		Investment	Net Income
Park	Parking Lot	$ 200	$ 22
B1	1-story building	$4,000	$600
B2	2-story building	$5,500	$720
B3	3-story building	$7,500	$960

income minus the investment cost. For the parking lot, assuming a 10% discount rate, the NPV would be $22,000/10% − $200,000 = $20,000. Table 9.12 shows the rate of return and the present worth for these four options.

Table 9.12   IRR and NPV for the Mutually Exclusive Options for Developing a Site

Project	Internal Rate of Return	Net Present Value
Parking lot	11.0%	$ 20
1-story building	15.0%	$200
2-story building	13.1%	$170
3-story building	12.8%	$210

If our hurdle rate equals our discount rate of 10%, then all of the projects are acceptable, whether we consider the IRR or the net present value. However, if we have to choose just one of these, then we have a problem. Considering only the return on investment, the best choice appears to be the 1-story building with its 15% return. However, the 3-story building has a higher NPV. Which is really the best project? In your presentation to the board of directors, do you recommend the 1-story building because the company always uses IRR to rank projects? Do you recommend the 3-story building because the textbooks always recommend maximizing NPV? Do you accept a suggestion to compromise on a 2-story building? What should you do?

In dealing with these questions, the analysis must be done one step at a time, beginning with the option that requires the least investment (in this case, the parking lot). Since the IRR of this project exceeds the hurdle rate of 10%, it is acceptable. The question now concerns the additional benefits that might be obtained from additional investment in this site. The 1-story building requires an additional $3.8 million dollar investment in order to gain an additional $578,000 in annual income. The rate of return for this incremental investment is therefore $578/$3,800 = 15.2%, which is well above the hurdle rate. The NPV for this building is 10 times greater than the NPV for the parking lot, so both measures indicate that the 1-story building would be a good investment. Now we need to consider the benefit to be gained by the additional investment required to go from 1 to 2 stories. The additional investment of $1.5 million produces additional net income of $120,000 per year, so the ROI for the increment is only 8%, which is less than our hurdle rate of 10%. Thus the 2-story building is not as good as the 1-story building. Although the IRR for this building is 13.1%, which is well above the hurdle rate, the incremental return for the additional $1.5 million is unacceptable. If we look only at the NPV, we immediately reject the 2-story building because its NPV is $30,000 less than the NPV of the 1-story building.

Now we proceed to the fourth and final option, the 3-story building. We compare this building to the best of the previous options, namely, the 1-story building. The incremental investment in this case is $3.5 million and the incremental net income is $360 thousand, so the incremental return is 10.3%—just over the 10% hurdle rate. Therefore the incremental investment is in fact justifiable. Once again, the NPV immediately gives the same result: the 3-story building has the highest NPV and therefore is the preferred investment.

Conceivably, the board of directors will be unwilling to commit $7.5 million to this site. If so, someone who hadn't followed the logic very closely might suggest cutting back to the 2-story building, which after all has an IRR of 13.1% (perhaps snidely noting that this figure is higher than the 12.8% for the option recommended by the junior analyst). At that point in the meeting, you have to stand your ground: if the board is unwilling to commit $7.5 million, then they should stay with the 1-story building because it has a higher NPV than the 2-story building. And, if necessary, you can explain that the incremental $1.5 investment required for the 2-story building would be better invested in another of the company's projects.

The procedure illustrated in Example 9.9 can be used with any set of mutually exclusive investment alternatives:

1. Rank the alternatives in increasing order of investment required.
2. Estimate the IRR for the each alternative.

3. Choose as a base case the first alternative whose IRR exceeds the hurdle rate.
4. Compare the next alternative (i.e., the alternative with the next-highest investment requirement) to the base case:
   a. Calculate the IRR for the incremental investment.
   b. If the incremental IRR is unacceptable, consider the next alternative and repeat this step.
   c. If the incremental IRR is acceptable, make this alternative the new base case and repeat this step until either the capital budget is reached or all alternatives have been tested.

This process finds the highest investment that can be justified among the competing projects. It also indicates the project with the highest NPV—which is why it is desirable to estimate the NPV even when you must present results in terms of IRR.

9.10   DEALING WITH UNEQUAL LIVES OF COMPETING PROJECTS

Competing projects may well have different project lives. If so, then several approaches can be taken to ensure that comparisons are made in a reasonable manner.

One possibility is to choose a longer period that is an integral multiple of the lives expected for each of the projects. It can then be assumed that the projects will be repeated two or more times over the course of extended period of analysis. For example, if competing projects have lives of 3 years or 4 years, then each project can be analyzed over a period of 12 years because this time frame would involve four cycles of the 3-year projects and three cycles of the 4-year projects. This approach could lead to some extraordinarily long life cycles when considering many projects with many different lives. For example, if projects have lives of 5, 7, or 10 years, then a 70-year project life is needed to have an integral number of cycles for each project. The problem with such a long project life is that technology, population, related development, and prices are likely to change so much that very long-term estimates will be questionable. It is unreasonable to use a 70-year horizon to compare options that all have lives of at most 10 years.

A second approach is to use the annuity value rather than the net present value. This approach assumes that any of the projects either can be extended at the same (or a similar) annuity value or can be replaced by better projects. If the projects have similar but not identical lives, the differences will not be dramatic:

> *Fundamentally Equivalent Annual Cost is a robust measure regardless of the alterations from the original project and its identical repetition assumption. . . . In reality, projects often do not repeat, but are rarely divested during their first life and dramatic cost change occurs only in the long run.*[3]

For typical projects, where the effect of unexpected early termination is minor and discount rates exceed 10%, the equivalent annual cost is reasonable to use even though projects have different lives. For riskier projects or projects with great uncertainty in cash flows, sensitivity analysis must be done to consider the effect of early termination and variable cash flows on the equivalent annual cost.[4]

A third approach is to include a residual value for each project at the end of the analysis period. This approach assumes that it is possible to estimate residual values; although it may be feasible, it may also be much more trouble than it is worth.

---

[3] Ted. G. Eschenbach, Robert B. Koplar, and Alice E. Smith, "Violating the Identical Repetition Assumptions of EAC," 1990 *International Industrial Engineering Conference Proceedings*, Institute of Industrial Engineers, 1990, 99-103.

[4] Ted G. Eschenbach, Robert B. Koplar, and Alice E. Smith, "Violating the Identical Repetition Assumption of EAC," *International Industrial Engineering Conference Proceedings*, Institute of Industrial Engineers, 1990, 99–103.

A fourth approach is simply to use a long enough time period that any differences will be minimal. If discount rates are greater than 10%, then what happens after 20–30 years will have minimal impact on NPV, as discussed earlier.

As always, it is important to use common sense. When in doubt, do some sensitivity analysis using different time periods to determine to what extent, if any, the choice of the period of analysis is causing differences in rankings among the alternatives. There is no "right" method that must be followed.

9.11   SPLITTING A PROJECT INTO PIECES FOR DIFFERENT PARTIES

This chapter so far has considered the perspective of entrepreneurs, developers, companies, or agencies as they evaluate their options for undertaking construction projects. The discount rates and hurdle rates that they use will reflect their own investment opportunities, their own cost of capital, and their own perceptions of the risks associated with the projects they are examining. If the projects are funded entirely by cash on hand, then this is the only perspective that matters.

More commonly, as discussed in Chapter 8, projects and companies are leveraged. Financing a project is possible only if a major portion of the money required for the investment can be raised from outside investors. If this money is a small portion of the total funds sought by the company, then the cost of the capital required (i.e., the interest rate on loans or bonds and the price per share of stock sold) can be assumed to relate to the overall financial strength of the company or agency. The discount rate used in calculating the present worth and the hurdle rate would be at least as high as the organization's weighted average cost of capital, and the financing of any particular project would be a small part of the overall financial management of the company or the agency.

Additional analysis is necessary if the project is undertaken as a stand-alone activity of a new company, if the project requires funding that is tied to its actual results (rather than to the overall financial strength or the organization), or if the project is a major departure from other activities of the company. In these cases, it is necessary to consider how the financial markets and potential investors will view the risks of the project.

The project may require bank loans that are secured by the expected rents, tolls, or other proceeds of the project. The value of the company's stock could be related directly to the success of the project, taking into account the interest payments that must be paid to banks or bondholders before paying any dividends to stockholders. In these situations, it is necessary to consider the different perspectives of the potential investors.

A mortgage is a loan that is secured by a lien on the property. If mortgage payments are not made in a timely fashion, the mortgage holder has the right to foreclose on the property. Since the loan is backed by property, the mortgage is less risky than an unsecured loan, and the interest payments on a mortgage will be lower than the interest on an unsecured loan. It is possible to have multiple mortgages on a property. If so, then the mortgage agreements will state the order in which payments will be made if there is insufficient cash to make all of the contractual payments. The first mortgage generally has priority over the second or any other mortgages, meaning that the holder of the first mortgage has first call on the cash flows of the company. The risk of not getting paid is therefore higher for the holder of a second mortgage than for the holder of the first mortgage.

After the payments are made on secured loans, the next priority is to make payments on unsecured loans and to pay interest on bonds. If a company is unable to make such payments, then it can be forced to declare bankruptcy. A bankrupt company can in many cases suspend mortgage payments, interest payments, taxes, and other fixed charges in order to reduce the outflow of cash while attempting to reorganize.

After all fixed charges and taxes are covered, whatever cash is left over can be paid out as dividends to stockholders or reinvested in the company. This portion of the cash flow varies with the success of the company or the project; the higher the fixed charges as a total proportion of expected cash flows, the more uncertain the prospects for the company. The value of the company to the owners depends upon this portion of the cash flow: the higher and the more reliable the cash flow, the greater the value of the company.

9.12   SUMMARY

Maximizing Net Present Value

The equivalent worth methods provide the best way to compare alternatives. If the net present value is positive, then a project is worth pursuing, at least from the financial perspective. If the NPV is negative, then it is not worth pursuing. If the NPV is positive, then any future values and annuity values will also be positive, so any of these measures can be used to determine whether a project is worthwhile from the financial or economic perspective. Moreover, each of these measures will produce the same ranking of independent alternatives and the same choice among mutually exclusive alternatives.

Importance of the Project Life

When comparing alternatives, analysts need to choose the time period with some care. In general, a period of 20–30 years is sufficient because the discounted costs and benefits of more distant benefits will add very little to the present worth of a project. The choice of a time period should not determine the outcome of the analysis. If costs and benefits have both reached a steady state by the end of the analysis period, then there is no reason to worry about the choice of the project life, so long as the discounted cash flows from the excluded years contribute little or nothing to the present worth. If either costs or benefits are expected to rise or fall sharply just after the end of the analysis period, then a longer period is needed. For example, if extensive rehabilitation is anticipated in around 22–25 years, then using a 20-year life could be very misleading and a 30-year life would be better. In some cases, where extraordinary costs or benefits will occur in the far distant future, much longer time periods should be considered. For example, the costs of dismantling an obsolete nuclear power plant and the ultimate costs of safely disposing or sequestering spent nuclear fuel should be included in the analysis, even if such costs are expected only after 40 or more years of operations.

Does Discounting Ignore Future Catastrophes?

Some critics complain that discounting cash flows means that the costs of future catastrophes can be ignored when evaluating projects. However, even the discounted costs of major floods, earthquakes and other catastrophes can be enormous and certainly large enough to justify engineering, regulatory or other means of avoiding or mitigating the consequences of future catastrophes.

Internal Rate of Return

Companies commonly use the internal rate of return, rather than NPV, to rank competing proposals for projects. The IRR is useful because it can be calculated without reference to any predetermined discount rate. It therefore appears to provide an objective assessment and an obvious means of ranking independent alternatives. However, using this measure to rank projects can lead to three potential problems:

- If cash flows are highly variable, with multiple periods where cash flows are negative, then the methods used to estimate IRR may come up with two values.
- This method implies that all positive cash flows can be reinvested at the IRR over the life of the project. In fact, cash obtained from the project may have to be invested in ways that have much different returns.

- This method does not provide the correct rankings for mutually exclusive alternatives. It is necessary to consider the incremental return for incremental investments to determine which alternative is best.

External Rate of Return

The IRR can be modified to deal with the first two problems by using an "external" rate of return for converting all periods with negative cash flow to an equivalent present value and converting all periods with positive cash flow to an equivalent future value. The rate of return is then uniquely defined as the annual return that will cause the present value of the costs to grow into the future value of the benefits. This approach will still require analysis of incremental returns for incremental investments in order to obtain the correct selection among mutually exclusive alternatives.

Constant Dollar versus Current Dollar Analysis

Projects can be evaluated in terms of constant dollars or current dollars. In a constant dollar analysis, the effects of inflation are ignored both in selecting a discount rate and in predicting cash flows, In a current dollar analysis, the discount rate is higher, taking into account expectations concerning inflation., and the analysis is based upon actual cash flows. Two mistakes must be avoided:

- Using the real discount rate in a constant dollar analysis will discount future cash flows too heavily, thereby tending to favor small projects with immediate benefits.
- Using a discount rate that does not consider inflation in a current dollar analysis will over-emphasize future cash flows, thereby tending to favor large projects that eventually produce long-term benefits.

Choosing among Independent Investment Options

Any alternative that has positive NPV is worth pursuing, at least from a financial perspective. If the NPV is positive, then any equivalent annuity or future value will also be positive. The rankings of alternatives based on NPV, annuities, or future value will all be the same.

If the internal rate of return is used to rank projects, then the higher the IRR the better. Any project with positive NPV will also have an IRR greater than the discount rate used to calculate the NPV. A company may rank its investment options by comparing the IRR of each project to a hurdle rate. The hurdle rate will be at least as high as the company's weighted average cost of capital and is, in effect, the company's minimum acceptable rate of return (MARR).

Choosing among Mutually Exclusive Projects

If selecting one option precludes other options, then the options are mutually exclusive. The basic rule is to choose the option with the highest NPV (which will also have the highest annuity value and the highest future value),

If the internal rate of return is used, then care must be taken in ranking projects, because the project with the highest IRR is not necessarily the best project. Constructing a smaller project with a higher IRR may preclude a larger project with a lower, but still acceptable IRR. It is therefore necessary to follow a well-defined procedure in determining which project is best.

1. Rank the mutually exclusive projects in order of increasing investment requirements.
2. Determine the IRR for each project.
3. Starting with the smallest project, select the first project with an acceptable IRR as the base project.

4. Calculate the incremental costs and incremental benefits for each larger project.
5. Calculate the incremental rate of return for each larger project.
6. Select the first project larger than the base project that has an acceptable incremental rate of return. This becomes the new base project.
7. Repeat the analysis of incremental costs relative to the most recent base project.
8. If there are no projects for which the incremental benefits justify the incremental costs, then the most recent base project is the best project.

Dealing with Unequal Lives of Competing Projects

There are several options for comparing projects that have different expected lives:

- Use a time period for evaluation that is an integral multiple of the lives expected for each of the projects under consideration, e.g., use a 20-year time horizon for comparing a project with a 5-year life to one with a 4-year life. This approach may lead to unrealistically long time periods for evaluation.
- Compare projects based upon their annuity values. This approach will often be reasonable for projects with similar lives.
- Compare projects over a fixed time period, and assign residual values for each project as of the end of that time period. This approach is feasible only if residual values can readily be estimated.
- Compare projects over a long-enough time period that more distant cash flows will have a minimal effect on the analysis.

Sensitivity analysis can be used with any of these approaches to determine whether or not differences in the evaluation period cause differences in ranking of the project.

Splitting a Project into Pieces for Different Parties

Each party to a project will share some of the costs and some of the benefits. The risks to each party depend upon a comparison of their costs and benefits, and upon the risks associated with those costs and benefits. Construction companies want to be paid for their efforts, but are little unconcerned about the long-run financing of a project. Bankers need to know whether the expected revenue will be sufficient to cover monthly mortgage payments, but are otherwise unconcerned about the overall profitability of the project. Investors need to know whether their share of the profits will be sufficient to justify their share of the investments, whether they are purchasing bonds to earn interest or purchasing stock in hopes of dividends and capital gains. Owners consider what they must put into the project and what prospects they have for making money after the construction companies, the banks and the investors have taken their share of the revenues.

DISCUSSION: THE LIMITS OF FINANCIAL ANALYSIS

This chapter has dealt almost exclusively with financial issues. In the private sector, financial performance is in fact the most important aspect of project evaluation. In the public sector, where projects are undertaken to meet public needs rather than make a profit, economic, social, environmental, and sustainability issues also are relevant. To the extent that economic factors can be expressed in monetary terms, it is possible to use the same methodologies to calculate the NPV and the IRR. However, these global measures are not the only things to consider when evaluating any complex project.

Other economic issues include distributional equity (who wins and who loses), regional economic impact (the use of local labor and resources and the multiplier effect on the local economy), and nonfinancial externalities (environmental and social impacts and the need for remediation). Any large project has an impact on the public, and many costs and benefits are likely to be difficult to quantify and even more difficult to value. In some cases, nonquantifiable factors are the major issues in project evaluation. Many of these issues have been discussed in earlier chapters, and some are covered in subsequent chapters.

In conclusion, despite spending half of this text focusing on financial matters, we have to conclude that financial feasibility is essential for any project—but that financial feasibility may have little or nothing to do with project desirability. Being able to get money to build something is much different from deciding whether something should be built. Financing difficulties may preclude certain highly desirable projects, yet encourage other clearly undesirable projects.

Engineers, managers, planners, and politicians have some personal responsibility for pursuing desirable projects that are financially feasible. Project evaluation depends upon properly presenting estimated costs and benefits and disclosing assumptions about discount rates, project lives, and the types and distribution of costs and benefits. It is not enough to show that a particular project can be justified; it is also necessary to show that the project is better than the available alternatives.

ESSAY AND DISCUSSION QUESTIONS

9.1 What are the advantages and disadvantages of using the internal rate of return as a means of comparing alternative projects?

9.2 Company managers are often urged to "maximize the net present value of cash flows." Why? Is this good advice? Is it also true for the public sector?

9.3 It is easy to say, "Choose the project with the highest NPV," but how many possible projects are there likely to be? Is it really feasible to "list all the options and select those with positive NPV for further analysis? Let's consider a common situation. Many cities are built along the banks of a river, necessitating construction of bridges or tunnels or the use of ferries to link the two sides of the river. Suppose a city already has four bridges, a tunnel, and two ferry services providing transportation across the river that bisects the city. Various proposals have been made for increasing capacity:

- Increase bus service so as to reduce automobile traffic and relieve congestion.
- Charge tolls on some or all of the existing bridges.
- Expand some or all of the existing bridges.
- Construct new bridges, of varying sizes, at 10 possible locations.
- Construct new tunnels, of varying sizes, at 3 possible locations.
- Subsidize additional ferry service for several routes along and across the river.

Identify some options or sets of options that can be considered to be independent. Identify some options or sets of options that are mutually exclusive. How many independent sets of options could be identified? How would you structure the options or sets of options that could be considered?

PROBLEMS

9.1 A dispute has broken out in a board meeting of a real estate investment firm. Mr. Park advocates developing a site as a parking lot that would require a minimal level of investment, produce revenues of close to $1 million per year, and have an IRR of more than 40%. Mr. Macy prefers a mixed development with several small stores at street level and apartments on the upper levels. He argues that the project will have a NPV of $6 to $10 million, using the company's hurdle rate of 15%. "Fine," says Mr. Park, "but your IRR is still less than 20%, so my project is better." Which project would you support?

9.2 A company has a hurdle rate of 15% and a capital budget of $10 million. Which of the projects shown in the table below can be supported? Which projects could be supported if the capital budget were $15 million?

Problem 9.2: IRR Estimated for Various Projects

Project	Investment	Estimated IRR
A	$1 million	20%
B	$2 million	19%
C	$0.7 million	18%
D	$0.3 million	18%
E	$5 million	17%
F	$2 million	16%
G	$1 million	15%
H	$3 million	12%
I	$1 million	10%

9.3 Three vice presidents of a manufacturing company have once again advocated their pet projects. The VP Engineering wants to invest in energy efficiency, and he claims his project will save the company $2 million per year for 20 years. The VP Sales wants to invest in a new manufacturing operation that she claims has a NPV of $10 million. The VP Operations wants to invest in new equipment that will provide benefits that will have a future value of $30 million in 10 years. All departments have used standard techniques for preparing their estimates of the net benefits, and all have used the company's hurdle rate of 10% to discount cash flows. Which proposal is the best?

9.4 A company is trying to decide which type of equipment to acquire. Management expects to get 1000 hours of use per year for the equipment over a period of 5 years. The purchase price, operating costs, and expected life of the equipment are shown in the following table. Which should they purchase? The company's MARR is 10%.

Problem 9.4: Characteristics of Four Types of Equipment

Type of Machine	Purchase Price	Operating Cost/Hour	Expected Life (hours of service)	Expected Life (years of service)
Workman	$100,000	$20	10,000	10
Master	$120,000	$18	50,000	20
Avatar	$150,000	$15	100,000	30

9.5 Consider another company facing the same choice of equipment presented in the previous problem. This company expects to use the equipment for 2000 hours per year indefinitely. Calculate the equivalent uniform annual cost (EUAC) of the purchase price plus operating cost over the expected life of each option. Which is the best option?

9.6 Still considering the choices of equipment presented in the previous two problems, under what circumstances would the "Master" option be justifiable?

9.7 You have been asked to evaluate four projects (see table). The NPV of the investment and the equivalent uniform annual net benefits (not including the investment) are given. You are trying to decide if the annual net benefits are large enough to justify any of the investments. The minimum acceptable rate of return for your firm is 10%.

Problem 9.7: Investment and Benefit Data for Four Projects

Project	Investment (NPV as of time zero)	Equivalent Annual Net Benefits	NPV (using capital worth method)	IRR
A	$1 million	$130,000		
B	$2 million	$190,000		
C	$2.6 million	$286,000		
D	$4 million	$480,000		

a. Calculate the NPV and the IRR for each project, and fill in the table. In calculating the NPV, assume the annual net benefits continue forever, so that you can use the capital worth method.

b. Assume that these projects are independent, so the firm could carry out any or all of them. Which projects are justified using the NPV method? Which are justified using the IRR method?

c. Assume that these projects are mutually exclusive alternatives for developing a specific site. Which project is best? Why?

9.8 You have four options for constructing an apartment building on a site. Your MARR is 10%. Which of the following options are acceptable, and which option is the best? (Assume that net rents continue indefinitely, with no inflation.)

a. *5-story building, no frills*: $1 million investment, annual net income $0.11 million

b. *5-story building, with pools and gardens*: $1.5 million investment, annual net income $0.12 million

c. *10-story building, no frills*: $2 million investment, annual net income $0.21 million

d. *10-story building, upscale*: $2.5 million investment, annual net income $0.25 million

e. *15-story building, upscale*: $3.3 million investment, annual net income $0.35 million

9.9 You are a summer intern working for a hot-shot project evaluation firm, and your supervisor admits to having a problem. At first he thought his project was really a terrible idea because it had only a 2% IRR; but then he discovered that it also had an IRR of 18%, which is well above the firm's hurdle rate of 10%. He checked his numbers, and the calculations (shown below) were in fact correct. He also did a few side calculations using the firm's hurdle rate of 10% (also shown below). Explain the problem,

Problem 9.9 Cash Flows for a Project with Two Different IRRs

Discount rate:	1.89%							
Year	1	5	10	15	20	25	50	Total
Cash flow	−100	180	−50	−200	75	2000	−3000	−1095
NPV	−98.1	163.9	−41.5	−151.1	51.6	1253.3	−1178.1	0.0
Discount rate:	**18.24%**							
Year	1	5	10	15	20	25	50	Total
Cash flow	−100	180	−50	−200	75	2000	−3000	−1095
NPV	−84.5	77.9	−9.4	−16.2	2.6	30.3	−0.7	0.0

complete the analysis, and tell your supervisor whether this is a good project.

Additional calculations for these cash flows:

- Present worth of construction (year 1) plus rehabilitation (years 10 and 15) is $183.62 million, assuming MARR of 10%.
- Future worth of all revenues (years 5, 20, and 25) will be $36.098 billion at the end of year 50, assuming MARR of 10%.

9.10 The project described in the previous question is rather strange, at least in terms of its cash flows. Let's assume that the numbers shown here are acknowledged to be rough estimates, although the general trends are likely to be similar to what is shown. Let's also assume that some people believe it is worth doing because of the great benefits over the first 25 years, but others are concerned about the long-term costs. What additional analyses would you suggest before reaching a final decision about whether to recommend proceeding with this project?

9.11 Highway safety can be increased by improving the roads, the vehicles, or the drivers, as shown in the following examples:

a. You are the safety officer in a state Department of Transportation, and you have a budget for improving highway safety. Some of your options are shown in Table1 below as "Road Options R1, R2, and R3"). For each of these options, calculate the cost effectiveness for reducing fatalities.

b. You are a safety engineer for an automobile manufacturer. Some of the options for improving automobile safety are shown in Table 2 below as "Auto Improvements A0, A1, and A2." For each of these options, calculate the cost effectiveness for reducing fatality rates.

c. A fatal accident recently occurred at one of the dangerous intersections in your state. A Chevrolet skidded on a patch of ice, ran a red light, and crashed into the passenger side of a 2005 Ford. The driver, who was protected by an air bag, was unhurt; but the front-seat passenger, who was protected only by a seat belt, was killed. The driver of the Chevrolet was unhurt because his air bag inflated; the car did not have antilock brakes (and the driver was quoted as saying, "I should've got the antilock brakes!"). Because of this accident, an influential politician has strongly criticized the Department of Transportation for not eliminating all of the dangerous intersections, and he would like to divert all safety funds to this task. Since you are the safety specialist in your state's DOT, you have been asked to respond. In light of the opportunities shown in Tables 1 and 2 for improving highway and automobile safety, explain why you agree or disagree with the proposal.

Problem 9.11, Table 1: Highway Safety Improvements and Their Effects on Fatalities

Road Option	Cost	Expected Reduction in Fatalities
R1: Ensure better enforcement of speed limits on 100 miles of interstate highways.	$1,000,000 per year	2.0 per year
R2: Provide wider shoulders on 100 miles of major rural roads.	$20,000,000	1.0 per year
R3: Eliminate dangerous intersections by building bridges.	$5,000,000 per intersection	0.05 per intersection per year

Problem 9.11, Table 2: Automobile Safety Improvements and Their Effect on Automobile Fatality Rates

Auto Improvement Option	Cost	Expected Reduction in Fatality Rate Over the 100,000-Mile Life of a Typical Car
A0: Seat belts plus air bags for driver (already required)	$100 per vehicle	0.002
A1: Front plus side air bags for driver and passengers (currently optional)	$400 per vehicle	0.0002
A2: Antilock brakes (currently optional)	$400 per vehicle	0.0005

Chapter 10

Rules of the Game: Taxes, Depreciation, and Regulation

The zoning law of 1916—the nation's first—regulated the bulk of buildings, their height, and their uses. It divided the city into three zones—residential, business, and unrestricted—and empowered the Board of Estimate to regulate the use, height, and bulk of every building on every street in the city, depending upon what zone the block was in.[1]

CHAPTER CONCEPTS

---

[1] New York City implemented zoning to avoid turning Manhattan into dark canyons with skyscrapers towering above and keeping light and fresh air away from city streets. John Tauranac, *The Empire State Building: The Making of a Landmark* (New York: St. Martin's Griffin, 1995), 55.

10.1   INTRODUCTION

Government policies affect how projects are conceived, what kinds of projects can or cannot be implemented, where projects can or cannot be built, how projects can or cannot be constructed, and how successful they will be once they are implemented. This chapter examines how policies such as taxes, zoning, building codes, and environmental regulations affect project evaluation.

These policies are in effect "rules of the game" for limiting the kinds of developments that can be pursued and influencing how the players tabulate their scores (i.e., their profits and their ability to complete projects). The players—real estate companies, entrepreneurs, public agencies, infrastructure operators, investors, banks—still have to figure out what they want to do and how to do it, but they must abide by the rules that have been established. Changing the rules will change the way the "game" is played, and the rules can be adjusted to promote projects that are believed to provide economic, social, environmental, or sustainability benefits for the public.

Taxes are relevant to project evaluation because taxes affect cash flows. If the goal is to maximize the net present value of cash flows, then it is necessary to consider taxes. Moreover, local, state, and federal governments may impose taxes or offer tax credits to discourage or promote certain kinds of development, so it is important to be able to comprehend the effects of tax policy on project design and evaluation. Different kinds of taxes may apply to the profits from constructing, operating, and selling infrastructure, so it is necessary to understand how tax laws categorize each type of expenditure and each type of revenue. Accounting rules established by law or by regulation determine what kinds of expenses are treated as current expenses and what kinds are treated as capital expenditures. Arcane rules may determine whether money spent on rehabilitating infrastructure is treated as operating expense—which is fully deductible as an expense in the current year—or a capital expense that can be deducted from taxable income only over a period of many years. Section 10.2 shows how taxes affect cash flows and provides examples of how tax policy can be used to promote certain types of projects.

Zoning is the major tool used by local governments to guide land use. Zoning restrictions limit the types of development that may be pursued in certain locations, such as restricting one area to residential use while specifying another area as suitable for industrial use. Zoning restrictions may also limit the size or height of buildings or the location of buildings on a site. Section 10.3 provides examples of the effect of zoning on the potential value of a site and the opportunities for projects.

Building codes define what types of construction materials, designs, and methods can be utilized. Regulations can be established to reduce risks during construction or operation or decommissioning of a project. New technologies, such as the use of plastic pipes for plumbing, had to be approved for use in building codes before they could be widely used. In transportation, governments may create design standards for highways and bridges that serve a similar function (i.e., promoting safety during construction and operation). Likewise, governments may establish standards for the construction and operation of water resource systems and for other types of infrastructure. Section 10.4 describes the logic underlying safety standards, along with some examples of how these policies affect project design.

Environmental restrictions restrict the nature, location, and cost of projects. Land use regulations may include restrictions on development in or near wetlands or waterways. The types of materials used in construction may be subject to restrictions, such as laws prohibiting the use of asbestos because of its link to lung cancer. Restrictions may also limit the types of work that can be undertaken at night, so as to limit the disruption to neighborhoods. Contractors may be required to take special precautions to prevent dust and runoff from construction sites from contaminating nearby areas. Section 10.5 shows how environmental policies can affect project evaluation.

10.2   DEPRECIATION AND TAXES

The previous chapters have introduced two quite different ways of thinking about financial performance. In Chapters 7 to 9, this text repeatedly recommends evaluating the financial performance of projects by calculating the net present value of all of the cash flows associated with that project. That is a pretty clear statement of how to consider financial performance—but it is entirely different from the way that corporations describe their performance to their stockholders, investors, and tax collectors! Chapter 9, in discussing the internal rate of return, introduces a rival way of looking at financial performance, one that centers on rate of return on investment rather than NPV of cash flows. After resolving some methodological difficulties, that chapter concludes that the IRR can be used, with care, to reach the same conclusions as the NPV analysis concerning which project or projects are worth pursuing.

The logic introduced in Chapter 9 is indeed the proper way to evaluate the financial aspects of projects—but it has little or nothing to do with taxes or a company's financial statements. Taxes and financial statements are structured according to strict guidelines known as **generally accepted accounting principles (GAAP)**. By following well-defined accounting rules, it is possible to define terms such as profit and return on investment (ROI). Profit is a critical measure, because **income taxes** are stated as a percentage of profits. Profit, ROI, and other measures used in financial statements are critical because investors and analysts use them in judging whether to invest in a company. The financial markets rely upon the validity and the comparability of data produced by companies in their financial statements. Therefore, if a company sells stocks or bonds, it is required to use GAAP in preparing those statements as well as their tax returns.

Since taxes are large cash flows, they cannot be ignored when evaluating projects. And since the amount of taxes to be paid depends upon accounting rules, it is necessary to understand some basic concepts of accounting. One of the most important rules is that capital expenditures cannot be treated as a current expense, but instead must be spread out over many years as a depreciation expense. In many infrastructure projects, depreciation is one of the largest expenses, so the rules that determine how to calculate depreciation are important. Section 10.2.1 shows various methods that may be used to determine depreciation expense. Section 10.2.2 shows how to convert pretax cash flows into after-tax cash flows, taking into account depreciation and income taxes. Sections 10.2.3 and 10.2.4 illustrate how income taxes and accounting rules might affect a developer. Section 10.2.5 shows how real estate taxes and other local taxes might affect project evaluation.

10.2.1 Depreciation

Depreciation is a method for converting capital expenditures into annual operating expenses. The main reason for using depreciation is that a capital expenditure is entirely different from an operating expense. Operating expenses for a building include such things as electricity for lights, oil for heat, and wages for the people who manage and maintain the building. At the end of each week or month, the owner of the building knows how much was spent on electricity, oil, and wages: the lights worked, the building was warm, the rooms and hallways were cleaned, and rents were collected. The money was spent, the work was done, and the accounting is complete.

Capital expenditures are entirely different. The building may have cost $10 million to construct, and when the construction is complete the owner may have a mortgage for $10 million—but the owner also has a building. Now, the building may or may not be worth $10 million, because the value of the building depends upon the real estate market, the condition of the building, and the annual rent payments, not the cost of the building. However, the accounting assumption is that if the building cost $10 million to construct, then the building is an asset worth $10 million when it is put into service. The owners may have spent $10 million, but they have created an asset worth $10 million, so they have not had any loss in value.

The same concept can be applied to the purchase of a car for $20,000, the construction of a bus terminal for $20 million, or the construction of a vast pipeline for $2 billion. The money may have been spent, but an asset has been created, and accountants will record the book value of the asset as being equal to the investment cost. So right at the beginning of the life of the asset, we have an accounting assumption that is accepted even though it most likely is wrong. The car may be worth only $15,000 as soon as we drive it out of the dealer's lot; the bus terminal may be worth only $10 million to anyone other than the bus company, while it may be worth $30 million to the bus company. We don't quibble about this discrepancy between book value and real value—not because we are so flexible in our thinking, but because the tax collector tells us that we will use the accountant's book value as one factor in computing the taxes we owe. If you want to use the real value of the asset in your own internal reporting, that is fine; just don't confuse the book value and the real value when doing your taxes. Also, although we really do know what we paid for the car, the bus terminal, and the pipeline, we probably do not know what they really are worth today. So it is convenient for us, as well as the tax collector, to use the accounting assumption.

Now we have a place to start for figuring out depreciation, namely the book value of the asset. Perhaps now we can try to determine how much the asset deteriorates each year, so we can use the actual decline in life as the amount of depreciation. This turns out to be a difficult task. What is the expected life of a car? How much of the life of a car is consumed by time? By usage? By exposure to rain and snow? Some engineering analyses can answer questions like these, but the tangle of assumptions and analyses quickly becomes quite thick. For internal purposes, say for a rental car company, it might be a very good thing to understand the life of cars based on the kind of usage they receive and whether they are based in the heat of Florida or the snow and ice of Minnesota. For most companies, however, these likely are difficult calculations with little or no benefit to management; a plan to replace company cars after 5 years is sufficient for them. However, all companies have to use an accepted methodology to account for depreciation of their cars.

At this point, we need three more accounting assumptions to make it easy to estimate depreciation:

1. The life of the asset
2. The salvage value of the asset (which does not include the value of land, since land is generally assumed not to depreciate in value)
3. The depreciation rate over the life of the asset

The life of the asset could be defined based on some sort of study of past experience—or it could be defined by the tax collector. A car, for example, might be assumed to have a life of 10 years, while a bus terminal or a pipeline might be assumed to have a life of 30 or 50 years. The salvage value is the remaining value of the asset at the end of its life. For a car, the salvage value might be the scrap value of the car, which might be assumed to be 5% of the original purchase price. For a bus terminal or a pipeline, the salvage value might be assumed to be the book value of the land (i.e., the purchase price of the land required for the project) or a percentage of the investment cost. The accounting principle is that depreciation causes the book value of an asset to decline from the original investment cost to its salvage value at the end of the asset's life. The depreciation could most easily be assumed to proceed at a constant rate over the life of the asset:

$$\text{Annual Depreciation} = (\text{Investment} - \text{Salvage})/\text{Life} \qquad \text{(Eq. 10.1)}$$

If the life of a $4 million asset is 10 years and the salvage value is $1 million, then the asset will depreciate in value from $4 million to $1 million over its 10-year life, for an average depreciation rate of $0.3 million per year (Figure 10.1). This is called **straight-line depreciation**, because the book value of the asset declines in a straight line from the initial investment cost to the salvage value.

Other approaches to depreciation can be imagined. One might argue that depreciation should be greater at first, to reflect the immediate loss in value of cars and some other assets. Once we have accepted that the accounting assumptions may be tied to convenience rather than reality, it is easy to come up with some alternatives. For example, instead of assuming a fixed amount of depreciation over the life of the asset, it is possible each year to take a fixed percentage (P%) of the remaining book value over the life of the asset, without considering the salvage value at all.

$$\text{First-Year Depreciation} = (\text{P\%})(\text{Investment}) \qquad \text{(Eq. 10.2)}$$

$$\text{Reduced Book Value at End of First Year} = (1 - \text{P\%})(\text{Investment}) \qquad \text{(Eq. 10.3)}$$

$$\text{Second-Year Depreciation} = (\text{P\%})(1 - \text{P\%})(\text{Investment}) \qquad \text{(Eq. 10.4)}$$

$$\text{Year N depreciation} = \text{P\%}(\text{Book value at end of year N} - 1) \qquad \text{(Eq. 10.5)}$$

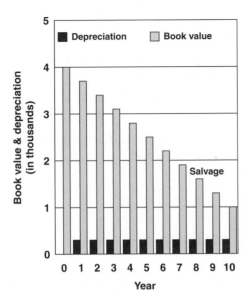

Figure 10.1 Straight-Line Depreciation.

These equations can readily be used to create a table in a spreadsheet that shows the initial book value in each year, the amount of depreciation allowed in that year, and the final book value.

With the same example used above, suppose the book value was depreciated by twice the amount allowed under straight-line depreciation, which would be 20% of the remaining value each year for 10 years. This method is called **double declining balance depreciation** because the amount of depreciation allowed in the first year is double the amount allowed under straight-line depreciation. The book value would decline by (0.20) $4 million = $800,000 in the first year. In the second year, the initial book value would therefore be $3.2 million, and it would decline by (0.20) $3.2 million = $640,000 in the second year. Each subsequent year the amount of the depreciation declines by a smaller amount, as illustrated in Figure 10.2. For the first 5 years, the annual depreciation is higher using the double declining balance; but after that, the fixed depreciation of $0.3 million under the straight-line approach is better.

This method can be used with other percentages, and the general method is known as **declining balance depreciation**.

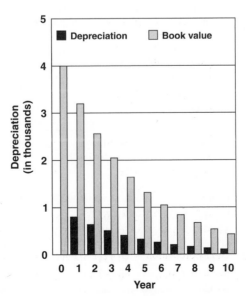

Figure 10.2 Double Declining Balance Depreciation.

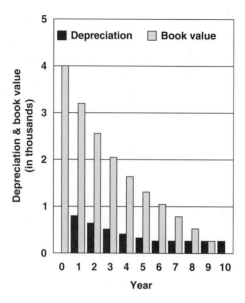

Figure 10.3 Double Declining Balance Reverting to Straight-Line Depreciation in Year 6.

A related approach is to use the declining balance approach only so long as that approach results in depreciation greater than the straight-line approach. Using this method, it is necessary to compare the depreciation for the declining balance approach with the depreciation that would be obtained by switching to straight-line depreciation of the remaining book value over the remaining life. Once the straight-line approach provides greater depreciation, switch to that approach and continue over the life of the project. Once again, this method can readily be examined in a spreadsheet:

1. Start by using the declining balance in the first year.
2. For the second year, calculate depreciation using two methods:
   a. Continue using the declining balance method.
   b. Determine the depreciation of the book value at the beginning of the year using the straight-line method (see Eq. 10.1).
3. If the depreciation under the straight-line method is higher than that allowed under the declining balance method, then switch to that method; if not, then continue using the declining balance method.

This method is illustrated in Figure 10.3. Table 10.1 compares the book value and the annual depreciation for the three methods. Like the double declining balance method, this approach provides

Table 10.1   Three Methods of Depreciating an Asset With Initial Book Value of $4 Million and a Life of 10 Years ($millions)

Year	Straight Line Book Value at Beginning of Year	Depreciation	Double Declining Balance Book Value at Beginning of Year	Depreciation	Double Declining Balance Reverting to Straight Line Book Value at Beginning of Year	Depreciation
1	$4.000	$0.300	$4.000	$0.800	$4.000	$0.800
2	$3.700	$0.300	$3.200	$0.640	$3.200	$0.640
3	$3.400	$0.300	$2.560	$0.512	$2.560	$0.512
4	$3.100	$0.300	$2.048	$0.410	$2.048	$0.410
5	$2.800	$0.300	$1.638	$0.328	$1.638	$0.328
6	$2.500	$0.300	$1.311	$0.262	$1.311	$0.2621
7	$2.200	$0.300	$1.049	$0.210	$1.049	$0.2621
8	$1.900	$0.300	$0.839	$0.168	$0.786	$0.2621
9	$1.600	$0.300	$0.671	$0.134	$0.524	$0.2621
10	$1.300	$0.300	$0.537	$0.107	$0.262	$0.2621
	$1.000		$0.429		$0.000	

a way to accelerate depreciation of an asset, thereby allowing larger tax deductions during the early years of an asset's life.

One might think that these rather arbitrary methods would require a great deal of bookkeeping, plus considerable trouble figuring out what the life of an asset is. That is true. And the accountants, legislators, and tax collectors in the United States have therefore come up with an even simpler approach for accelerated depreciation of assets. The U.S. Internal Revenue Service allows companies to use what is called the Modified Accelerated Cost Recovery System (MACRS), which was introduced in 1986. This system divides assets into six categories, defines an asset life for each category, and assumes the salvage value is zero. Another simplifying assumption is that the first and last years of an asset's life are assumed to be exactly 6 months, so it is not necessary to track the actual date that assets were put into service. Since the options are limited—standard lengths for the first and last years, no salvage value, and no need to determine lives—bookkeeping is simplified, and the choice of an asset life need not be justified. As noted above, the term *accelerated* means that the lives in the MACRS are generally shorter than those previously used, so that companies can take more depreciation sooner. The option of using straight-line depreciation is still available for some assets.

10.2.2   Income Taxes

Individuals and companies pay income taxes that are calculated as a percentage of annual income or profits. This percentage is known as the tax rate. Income taxes may be progressive, meaning that the marginal tax rate increases as income or profits rise. The maximum tax rates vary from country to country, and a country may raise or lower tax rates from time to time. In the United States, additional income taxes are imposed by most of the states and some of the largest cities. Typical tax rates for large companies are 34% for federal tax and 6% for state tax. Federal tax rates are 10% or less for individuals with little or no income, but the rates rise to 35% for individuals with high incomes.

State and local taxes are deductible expenses when calculating federal income taxes, so the effective income tax rate is

$$\text{Effective Income Tax Rate} = SR + LR + FR(1 - SR - LR) \qquad \text{(Eq. 10.6)}$$

where

$SR =$ state income tax rate

$LR =$ local income tax rate

$FR =$ federal income rate

Consider a company that pays federal taxes at a rate of 34% plus state taxes of 6% and a city tax of 1%. Their effective income tax rate is

$$\text{Effective Income Tax Rate} = 6\% + 1\% + 34\%(1 - 0.06 - 0.01) = 39.55\%$$

The previous chapters have introduced the concept of a minimum acceptable rate of return (MARR) and a hurdle rate for the internal rate of return (IRR). Both of these financial objectives can be calculated before or after income taxes. The relationship between the before-tax and **after-tax MARR** and IRR can be approximated as follows:

$$\text{After-tax MARR} = (1 - \text{Effective Income Tax Rate})(\text{MARR}) \qquad \text{(Eq. 10.7)}$$

$$\text{After-tax IRR} = (1 - \text{Effective Income Tax Rate})(\text{IRR}) \qquad \text{(Eq. 10.8)}$$

EXAMPLE 10.1   Comparing After-Tax Returns for Corporate Bonds and Tax-Free Municipal Bonds

In the United States, tax laws allow state and local governments to sell bonds whose interest is not subject to federal tax. Investors living in the state that issues such bonds pay no state taxes on the interest received from such bonds. Suppose a group of investors are considering municipal bonds instead of buying an AA-rated corporate bond that pays 6% interest. Would these investors be willing to buy a municipal bond offered by their local port authority that pays 4.0% interest, or a municipal bond offered by a neighboring state that pays 4.125%? The investors agree that all three bonds are very high quality, because all three are rated double-A by bond-rating services. Some of the investors are in the 35% tax bracket, while others are in the 15% tax bracket; all of them pay 6% for state income taxes.

First consider the high-income investors. Their effective income tax rate is calculated as follows:

$$\text{Effective Tax Rate} = 6\% + 35\%(1 - 0.06) = 38.9\%$$

For these investors, the 6% interest received from the corporate bond will be taxed at the effective rate of 38.9%, so that they will receive only $6\%*(1 - 0.389) = 3.67\%$ interest after deducting federal and state taxes. They therefore would prefer the tax-free interest of 4% that they could obtain from buying the local bonds offered by the port authority. The bonds offered by the nearby state will be subject to the 6% state tax, so their after-tax interest rate would be $4.125\% (0.94) = 3.88\%$. The best choice for this investor is to buy the local port authority bonds that pay 4.0% interest and are free of both state and federal tax.

For the lower-income investors, the corporate bonds provide a better option because these investors have a much lower **effective tax rate**:

$$\text{Effective Tax Rate} = 6\% + 15\%(1 - 0.06) = 20.1\%$$

Their after-tax yield on the corporate bonds therefore is $6\% (1 - 0.201) = 4.79\%$. This is well above either of the other options.

This example illustrates how the tax code can encourage certain investors to buy bonds to finance what legislators believe to be desirable projects. If the interest on municipal bonds were not deductible, then municipalities and states would have to offer much higher interest rates to attract investors. Changes in tax rules make it possible for state and local governments to pay less interest while at the same time allowing purchasers of their bonds to earn higher after-tax interest than they can obtain from corporate bonds with much higher interest rates.

10.2.3   Tax Credits

A tax credit is an amount that can be used to offset income taxes that are owed. Governments may offer tax credits to promote investment in certain kinds of activities, such as education, housing for the elderly, rail transportation, or alternative energy programs. Legislation defines the type of expenditures that would qualify for the program and the amount of the tax credit. Infrastructure programs can be promoted via an investment tax credit (ITC) that could be used to reduce taxes during the year of the investment. Since infrastructure will be depreciated over a life of many years, the investment tax credit can be a major boon to developers during the first year of their projects.

EXAMPLE 10.2   Using an ITC to Promote Investment in Low-Income Housing

Suppose the federal government has decided to provide a 20% investment tax credit to promote investment in housing for low-income families. How much would this tax credit reduce the cost of a qualifying $4 million investment?

The tax credit is 20% of the qualifying investment, which would be 20% of $4 million, or $800,000 in this case. This tax credit could be used to offset taxes, so the benefit to the project would depend upon the timing of both the investment and the payment of taxes. If the investment occurred in year 1, and if the company adjusted its estimated tax payments to reflect the tax credit, then the benefit would also be received during year 1, reducing the cost of the project from $4 million to approximately $3.2 million.

The tax credit is a benefit only if the company actually pays taxes that are at least as great as the credit. If the company pays only $200,000 per year in federal taxes, then the tax credit has to be taken over a period of 4 years. In this case, the tax credit is very much like an annuity of $200,000 per year for 4 years. At a MARR of 10%, the tax credit is worth

$$\text{Value of Tax Credit} = \$200,000[P/A,10\%,4] = \$200,000(3.17) = \$634,000$$

In other words, the tax credit is most valuable to profitable companies that are able to use the credit to reduce their current taxes.

10.2.4   How Depreciation and Taxes Affect Cash Flows

The previous sections have illustrated how accounting rules and the tax code define what expenses can be deducted from revenue to calculate income and what portion of income must be paid as income tax. We have learned that the rules and regulations related to depreciation have more to do with policy and politics than they have to do with the actual deterioration of an asset. Since depreciation is such an important policy tool, it is worth considering in some detail how depreciation affects projects. There are four things to consider:

1. Depreciation converts investments into recurring expenses. The investment involves actual cash expenditures (possibly using borrowed money) that are not deductible for income tax purposes, while depreciation is a noncash expense that is deductible.
2. Accelerated depreciation allows a business to increase depreciation during the early years of an asset's life. This accounting change has the effect of reducing profits but increasing cash flow.
3. Depreciation can be treated as an expense even if it is clear that the asset is actually increasing in value, as is often the case with buildings.
4. If an asset is sold, the new owner can depreciate the asset again, using the purchase price as the initial book value and using a life and salvage value allowed by accounting rules and tax regulations.

Table 10.2 shows the tax advantages of the three options for depreciating the asset that was included in Table 10.1 above. It is assumed that the business's effective tax rate is 40% and that the business indeed has taxable income remaining, whichever method is used. For straight-line depreciation, the annual tax benefits equal $120,000, which is 40% of the annual depreciation of $300,000. Over 10 years, the tax benefit is $1.2 million, and the net present value of the tax benefit, discounted at 10%, is $0.737 million. With the double declining balance method, the annual tax benefits are initially much higher, but they decline over the 10-year life of the asset. The total tax benefit is greater, with an NPV of $1.023 million. The third method provides even more tax benefits because this method fully depreciates the asset (i.e., no salvage value), and the NPV of the tax benefit is $1.096 million. The accelerated methods of computing depreciation thus could increase the value of this asset by $0.36 million—which is 9% of the initial construction cost. This is a good reason for developers to be extremely interested in the nuances of the tax code!

Now think about what happens if the owner has no taxable income. In that case, the depreciation has no tax benefit, however it is calculated. The owner in such a situation might consider selling the asset to

Table 10.2   Tax Benefits of Depreciating an Asset with Initial Book Value of $4 Million and a Life of 10 Years (in millions)

Year	Straight Line		Double Declining Balance		Double Declining Balance Reverting to Straight Line	
	Depreciation	Tax Benefit	Depreciation	Tax Benefit	Depreciation	Tax Benefit
1	$0.3	$0.120	$0.800	$0.320	$0.800	$0.320
2	$0.3	$0.120	$0.640	$0.256	$0.640	$0.256
3	$0.3	$0.120	$0.512	$0.205	$0.512	$0.205
4	$0.3	$0.120	$0.410	$0.164	$0.410	$0.164
5	$0.3	$0.120	$0.328	$0.131	$0.328	$0.131
6	$0.3	$0.120	$0.262	$0.105	$0.2621	$0.105
7	$0.3	$0.120	$0.210	$0.084	$0.2621	$0.105
8	$0.3	$0.120	$0.168	$0.067	$0.2621	$0.105
9	$0.3	$0.120	$0.134	$0.054	$0.2621	$0.105
10	$0.3	$0.120	$0.107	$0.043	$0.2621	$0.105
Total		**$1.200**	**$0.086**	**$1.428**		**$1.600**
NPV		**$0.737**		**$1.023**		**$1.096**

someone who can use those tax benefits—and who will allow the seller to lease the building on a long-term lease. Or, in what is almost the same thing, the business can simply lease assets from companies that do get tax benefits from ownership and pass them on to the business in the form of low lease rates.

Let's return to the situation where the owner does have plenty of taxable income and can therefore take advantage of the tax benefits. In year 11, a change may be needed because no further tax benefits can be obtained from the fully depreciated asset. The owner's taxes will therefore increase by $120,000 per year. This conceivably is a good time to sell the asset (if it still has useful remaining life), so that someone else can capture tax benefits from a new process of depreciation.

Depreciation is especially important in real estate because buildings are very long-lived assets that can increase in value if they are well-maintained or in a good location. If the building's cost can be covered by a mortgage, and if rents are sufficient to cover the mortgage payments and upkeep, then the tax advantages of depreciation can be important to the financial structure of the project. For example, consider a skyscraper that is constructed at a cost of $450 million and has a salvage value of $50 million. If this building is depreciated using straight-line depreciation over 40 years, then the depreciation will be $10 million per year and the tax advantage will be $4 million per year for a business with an effective tax rate of 40%. For someone with a discount rate of 10%, this tax benefit is worth approximately $4 million/0.10 = $40 million.

When a capital asset is sold, accountants and tax collectors will treat the difference between the sale price and the depreciated basis for the asset as a **capital gain or loss**.

$$\text{Capital Gain} = \text{Sale Price} - \text{Book Value} \qquad \text{(Eq. 10.9)}$$

The accounting logic (or perhaps we should say the "accounting fiction") is that the value of the asset actually equals the depreciated basis that is reported to the tax collector and to shareholders up until the moment the asset is sold. At the time of sale, the asset instantly either gains or loses value. Some assets, like automobiles, lose much of their value as soon as they are first put into service, so their depreciated basis at first overstates the value of the asset. Other assets—notably real estate—not only depreciate less rapidly than assumed by the accounting rules, but are very likely to increase in value. Capital gains are typically taxed at a lower rate than income, so both companies and individuals are well-advised to receive cash payments as capital gains rather than as income.

Depreciation, therefore, is an accounting technique that protects some cash receipts from being taxed as income while allowing taxes on any increases in value to be deferred until the asset is sold—and then taxing the capital gain at a lower rate.

EXAMPLE 10.3 Depreciation and Capital Gains

Consider the $4 million asset described in Tables 10.1 and 10.2. As shown in those tables, the depreciation on this asset can save the owners between $0.7 and $1.1 million, depending upon the method used to depreciate the asset. Now suppose that the asset has actually increased in value and is sold at the end of year 10 for $5 million. The capital gain will be the difference between the sale price and the depreciated basis. If straight-line depreciation was used, then the capital gain will be $4 million with straight-line depreciation, $4.571 million with double declining balance, and $5 million with double declining balance reverting to straight-line depreciation. If the capital gains tax rate were 20%, then the capital gains tax will range from $0.6 million to $1 million. Assuming a 10% discount rate, these taxes will be discounted by a factor [P/F,10%,10] = 0.3855. The NPV of the capital gains tax is therefore

$$\$0.6 \text{ million}(0.3855) = \$0.213 \text{ million}$$

with straight-line depreciation and

$$\$1 \text{ million}(0.3855) = \$0.3855 \text{ million}$$

with double declining balance reverting to straight-line depreciation. The combined benefit of the tax policy is as shown in Table 10.3. Note that without depreciation, the income taxes increase during each year, as shown in Table 10.1; but the capital gain will be only

Table 10.3   Depreciation Benefits and Capital Gains Taxes for an Asset with Initial Book Value of $4 Million, a Life of 10 Years, and a Sale Price of $5 million (in millions)

	Straight Line	Double Declining Balance	Double Declining Balance Reverting to Straight Line
NPV of tax benefits of depreciation (Table 9.1)	$0.737	$1.023	$1.096
NPV of capital gains tax	($0.213)	($0.352)	($0.386)
Net benefit to owner	$0.524	$0.671	$0.710
NPV of capital gains tax if depreciation had not been allowed	($0.077)	($0.077)	($0.077)

$1 million (i.e., the $5 million sale price minus the $4 million initial cost). The tax on the capital gain would be $0.2 million in year 10, with an NPV of $0.2 million (0.3855) = $0.0771. In other words, the ability to depreciate an asset that is actually expected to increase in value provides a valuable benefit to owners and a substantial incentive to invest in such an asset.

This example illustrates how the present worth of the tax savings created by depreciation can be far greater than the tax liabilities associated with the eventual capital gain. From the owner's perspective, the tax treatment of real estate and other assets that are likely to increase in value can be extremely favorable. From the public's perspective, the favorable treatment of real estate may also be quite acceptable, either as a way to reduce the costs of providing rental housing or as a way to help attract investments in retail outlets, office buildings, and other businesses that provide jobs and increase regional income.

10.2.5   Local Taxes

Depreciation, while important, is typically governed by accounting standards and federal tax regulation. Both of these are slow to change, and neither of them is susceptible to manipulation by local governments. Local governments therefore are much more interested in the tax incentives they control—namely, local sales taxes, local income taxes, and real estate taxes. Land use regulation, another aspect of local control, is discussed in the next section.

Local taxes can be used in two ways:

1. Specific taxes can be targeted to specific projects (e.g., sales taxes are sometimes used to finance transit projects).
2. Real estate taxes can be reduced to make certain kinds of development more attractive (e.g., a city might encourage a manufacturer to locate a new plant by offering a reduced real estate tax rate for a period of years).

As was the case with depreciation and capital gains, the key consideration is the effect of changes in tax laws on the cash flows associated with potential projects. If taxes are increased to finance public projects, then the public authorities must satisfy voters that the benefits of the proposed projects will be worth the increase in taxes. Specific proposals often are presented to voters in a referendum, and voters may decide whether to spend extra money to build schools, improve roads, construct bridges, or create a new sewerage system.

Tax deals designed to attract businesses or to encourage specific projects may not receive the same scrutiny, because it is usually easier politically to provide tax breaks that promote a project than it is to raise taxes on voters in order to finance a project. Still, it is well worth considering whether the alleged economic benefits justify the tax concessions that are used to attract development. Large national and international corporations can encourage competition among regions and even countries, and they may eventually locate where local governments are willing to forgo almost all taxation. It is possible—but not desirable—for local governments to forgo tax revenues even though they commit to considerable spending for local infrastructure and services related to the new project (e.g., new roads, increased road maintenance, increased police protection, and general increases in education and other public services related to population growth).

10.3 LAND USE REGULATIONS

If there are no limitations on land use, then developers may attempt to construct whatever provides them the greatest financial benefits. In some locations, this could be high-rise office buildings; in others it might be new factories, a fast-food restaurant, or a recycling facility. Some cities make little or no attempt to limit development; they allow developers and landowners to do whatever they like with their land. Other cities use elaborate plans for guiding development, and they zone different parts of the city for different types of development.

Rationales for zoning include separating incompatible land uses, protecting scenic or environmentally sensitive areas, preserving land for special uses, and limiting the density of development to preserve the aesthetic or social character of the community. Opponents of zoning may have little faith in the ability of planners to direct development, they may believe that landowners should be able to do whatever they want with their land, or they may fear that greed, politics, and corruption will overrule fairness and common sense in planning.

It is beyond the scope of this text to delve into the pros and cons of planning and zoning. However, it is central to our text to understand how zoning and other land use regulations affect the value of land and the potential for projects.

Chapter 9 provided examples of how to evaluate mutually exclusive projects. That chapter espoused the principle of selecting the project with the highest net present value, which would result in the maximum amount of profitable investment in any particular site. If a site is subject to zoning regulations, then the most profitable developments may well be prohibited. If so, then landowners or developers may seek a zoning variance that would allow them to undertake a different kind of project or a larger project. To secure a variance, whoever owns a plot of land may apply to the zoning board or whatever authority controls the land use regulations. If the land is rezoned, then the more profitable project will be feasible, so the value of the land increases—perhaps considerably. If land formerly zoned for single-family houses on 1-acre lots is rezoned to allow apartment buildings, then land values could double or triple. If the land is rezoned to allow construction of a large mall, then a few acres of land could suddenly rise in value from less than $100,000 per acre to $10 million or more! The need for an open, public process is evident since the opportunities for corruption are obvious.

For example, what would it be worth to get zoning board and planning commission approval for a grand development in an area currently zoned for single-family housing or in an area formerly developed for light-density retail or sports fields? Quite a lot, actually, and the extra value can spur some highly dubious activities involving politicians as well as developers (see Box 10.1).

Zoning restrictions may also limit the size or height of buildings or the location of buildings on a site. To preserve the architectural integrity and prominence of their historic structures, Washington, D.C., and Paris are among the cities that prohibit the building of skyscrapers in the center of the city. New York City, after allowing hundreds of skyscrapers to be built in the lower portions of Manhattan Island, found that these immense structures can transform streets into dark canyons. To prevent further crowding of the airspace, the city instituted zoning laws in 1916 that allowed buildings to be constructed right out to the sidewalks, but required **setbacks** as the buildings rose higher:

> On 25 percent of the plot, buildings could rise as high as technology and the will of the developer were willing and able to take them. The law divided the city into zones that allowed buildings to rise a different multiple of the width of the street before requiring a setback. In areas zoned for the most intensive use, . . . buildings could rise straight up from the building line two and a half times the width of the street before setbacks were required. . . . In some specialty zoned retail districts buildings could rise only one and a quarter times the street width, but in most of Manhattan buildings could rise one and a half times the width of the street.[2]

---

[2] Tauranac, *The Empire State Building*, 56–57.

Box 10.1   Planning, Rezoning, and Redevelopment

When public planning commissions must vote on whether to rezone land for more lucrative developments, obvious opportunities arise for political maneuvering, shady financial deals, double-dealing, and worse. Consider the scandal associated with a proposal to create a colossal sports and gambling complex in the site of a defunct racecourse situated within a large public park near Dublin, Ireland. This 300 million-pound proposal, in addition to creating a variety of sporting facilities, sought to create what would have been the largest casino in Europe. Proceeds from gambling would provide the cash to fund the rest of the project.

The gambling scheme failed, despite much political maneuvering, transfers of envelopes of cash, lavish "corporate hospitality" at football matches, and offers of lucrative consulting contracts to political opponents. The scheme failed because of well-organized opposition that gathered 20,000 signatures from local residents against the introduction of gambling into one of the wealthier suburbs of Dublin. The promoters of the scheme had no option but to sell out to developers with plans for housing, public parks, and community facilities that were more acceptable to the public.

Among the promoters of the original gambling scheme was Mr. Bertie Ahern, the taoiseach (prime minister) of Ireland. After much public haggling over his involvement, which apparently included more than $1 million in dubious transfers to and from his bank accounts, Mr. Ahern announced his resignation on April 2, 2008, while still denying that he had received a corrupt payment.

Frank McDonald and Kathy Sheridan, *The Builders* (Dublin: Penguin Press, 2008), 54–58.

Using setbacks enabled the city to benefit from the density of skyscrapers without having huge buildings block all of the light and air from reaching the street levels. Also, by allowing towers of any size on a quarter of the lot, this zoning law allowed for very tall buildings. The setbacks required by this zoning law were reflected in the design of the Empire State Building, the first 100-story building in the world.

Figure 10.4 Typical Sidewalk View in Downtown Vancouver, British Columbia. Despite the prevalence of high-rise buildings, downtown Vancouver enthralls visitors with many tiny parks, sculptures, and waterfalls that make walking to and along the waterfront a delightful experience.

Another approach to zoning is to limit the **floor area ratio** (**FAR**), which is the ratio of the sum of the total floor area of a building to the size of the site. A floor area ratio of 14, for example, allows a 14-story building that covers the entire lot or a 28-story building that covers half of the site. The same FAR allows a mixture of buildings on a site. For example, with the same FAR of 14, it is possible to build a 40-story building on a quarter of the lot and a 6-story building on two-thirds of the site while leaving the remaining portion of the site as open space: $40(1/4) + 6(2/3) = 14$.

In suburban and rural locations, zoning is frequently used to limit the density of development by specifying a **minimum lot size** and limiting development to single-family residences for large portions of the region. The minimum lot size in a suburban location may be a quarter of an acre or less while the minimum lot size in a rural location is sometimes set as high as 5 acres. These restrictions are certainly effective in limiting the total number of housing units, but they may also lead to sprawl (i.e., the conversion of large portions of farmland and open space to residential uses).

A different approach to limiting density is known as **cluster zoning**, in which the density of development is still stated as the number of units per area. However, the housing units (or other structures) can be clustered in a small portion of the site, leaving large portions of the site as open space. Cluster zoning can be effective in minimizing the environmental impact of development because it not only leaves much of the site as open space but also allows utilities and transportation services to be more efficient, since people and houses are closer together.

10.4   BUILDING CODES AND OTHER SAFETY STANDARDS

Governments establish building codes to reduce the risks associated with structural failures, accidents, fires, floods, earthquakes, and other natural or man-made hazards. Minimum standards for construction are needed to protect people from unsafe or unsanitary conditions in their buildings. As building technologies change—and as accidents or natural disasters identify problems—building codes may be updated.

Here are some examples of regulations designed to reduce the risks of natural hazards:

- Design standards and inspection requirements for dams
- Requiring brick firewalls between attached structures in order to limit the possible spread of fire
- Requiring sprinkler systems and fire escapes for schools, apartment buildings, and office buildings
- Higher standards for structures constructed in earthquake zones
- Prohibitions regarding the use of lead pipes, asbestos insulation, and other materials that are known to cause health problems

Here are some regulations designed to improve safety:

- The use of nets and other precautions when constructing bridges and high-rise buildings
- Requiring railings for steps on residential housing
- Requiring numerous safeguards for nuclear power plants

10.5   ENVIRONMENTAL REGULATIONS AND RESTRICTIONS

10.5.1   Regulations

Environmental regulations may affect what, how, where and when something is built. Examples include

- Minimum setbacks from rivers, ponds, lakes, and wetlands
- Restrictions on filling wetlands

Figure 10.5 Landscaping for a Mall in Front Royal, Virginia. Landscaping here features trees and shrubs along the roads and parking lots and preservation of wetlands between the mall and the main access road from Interstate 66. The zoning code provides guidelines for development of "highway corridors" along the major entrances into the town and "strives to ensure that such development is compatible in use, appearance, and functional operations with the Town's economic development policies and action strategies."

- Restrictions on development in floodplains
- Designation of areas as wilderness

Minimum setbacks from water are desirable for several reasons. Maintaining a natural buffer will help limit water pollution resulting from runoff of rainwater and allow a passageway for wildlife (Figure 10.5). In cities, pedestrian access to waterways and bodies of water is an important consideration, and many cities have developed linear parks along rivers, lakes, or the ocean.

Restrictions on filling wetlands recognize the importance of wetlands for many species of animals and plants as well as the ability of wetlands to act as a natural buffer for holding water following heavy rains or snowmelt. When building roads or other transportation routes, it is sometimes necessary to cross wetlands; but routes might be chosen to limit the area of wetlands that would be filled in, and some construction techniques (e.g., use of bridges rather than a causeway) might limit disruption to the wetland. It may also be possible to create new wetlands alongside the highways, so that the net impact of the highway construction can be mitigated in some locations.

Some states have created a system that requires highway construction and other development projects to cover the costs of preserving and protecting wetlands. When new roads are constructed, the state department of transportation is required to (1) limit the destruction of wetlands, if possible; (2) create new wetlands, if feasible; and (3) make a payment into a special fund for any wetlands that are destroyed. A separate agency uses the money from this special fund to acquire and preserve wetlands elsewhere in the state as well as create new wetlands in locations where they are desirable. This process can be structured to preserve much more the of wetlands area than is destroyed.

10.5.2   Easements

Owners may choose to put conservation easements on their property to limit future development. Easements may prohibit any development, or they may allow limited development such as the construction of a single house on a specified portion of the site. They may allow activities such as agriculture, forestry, sports (e.g., hiking or skiing), or they may require the land to remain in or revert to a wild state. An easement is a legal document. When an owner puts a restrictive easement on property, the value of that property will decline. The decline in value can, in some instances, be considered to be a charitable donation because some legislation encourages landowners to provide easements aimed at preserving open space or protecting wildlife. The charitable donation is an amount that can be deducted from income when calculating how much is owed in income tax, so there can be a financial benefit to an individual or a company for placing an easement on a property. The amount of the deduction must be verified by qualified assessors who determine the value of the property with and without the easement, based on recent sales of similar properties.

EXAMPLE 10.4   How the Tax Code Can Promote Preservation of Open Space

The Smiths, a recently retired couple, own an old farmhouse with 200 acres of fields, wetlands, and forests.[3] They purchased the property many years ago for $175,000. They are planning to sell their property in a couple of years so that they can live closer to their children and grandchildren. A developer has offered $1.5 million for their property, which he feels would be an excellent location for a large number of vacation homes and condominiums. However, the couple would prefer to preserve their land as open space for the benefit of future generations. Moreover, they would also like to keep the property as a place to spend family holidays and vacations.

A local conservation group is interested in conserving the site. They have suggested that the couple donate an easement on the property, which they could then use as a charitable donation that would offset their taxes. The value of the charitable donation would be the difference between the current market value of the property and the value of the property if it were restricted to farming and forestry. A real estate expert assessed these values and concluded that the current value was indeed $1.5 million, while the value would be only $0.5 million if an easement prohibited any further development of the land. The difference of $1 million would be the value of the easement.

The conservation group suggested that the couple donate an easement on the land, thereby obtaining charitable deductions and income tax savings for many years. The land would remain open and they, their children, and grandchildren could still use the property as a vacation home. The conservation group illustrated the financial benefits that would be gained by a couple with income of $200,000 per year and a marginal tax rate of 34%. If the charitable donation of the easement were spread out over 20 years, then the annual charitable deduction would be worth $50,000 and the annual tax savings would be $50,000 (34%) = $17,000. The value of the tax savings to the couple would be nearly $200,000:

$$\text{Value of Tax Savings} = \$17,000[P/A,6\%,20] = \$17,000{}^*11.47 = \$195,000$$

The Smiths were impressed, but noted that their income was only half that amount. Still, they estimated that they could save $10,000 per year for 10 years, which would be a little less than they needed to be able to make the move while keeping the old farmhouse as a vacation home.

The conservation group then suggested subdividing the property into three pieces: a 10-acre plot with the old farmhouse, another 10-acre plot that could be sold to raise some immediate cash, and 180 acres of conservation land. The tax benefits would be 5–10% less, but the 10 acres could likely be sold for $75,000–$100,000.

This adjustment to the plan turned out to be satisfactory to the Smiths, so they subdivided the property, donated the easement to the local conservation group, and sold the 10-acre lot.

The chair of the conservation group noted that a similar situation had arisen a few years earlier; in that case the owner of a farm had very little taxable income, so the tax benefits of donating an easement were of no use. Moreover, the owner needed a substantial amount of cash to cover the loss of his farming income when he retired. The owner could have sold the farm to a developer for $1 million, but he hated to see his farm turned into condominiums. In that case, the conservation group was able to raise funds from the local community to purchase an easement for $400,000—much less than the developer was offering for the farm, but enough to help the farmer in his retirement.

10.6   SUMMARY

Depreciation and Taxes

Income tax is based on net income, the difference between income and expense. Income taxes may be levied by local, state, or federal governments. Local and state taxes may be deductible from income when calculating federal income tax. The effective tax rate combines the effects of local,

---

[3] This fictional example is based upon actual recent examples of conservation efforts sponsored by local conservation groups in New Hampshire.

state, and federal taxes. Assuming that local and state taxes are deductible, the effective tax rate equals

$$\text{Effective Income Tax Rate} = SR + LR + FR(1 - SR - LR) \qquad \text{(Eq. 10.10)}$$

where

> SR = state income tax rate
> LR = local income tax rate
> FR = federal income rate

The after-tax MARR equals the pretax MARR multiplied by $(1 - \text{effective tax rate})$. Since companies and investors are concerned with after-tax cash flows, their decisions concerning where to invest, when to invest, and how much to invest may be affected by tax policy. For example, governments may allow tax credits to promote certain types of infrastructure investments.

Since capital investment is not an expense, investment does not affect profit or income tax. However, capital assets decline in value over their lifetime, and the depreciation in the value of capital assets can be treated as an expense. Depreciation is a noncash expense that represents the actual or assumed decline in value of such assets (but remember that land cannot be depreciated).

Depreciation, although it is a noncash expense, is deductible from income and therefore can reduce income tax payments—and affect the perceived value of a project to investors. Tax laws and accounting rules determine the methods that can be used to depreciate an asset. One commonly used method is straight-line depreciation, which assumes that a capital asset declines in value at a uniform rate over its life. Other commonly used methods include double declining balance and asset class depreciation, which allow higher rates of depreciation during the early years of an asset's life. Accelerated depreciation results in lower profits, which may seem undesirable, but it also results in lower taxes and an increase in the net present value of a project. Public policy may therefore allow accelerated depreciation as a means of encouraging certain types of investments.

Land Use Regulations

Zoning limits the type of development that can be undertaken in a city or town. Certain areas may be zoned for residential while others are zoned for commercial or industrial. Zoning may require a minimum lot size, designate how close a building can be to property boundaries, or limit the maximum extent of development—for example, by specifying a maximum floor area ratio (FAR). Since the value of land depends largely upon the potential for development and the character of the surrounding areas, zoning may either enhance or depress land values. For example, prohibiting obnoxious development within a residential neighborhood may increase the value of the homes in that area, but prohibiting more intensive development (e.g., apartment buildings or smaller lot sizes) may depress land values.

Developers often seek zoning variances in order to initiate more intensive uses of the land, or what they believe to be better (and certainly more profitable) uses of the land. Where rezoning would cause a dramatic increase in value, but equally dramatic changes in a neighborhood, there is potential for great local political conflict and controversy.

Building Codes and Other Safety Standards

Building codes define what materials and methods can be used in construction, and they may mandate better construction techniques and materials in areas prone to natural hazards, especially earthquakes or flooding.

Safety standards may be part of building codes. For specialized facilities, such as nuclear power plants, safety standards may require great care (and great expense) in design, construction, and operation. As technology advances, the logic underlying building codes and safety codes may need to be revised to allow cheaper or more effective methods and designs (e.g., plastic pipes for plumbing). As knowledge of environmental impacts and risks increases, building codes may need to be revised to avoid dangers that were previously unknown (e.g., the use of asbestos for insulation is banned because this material is linked to lung cancer).

Environmental Regulations and Restrictions

Environmental regulations may affect what, how, where, and when something is built. Examples include

- Minimum setbacks from rivers, ponds, lakes, and wetlands
- Restrictions on filling wetlands
- Restrictions on development in floodplains
- Designation of areas as wilderness

Legislation or regulation may require infrastructure projects to mitigate their effects on the environment. For example, transportation projects that disrupt wetlands may be required to pay money into a special fund used to protect or create other wetlands.

Landowners may choose to accept restrictions on their own property by creating an easement. For example, an easement could allow a right-of-way (e.g., for power lines or access to other properties), allow public access, or allow or prohibit certain uses of the land. Conservation easements are commonly used to promote preservation of open space. Such easements may prohibit further development or allow only agricultural or forestry activities. If landowners donate an easement on their property to a qualified charitable organization, then they can treat the assessed reduction in the property's value as a tax deduction.

ESSAY AND DISCUSSION QUESTIONS

10.1 Investment tax credits, bonds with tax-free interest, and accelerated depreciation are mechanisms used by governments to promote development that is believed to be in the public interest, such as housing for the elderly, expansion of port facilities, or construction of windmills to generate electricity. Each of these mechanisms is a variation on a single theme: allow developers to reduce their taxes if they devote some of their resources to projects that are in the public interest. Since no tax revenues need to be spent on the projects, these mechanisms are effective ways to promote development that is in the public interest. Do you agree or disagree?

10.2 Accelerated depreciation allows developers to make more money and avoid paying taxes. The rules quickly become so complex that few people can understand them, and it is therefore possible for politicians to give tax breaks to big business (and to their campaign contributors) without incurring any criticism from the public. Depreciation is indeed a legitimate expense, but only if it is based on sound engineering and economic principles. Allowing accelerated depreciations simply siphons money from the public treasury. Do you agree or disagree?

10.3 Compare and contrast cluster zoning and traditional zoning that provides for a minimum lot size. What environmental advantages can be gained by using cluster zoning? What economic or social advantages can be gained by using cluster zoning?

PROBLEMS

10.1 A company builds a new manufacturing plant at a cost of $250 million. The life of the plant is 40 years, and the salvage value is estimated to be $50 million. The plant goes into operation on January 1.

a. What is the straight-line depreciation in year 1?
b. What is the straight-line depreciation in year 10?
c. What is the first-year depreciation assuming double declining balance?
d. What is the depreciation in year 10 assuming double declining balance?

10.2 A developer pays federal income tax at a marginal rate of 27% and state tax of 6%. What is his effective tax rate?

10.3 An investor with a pretax MARR of 15% is in the 35% bracket for federal income tax and also pays incomes 5% for state income tax and 1% for city income tax. What is this investor's after-tax MARR?

10.4 A young couple has received a gift of $5,000, which they would like to use to buy shares in a mutual fund. They are in the 15% bracket for federal income tax, and they pay no state or local income taxes. Should they purchase shares in a tax-free municipal bond fund that pays 4% annual interest or in a high-quality corporate bond fund that pays 6% interest? Both funds are highly recommended, and both have had similar fluctuations in price; the only difference between them is that the interest from the municipal bond fund is tax deductible, whereas the higher interest from the corporate bond fund is not.

10.5 A retired couple has received a gift of $5,000, which they would like to use to buy shares in a mutual fund. They are in the 30% bracket for federal income tax, and they pay 5% state and 1% local income taxes. Should they purchase shares in a tax-free municipal bond fund that pays 4% annual interest or a high-quality corporate bond fund that pays 6% interest? Both funds are highly recommended, and both have had similar fluctuations in price; the only difference between them is that the interest from the municipal bond fund is tax deductible, whereas the higher interest from the corporate bond fund is not.

10.6 You have an opportunity to acquire an office building for $1 million. The annual net revenue from operations is expected to provide an after-tax profit of 9% annually on this million-dollar investment. If you like, you can arrange to finance 80% of the purchase price with an 8% line of credit from your bank. They will let you borrow $800,000 and just pay the interest each year indefinitely. Your effective tax rate is 40%, and there is no inflation.

a. What is your annual before-tax profit if you do not borrow any money?

b. What is your annual before-tax profit if you do borrow the $800,000?

c. What is your annual after-tax profit if you borrow the $800,000?

d. What is the ROI on your equity investment of $200,000 if you borrow the money?

e. Comment (one sentence): What do you gain and what do you lose if you borrow?

10.7 What's the best tax deal? A city wants to build a new bridge to relieve congestion. The mayor wants a private company to build and operate the facility, which is expected to cost $100 million to construct and $1 million per year to operate. If demand projections are correct, about 7 to 8 million vehicles per year will be using the bridge. The average savings will be about 10 minutes for the average user, and the average value of time for users is estimated to be on the order of $10 per hour.

a. You are considering bidding on the project. To bid, you need to state what toll you will charge and agree to pay all investment and operating costs. What toll will you need to charge just to cover your investment costs plus your operating costs, assuming that your cost of capital is 10%? What toll would you have to charge to achieve a pretax ROI of 15% for this $100 million project? For simplicity, assume that the company spends $100 million of its own money to build the bridge.

b. Suppose your company pays federal taxes at a marginal rate of 34% and state taxes at a marginal rate of 10%. What would your after-tax cash flow be, assuming that (1) you paid for the bridge in cash (i.e., you are not yet not considering financing options); (2) you charge a toll of $2; and (3) you can depreciate the bridge over 40 years, assuming a residual value of $20 million. What is your after-tax ROI?

c. Public support could help improve your ROI and your willingness to undertake this project. Would you rather have the mayor urge the city council to agree to (1) cover 5–10% of the construction costs, (2) pay 25–50% of the annual operating costs, or (3) guarantee the loans so that you could lower your cost of capital by 1% or more?

d. The mayor is also willing to allow you to charge whatever toll you like instead of providing any of the concessions discussed in part c. Explain why this is or is not a good option.

10.8 You are considering buying a corner lot near the center of the city. You would like to build a 10-story office tower, but the lot is zoned for residential development only, with a maximum height of 3 stories. The property is assessed at $400,000 based on the current zoning, but the seller wants $2 million. If you can get a zoning variance, you will be able to construct your office tower. Based on similar completed projects, you estimate the construction cost to be $20 million, not including the cost of buying the land. You anticipate that construction can be completed within a year, after which you will begin to receive lease income. Based on your study of market conditions, you anticipate annual net income (i.e., rents minus operating expenses and depreciation) to be $3 to $4 million. Your pretax MARR is 12.5%. If you can get the zoning variance, would you be willing to buy the lot for $2 million? Why or why not? What is the maximum you would be willing to pay? Why?

10.9 Look up the zoning code for your city or town, and identify at least five ways that the code seeks to balance environmental and economic considerations. (You can probably find the zoning code online by entering the phrase "town name & zoning" into a search engine.) Do you think that the zoning code does a good job of protecting the environment?

CASE STUDY: LESS TIME IS MORE MONEY

The "Time is Money" problem in an earlier chapter explored the financing of a skyscraper. That problem did not explicitly consider the effect of depreciation or taxes, because it was primarily concerned with illustrating the basic financial issues, the effect of leverage, and the differences in discount rates used by developers, bankers, and potential purchasers. This problem shows how being able to depreciate a building over a shorter time translates into a higher value for the building, which may be enough to make an unacceptable project worth pursuing.

Consider the effects of depreciation on a similar proposed project:

- The purchase price of the land is $50 million.
- The completed cost of the building is $350 million as of the time that the building is occupied and the owners begin to receive lease payments.
- The annual net revenue from leases (total lease income minus all of the owner's expenses other than depreciation) is $40

million. The annual accounting profit from the building will be $40 million minus depreciation.

- The building is depreciated using straight-line depreciation over a period of 40 years. The salvage value of the structure is $30 million.
- The owner's effective income tax rate is 40%.
- The owner's pretax MARR is 15%.
- If the owner sells the building, there will be a 20% capital gains tax on the difference between the sale price and the depreciated basis for the building.

Questions

a. How important is depreciation to this owner?
   i.   What is the annual accounting profit?
   ii. What is the NPV of the owner's cash flows over the first 20 years of the project? Over the 40-year life of the project? (Do not include the cost of the land or construction.) Is this enough to justify the cost of the building?
   iii. If the owner sells the building for $900 million at the end of 20 years, what is the capital gains tax? What is the NPV of the sale price minus the capital gains tax? How does this affect the overall value of the project?

b. Suppose that a change in the tax code allowed the owner to depreciate the building over 20 years rather than 40 years:
   i.   What is the annual accounting profit?
   ii. What is the NPV of the owner's cash flows over the first 20 years of the project? Over the 40-year life of the project?
   iii. If the owner sells the building for $900 million at the end of 20 years, what is the capital gains tax?
   iv. How much does the change in depreciation increase the NPV of the building and the desirability of the project?

Part **III**

Developing Projects and Programs to Deal with Problems and Opportunities

P art I of the text is an overview of many of the processes needed to assess the performance of infrastructure projects and to evaluate alternatives for improving performance. Inevitably, many aspects of performance will have to be considered, and many possible impacts on society or the environment must be minimized or mitigated. Part II shows how engineering economics provides the methodologies needed to compare projects in terms of financial and economic impacts, including the effects of taxes and depreciation. Part III goes into greater detail on various topics that are relevant to projects and programs aimed at improving infrastructure performance, as suggested by the following figure.

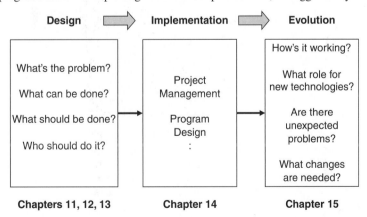

Structure of Part III

Chapters 11, 12, and 13 relate to the preliminary design of infrastructure projects, which includes identifying needs, developing alternatives, deciding who should be involved in the project, and reducing the risks associated with the project. Most of the material in these three chapters builds upon concepts that were introduced in Part I, which provides a broad framework for considering ways to improve infrastructure-based systems. Given this framework from Part I, as well as the engineering economics methods covered in Part II, we can now consider some of the more detailed aspects of project evaluation.

Eventually, the decision is made to proceed with a project. The emphasis then shifts from design to implementation, and an entirely different set of concepts and methodologies are needed. Chapter 14 introduces the basic terminology and concepts of project management and illustrates some ways to manage projects effectively, including a case study of Boston's "Big Dig." That chapter also discusses how programs might be established to provide guidelines for undertaking a series of related projects, such as the Interstate Highway Program.

Infrastructure projects, even extremely large projects, are typically just a minor element in the evolution of a broader, infrastructure-based system. Over time, needs and technologies change—as does economic geography. Systems need to be maintained, expanded to keep up with growth, and modernized as technological capabilities improve. Eventually, entirely new systems may emerge to compete with and perhaps eventually replace existing systems. Chapter 15 describes the long-term movement toward more sustainable infrastructure. It emphasizes the importance of recognizing the history of systems, the current context of infrastructure systems, and trends that may, over several decades, dramatically affect people's views on how best to meet society's needs for transportation, water resources, electrical power, clean air and water, housing, and other services provided by infrastructure-based systems.

Chapter 11

Developing a Strategy to Deal with a Problem

I hope that ultimately it will prove feasible to build a sea-level canal. Such a canal would undoubtedly be best in the end, if feasible, and I feel that one of the chief advantages of the Panama Route is that ultimately a sea-level canal will be a possibility. But, while paying heed to the ideal perfectibility of the scheme from an engineer's standpoint, remember the need of having a plan which shall provide for the immediate building of the canal on the safest terms and in the shortest possible time.[1]

CHAPTER CONCEPTS

---

[1] President Theodore Roosevelt's instructions to the International Board of Consulting Engineers that was assembled to consider the principal problems in constructing the Panama Canal, September 1905.

11.1 INTRODUCTION

As stated in Chapter 1, project evaluation is an art, not a science. While Chapters 7 and 9 have introduced quantitative methods for discounting cash flows and ranking alternatives in terms of financial matters, Chapter 8 emphasizes that the choice of a discount rate requires considerable judgment by any of the parties that may be evaluating a project. Moreover, predictions of financial costs and benefits are seldom precise, and predictions of social and environmental impacts are even more difficult to quantify. Assumptions can and must be made to allow comparisons of alternatives, but these assumptions are always prone to error—sometimes egregious error. Many real estate projects that were initiated in 2006 and 2007 anticipated continuing increases in demand and ever-rising prices; hopes for these projects were dashed when the real estate market collapsed soon afterward.

Figure 11.1 summarizes the major steps in project evaluation that were identified in Chapter 1. This chapter presents various methods that can be used at various stages in this process.

The first step, defining the problem, may be the most critical, because the greatest uncertainties and the greatest opportunities are encountered when projects are first considered. It is therefore vital to encourage broad conceptual thinking at the very beginning of a project. What exactly are the needs to be addressed, what investment options are available, and what other kinds of approaches can be used to address these needs? What economic, technological, or social changes might take place over the lifetime of a project? Section 11.2 presents strategies for developing a good **statement of needs**, which can then be used to identify **project objectives**. Given a clear statement of needs and objectives, it is possible to consider how best to meet those objectives. Section 11.3 presents **brainstorming** as a means of generating a wide range of alternatives. Two examples illustrate how a poor statement of needs can lead to incorrect solutions. Defining highway congestion as a lack of highway capacity will restrict consideration of transit alternatives as well as strategies for reducing highway demand. Consultants in Bolivia, in evaluating an expensive rail project, failed to recognize that inland waterways are much cheaper than rail in moving soybeans to ocean ports.

Section 11.4 provides methodologies for ensuring that the evaluation process is robust, which means that the selection of alternatives and final design decisions are not narrowly based on fixed assumptions. **Sensitivity analysis** (Section 11.4.1) involves a systematic examination of the impact of changes in any and all of the major assumptions underlying a study. This is an analytical technique that will indicate which factors are most important to the success or failure of a project. **Scenario analysis** (Section 11.4.2) takes a different approach to dealing with uncertainty by looking at possible ways in

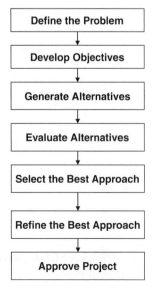

Figure 11.1 Steps in Project Evaluation.

which future conditions might evolve. A **scenario** is a set of internally consistent assumptions that together provide a vision of a possible future within which a project will be implemented. Each scenario includes a consistent set of assumptions about the factors that might affect a project's success, such as the rate of economic growth, the price of energy, or the worldwide response to global warming. Sensitivity analysis produces precise mathematical results; scenario analysis requires consideration of what is possible and how best to prepare for **alternative futures**, as illustrated by a case study of **Southern California Edison** (Section 11.4.3). **Probabilistic analysis** (Section 11.4.4) may be useful or necessary in making initial estimates of project costs or benefits. Experience may provide the basis for estimating the likely range of costs or benefits and developing strategies for bidding on a project.

While project evaluation as structured in Figure 11.1 may appear to be a linear process, it is actually an iterative process (Section 11.5). At the beginning of an evaluation, the objectives may be poorly stated, additional ideas may emerge from preliminary analysis, and review of preliminary results may indicate where further analysis or a broader perspective would be useful.

The chapter closes with the **Pearl River Delta case study** (Section 11.6). This case study illustrates how an experienced interdisciplinary team used an iterative process to define the needs and objectives of a major infrastructure project. What seemed to be a question of how to pay for a bridge turned out to be a question of how the construction of one or more bridges could influence traffic management, economic development, open space, water quality, and other factors within the populous Pearl River Delta region of China. The context in which such a project is initially viewed may dramatically limit or expand the long-term costs and benefits to an entire region of a country.

Figure 11.2 shows how the material in this chapter fits within the overall framework of project evaluation. The end result is the selection and approval of a particular project or set of projects that are then implemented. At that point, the emphasis shifts from project evaluation to project management, which is introduced in Chapter 14.

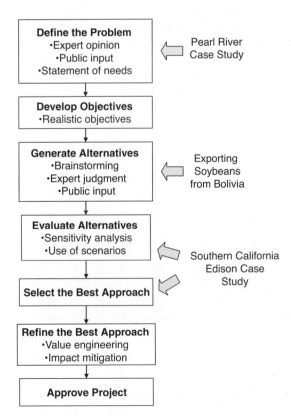

Figure 11.2 Aspects of Project Evaluation Addressed in This Chapter.

Notice the similarities between project design and project evaluation. Neither of these processes can be readily quantified or reduced to a simple set of rules or guidelines. Both of these processes benefit from broad thinking about what the problems and opportunities really are, how they can best be addressed, and what alternatives might be suggested. Project design, at a basic level, concerns engineering and architectural matters—what is to be built, where it is to be built, and how it will be built. Project evaluation, at a fundamental level, concerns the feasibility of the project, encompassing the financial, economic, social, and environmental factors that will ultimately determine whether the project should be undertaken. Both design and evaluation rely upon the intelligence, insight, creativity, and common sense of the people who undertake the process.

11.2   IDENTIFYING NEEDS AND OBJECTIVES

11.2.1   Needs

Infrastructure-based systems address fundamental needs of society, such as mobility, availability of safe drinking water, safe and secure housing, communication, or public health. These are all general concepts, and improvements in any of these areas would be welcomed by almost everyone in any society at any stage of historical or technological development. The types of improvement that are possible and the degree to which these needs are felt to be satisfied, however, has certainly changed over the course of history and certainly differs today for different groups of people and different regions of the world. For example, it is no longer acceptable to dump raw sewerage into lakes, rivers, or the ocean, nor will society acquiesce to the loss of the recreational uses of its waterfront because of unhealthy water quality. In Boston and elsewhere, major projects have updated sewerage treatment facilities, prevented storm runoff from overflowing the sewer systems, and made formerly polluted waters fit for recreational use (Figure 11.3).

The motivation for major infrastructure projects is generally based on a perception that needs are not being met as well as they could or should be. Impetus for change could come from four directions:

1. Dissatisfaction with the existing system
2. Ideas for improving the operation and management of the existing system, including techno-logical innovations aimed at improving system performance
3. Invention of technologies or access to previously unavailable resources that allow entirely new approaches to satisfying the same needs.

Figure 11.3 Deer Island Sewerage Treatment Plant. This facility, which is situated on a peninsula extending into Boston Harbor, was part of a multibillion-dollar effort to clean up the harbor to protect aquatic life and preserve opportunities for sailing, swimming, and other water sports. The effort was mandated via enforcement of federal water quality regulations, and it was funded in part by increases in the water and sewerage rates paid by residents and businesses.

4. Court orders that require a company or a public agency to provide better service or to reduce environmental impacts of their activities

The existing system may be perceived as being too costly, ineffective, too small, too slow, or too decrepit. The perception of problems leads people—users, owners, employees, managers, regulators, or innovators—to identify a *need for system improvement*.

Even if the system is operating satisfactorily, its managers, engineers, and consultants will continually be looking for ways to improve it, and inventors will be looking for new technologies that they could develop for the system. Thus, there is always the **possibility for system improvement**.

There will always be alternative ways to satisfy societal needs, and no infrastructure-based system is immune from competition. New technologies, potential economies of scale, new managerial capabilities, access to a more highly educated labor force, or greater societal concern with social and environmental impacts may lead to the introduction of a different kind of system. At some point, the **opportunity for introducing a new system** may be perceived as a **need for the new system**.

A well-conceived statement of needs leads to consideration of many different kinds of options for satisfying these needs. A poorly conceived statement of needs unnecessarily restricts the evaluation process and can ultimately lead to ineffective or even damaging projects. In general, statements that emphasize performance are superior to statements that refer to specific technologies.

11.2.2 Needs versus Desires

The statement of need should recognize that the level of "need" depends in part upon cost and capabilities. We may want to have much more than we have; whether we actually need any more depends in part upon whether we can afford it. There is also a hierarchy of infrastructure needs. Making clean water available for household consumption is a basic public health need. Making sure that buildings and bridges will not collapse is a basic need related to public safety. Relieving congestion to reduce commuting time would be nice, but this need is not on a par with providing clean drinking water or safe buildings. If the costs of congestion can be quantified, and if cost-effective approaches to reducing congestion can be used, then it may make sense to proceed with one or more of them.

11.2.3 Objectives

Objectives can be based on needs, but the linkage between needs and objectives may be well defined. Objectives are more specific statements of what is to be achieved, and objectives can be formulated even when there are conflicting views as to what is needed. An objective to build roads can be applauded by commuters who believe they need roads and by contractors who believe they need the work. Debate over whether there is a need for something can be endless; agreeing upon an objective is perhaps an easier task. Once it is established by a public agency or a private company, an objective is definite. We might not agree with a stated objective, but we cannot dispute that it has in fact been stated!

Objectives can be stated qualitatively, but they may not do much more than indicate the desired direction of change. Quantitative objectives are more difficult to formulate, but they provide better targets for managing system performance and measuring the success or failure of projects. The difference is evident by comparing the objectives as stated in the two columns of Table 11.1. Note that the quantitative statements may suggest different strategies for achieving the general objectives and that they may considerably influence who benefits from projects that address the objectives. Qualitative objectives may result when there is disagreement about the needs or about how to interpret the objective; parties may agree to disagree about the specifics, yet accept the qualitative statement.

Table 11.1   Qualitative versus Quantitative Objectives

Qualitative	Quantitative
Reduce congestion.	• Reduce congestion by 10% within 3 years, as measured by average commuting time for the metropolitan region. • Reduce congestion on the urban interstates by 5% and reduce congestion on the major arterials by 20%, as measured by monitoring average travel speeds on the region's major commuting routes.
Increase the use of renewable energy resources.	• Increase the use of renewable energy resources by 5% per year over the next 10 years. • Generate at least 20% of the region's electricity from renewable energy resource within the next 10 years.
Improve access to clean drinking water in developing countries.	• Ensure that clean drinking water is available within every village in rural regions of a particular country within the next 5 years. • Increase the percentage of the population with access to clean water within 50 yards of their homes from 35% to 75% within 10 years.
Reduce homelessness.	• Provide sufficient shelter in the city for up to 5000 homeless people during the upcoming winter. • Upgrade mental health plans to ensure that patients with designated mental problems are no longer released prematurely from state hospitals unless they have a place to live.

EXAMPLE 11.1   Water Supply Needs and Objectives

As population grows and as demand for water increases, existing supplies of water may appear to be inadequate for a city. If something isn't done, then forecasts may indicate that demand will outstrip supply in the not-too-distant future. The problem could be discussed by local governments, by administrators and engineers at the city's water department, and by columnists in local newspapers. The framing of the problem is likely to determine the city's response to the situation. Consider these ways of presenting the problem:

- The city needs more water.
- The city needs to restrict the use of water to avoid future shortfalls.
- The city needs to ensure that sufficient water of appropriate quality is available for personal, industrial, recreational, agricultural, and other uses.
- The city, its residents, and local companies must work together to ensure that water is not wasted and that long-term, cost-effective strategies are undertaken to manage water supply and demand.

The statement that "the city needs more water" will lead to an examination of ways to get more water, which could involve expanding the reservoir, constructing pipelines to get water from another watershed, or building a desalinization plant to obtain fresh water from the sea. An objective related to this perceived need might be to "increase the water supply by 20% within 5 years." Such an objective would lead directly to consideration of major water resource projects.

The statement that "the city needs to restrict the use of water" could lead to a pair of objectives: "reduce consumption of water by at least 5%" and "reduce consumption by 20% whenever the governor declares that we are in a drought." These objectives could be achieved by regulations such as limits on watering lawns and washing cars as well as restrictions on water use by the parks department or other city agencies. Efforts could be initiated to find out whether industries are wasting water and whether certain types of activities should be banned. This statement might also suggest a pricing option: raise the price of water so that individuals and businesses would have an incentive to conserve water.

The statement that "the city needs to ensure that sufficient water is available" for multiple uses suggests that different strategies might exist for different types of uses. Farmers who use water for irrigation do not need to use potable water for that purpose, nor does the parks department need to use potable water to keep the grass green in public open spaces. This statement of needs might encourage reuse as well as conservation of water.

The final statement emphasizes the importance of conservation and cost effectiveness. This statement does not put the burden solely on the government, and it indicates that unlimited access to water should not be taken for granted.

EXAMPLE 11.2   Rush-Hour Congestion Needs and Objectives

The ability for residents to commute to and from their jobs is a basic transportation need for a region. Rush-hour congestion clearly frustrates commuters, and they might believe that "we need more highway capacity." If this is how the problem is perceived by the public, the mayor, and government agencies, then projects that increase highway capacity are what will be considered. A broader statement of the problem would be "we need to reduce rush-hour congestion." Adding highway capacity remains an option, but there are many others. Using tolls to manage traffic flows, limiting downtown parking spaces, constructing express lanes for buses and high-occupancy vehicles, encouraging companies to stagger their work hours, and improving traffic management are all alternatives that might help reduce rush-hour congestion. Over the long run, strategies that promote dense development along transit corridors may encourage the use of transit and reduce the rate of growth in demand for highways.

As shown above in Table 11.1, the need to reduce congestion can be translated into specific objectives for reducing travel times on specific roads or reducing average commuting times. These two measures are subtly different. Measuring travel times on roads will show how delays increase or decrease in specific locations; these measures are logically consistent with the commonsense view of congestion. Measuring commuting times might give a similar result, except that people may be moving to the suburbs from the city; in that case, their commute will be longer even if average speed does not change on any roads. Also, some commuters might shift from an unreliable 20- to 40-minute auto commute to a reliable, less stressful 40- to 45-minute commute using commuter rail; their average commuting time would increase, but they are happier with their commute. Longer commuting times do not necessarily mean more congestion.

11.3   IDENTIFYING ALTERNATIVES

Given a clear statement of needs or of a problem, it is possible to consider the alternatives for dealing with the problem. The number of alternatives that can be investigated in detail is limited, so it is important to take care in deciding what to investigate. If only a few options can be considered, then it is important to consider distinctly different options in order to get the most information possible out of the analysis. It may be possible to identify the major competing alternatives or to select alternatives that clearly cover the major design options. The options may consider a preferred alternative, although it might be prudent to reserve judgment about what is preferred until some analysis has been done. The options should also consider a "do nothing" or "business as usual" option that depicts what would happen if no major project is undertaken.

Here are some pitfalls to avoid, if possible:

- *The Good, the Bad, and the Ugly:* There may be pressure to analyze only the option that has the most powerful supporters. To make a pretense of objectivity, a few other options will be thrown in—with care taken to ensure they are totally unrealistic. This analysis clearly identifies which option is good, but it provides essentially no insight into what realistic alternatives might have been.
- *Tweedle Dum and Tweedle Dee:* If options are too close to each other, then the analysis will not find much difference between them, and the assessment and evaluation efforts will be largely irrelevant. It will be better begin a study of river crossings by considering a bridge versus a tunnel versus a bridge/causeway than to start by looking at a cable-stayed bridge versus a suspension bridge. Don't let the difference between options be so small that there really is none. Sensitivity analysis, discussed in the next section, provides an opportunity to look at small changes in design.
- *Overkill:* Sometimes new technology opens up an opportunity to make substantial improvements in performance, and it is possible to show great benefits even if a high-priced option is introduced. However, nearly all of the benefits may be achievable with a much less expensive version of the new technology. The options should therefore include a couple of technological options, not just the most sophisticated option.

The remaining portions of this section present four useful methods for generating alternatives:

1. Brainstorming
2. Systematic identification of options
3. Public input
4. Best practices and expert opinion

Whatever method is used, common sense is useful in structuring a set of meaningful options and avoiding pitfalls such as those listed above. Since evaluation is an iterative process, it may be useful to structure the alternatives to maximize the knowledge gained from the analysis. The alternatives can be revised in subsequent analyses, and the alternative that is finally adopted is likely to be a composite of two or more of the alternatives that were initially investigated.

EXAMPLE 11.3   Alternatives for Exporting Soybeans from Bolivia

The eastern region of Bolivia, like the adjacent portions of Brazil, includes some of the most productive agricultural lands in the world. Soybeans are one of the most profitable crops in this region because of worldwide demand for soy products and because the South American harvest is 6 months offset from the more abundant harvest in the Northern Hemisphere. Due to a 1000-mile gap in the Bolivian rail system, soybeans have to be exported via Brazil. If this gap were filled by a new railroad, then soybeans could be hauled over the Andes to a Pacific port for export to Asia. This strategy would save thousands of miles for the ocean trip, not to mention avoiding the delay and cost of going through the Panama Canal. The idea of constructing a rail line to unite the agricultural east with the central and western portions of the country has been discussed for more than 75 years. Known as the "Interconnection," this rail line has been the dream of many an engineer and many a railroad president.

A transportation consulting firm conducted a preliminary study comparing the cost of moving soybeans by rail from eastern Bolivia to Brazilian ports on the Atlantic to the cost of moving them via Bolivia to Chilean ports on the Pacific. The study showed that the Bolivian route would be competitive, so the government continued to explore the issue. Another consulting team was asked to visit exporters and transportation officials in both the Santa Cruz region of Bolivia and in the neighboring states of Brazil. The team updated the analysis of the prior study and confirmed that the Bolivian rail route was indeed competitive with the rail route to the Brazilian ports on the Atlantic. However, Brazilian rail officials described their plans for building a new rail line that would connect to a port on the Amazon, which would shorten the rail trip by more than a thousand miles with little change in the ocean shipping cost; they expected this route to be used for exporting soybeans from much of the region. With this new route, it was no longer as clear that the Bolivian Interconnection would be able to attract a substantial amount of soybean traffic.

Interviews with exporters were even more discouraging, for they pointed out that most of the soybeans were moved south on the Paraná River to Argentina, where they were loaded onto ships for export around the world. The barge movements along the river were much cheaper than the rail movements, and the main costs of ocean shipping related to loading and unloading. The savings from the shorter ocean trip were nowhere near enough to cover the increases in cost that would result from using rail to reach a Pacific port.

Failing to include barge transportation was a major defect in the original study. Had government officials not been so concerned with competing with Brazil's rail system, they might have recognized the need to investigate the inland waterway option. Had the consultants viewed their task as studying choices faced by exporters rather than as studying relative costs of using different rail routes, they would quickly have realized that the barge service was superior (i.e., much cheaper) than any of the existing rail options.

In this case, the do-nothing option prevailed, and Bolivia did not attempt to build a railroad up the eastern slope of the Andes.

11.3.1 Brainstorming

Brainstorming can be an effective way for a group of people to come up with a great many ideas in a short time. The process involves posing a question to a group of people of varied backgrounds who will then respond quickly with whatever ideas they think are relevant. In a brainstorming session, the objective is to generate ideas, not debate or evaluate them. It doesn't matter (at this stage) whether the ideas are good or bad. By merely compiling the ideas, the group can quickly gather a lot of input; and the rapid identification of ideas or options may focus people's thoughts and stimulate even more ideas. When people know that their ideas will not be challenged or dismissed, they are more likely to speak up.

Table 11.2 Two Possible Ways to Structure
Discussion Groups

Skills and Interests	Role
Design	Developer
Location	Owner/Investor
Operation	Abutter
Construction process	Government official
Safety	Environmental group
Economic impact	Chamber of Commerce

Here are some keys to effective brainstorming:

- Have a clear statement of a need or a problem.
- Assemble a group of knowledgeable people with an interest in the problem.
- Provide sufficient background information so that everyone in the group has at least an overview of the problem.
- Allow a short time for people to gather their own thoughts about the problem.
- Have one or two facilitators who will ask for ideas and write down whatever ideas are suggested.
- Allow enough time for everyone to present all of their ideas.
- Rearrange, combine, condense, and refine all of the ideas that have been generated.
- Allow enough time for everyone to review the list of ideas and make sure that the final list of ideas reflects what was actually presented.

Various techniques can be used to provide more structure to the process or to allow brainstorming to work effectively with a large group of people. Brainstorming works well in small groups, where everyone feels comfortable in speaking. Limiting the size of the group to 10 to 20 people is generally a good idea. If many people are interested in an issue, then they can be divided into smaller groups to stimulate participation. The ideas generated by each group can then be consolidated by coordinators.

People can be invited to participate in a brainstorming session based on the nature of their skills and interests or of their role with respect to a possible project. Table 11.2 shows an example of how this might be done for discussion of an infrastructure project. Groups could be structured so that people with the same skills or the same role could meet to discuss a project. After an hour or two of discussion, each group could summarize what they believe to be the key issues. The groups could then be restructured to have either a mix of skills and interests or a mix of roles.

11.3.2 Systematic Identification of Options

In many cases, the options will be well defined, and it will be possible to define the options that are available. The following types of options will be available for improving the performance of infrastructure:

- Adjust prices to manage demand.
- Improve productivity of the existing system through better management.
  - Obtain better coordination among different parts of the system.
  - Adjust priorities given to different aspects of performance.
- Repair or rehabilitate elements of the system.
- Make investments that allow more productive use of the existing system or network.
- Make investments that expand an existing system of network.
- Make investments that create new types of infrastructure.

The timing and location for any of these options can be varied systematically. Alternatives can also be structured to explore opportunities for implementing a project in stages.

EXAMPLE 11.4 Alternatives for Meeting Future Demands for Electricity

What options might be pursued to ensure an adequate supply of electricity to meet future demand? The framework just presented can be used to identify a wide range of options:

- *Adjust prices to manage demand:* Raise prices to promote conservation; establish differential prices by time of day to reduce peak demands on the system.
- *Improve productivity of the existing system through better management:*
  - ○ *Obtain better coordination among different parts of the system:* Encourage customers to use energy-efficient appliances; offer customers opportunities to purchase energy-efficient light bulbs.
  - ○ *Adjust priorities given to different aspects of performance:* Pay more attention to meeting future energy needs and less attention to minimizing current costs.
- *Repair or rehabilitate elements of the system:* Upgrade the least efficient power plants.
- *Investments that allow more productive use of the existing system or network:* Improve coordination among different power suppliers (i.e., improve the capacity or flexibility of the electrical grid).
- *Investments that expand an existing system of network:* Add power plants and extend transmission lines.
- *Investments that create new types of infrastructure:* Develop renewable sources of energy; develop nuclear power plants.

11.3.3 Public Input

Public hearings provide an opportunity for developers and government officials to present proposals to the public or receive input from the public concerning needs for projects. Public input is useful in identifying potential social and environmental problems associated with a project. Involving the public early on helps raise controversial issues while there is still time to deal effectively with them. Failure to involve the public may lead to bitter confrontations at a stage when it is difficult and expensive to modify a proposed project to mitigate social or environmental impacts.

Another way to obtain public input is to work closely with existing organizations that have an interest in the proposed project and its potential impacts. This approach may involve conservation groups or public interest legal organizations that are interested in ensuring the project is structured to minimize impacts on the environment or on the neighborhoods most affected by the project. Chapter 5 includes two case studies in which public input helped to clarify objectives and alternatives for efforts aimed at promoting sustainable development.

11.3.4 Best Practices and Expert Opinion

For well-established infrastructure-based systems, there will likely be identifiable examples of best practices for improving performance. Experts, who are familiar with many such examples, can likely identify the options available for improving infrastructure performance in their area of expertise.

11.4 DEALING WITH UNCERTAINTY WHEN ASSESSING ALTERNATIVES

After identifying a set of alternatives, the next step is to assess the costs and benefits associated with each option. The assessment should consider the financial, economic, social, and environmental impacts, using the methods described earlier in this text along with whatever engineering and institutional analyses are appropriate for the project. This stage of the analysis produces a great deal of detail for each alternative, and all of it must be summarized and presented to decision makers for their consideration.

Analysis, evaluation, and selection are different activities that may well be carried out by different people. The analysis is—or should be—a rational, objective determination of the costs and benefits associated with each alternative. If the results are challenged, then further analysis may be needed to obtain more credible results. However, the analysis always involves some uncertainty, and it always requires assumptions—many of them trivial or uncontroversial, but some of them likely to be important and subject to debate.

Sensitivity analysis, scenario analysis, and probabilistic analysis are techniques that can be helpful in dealing with risks, uncertainty, and shaky assumptions. These techniques are addressed in the next three sections.

11.4.1   Sensitivity Analysis

Sensitivity analysis is a means of determining the importance of assumptions used in the analysis. In concept, sensitivity analysis is quite simple:

1. Prepare a base case using the best estimates for all variables needed to estimate impacts.
2. For the most important or least understood variables, repeat the analysis for a reasonable range of possible values.
3. Create tables or charts that illustrate the sensitivity of the results to the variables that prove to be the most important.

For example, in considering whether to construct an office building, it would be worth doing some sensitivity analysis on the following factors:

- Size of the building (number of stories or amount of rentable space)
- Construction cost
- Construction time
- Average rental payments
- Occupancy rate
- Interest rate on the construction loan
- Interest rate on the permanent mortgage

In practice, sensitivity analysis is an art that requires good judgment and the ability to interpret and display results. Judgment is necessary in selecting variables, in determining the range of values to consider, and in deciding how to present the results. Judgment is necessary primarily because of the great many factors that might be examined and the possibility of considering changes in multiple factors—there are too many possible combinations to consider. Even if the analysis is feasible, a mindless analysis of all combinations of all the factors would create a stack of results too thick to contemplate. For example, for just three possible values for each of the seven factors listed above, there are $3^7 = 2187$ cases to consider. This rapid combinatorial increase in options that perhaps should be considered is sometimes referred to as the **curse of dimensionality**. It is not feasible to analyze all possible combinations; so an alternative approach, one based on logic and common sense, is necessary to structure an effective sensitivity analysis.

To begin a sensitivity analysis, consider one variable at a time. For each variable, choose a small number (from 3 to 10 is usually reasonable) of values that cover the likely range of values that might be encountered. Example 11.5 shows how a sensitivity analysis might be structured for a proposed office building.

EXAMPLE 11.5   Sensitivity Analysis for a Proposed Office Building

For the variables related to an office building, appropriate values might be as shown in Table 11.3. The base case corresponds to the preliminary plan for the building, which calls for a building of about 50 stories that would be constructed over a period of 2.5 years at a cost of about $2 billion. The financial plan for the building assumes 80% occupancy and an average rental rate of $50/square foot per year. Note that the values selected for the sensitivity analysis are based on some thought concerning the variable as well as the nature of the variable:

Table 11.3 Structuring a Sensitivity Analysis

Variable	Base Case	Values for Sensitivity Analysis
Size of the building	50 stories	40, 45, 50, 55, 60
Construction cost	$2 billion	$1.9, $2, $2.1, $2.2 billion
Construction time	30 months	27, 30, 33, 36
Average annual rent	$50/square foot	$24, $30, $35, $40, $45, $50, $55, $60
Occupancy rate	80%	60%, 70%, 80%, 90%, 95%
Interest rate on the construction loan	10%	9%, 9.5%, 10%, 10.5%, 11%, 11.5%, 12%
Interest rate on the permanent mortgage	8%	7.5%, 8%, 8.5%, 9%, 9.5%, 10%, 11%, 12%

- *Size of the building:* This value is varied between 40 and 60 in increments of five stories on the assumption that the project under consideration is "about 50 stories" and that two values on either side of 50 stories will be sufficient to illustrate the importance of size.
- *Construction cost:* The rough budget is $400 million. Costs are more likely to be over rather than under that figure, but we are pretty sure the cost will be no more than $440 million.
- *Construction time:* This value is expected to be 30 months. It could be a bit shorter if all goes well, or somewhat longer if problems arise with weather, materials, contractors, or delays.
- *Rental payments:* Current annual rents for similar buildings are about $50/square foot, so it seems reasonable to try a couple of values either side of that. However, current rental rates are very high compared to only a few years ago, so it will be prudent to consider some sharply lower values as well.
- *Interest rates:* Interest rates could be higher unless some tenants sign long-term leases very early, so we need to consider that rates might go quite a bit higher than we are planning on.

As illustrated in this example, sensitivity analysis most decidedly is not a mindless calculation of "base case plus or minus 5, 10, 15, and 20%."

Example has 11.5 illustrated why it is important to think carefully about how best to structure a sensitivity analysis. It is better to consider a reasonable range of variation in each key variable, rather than assuming the same changes of 5, 10, 15, or 20% for all of these variables. Example 11.6 illustrates the problems that may arise from using this kind of systematic but rather mindless approach. First, the specified percentage changes will not be of equal likelihood or of equal importance for each variable. Second, it often is more useful to consider results for reasonable increments in each variable (e.g., increments of 1 or 2 months in construction time rather than increments of 5 or 10%. By showing the percentage change, all of the results can be displayed on a single graph—but the value of that graph is undermined by the qualitative differences in interpreting the results for each variable.

EXAMPLE 11.6 Sensitivity Analysis for a Proposed Office Building (continued)

Figures 11.4 and 11.5 present the results of sensitivity analyses that relate to the same office building considered in the previous example. Figure 11.4 charts the results of a "mindless" sensitivity analysis in which three variables—rents, construction cost, and the interest rate on the permanent mortgage—are varied by plus or minus 10, 20, or 30%. The three lines plot net present value (NPV) of the project for each of the values considered. The NPV for the base case is $800 million, and the NPV of the construction cost is $2 billion. In Figure 11.4, it appears that the most important factor is the construction cost; the least important factor is the interest rate on the loan. However, this chart shows a variation in construction cost of plus or minus 30%, whereas in the previous example it was assumed that costs would be no more than 5% lower or 10% higher than budget. The range of uncertainly related to construction cost is therefore not $1.4 to 2.6 billion as shown in Figure 11.4, but $1.9 to $2.2 billion as shown in Table 11.3. For rents, we have the opposite problem: the chart shows results for only a 30% decline, but the table suggested that a 50% decline is possible. If the trend indicated in the plot continues, then the NPV will

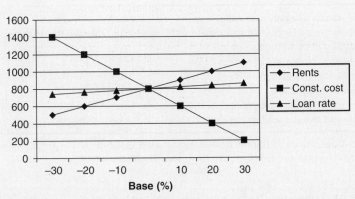

Figure 11.4 Showing Sensitivity Results for Multiple Variables on a Single Chart (sometimes called a spider plot).

drop to about $250 million if rents fall to $25/square foot. Thus, if we believe the logic underlying Table 11.3, a decline in rents is actually more of a potential problem than an increase in construction costs. Figure 11.4 therefore is misleading with respect to the relative importance of uncertainty in rents and construction costs.

Interpreting the sensitivity to interest rates is complicated in Figure 11.4 for a different reason: it is confusing to talk about a 10 or 20% decline in interest rates. The **spider plot** seems to show that the results are not that sensitive to interest rates, but the more carefully designed sensitivity analysis in Figure 11.5 shows a different result. Here the interest rate is shown on the x-axis, and it is apparent that the NPV of the project drops quickly if interest rates go above 9%.

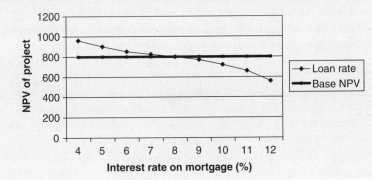

Figure 11.5 A More Effective Way to Illustrate the Sensitivity to Interest Rates.

The need for judgment and common sense in selecting variables for and in displaying results of a sensitivity analysis should be evident from Examples 11.5 and 11.6. A spider plot such as shown in Figure 11.4 may be useful, but only if the percentage variations make sense for each of the variables included in the plot. Care must also be taken in displaying results when multiple variables are changed simultaneously, and it is important to explain which variables were changed, by how much, and why this particular set of changes was felt to be important. When considering multiple variables, it is best to develop a coherent story that can be readily illustrated with a few tables or graphs rather than a mammoth printout showing all possible combinations of multiple variables. If you start explaining how and why you are changing multiple variables, you are getting very close to scenario analysis, which is the subject of the next section.

In summary, sensitivity analysis involves systematically varying key parameters to determine their effects on performance. Sensitivity analysis is an art, not a science. Judgment is required in deciding what variables to consider, what to vary, how much to vary them, and how to display and interpret the results.

11.4.2 Use of Scenarios

When evaluating a project, it is essential to remember that the future is uncertain, that events may take decidedly different paths, and that whatever project is implemented, it will have to live with whatever

Table 11.4 Steps for Developing Scenarios

Identify the focal issue or decision.
Identify key local forces that will influence the success or failure of the decision.
Identify key external forces.
Rank by importance and uncertainty.
Select scenarios defined by drivers.
Flesh out the scenarios.
Determine the implications of each scenario.
Identify and monitor leading indicators.

Source: Peter Schwartz, "Steps to Developing Scenarios," appendix to *The Art of the Long View: Paths to Strategic Insight for Yourself and Your Company* (New York: Doubleday, 1996).

really does take place in the future. It is therefore useful to consider scenarios that encompass different—but coherent—visions of the major factors likely to affect a project.

Peter Schwartz has identified steps that can be taken for developing scenarios (Table 11.4).[2] He recommends building scenarios around the factors most likely to affect the decision at hand. In other words, begin by thinking about the proposed project or the need you are trying to address, and then think about the factors and forces that will affect that need. Schwartz distinguishes between local forces, which directly affect a particular project, and the macro forces that affect the overall environment of the project. For instance, the success of a new office building depends upon the local real estate market and the size and location of office buildings that might be built in the future. These local forces are in turn influenced by national economic trends and the rate of population growth in the region. Identifying the driving forces requires the most research.

> *In order to adequately define the driving forces, research is usually required. Research may cover markets, new technology, political factors, economic forces, and so on. The scenario planner is searching for the major trends and the trend breaks. The latter are the most difficult to find; novelty is difficult to anticipate.*
>
> Peter Schwartz, *The Art of the Long View*

The most important elements to consider are those that are most uncertain and most likely to affect the decision or the evaluation. Demographics are likely to be about the same for all scenarios, but energy prices might be much higher in some scenarios.

Structuring of a set of scenarios is the next step. Schwarz calls this one of the most critical steps because it requires choosing a limited number of scenarios that effectively cover the range of possibilities that might be encountered. If too many scenarios are used, they will be too difficult to flesh out; and it will be difficult to interpret and compare the results of the scenario analysis. What is needed is a small number of scenarios that clearly illustrate the differences among the different strategies that might be pursued. Schwartz gives an example of four scenarios based on two key drivers related to the success of investments by an automobile manufacturer (fuel prices and protectionism):

1. High fuel prices in a protectionist environment (i.e., government regulations restrict imports in order to favor local manufacturers)
2. High fuel prices in a global economy
3. Low fuel prices in a protectionist environment
4. Low fuel prices in a global economy

---

[2] Peter Schwartz, *The Art of the Long View: Paths to Strategic Insight for Yourself and Your Company* (New York: Doubleday, 1996).

Each of these scenarios can be fleshed out by considering appropriate values for the key local and external forces that were identified in the earlier steps. The next step is to consider the proposed decision in the context of each scenario. If a decision looks good only in one scenario and bad in many, then that is likely to be a poor decision. For example, to consider the automobile situation, continuing to produce expensive, gas-guzzling cars looks good only in the third scenario since either global competition or high fuel prices will favor other kinds of cars.

Finally, Schwartz recommends monitoring the key factors in order to figure out how the world is evolving (i.e., which scenario is actually unfolding). Over time, new factors will evolve, new decisions will require another set of scenarios, and a well-informed executive will be able to make better decisions.

Scenarios can help provide structure to complex systems problems. For example, deteriorating air quality threatens public health in many of the world's megacities. Air quality is a particular problem in Mexico City, whose population of 20 million is second only to that of Tokyo. A study of strategies for improving air quality in Mexico City developed three scenarios based on five key elements, as shown in Table 11.5.[3] The "Divided city" scenario envisioned the combination of high population growth and low economic growth leading to higher income disparity and separate cities for the rich and the poor. A different future was envisioned in the "Changing climates" scenario, including greater concern for environmental issues and income inequality within a context of moderate economic growth and better land use planning. The "Growth unbounded" scenario envisioned unrestrained land use and rapid sprawl related to stronger economic growth. This study considered 28 strategies ranging from vehicle technology to transport management to investment in public transit. The study found that even the most aggressive strategies would be unable to "achieve sustained emission reductions in the case of unbounded population growth, auto ownership, and urban sprawl. Therefore a refocusing of efforts on long-term land use planning issues is recommended."

Table 11.5   Scenarios for Assessing Strategies to Improve Air Quality in Mexico City

Strategy	Population Growth	Economic Growth	Urban Sprawl	Income Disparity	Environmental Consciousness
Divided city	High	Low	Increased	High	Low
Changing climates	High	Moderate	Low	Low	High
Growth unbounded	Low	High	Increased	High	Low

Source: Ali Mostashari, Joseph M. Sussman, and Stephen R. Connors, "Design of Robust Emission Reduction Strategies," *TRR* no. 1889 (2004).

Clearly, a great deal of judgment is required to develop a good set of scenarios. But a great deal of judgment is required to make any important decision. The great value of scenario analysis is its ability to consider projects under different conditions, which at least removes the sense of certainty that might otherwise creep into the decision. The following section presents a case study of the use of scenarios by an energy company faced with the need to expand its production capabilities.

11.4.3   Case Study: Use of Scenarios by Southern California Edison

Southern California Edison (SCE) has used scenario planning to guide its investment strategy. SCE planners used a three-stage process: scenario development, implications of scenarios, and development

---

[3] The scenarios were developed by R. Dodder and cited by Ali Mostashari, Joseph M. Sussman, and Stephen R. Connors in "Design of Robust Emission Reduction Strategies for Road-Based Public Transportation in Mexico City, Mexico: Multiattribute Trade-Off Analysis for Metropolitan Area," *Transportation Research Record* no. 1880 (2004).

of strategies. Scenario development followed a process similar to that recommended by Schwartz. Eight drivers were identified:

1. Price of fuel and purchased power
2. Base case rates based on current operating and maintenance costs
3. Demand for electricity, which was assumed to be proportional to economic growth
4. Changes in environmental regulations
5. Open access to SCE's transmission
6. Customers generating their own power
7. Technical innovation (for the company and for consumer appliances)
8. Population growth
9. Generation shutdown

> *By focusing on plausible uncertainties and postulating alternative futures, scenario planning emphasizes the unpredictability of future events and their impact on operations, sales, prices, demand, and so forth. The process of constructing and analyzing the impact of scenarios forces the planners to delve into the dynamics, the cause-and-effect relationships that determine the future. The process identifies major weaknesses as well as major opportunities that exist under different scenarios. Consequently, management can prepare contingency plans to deal with threats and take advantage of opportunities.*
>
> Fred Mobasheri, Lowell H. Orren, and Fereidoon P. Sioshansi, "Scenario Planning at Southern California Edison," *Interfaces* 19, no. 5 (1989): 34.

SCE planners then created an initial set of 45 scenarios based on these eight drivers, which they clustered into groups from which they eventually selected 12 scenarios for analysis. The 12 scenarios were each defined by changes in production requirements (megawatts) related to the eight drivers. The scenarios were designed to span a production range from 5000 megawatts below their base case forecast to 5000 megawatts above base. Table 11.6 shows how four of the strategies were defined in terms of increases or decreases in production that were related to the eight drivers.

Taken together, these 12 scenarios indicated the extent to which SCE should plan for dealing with the need to increase or decrease production. SCE then identified ways to increase or reduce production capacity for each scenario by using strategies like these:

Table 11.6   Change in Production Requirements (in megawatts) for SCE for Selected Strategies

Driver	Economic Bust	Expanded Environmental Concern	Low-Cost Fuel	Economic Boom
Fuel and purchased power	−1000	—	1000	500
Base rates	−500	—	500	500
Economic growth	−3000	—	2000	3000
Environment	—	−1500	—	—
Open access	—	—	—	—
Self-generation	—	—	—	—
Technical innovation	—	—	—	500
Population growth	−500	−500	500	500
Generation shutdown	—	—	—	—
Total changes from base case	−5000	−2000	4000	5000

Source: Mobasheri et al. (1989, 38).

- Capacity could be increased by 900 megawatts (MW) by taking power plants out of standby reserve or reduced by as much as 1500 MW by putting additional power plants into standby reserve.
- Energy purchased from other companies could be increased by as much as 2000 MW.
- Energy management, such as peak-load pricing, could be used to add 500 MW or to reduce 1050 MW of production.

This analysis revealed problems that SCE would have to deal with under some of the extremely low-demand or extremely high-demand scenarios. For example, to deal with increased power needs, SCE identified projects that could be implemented quickly so as to increase capacity, and the company determined that it would need to enhance its ability for energy management.

11.4.4   Probabilistic Analysis

In some situations, using probabilistic analysis may help gain a better understanding of the risks associated with a project. To use probabilistic analysis, the following conditions are necessary:

- The most important factors that affect performance for the project have been identified.
- Algebraic expressions relate these key factors to performance.
- Past experience or special studies provide a basis for estimating
    - Probabilities associated with each key factor.
    - Interrelationships among the key factors.

Previous chapters have presented examples of algebraic expressions that relate key factors to performance. For canals, capacity and cost were related to such factors as the dimensions of the canal, the size of boats, the speed of the boats, the number of locks, the time required to go through the locks, the costs of excavating the canal and building the locks, the operating and maintenance costs of the canal, the interest rates on loans, and the costs of moving freight by competing modes of transportation. For office buildings, the net present value of the project was related to the costs and time for construction, land costs, rentable floor space, rental rates, occupancy rates, and terms of the construction loan and the permanent mortgage. For both canals and office buildings, some of the key factors will be more uncertain than others, and sensitivity analysis can be used to identify the factors most critical to project success.

Let's begin by considering construction cost from the perspective of an experienced contractor who is submitting a bid to construct an office building. Construction will involve a number of distinct activities, such as clearing and leveling the site, excavation, laying the foundation, erecting the steel frame for the structure, and so on. The contractor has good estimates of the time and cost associated with each of these major steps of the construction process. The contractor also knows how delays in one step might affect other steps of construction. The contractor thus has the information necessary to construct a detailed model of the construction process, showing the time and cost for each step as a discrete or continuous random variable. Such a model could be run numerous times to estimate the distribution of construction cost for the project

The contractor might also be willing to estimate the possible range of construction costs by assuming the cost per square foot of the proposed project will be similar to the costs of previous projects. Either way, the contractor could create a chart something like the one in Figure 11.6 to represent the anticipated range of construction costs.

In Figure 11.6, the x-axis is the cost per square foot of an office building; the y-axis represents the number of office buildings. This plot could be based on the actual costs of several hundred buildings or the results of several hundred runs of the model of construction costs, or it could be based entirely on the expert judgment of the contractor. Based on this distribution, the expected cost is $106/square foot, but the costs could be as low as $80 or as high as $128/square foot. What should the contractor do?

- The bid should be higher than the expected cost.
- The bid should include a contingency amount that will cover the most likely range of cost variation.

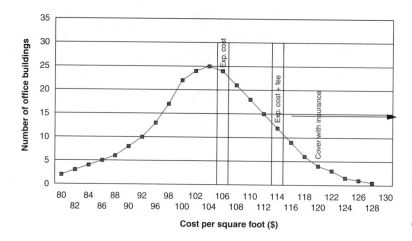

Figure 11.6 Bidding on a Project Based on the Estimated Distribution of Construction Costs.

- The bid should include a fee that, if necessary, could cover further cost overruns.
- The contractor should understand the factors that might lead to excessive costs and take steps to deal with those situations (e.g., take out insurance to cover the costs of major accidents).

> *When we submit a bid for constructing a new sports stadium, we are asked to submit a lump sum price with contingencies or a guaranteed maximum price. Often we need to bid before we have any agreements with our sub-contractors. We therefore have to bid based on our past experience, which allows us to estimate the expected cost and the likely range of costs around that estimate. We start with a bid that is above the expected cost, add something for contingencies, and then add a fee. If need be, the fee can be used to cover higher-than-expected costs. We then take out insurance to cover the most extreme costs. The worst case scenarios involve fatal accidents. A major accident in the construction of one stadium resulted in 3 fatalities, extensive work to recover from the accident, and a delay of one year. The added costs included a judgment of $100 million related to the fatalities, $25 million to fix the damage, and $25 million related to the delay in completion of the stadium.*
>
> Bob Wyatt, "Emerging Trends in the Development, Design and Construction of Stadiums for Professional Sports," presentation at MIT, December 7, 2000.

Using probabilistic analysis presents three major difficulties. First, the probabilities are not well understood, especially when considering the potential interaction among different variables. Second, the mathematics quickly becomes very complicated. Even if it is possible to do the math, it may be too difficult for analysts to explain the results to the decision makers—owners, developers, and investors. Third, the real questions for the decision makers are how to recognize and deal with possibilities, rather than quantifying probabilities. To identify and respond to strategic concerns, sensitivity analysis and scenarios may be sufficient, without trying to formulate or estimate probabilities. Probabilistic analysis is addressed in more detail in Chapter 13.

11.5   EVALUATING THE ALTERNATIVES

The systematic analysis described in the previous section provides the basis for comparing alternatives, identifying the strengths and weaknesses of various aspects of each alternative, and selecting the alternatives that deserve further study. In an iterative process, there is no clear point at which the iterations are complete. Eventually, though, the alternatives are narrowed down to just a few possibilities and a decision can be made as to which alternative is preferable (Figure 11.7). Attention can then be given to

Figure 11.7 Tren Urbano's San Piedras Station. This station, which was under construction in January 2004, is situated underneath an active retail district in San Juan, Puerto Rico. To improve transportation, this city at first considered options related to bus, light-rail, and heavy-rail transit. After deciding to proceed with a heavy-rail system, city planners had to select the route and then decide where to elevate the system, where to run at ground level, and where (as at San Piedras) it was worth the extra construction cost to run underground.

refining the preferred alternative and developing a staged implementation strategy. At this stage of the analysis, studies can be undertaken to consider modest variations in the size and scope of the project. Detailed studies can seek ways to reduce the costs or improve the performance of the study or to identify better ways to mitigate negative social or environmental impacts. Implementation strategies may consider options for staging the development or ways to allow room for further expansion. Design studies can also take into account options for eventual decommissioning and reuse of site and materials.

Parts I and II have devoted a great deal of attention to the issues and performance measures that should be considered in analyzing and evaluating alternatives. In addition, there will likely be a need for highly technical analyses and decisions that require extensive input from experts. The key technical issues will vary with different types of infrastructure, and such issues are beyond the scope of this text. While technical issues must eventually be considered, it is often clear that the major issues are not technical. Instead, the major issues are related to location, size, or timing as well as to political feasibility or sustainability. The case study in the next section shows that brainstorming, comparative studies, and expert opinion can be very useful in the early stages of a study—well before it is necessary to consider technical engineering matters at any depth.

11.6 PEARL RIVER DELTA CASE STUDY

The greatest mistakes in project evaluation are likely to be made very early by defining projects too narrowly or incorrectly. Likewise, the greatest contributions to the success of a project may come from people who view possibilities from a broad perspective. This case study shows how an experienced interdisciplinary team framed the issues and opportunities associated with a major infrastructure project within one of the world's largest multi-centric regions, namely the Pearl River Delta in China. The team participated in the earliest stages of public debates concerning the possibility of bridging the Pearl River Estuary. The basic idea had been around for at least two decades, but there was as yet no commitment to any particular plan. It was therefore possible to introduce new ideas, debate the justification of the project, explore related issues, and prepare a broader context for evaluating the project.

11.6.1 Background

The Pearl River Delta is one of the most densely populated regions in the world. Guangzhou, at the head of the delta, is a city of 15 million people; Hong Kong, on the eastern edge of the delta, is one of the most prosperous cities and largest ports in the world; the region includes Macau on the west side of the delta and a half dozen other major cities with more than a million people each. Overall, more than 40 million people live in this region. The institutional structure of the region is complex, since Macau and Hong Kong are special administrative regions (SARs) and Zhuhai and Shenzhen are special economic regions (SERs). While it is all part of China, the region has complex boundaries and customs regulations. At the time of the study, economic growth in the region was expected to continue its torrid pace as the various pieces of the region become more fully integrated.[4]

By 2002, proposals had originated both from Hong Kong and China to build a bridge and/or a tunnel across the Pearl River to integrate Hong Kong with the West Delta. The proposals sparked debates about the need and justification for such a project, the nature and location of the crossing, the financing of the project, and the potential social and environmental impacts. The Hong Kong 2022 Foundation asked an interdisciplinary team from MIT to provide guidance in identifying and assessing the major issues and constraints that should be considered in assessing any proposal for creating a link between Hong Kong and the west side of the delta. The foundation was not at that time interested in specific designs.

The team was headed by Professors Tunney Lee and Ralph Gakenheimer of MIT's Department of Urban Studies and Planning and Nien Dak Sze, chairman of AER, Inc., a consulting firm with experience related to major projects in Hong Kong. Fred Salvucci, senior lecturer in the Department of Civil Engineering, was a key member of team; his unique experience included two terms as secretary of the Executive Office of Transportation and Construction in Massachusetts at a time when the state initiated several major transit and highway projects. Salvucci and Lee both had a long history in transportation planning in Boston, going back to the early 1970s when Massachusetts decided to scrap the plans for ever more highways and instead develop a more balanced transportation system.[5]

The study revolved around a series of informal weekly or biweekly meetings that allowed a great deal of open-ended discussion about options, issues, strategies for projects, and schemes for presenting our ideas. Articles about the project, which appeared regularly in the Chinese press, were circulated to the team.

It quickly became evident, based on comparisons with similar projects successfully completed around the world, that it would be possible to build a bridge across the Pearl River. Depending upon the location of the project, a tunnel would be desirable to avoid any interference with shipping to and from the port of Hong Kong. It was also evident that the economic benefit would likely be significant, so that it would be possible to justify the project either from the perspective of the public sector (effects on GDP and regional integration) or the private sector (profitability based on toll-based financing). A major concern was that government agencies would move too quickly to begin construction of a project without clearly understanding the range of relevant issues and impacts. It would be easy to build and finance a bridge that would not be close to the best size, be in the best location, have the best design, integrate the best with other infrastructure projects, or produce the best environmental and social impacts.

The team members did not have the data, the analytical resources, or the inclination to conduct detailed traffic analyses or pursue any technical or economic analysis. Instead, they used international examples, their combined experience with major projects, their knowledge of transportation systems, and straightforward analysis to highlight what they felt were key issues.

- The bridge was feasible since bridges and tunnels of similar length had been constructed elsewhere in the world (Table 11.7).

---

[4] M. Enright, K. M. Chang, E. Scott, and W. H. Zhu, *Hong Kong & the Pearl River Delta: The Economic Interaction* (Hong Kong: The 2022 Foundation, 2003).

[5] Other members of the team included Ken Kruckemeyer (lecturer at MIT and former engineer with EOTC; expert in bridge design and neighborhood impacts of transportation projects) and Gerry Flood (expert in mapping and computer graphics). The team was supported by graduate students Yanni Tsippis, Dalong Shi, and Mark Schofield in MIT's Transportation program As a member of the team, my role was to provide support in two areas—project evaluation and freight transportation. As a participant, I was able to observe firsthand how discussions and ideas came up, mutated, and eventually became part of a consensus about what could or should be done and what should not be done.

Table 11.7   Examples of Long Bridges and Tunnels

Bridge or Tunnel	Length	Cost	Toll
Chunnel (rail tunnel connecting Great Britain and France)	50 km	$21 billion	$75
Lake Pontchartrain Causeway (causeway connecting New Orleans to points north of the city)	39	$0.06 billion	$1.50
Chesapeake Bay Bridge/Tunnel (connecting Norfolk, Virginia, with the Eastern Shore of Virginia)	28	$0.4 billion	$10
Oresund (Denmark–Sweden)	16	$2.4 billion	$32
Tokyo Bay Aqualine (connecting Tokyo with the relatively undeveloped eastern side of Tokyo Bay)	15	$11.7 billion	$20

- Since the bridge/tunnel would provide a much shorter route between two densely populated areas, the Pearl River Delta offered an excellent opportunity for constructing a bridge.
- The debate should consider the location of the bridge and the possibility of a Y-shaped bridge (i.e., one that links two cities on one side with one on the other side), a double-Y, or two separate bridges.
- The bridge should be considered as a key link in a multi-centric region with implications for traffic management, investment, and economic growth throughout the region.
- "More than a bridge"—the project should be reviewed in light of opportunities for such things as renewable energy (wind power or solar farms located on or near the bridge), development of existing islands, the creation and development of new islands, and the location of piers and new islands so as to promote river flow and prevent silting.

The research produced a highly polished report laden with pictures and figures with great visual appeal.[6] The report, which included poster-sized pullouts, was aimed at quickly conveying information and insights. The report was presented at a workshop held at Hong Kong University on March 25–26, 2003. The workshop was organized by the sponsor, the 2022 Foundation, as a means of promoting discussion among government officials, representatives of nongovernmental organizations (NGOs), and private sector business leaders from both China and Hong Kong. The study helped promote awareness of the wide range of benefits and options and of the importance of integrating the bridge project with other regional infrastructure planning efforts related to economic development, transportation systems, and the environment.

11.6.2   How the Team Did Its Work

Ideas require time to gestate. As the team members met regularly over a period of 8 months, they evolved an ever more complex view of the project, along with an increasingly coherent story to tell about the project. The process involved brainstorming, contemplation, debates, reconsideration of issues, introduction of new issues and perspectives, preliminary analysis, and more debates. The idea of using international comparisons came up at the team's first meeting in May 2002. The major options for the alignment were identified by August, and the possibility of building two bridges was broached in September. The team did not seriously consider the importance of viewing the bridge as a key link in a multi-centric region until early in 2003. The role of tolls was discussed in July 2002 and then revisited in February as part of a broader discussion of traffic management within the region. The team referred to environmental concerns at the outset, but eventually had more specific ideas about leaving more open

---

[6] C. D. Martland, R. A. Gakenheimer, K. E. Kruckemeyer, T. Lee, M. Murga, F. P. Salvucci, D. Shi et al., *Linking the Delta: Bridging the Pearl River Delta* (Hong Kong: The 2022 Foundation, 2003). Available online at www.2022foundation.com.

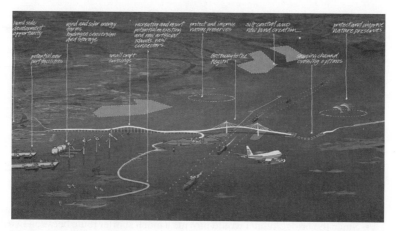

Figure 11.8 Schematic Representation of the Issues Related to Bridging the Pearl River Delta.

Source: *Linking the Delta: Bridging the Pearl River Delta* (Hong Kong: 2022 Foundation, 2003).

space on the east side, bypassing the coast when developing the west side, and integrating the bridge with efforts to clean up the estuary. The team also discussed the aesthetics of the bridge, including an idea that the bridge could be designed to resemble a dragon when viewed from the air—with the ability to have fireworks propelled from the dragon's mouth. The result was not something that any team member could have created individually, nor was it something that the entire group could have created in a short time.

While the team was developing its view of the project, the *South China Morning Post* published many articles containing information and opinions about the economic potential for the bridge, financing ideas, and environmental considerations, especially issues unique to the region. Major potential benefits of the project included integrating Hong Kong within the Pearl River Delta region and strengthening that region's ability to compete with other regions in China, notably Shanghai. Better connectivity would promote regional economic growth and help maintain the role of Hong Kong and the region as an international logistics center.

The results of the interdisciplinary study did indeed broaden the debate. One impetus for the study was a specific proposal by a Hong Kong businessman, Sir Gordon Wu, who had offered to build a bridge at no cost to the government. His company would build the bridge and finance the costs by pledging toll revenues to cover debt servicing. He indicated that the public sector would have to cover only the cost of a 9-km connecting road. The early articles referred to "the bridge" and "the route" and to whether a public subsidy would be needed. The team's report, as described above and illustrated in Figure 11.8, widened the debate and specifically emphasized the need to consider other uses for tolls (traffic management and financing for other parts of the system, not just financing for the bridge), the need to maximize the public benefits of the bridge, and the potential for tying the bridge project into comprehensive plans for improving the region's infrastructure and environment.

11.6.3 Lessons Learned from the Pearl River Delta Study

An interdisciplinary team will need to go far and fast in preliminary thinking about a project like this. It will not be possible to take the time to study all of the issues that arise, nor will it be necessary to go into tremendous detail at this early stage of project evaluation. Digressions on methodological issues or technical issues may divert time and resources from the task of understanding strategic issues that could affect the public debate.

The private sector may realize the gross feasibility of a project—so someone may push for an immediate start, and the public response may be, "Why not let them begin?" It must be emphasized that the public should be looking for the best project, not simply a good project or a profitable project. The benefits of a project may be great enough to "really do it right."

Spirited discussion may quickly give a greater appreciation of options—location, size, design, related infrastructure, timing, and so on.

A project may initially be regarded as a local project with specific costs and benefits. As discussion progresses, a study team will gain a greater appreciation of a wider range of costs and benefits affecting regional and national as well as local concerns.

11.7   SUMMARY

It may take years to complete and decades to pay for any major infrastructure project, and the economic, social, and environmental impacts of a project may be felt for a century or longer. No one can predict with any certainty what will happen to the economy, to technology, or to government policies over a period of just a few years, and the uncertainty of what might happen over a period of decades is much greater. The chapter presents two broad strategies for dealing with uncertainty:

1. Think strategically from the beginning: pay great attention to the statement of needs and objectives (Section 11.2) and to identifying alternative ways to meet those needs and objectives (Section 11.3).
2. Consider the variability inherent in any analysis of costs and benefits by conducting sensitivity analysis for each alternative, considering multiple scenarios for the future, and conducting probabilistic analysis concerning the effects of key factors on project outcomes (Section 11.4).

The chapter concludes with a case study that demonstrates how brainstorming and open-ended, interdisciplinary discussions can change the scope and context of a major project (Section 11.6).

Identifying Needs and Objectives

The greatest uncertainties arise at the beginning of the process, when people recognize shortcomings in an infrastructure-based system and believe that some sort of project might be desirable. How the need is specified and how the objectives for a project are defined will influence the types of alternatives considered. Poorly conceived statements of needs and objectives may lead to consideration of projects that really are not the best ways to address the real problems. Well-structured statements of needs and problems can lead to a wider range of possible solutions and a greater chance of success. Needs and objectives should be stated in terms of system performance rather than in terms of technology or size of the system.

In the Pearl River Delta case study, the need was originally defined as providing a new bridge across the Pearl River, and the solution appeared to be to allow a private company to finance, construct, own, and operate a toll bridge. When an interdisciplinary team looked at the needs for the region, adding bridge capacity eventually was seen as just part of the region's need for expanded transport capacity, and the construction of one or more bridges was found to be an important factor with respect to regional development strategies, development of renewable energy, and efforts to improve water quality in the region. The best approach to the problem was characterized as "More than a Bridge."

Table 11.8 presents some guidelines for thinking about options and alternatives. It is important to be flexible in thinking about a project and to be open to new possibilities. Over the long course of a project's life, many changes in social, economic, and competitive conditions can take place, and some thought at the outset may prove useful in designing projects that will be able to function more effectively throughout an uncertain future.

Identifying Alternatives

Once needs have been identified and objectives have been clarified, it is time to generate some alternatives for meeting those objectives. The number of alternatives that can be investigated in detail is limited, so it is important to take care in deciding what to investigate. The alternatives should include

Table 11.8 Guidelines for the Early Stages of Project Evaluation

---

- Address the grand issues:
  - Economic viability—is there any case at all?
  - Engineering—what are the options for capacity, staging, and flexibility?
  - Financial feasibility—is the project best done as a public project, a private project, or a partnership?
  - Political feasibility—who is likely to support or oppose the project?
  - Size—would a larger or smaller project be better?
- Consider comparable projects—this will provide an order of magnitude assessment concerning the viability of the project.
- Consider the possibility that benefits may be so great that you do not have to accept the minimal project.
- Plan for flexibility and capacity.
- Think at all scales—local, regional, and national.
- Think about aesthetics, and plan with an eye to style.
- Think about political issues.
- Guide the press—help them avoid taking too narrow a focus too soon; help them understand the grand issues.

---

a base case ("do nothing") that reflects what would happen if no new projects are undertaken. The other alternatives should be structured to include the major options known to be available.

Four useful methods for generating alternatives are introduced in this chapter:

1. Brainstorming
2. Systematic identification of options
3. Public input
4. Best practices and expert opinion

If input is received from experts and the public, and if brainstorming is used effectively, then a wide range of options will surely be suggested. The task will then be to select a few options that cover the range of suggestions and will provide insight into how best to proceed. Common sense will be useful in structuring a set of meaningful options and avoiding pitfalls such as "the Good, the Bad and the Ugly," "Tweedle Dum and Tweedle Dee," and "Overkill."

Since evaluation is an iterative process, the alternatives can be structured in such a way as to maximize the knowledge gained from the analysis. The alternative finally adopted is likely to be a composite of two or more of the alternatives that were investigated.

Dealing with Uncertainty when Assessing Alternatives

Methods described in prior chapters showed how to assess financial, economic, social, and environmental impacts. This chapter presented methods that are useful in understanding and dealing with the uncertainty that is embedded in the assessment.

Sensitivity analysis requires systematic variation of the key factors used in the analysis. The goal is to determine which factors are most important (i.e., to determine the sensitivity of the results and conclusions to reasonable changes in the main factors). Outputs such as project cost, annual revenue, return on investment, or net present value can be plotted for variations in assumptions concerning costs, construction time, financial factors, or demand. In practice, sensitivity analysis is an art that requires good judgment and the ability to interpret and display results. Judgment is necessary in selecting variables, in determining the range of values to consider, and in deciding how to present the results.

A scenario is a set of internally consistent assumptions that together provide a vision of a possible future within which a project will be implemented. Scenarios are broader in scope than sensitivity analysis, since they attempt to identify how various interrelated factors might change over the course of a project. The elements of a scenario represent the key factors that will affect the cost of the project, the demand for the project, or the public perception of the project. Elements of a scenario might include measures that reflect

- General economic conditions
- Responses of competitors to the proposed project
- Construction prices
- Energy prices

A set of scenarios can be created by varying the most important elements in a logical manner. A common practice is to consider a base case along with an optimistic and a pessimistic case. An example showed how Southern California Edison used scenario analysis in determining strategies for expanding its capabilities. SCE planners developed 12 scenarios based on variations in what they felt to be eight key drivers of performance, including fuel price, economic growth, population growth, and several technical factors related to electrical production. The exercise convinced SCE that it faced some situations where there would be problems with too much or too little supply, and the company took steps that allowed it to react more quickly to changing conditions.

In some situations, it may be possible to use probabilistic analysis to gain a better understanding of the risks associated with a project. To use probabilistic analysis, the following conditions are necessary:

- The most important factors that affect performance for the project have been identified.
- Algebraic expressions relate these key factors to performance.
- Past experience or special studies provide a basis for estimating
  - Probabilities associated with each key factor
  - Interrelationships of the key factors.

An example was shown in which a contractor used past experience to develop a simple model for estimating the range of costs for building sports stadiums. With this probabilistic model, he was able to bid on projects even before reaching agreements subcontractors. The bid would be sufficiently above the expected cost to allow for the normal range of cost overruns, and insurance would be used to cover the most extreme costs, such as would be associated with a fatal accident during construction.

Evaluating the Alternatives

Using an iterative process makes it possible to revise the alternatives, the assumptions used, and the methodology for assessing the alternatives. Scenarios and sensitivity analysis may help in understanding where an option must be improved or where strategies must be modified.

ESSAY AND DISCUSSION QUESTIONS

11.1 A city is worried about the damage caused by flooding. In the past 20 years, three floods have occurred and altogether resulted in 20 fatalities and destruction of 25 homes and small businesses. Much of the city's area is within the floodplain of the river, including some of the areas with the fastest rate of development. The leading newspaper has launched a campaign calling for investment in "long-planned and much needed flood control projects." The Chamber of Commerce also supports the flood control projects, since they would enable development in a vast area of marshy wastelands that abut the south side of the river. The cost of these projects is estimated to be $500 million. Is this flood control project what the city needs?

11.2 Structure a set of scenarios that could be used to evaluate alternatives for bridging the Pearl River Delta.

11.3 Go online to review the status of major transportation improvements in the Pearl River Delta. What major projects, if any, have been completed since 2003? What is the current status of plans for or construction of bridges across the Pearl River Estuary? To what extent are the ideas presented in the case study reflected in the current depiction of the issues in the press or in blogs?

11.4 Choose one of the following topics, and identify at least three possible statements of needs and objectives:

a. A large earthquake in a rural area damaged many buildings and caused dozens of deaths in remote villages. Fortunately, the epicenter was far from the nearest city, which experienced only minor damage. However, the citizens and politicians in that city were understandably concerned about what might have happened, and everyone would like to do something to reduce the risks of a less remote earthquake.

b. As people age, they eventually reach a point where they should no longer be driving an automobile; yet many elderly people require their cars to buy groceries, visit friends, and receive medical care.

c. A significant portion of the fertilizers used for large-scale agriculture end up in rivers and streams adjacent to the fields, eventually finding their way to the Gulf of Mexico and contributing to ruination of that aquatic environment.

11.5 Choose the best statement of needs and objectives that was suggested for one of the topics in the previous question. Gather a group of at least five people, and use brainstorming to identify alternative ways of meeting the needs and objectives.

PROBLEMS

11.1 A company wants to construct a new factory that will have a capacity of 1 million units per year. The factory is expected cost $100 million to construct, and it should be operating within two years. The product is expected to sell for $40/unit, which is well above the expected cost of $30/unit, including the EUAC of the investment in the factory and the production cost assuming the factory operates at 95% capacity. The NPV for the project is expected to be $20 million, and the IRR is expected to be 15%. You have been asked to structure a sensitivity analysis for this project.

a. What variables would you include in the analysis, and what values would you suggest considering for each variable?

b. How would you present the results? (Describe the tables or charts that you would use—you do not have to do any of the analysis.)

11.2 A company would like to build wind farms in rural areas of New England. As of 2007, the price of oil was so high that the company believed it would make a profit if it could receive an investment tax credit of 20% from the federal government. Other initiatives were under way in the region that would reduce the demand for fossil fuels, including expanded use of water power, the use of wood-powered electrical power plants, tax incentives aimed at conservation, and greater residential use of wood stoves for heating.

a. Identify at least four key elements that you would include in developing scenarios for evaluating the potential success of the wind farms.

b. Identify at least four scenarios based on the key elements that you have identified.

CASE STUDY: WE'LL CROSS THAT BRIDGE WHEN WE COME TO IT (PART II)

This case revisits the situation that was discussed at the end of Chapter 5. The earlier case introduced the actors, the alternatives, and the issues and asked you to identify an evaluation strategy that included selection of criteria and weighting issues. Some additional information has now been gathered, and you can proceed with a more detailed analysis that leads to specific recommendations for constructing a bridge. For your convenience, the material from Chapter 5 is included at the beginning of this expanded case. The case is designed for teams of students who represent the different actors involved in the ongoing debate about which bridge is best.

Situation
Cammitibridge, a prosperous city of a million residents, is located on the north side of the Cammiti River. It is adjacent to and to some extent hemmed in by a state forest to the northeast and a series of steep hills to the northwest (see map at end). The town is named for its historic bridge, first built in 1720 and most recently rebuilt in 1850. The bridge is a narrow, two-lane bridge that spans the river so as to connect Cammitibridge with the main road to the capital, Boslondale, some 100 miles away. The bridge connects to a highly scenic road that winds for several miles toward the west below sandstone cliffs along the edge of the river. Everyone in the city agrees that a new bridge is needed, for various reasons:

• Development in Cammitibridge is increasingly limited by a lack of vacant land close to the city center. On the outskirts of town, people are starting to move up the slopes of the hills, but there just isn't much room for expansion downtown or along the major roads serving the city.

• There is considerable traffic between Cammitibridge and Boslondale, so that the old bridge is often congested, irritating those who must use it and limiting opportunities for development on the south side of the river.

• The current bridge provides little access to the land available for development south of the river. A new bridge in a new location would open up this land for development.

• The local construction industry supports the construction of a new bridge, because of the opportunity for new activity and profits.

• A consortium of private citizens, led by Canwy Bildem, has even proposed that they will replace and expand the existing bridge at their own expense if they are allowed to charge a toll of no more than $2; at the end of a 30-year period, they would turn the bridge over to the city.

There are four competing options for the bridge:

• *Bridge 1:* Expand the existing bridge at a cost of $20 million. The plan would basically build a second two-lane bridge next to the existing bridge. Once the new bridge is operating, the old bridge would be rehabilitated to handle heavier trucks and then reopened.

• *Bridge 2:* Replace the existing bridge with a new four-lane bridge at a cost of $50 million. The old bridge would be torn down upon completion of the new bridge. An additional $5 million would be required to modify the roads to match up with the new bridge.

- *Bridge 3:* Build a new four-lane bridge that connects to the developable land to the southeast. Since the river is wider at this location, the cost would be greater. Initial estimates are that the bridge would cost $75 million if built at the narrowest location, while access roads would cost $20 million.
- *Bridge 4:* A variation on Bridge 3, this option would change the bridge location slightly to improve access to the city, while requiring a longer bridge to get over the swamps; the costs would be $90 for the bridge plus $10 million for access.

The bridge could be financed in several ways:

- The city could sell revenue bonds to the public and pay the interest on the bonds out of general tax revenue.
- The city could sell revenue bonds to the public and charge tolls on the bridge sufficiently high to pay the interest on the bonds.
- The city could authorize Mr. Bildem to proceed with his plan to build a toll bridge without using public money.
- The city could probably fund the required connections (but not the bridge) from its ongoing budget for road construction and maintenance.

The economic benefits from construction of the bridge fall into several categories:

- Reductions in travel time for people who currently use the bridge
- Increased opportunities for development south of the river, leading to higher land values, new jobs, and greater real estate taxes for Cammitibridge

Preliminary analysis suggests that

- Expanding or replacing the current bridge will have minor effects on traffic volume or development, since very little open land is suitable for development near the existing routes.
- Building a bridge at the east end of town will provide a spark to development of that region; total traffic across the river is expected to grow quickly if a new bridge is built in that location. Also, some traffic will divert to the new bridge, reducing congestion in the city.

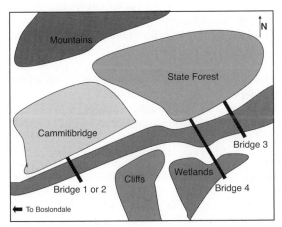

Schematic Diagram of Cammitibridge and Possible Bridge Locations

Table 1 summarizes the expected results from constructing each of the major options. It also shows the costs expected for constructing each bridge and for constructing new access roads. These costs can be assumed to be incurred at a uniform rate over the construction period. The travel time benefits represent the results of a network model showing the expected impacts on average commuting time for the city in the year 2015. The city normally considers travel time savings in its economic studies, using $10 per vehicle-hour to represent an average value of time to users. The traffic volumes are also for the year 2015; this year is currently used as a basis for traffic studies in the region (and, to simplify our analysis, we will assume that this year is in fact an average year over a 30-year planning horizon). Adding capacity for growth on the south side of the river is predicted to be helpful in relieving congestion in the city center; adding capacity to the existing bridge will prevent that location from becoming a major bottleneck for the city.

Assignment

The class will be divided into groups of 4 to 8 students. Half of the groups will analyze the choice of bridges from the perspective of the city's transportation department; the other groups will analyze the choice of bridges from the perspective of Mr. Bildem's company. Each group will have two subgroups; one will consider Bridges 1 and 2, the other will examine Bridges 3 and 4. Each subgroup will submit a joint analysis that answers the questions posed below for their bridges (i.e., everyone conducts a full set of analyses on at least one of the bridges; the students within each subgroup then compare and if necessary revise their analyses; each subgroup prepares a joint report on its bridges). The total group will prepare a one-page summary with its recommendations for the mayor and the public and attach four bridge reports as appendices. Group members can collaborate in any way they chose in preparing for their presentation.

Questions for the Public Sector Groups
You work for the city's transportation department. You have been asked to compare the NPV of the bridge options, taking into account the construction costs and the toll revenues over a 30-year period.

- What is the present value of the costs of building each bridge and its access roads using a discount rate of 8% per year (as required by city policy for infrastructure analysis)? To simplify the analysis, assume that all costs are incurred uniformly over the construction period.
- Assume that the city can sell 30-year bonds with an interest rate of 5% to cover all construction costs as well as interest on the bonds during the construction period (for example, at the end of year 2, 5% interest will be due for all of the bonds that had been sold at the end of the previous year). What is the total face value of the bonds that must be sold to cover all construction, access, and interest costs? (This will equal the total amount of outstanding bonds at the time when the bridge is finished).
- Assume that annual costs for toll collection are $250,000 and maintenance costs are expected to be 5% of the construction cost for the bridge and access roads. For each new bridge, what

Table 1   Expected Costs and Benefits Related to the Bridge (preliminary and subject to change)

	Bridge 1 Expand the existing bridge to 4 lanes.	Bridge 2 Replace existing bridge with 4-lane bridge.	Bridge 3 Build shorter bridge at east end of city.	Bridge 4 Build longer, more accessible bridge at east end of city.
Bridge cost	$20 million	$50 million	$75 million	$90 million
Access roads	—	$5 million	$20 million	$10 million
Construction time	2 years	3 years	4 years	4 years
Travel time savings for current users	4 minutes	4 minutes	5 minutes	7 minutes
Total users, west bridge	15 million/yr	15 million/yr	12 million/yr	11 million/yr
Total users, east bridge	—	—	5 million/yr	6 million/yr
Maximum toll	$2	$2	$2	$2
Increase in GRP	$10 million	$12 million	$20 million	$30 million
Population	1.1 million	1.1 million	1.15 million	1.2 million

toll would you charge to cover the interest on the bonds plus the costs of toll collections and maintenance, assuming traffic volumes as shown for 2015?

(*Note*: If Bridge 3 or 4 is built, tolls will not be charged on Bridge 1.)

Questions for the Private Sector Groups

You are the financial advisor to Mr. Bildem. You need to determine whether the NPV of future tolls is sufficient to cover the NPV of the construction costs. You also need to consider the possibility of getting financing (first a construction loan that would cover the costs of construction and then, once the bridge opens, a loan with lower interest rates that would be based on toll revenues).

- What is the present value of the cost of building each bridge and its access roads, assuming that you use Mr. Bildem's discount rate of 15%?
- Assume that Mr. Bildem can get a 10% line of credit for all construction costs for the bridge. Further assume that Mr. Bildem will pay only for the bridge, requiring the city to provide the access roads as its part of the deal. What will the total loan be when the bridge is open for construction? (For simplicity, assume all costs are incurred at the end of the year; the interest costs at the end of year n + 1 would equal 10% of the outstanding balance at the end of the prior year).
- You can refinance your loan once the bridge opens. Assume that the bank will let you borrow an amount that can be financed by an annual amount no greater than 50% of the average net toll revenues expected in 2015 (and the total loan cannot be greater than your construction costs).
  - Will the amount of the new loan be sufficient to cover the costs of construction loan?

  - If the loan is for 30 years at 8%, what will the annual payments be?
- To cover your loan payments and your share of operating costs (i.e., $250,000 plus 5% of the construction costs for the bridge), what toll would you have to charge for each bridge? Which bridge do you prefer to build? Why?

Preparation for Town Meeting

Everyone in the class has been appointed to the mayor's task force to evaluate bridge options from the perspective of the public. The people in the public sector groups are more concerned with overall economic impacts; the people in the private sector groups are more concerned with entrepreneurial opportunities for construction and development. At the meeting, you will hear reports from the head of the transportation department, Mr. Bildem, the mayor, and the head of the local high-tech development support group "Build Up or Shut Up." The group will then discuss and take a straw poll on three questions posed by the mayor:

1. Which of the bridges is the best for the city?
2. What is your recommendation concerning tolls?
3. Should the city accept Mr. Bildem's offer to build the bridge?

Agenda for Town Meeting

1. The head of Build Up or Shut Up introduces the need for a bridge and presents the criteria that should be considered.
2. Mr. Bildem presents the firm's preferred option.
3. The head of the transportation department presents city's preferred option.
4. The mayor introduces the questions and suggests criteria that should be considered.
5. General discussion.

Chapter **12**

Public-Private Partnerships

Well-structured and well-implemented PPPs offer the prospect of efficiency gains in the construction of infrastructure assets and the provision of infrastructure-based services and, therefore, also lower the government's costs in making these services available.[1]

CHAPTER CONCEPTS

Section 12.1 Introduction
Motivation for public-private partnerships

Relative strengths and weaknesses of public and private sector approaches to dealing with public needs

Situations where public agencies and the private sector might decide to work together on projects

Section 12.2 Principles of Public-Private Partnerships
Each partner must be satisfied with its share of the expected costs, benefits, and risks

Each project is a separate case

The partnership should be designed to deal with a particular situation

Section 12.3 Creating a Framework for a Partnership
Factors that will enter into the partnership

Options for private sector involvement in a public infrastructure project

Section 12.4 Determining How Much to Invest
Using their usual methods to determine the maximum amount each partner is willing to invest and what terms they prefer

Mechanisms for sharing risks

Negotiating the structure of the partnership

Sections 12.5 to 12.8 Case Studies to Illustrate Different Types of Partnerships
Complementary Strengths of Public and Private Partners: Tempe Town Lake

Public and Private Benefits: The Sheffield Flyover, Kansas City, Missouri

Maximize Ability to Undertake Projects: Toronto's Highway

Public Investment to Stimulate Economy: Province of Newfoundland's Investment in Offshore Oil Exploration

---

[1] Bernardin Akitoby, Richard Hemming, and Gard Schwartz, *Public Investment and Public-Private Partnerships* (Washington, DC: International Monetary Fund, 2007), p. 9.

12.1   INTRODUCTION

As demonstrated in examples throughout this text, public agencies, developers, companies, investors, and the general public have different perspectives on projects. The public sector is much more focused on identifying and satisfying public needs, whereas the private sector is much more concerned with achieving financial benefits. The public sector is led by elected officials who must justify their actions and decisions to voters, and these officials are involved in many aspects of society, from education to provisions of parks to water resources and waste management. The division of labor in the private sector produces leaders who are much more sharply focused on specific problems or types of activities. Some are concerned with finance, others with planning and design, and others with operations and management of specific types of infrastructure or industries. The public sector has the ability to raise money from taxation and, at the national level, to adjust the supply of money and the availability of credit for financing projects. The private sector offers entrepreneurs a chance to get rich by designing, promoting, constructing, and operating particular projects. Elaborate systems have evolved that enable developers to obtain the funds necessary to build projects, and people or companies who are willing to take risks have a chance to achieve great rewards. The public sector has the ability to guide development through tax policy, zoning, building codes, and provision of infrastructure. The public sector also has the responsibility for public heath, public safety, and environmental protection. People who are successful in the public sector have a chance to attain great influence over policy and development.

Few if any projects do not require some sort of cooperation between the public sector and the private sector. Even the construction of a small addition to your house or a dock for your boat will require a building permit and perhaps a wetlands permit. The construction of any large building is possible only if the developer secures numerous approvals from various government agencies related to the location, design, building materials, and construction processes. Likewise, when a government agency sets out to build a school or a road, that agency typically employs private sector contractors and subcontractors. These interactions between the public and private sector do not constitute a partnership. Building a house or a dock or an office building is clearly a private project, even though it must be sanctioned by government and follows rules and regulations specified by law. Likewise, building a school or a road is clearly a public project; even though the work is actually done by the private sector, the project is defined and funded by the public sector.

Figure 12.1 summarizes what is needed to have a successful project in the private sector. First, the decision to proceed depends upon having a positive net present value (NPV) or an internal rate of

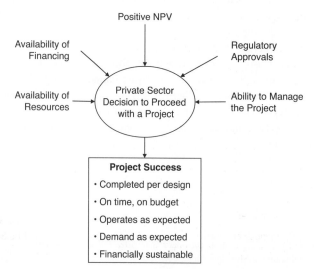

Figure 12.1 Overview of the Private Sector's View of a Successful Project.

return (IRR) that exceeds the company's hurdle rate for new investment. The project then needs to have regulatory approvals, which could relate to zoning, environmental impact assessment, site management, building codes, and any other mandated regulations. If financing is available, and if the company has access to the resources and managerial expertise to manage the project, it can proceed to implementation. But that is just the beginning. Whether the project is successful depends upon the ability to complete the project as designed, within the specified budget, and on schedule. Once the project is completed, ultimate success will depend upon how it is utilized; in particular, it must function safely and efficiently, and the demand must generate revenues that eventually will be sufficient to cover both operating and investment costs.

Figure 12.2 illustrates the broader perspective of a public agency. The first major difference is that the project can no longer be considered merely as a financial effort. Instead, economic, environmental, and social impacts must be considered and incorporated within a benefit-cost analysis. If taxpayers are willing to support a project because of economic or other benefits, then taxes can cover some of the costs, and the project does not require a positive NPV. The public sector will face similar regulations regarding land use and environmental impact assessment, but there is a major difference. Laws and regulations can be used to limit or restrict what either a private company or a public agency might do, but political actions can defeat proposed public projects. A private project must meet the regulations, but a public project must also satisfy the elected officials and ultimately the public. The second major difference is that the definition of success is much broader, because economic, social, and environmental factors will also be of direct interest to a public agency. To sustain a project over the long term,

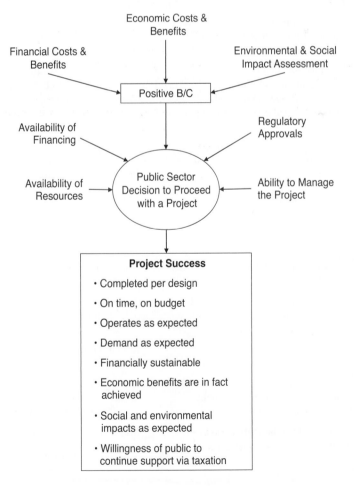

Figure 12.2 Overview of the Public Sector's View of a Successful Project.

especially if tax dollars are required to support operations, public officials and the public must be convinced that those tax dollars are worth spending.

Public-private partnerships arise when a public agency or agencies work with the private sector on a project to build, finance, and/or operate a facility. The public and private sectors are partners because they share in the risks and rewards of the project, and they share in the design and ownership of the project.

Some public-private partnerships arise in order to take advantage of the differing strengths of the two sectors. Table 12.1 compares some strengths of the public and the private sector as related to the implementation of infrastructure projects. Broadly speaking, the public sector must be more concerned with social, economic, and environmental issues while the private sector is more concerned with finance. The public sector has greater ability to identify public needs and to allocate funds to address those needs in a reasonably fair manner; the private sector can quickly respond to perceived needs without worrying so much about what is fair or what is most important to society as a whole. The public sector can raise money via taxes and the sale of low-interest bonds, and it has real powers related to land use. The private sector can raise money from investors, and it can obtain very large sums for clearly profitable projects. The comparisons could go on and on. Some people would add that the private sector is more efficient or more honest, but there are certainly examples of very efficient and inefficient operations in both sectors—just as honesty, wisdom, fraud, and foolishness can be found in both. The point is that the two sectors have different strengths and that some projects will be better designed, constructed, and operated if the two sectors work together.

Section 12.5 presents a case study of a public-private partnership (PPP) that takes advantage of the relative strengths of each party. The City of Tempe, Arizona, needed better flood control, it wanted to clean up an area largely used for waste disposal, and it hoped to encourage economic development. An innovative arrangement was developed whereby the city assembled a large parcel of land around the Salt River; the city then rezoned the land for higher-density development and sold plots to developers with the stipulation that they pay special fees and taxes that would ultimately pay for the creation of a 5-mile-long lake and extensive park land. The combination of open space and nearby high-density development achieved substantial public benefits, while developers had the opportunity to make profits that were based on the proximity of their land to what would become a very attractive site.

Table 12.1 Complementary Strengths of Public and Private Sector

Area	Public Sector Strengths	Private Sector Strengths
Identifying societal needs	Political process can establish priorities among competing needs	Private firms can respond quickly to perceived opportunities
Design	Establishment of building codes and construction standards	Development of new techniques or designs
Construction	A stable workforce can be assembled for continuing needs, such as road maintenance	Numerous companies exist or can be created for all types of construction
Finance	Access to tax revenues and ability to issue bonds with low (possibly tax free) interest rates; possible difficulties in raising fees to keep pace with inflation	Ability to raise capital for projects; recognizing the level of risk involved; greater freedom in raising fees to keep pace with inflation
Politics	Political approval may require more complete assessment of alternatives and a fairer distribution of costs and benefits	Insulation from politics may result in more objective assessment of projects
Land use	Able to assemble large tracts of land for public purposes using eminent domain	Owners can use their land any way they like, so long as they follow zoning and other regulations
Labor	May be required to use local labor or union labor, resulting in higher costs for construction or operation	May be able to use more efficient procedures or nonunion labor, resulting in cheaper construction

A second reason for public-private partnerships is that there are significant public and private benefits for a potential project. Taken together, the benefits might be enough to justify the project when neither the public nor the private sector is willing to tackle the project on their own.

Section 12.6 presents a case study of this type of PPP. In Kansas City, private railroads worked with public agencies to construct a flyover, which is a railroad bridge that takes one rail line over another rail line. The railroad benefited because the north-south trains no longer interfered with the east-west trains; the city and its residents benefited because delays at rail–highway grade crossings were eliminated or sharply reduced. Public financing reduced the costs of the project, and per car fees collected from the railroads will eventually pay for their portion of the project.

A third reason for PPP is that the public sector wishes to maximize its ability to undertake projects by gaining access to the financial markets. The ability of a government agency to raise money may be limited by budgets or by the credit rating of the agency, or a government may be unable to use its assets as collateral to raise money. The government can retain control over the design and location of infrastructure while leasing or selling facilities to a private company. The financing of the project can be based on fees related to the operation of the facility.

Section 12.7 provides an example of this type of PPP. The City of Toronto built a toll road that bypasses a portion of its downtown, and city planners wished to extend this road. Since the city did not want to issue bonds or raise taxes to pay for the road, it decided to sell the road to a company that would operate it and extend it—at the company's expense. The city sold the road for more than it had cost to build it, and Toronto thereby was able to have a private company extend the road with no further government expense.

A fourth reason for a public-private partnership involves a government interest in assisting new industries. While the economic potential may be evident, and companies may be willing to undertake projects, the companies may be unable to secure financing. The government may then decide to assist the company by making loans or even by making an equity investment. Section 12.8 provides an example of this type of partnership. When the private sector was unable to raise the financing necessary to start offshore oil drilling in Newfoundland, the provincial government decided to make a substantial investment in return for partial ownership of the new enterprise.

In summary, there are various reasons for PPP:

- *Complementary strengths:* The partnership builds upon the complementary strengths of the public and private sectors to complete a project that could not be undertaken—or done as well—by either sector.
- *Public and private benefits:* The partnership is created because the project has substantial public and private benefits, and cooperation allows both the public and private sector to satisfy their objectives more effectively than they could by acting alone.
- *Expanded public capabilities:* The partnership allows private capital to be accessed for undertaking public projects.
- *Economic development:* The partnership is created to exploit opportunities for economic development that will ultimately be profitable for the private sector while increasing jobs and economic activity within the region.

As illustrated by Figure 12.3, the overall structure of the project incorporates all of the elements that are of importance to either sector. The private sector still seeks a positive NPV and successful completion and operation of its portion of the project. The public sector still seeks a positive ratio of benefits to cost and successful completion and operation of its portion of the project, assuming negative social and environmental impacts are no worse than anticipated. Both sides of the partnership will need to be satisfied and financially successful if the project is to be a long-term success.

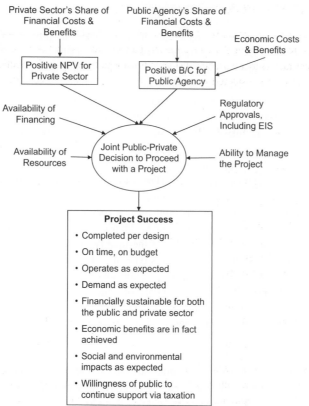

Figure 12.3 Overview of a Successful Public-Private Partnership.

12.2 PRINCIPLES OF PUBLIC-PRIVATE PARTNERSHIPS

The most important principle of public private partnerships is that *each side must bear an appropriate portion of the cost, benefits, and risks.* Private companies are unwilling to increase their investment in projects in order to provide public benefits. Public agencies are unwilling to make investments unless there is a sufficient public benefit. Private companies are concerned with financial performance (i.e., cash flows, profitability, and return on investment). Public agencies are concerned with cash flow only to the extent that cash is needed to implement a project; they may be willing to make substantial investments that are justified by socioeconomic benefits, such as an increase in economic activity or new jobs, reduction in congestion, or improvement in the environment or in public health.

A second principle for private companies is that *each project is a separate case.* After cooperating with a public agency on one project, a private company does not want to be forced to cooperate with that agency or other agencies on similar projects or other types of projects in the future.

A third principle is that *the partnership should be designed to deal with a particular situation*, since public participation can take many forms:

- Financial input and ownership
- Leasing publicly constructed facilities
- Authorizing private construction and operation of a facility that will ultimately revert to the government after a specified period

Which form of participation is most appropriate will depend upon the circumstances and the aims and capabilities of both the public and private partners. What was best in one situation is not necessarily what should be used in a subsequent situation.

12.3   CREATING A FRAMEWORK FOR A PARTNERSHIP

For parties to come together in a partnership, all parties must perceive that their potential benefits justify the costs and risks that they incur. As in creating any partnership, negotiation is possible and necessary. Section 12.3.1 describes factors that must be considered for any partnership, while Section 12.3.2 illustrates various options that might be considered for a public-private partnership to construct a highway bridge.

12.3.1   Key Questions Regarding the Partnership

Before a project can begin, the public agency and its partners must negotiate an acceptable way to share the costs, the risks, and the potential rewards. All parties must believe that their benefits are worth the risks and the costs that they incur, or they will be unable to come to an agreement. Key questions about the partnership include the following:

- Who pays how much for what portion of the project?
- What risks are accepted by each partner?
- Who controls design?
- Who controls construction?
- Who controls operations?
- Who owns what portion of the project?
- Will ownership of the project change over the life of the project?

Key differences in the public and private perspectives include the following:

- *Financial versus economic return:* The private sector requires a minimum level of financial performance, but the public agency may seek economic benefits such as relief of congestion, potential for development, jobs, and increases in regional product.
- *Cost of capital and access to capital:*
  - The cost of capital will generally be much greater for the private sector than for the public sector. The public sector can raise funds by selling government-guaranteed, tax-free bonds that may have very low interest rates, assuming the government agency has a good credit rating.
  - The private sector can raise funds for risky projects by selling stock, thereby giving investors partial ownership in the company and a chance to make a great deal of money if the project succeeds—as well as a chance to lose everything if the project fails.
  - The financial markets recognize and accept the possibility that a project will not be successful and that companies may be forced into bankruptcy; it is possible and not unusual for companies to fail, but public agencies are expected to endure despite failure.
- *Institutional and organizational flexibility:*
  - Public agencies control many of the policies and institutional requirements that constrain or limit a project; a public agency is likely to be in a better position to obtain approvals for such things as zoning variances, building permits, and environmental approvals because the public agency generally is viewed as working in the public interest.
  - Private companies likely are more flexible in creating organizational structures to deal with the design, construction, and operation of projects, especially projects that would be much different or much larger than what the public agencies have been involved in.
- *Time frame:* The time frame of private companies is heavily influenced because of discounting, which reduces present value of both positive and negative impacts that may occur in the distant future; public agencies (a) typically have a lower discount rate, and (b) typically have ethical commitments, constitutional outlooks, and other factors that lead them to have much greater concern for what happens in the very long run.

These four major differences in interests and perspective suggest ways that PPP can be productive. The role of the public agency may be (a) designing the project so that it provides economic, social, or environmental benefits to the public; (b) providing access to low-cost sources of capital; (c) securing project approvals from the various agencies involved in the project; or (d) ensuring that the project provides long-term, positive benefits to the public. The role of the private sector may be (a) designing the project so that it generates cash flows sufficient to justify the necessary investments, (b) providing access to financial markets that are willing and able to provide funds for very large or somewhat uncertain projects, (c) creating new organizations to help in designing, constructing, and operating a project, and (d) ensuring that the project does not take too long a perspective.

12.3.2 Possible Structures for a Bridge Project

Public Financing and Ownership

A bridge could be built as a public project, a private project, or a PPP. However it is financed, the costs are as illustrated in Figure 12.4. There is a large construction cost and continuing costs for maintenance and operations. For convenience, the construction cost of $1 million is pictured as occurring in the first time period, while annual costs of operations and maintenance are shown as a constant $70,000 per year.[2]

If the bridge is built as a public project, then there are several options for financing. The bridge could be viewed as part of the highway system, and whatever funds are used to construct highways can be used to pay for the construction of the bridge. For example, the federal or state government may have a highway trust fund (HTF)[3] that uses income from fuel taxes and registration fees to pay for authorized additions to the highway network (Figure 12.5). If the bridge is approved as a project that can be

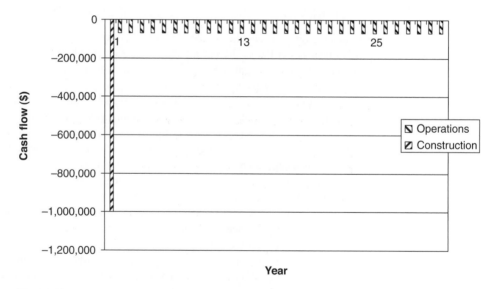

Figure 12.4 Costs of Constructing and Operating a Bridge.

---

[2] While construction may take longer than a year and maintenance costs are likely to rise over time, we know from Chapter 7 that it is possible to convert whatever costs are anticipated into an equivalent construction cost at time 0 and an equivalent uniform annual cost for operations and maintenance such as those shown in this figure.

[3] The federal HTF was created in 1956 as a mechanism for financing the construction of the Interstate Highway System. Fees and taxes on fuel and heavy trucks provided sufficient revenue to cover the federal government's 90% of the construction costs. Subsequent legislation allowed small amounts of the fund to be diverted to transit and intermodal projects. The federal fund can be used only for construction, not for operations or maintenance, which remain state responsibilities. States have similar funds, and fuel taxes again provide the major source of revenue. For the complete history of the HTF and the Interstate Highway System, see Tom Lewis, *Divided Highways* (New York: Penguin Books, 1997).

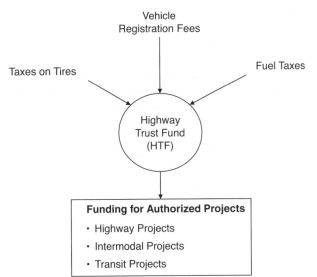

Figure 12.5 Structure of a Highway Trust Fund. Money collected from various fees and taxes is used to fund authorized projects, sometimes including transit or intermodal projects as well as highway projects.

supported by the HTF, the design and construction of the bridge can be begun. This is the basic approach used in the United States to create the Interstate Highway System and many state highways. State and city governments may also use tax revenues to support highway projects, and they can sell bonds to raise some of the funds required for construction.

Figure 12.6 shows the cash flows for financing a bridge, assuming that funds from the state's HTF cover half the construction expense while bonds are sold to cover the other half. Funds from the state's Department of Transportation (DOT) cover the annual operating costs of $70,000 and the interest payments of $20,000 on the bonds. After 30 years, the bonds will have to be redeemed, presumably by issuing new bonds.

If there are no tolls on the bridge, then the bonds are backed by credit of the state or local governments. If there are tolls, then the bonds would be backed by the expected toll payments. Once the bridge is constructed, it clearly belongs to a particular government agency, and that agency or another agency is responsible for maintaining and if necessary rehabilitating the bridge. If bonds were sold to pay for the bridge, then those bonds may affect the credit rating of the city or state. If the project

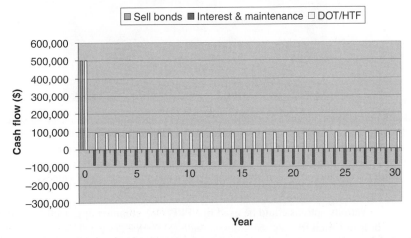

Figure 12.6 Cash Flows from the Sale of Bonds and Money from the State's HTF. These funds cover the initial costs of the project, plus continuing maintenance costs and interest payments on the bonds.

were funded out of tax revenues, then it may have been necessary to defer work on schools, water resource projects, or other government projects or activities. Also, some cities or states have limits on the total debt that they can incur. Borrowing to pay for the bridge therefore may limit their ability to borrow for some other purpose. Hence, there may be strong incentives to use tolls to finance the bridge.

Examples 12.1 to 12.3 provide a more detailed look at financing a bridge as a state highway project or as a toll bridge. The first question (Example 12.1) is whether the bridge is on a route that would qualify for funding as part of the state highway system. If not, the next question (Example 12.2) is whether high enough tolls can be charged to cover the costs of interest and operations. If so, then the question is whether the city should (a) build the bridge and collect the tolls, or (b) create a public-private partnership to build and operate the bridge (Example 12.3). The city could authorize the bridge and provide the connections to local roads while a private company could raise funds to pay for the construction costs. The private company would then charge tolls so as to earn a return on its investment. This approach works only if the expected value of the tolls is sufficient to provide an adequate return on capital (e.g., more than enough to cover the interest due on bonds and annual operating expense). The limit on the toll would be the value that users would place upon using the bridge. The toll could be quite high for a bridge that provided a much shorter or less congested route (in the previous chapter, Table 11.3 cites three tunnels and bridges with tolls greater than $10). If other bridges are located nearby, then the presence or absence of tolls on those bridges will affect what can be charged on the new bridge. With this approach to building the bridge, the costs would be borne by the private company, not by any public agency, so the construction of the bridge would not affect any public budgets or capital plans. On the other hand, if the bridge is built privately, then the design and capacity of the bridge, as well as the level of tolls charged, would be determined by the private company—possibly with an eye toward maximizing profits rather than maximizing public benefits. There could be intense public opposition to allowing a private company to charge what might be viewed as exorbitant tolls in order make excessive profits on an ugly bridge with limited capacity. Thus, there would likely be political pressure to retain some aspect of public control over the project.

Figure 12.7 Toll Booth on the West Virginia Turnpike. Most U.S. turnpikes were constructed by state governments before the Interstate Highway Program began. Tolls cover (1) the interest and redemption costs related to the bonds sold to finance the roads and (2) the costs of maintenance and operations. Tolls can also be used to manage congestion by charging higher rates either at peak hours or for an express lane. Some states have considered privatizing their toll roads to capture the value of their investments and use it either to finance future transport system improvements or reduce the state's indebtedness.

Various options could be used in a PPP. One common approach is for the public agency to seek bids in which the key variables include (a) bridge design, (b) tolls to be charged, and (c) length of time the private company will operate the bridge. The bridge would be owned (or eventually be owned) by the public agency, but it would be operated for an extensive period before it was turned over to the public agency. The public role could be to retain control over the size, design, location, and purpose of the bridge; to ensure that the tolls are reasonable; to provide some financial security for the private company by guaranteeing some sort of minimum annual payment if traffic volumes do not rise as expected; or to provide assurance that a competitive project will not be built within some specified period of time.

EXAMPLE 12.1   Can a Bridge Be Justified as a State Project?

A new bridge has been proposed that would reduce the travel time and cost between two rapidly growing regions in the rural portion of a state. The bridge, which is strongly supported by local officials, would create a route that will save each user an average of 10 miles and 15 minutes. The bridge is expected to cost between $40 and $60 million to construct, and annual maintenance and operating costs are expected to be $4 to $5 million. The route is currently used by 10,000 vehicles per day, and a preliminary study indicates that nearly all of this traffic would use the new bridge. About 80% of the vehicles on this route are automobiles, and nearly all of the rest are trucks. Local officials would like the state to fund the new bridge, which they would like constructed within the next 2 to 5 years. Can this bridge be justified as part of the state's transportation investment plan?

To answer this question, it is necessary to consider the state's transportation budget and the nature of other projects competing for state funds. The state has a prioritized list of transportation projects. The priorities are based on a formula that recognizes the benefits of reducing congestion, improving safety, reducing travel times, and promoting economic development. For this example, assume it is apparent that the proposed bridge would not have a very high priority. There are many projects involving bridges and road rehabilitation for much more heavily traveled routes in more densely populated areas of the state, whereas the existing route, though long, has very few accidents and essentially no congestion. In short, this is a low-priority bridge, and there is no immediate way to dramatize the need for it. Moreover, the state's highway trust fund is substantially underfunded, primarily because fuel taxes have not been increased for nearly 20 years. The trust fund can barely provide enough funds for high-priority projects, and medium-priority projects have been set back 10 or more years in the state's investment plan. In short, local officials cannot expect to have the state pay for the proposed new bridge.

EXAMPLE 12.2   Can the Bridge Be Financed with Tolls? (continuation of Example 12.1)

Would it be possible for a private company to build the bridge? If so, is that a good idea for the region? The economic value of the bridge is the time and cost saved by those who use the bridge to shorten their travel distances, plus additional benefits related to economic development that is likely to result from the increase in mobility provided by the new bridge.

In this example, assume that the major economic benefit comes from a reduction in travel expense for those who use the new bridge. The state DOT estimates the marginal cost per mile for driving an automobile to be $0.20, taking into consideration the cost of fuel and the wear and tear on the vehicle. The marginal cost per mile for driving a truck is on the order of $0.50. The average value of the time saved is on the order of $10 for automobile passengers and $20 for trucks.

Would a private company be able to finance the bridge by selling bonds backed by toll revenues? The first step is to estimate the annual revenue that must be raised by the tolls. If the bridge is financed by selling corporate bonds, the interest rate would have to be about 8%. The interest costs are thus be 8% of the construction cost, or $3 to $5 million.[4] The total annual cost, including maintenance and operations as well as interest, will therefore be $8 to $10 million.

The next step is to estimate the potential annual revenue. The toll can be no higher than the economic benefit of using the bridge. Using the DOT cost numbers, the average benefits per user can be estimated:

$$\text{Auto Benefits} = 10 \text{ miles}(\$0.20/\text{mile}) + 0.25 \text{ hours}(\$10/\text{hour}) = \$4.50$$
$$\text{Truck Benefits} = 10 \text{ miles}(\$0.50/\text{mile}) + 0.25 \text{ hours}(\$20/\text{hour}) = \$10$$
$$\text{Weighted Average Benefits per Vehicle} = 0.8(\$4.50) + 0.2(\$10) = \$5.60$$

If 10,000 vehicles used the bridge per day, the annual economic benefits would be as follows:

$$\text{Annual Benefits} = \$5.60/\text{veh.}(10,000 \text{ veh.}/\text{day})(365 \text{ days}/\text{yr}) = \$20 \text{ million per year.}$$

---

[4] If the construction cost is at the low end of the estimates, then the annual interest will be 8% of $40 million; at the high end of the estimates, the interest will be 8% of 60 million. These are all estimates, so all that can be said is that the interest payments are likely to be $3 to $5 million per year.

In other words, the annual economic benefits appear to be at least double the annual costs for interest and operations, even if the bridge costs are at the high end of what is anticipated. A toll of $3 for automobiles and $6 for trucks would be sufficient to cover annual costs. Moreover, since the regions served by the bridge are rapidly growing, traffic volumes and toll revenues can be expected to rise. Thus it does appear to be feasible for a private company to build the bridge using money raised by selling bonds and paying the interest on the bonds with tolls that users would be willing to pay.

EXAMPLE 12.3   Should the Bridge Be Built as a PPP? (continuation of Example 12.2)

Returning to the same bridge example, is there likely to be public opposition to allowing a private company to construct the bridge? If so, would the project be better structured as a PPP?

First, there could well be public outcry against allowing a private company to own and operate the bridge. The analysis has shown that a toll of $3 per car and $6 per truck would be more than sufficient to cover the 8% interest rate that the private company would pay on its bonds. However, the same analysis shows that the toll could be nearly 50% higher and still attract most of the traffic. If the bridge is as critical as the local officials believe, and if the region continues to grow as expected, then traffic volume—and toll revenues—can be expected to increase substantially over the life of the bridge.

The logic for public involvement is that interest costs could be lowered and that tolls could be controlled. A regional authority could be created that would own the bridge, and this authority could seek a partner or partners to construct and operate the bridge. With public backing, it would be possible to get lower interest rates by selling tax-free municipal bonds to fund the project. Even if the regional authority were unwilling or unable to sell bonds to finance the project, that agency could still seek bids for constructing and operating the bridge. It could also stipulate that the bridge (and the toll revenues) would revert to the regional authority after a period of 20 or more years.

12.4   DETERMINING HOW MUCH TO INVEST

To analyze a potential PPP, each party uses the same tools and techniques it uses for any other project. Each party has to estimate the cash flows associated with the project, and the public agency also has to estimate the expected economic, social, and environmental costs and benefits.

The private partner has to be satisfied that its NPV is likely to be positive or that its internal rate of return is likely to be acceptable, based on reasonable assumptions about the factors that will influence the outcomes of the project. The private partner is not interested in increasing its investment in order to allow greater public benefits, but may be willing to agree to a modified project so long as the public sector agrees to pay for any added costs.

Likewise, the public partner has to be satisfied that the benefit-cost ratio is adequate, based on its perception of the possible outcomes of the project. Both the benefits and costs can include nonfinancial factors such as improvements in congestion or air quality or access to public services. They can also include the economic benefits for the regions, such as job creation or the impact on average income or gross regional product.

Each partner will be able to determine the maximum amount they are willing to contribute to the project; if the combined amounts they are willing to pay exceed the cost of the project, then it is feasible. The parties will naturally try to find an arrangement whereby their actual contribution is less than what they are willing to pay; and this process could take some time. Note that it is not necessary for one party to explain its reasoning or provide its assumptions about unit costs or expected outcomes to the other party. The negotiations can proceed based on what each partner is willing to contribute and what arrangements are made for sharing the risks associated with the project.

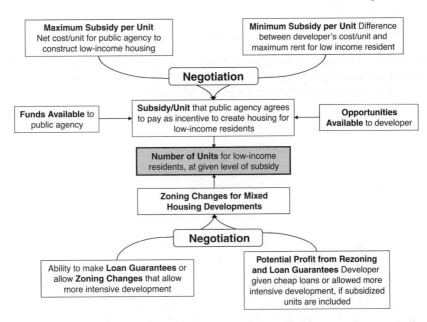

Figure 12.8 Key Factors in Negotiating for Low-Income Housing.

Example 12.4 shows how public agencies might negotiate with private to create more housing that is affordable for low-income residents. The key factors are as shown in Figure 12.8. From the public agency's perspective, there are four key issues:

1. The amount of money available for constructing or subsidizing housing for low-income residents
2. The cost per unit of constructing public housing for low-income residents, which places an upper limit on the subsidy per unit that the agency is willing to pay
3. The feasibility of changing zoning rules to allow more intensive development, which will determine how much of an incentive can be offered to developers
4. The ability to guarantee the developer's loans if the developer agrees to reserve a certain number of units for low-income residents

From the developer's perspective, the key factors are as follows:

- The cost per unit for construction, which must be covered by the combination of rent from the low-income occupant and the subsidy from the public agency. This determines the minimum subsidy the developer will accept.
- The ability to make more money if the developer is allowed to construct more units on a given plot of land, which would allow the developer to include more subsidized units or to reduce the minimum subsidy required
- Whether or not the benefits of lower interest rates are sufficient to offset the lost rents if some units are reserved for low-income residents
- Other opportunities for development: is it worth the developer's while to negotiate, or should the company simply build what is currently allowed on the sites it owns, or should it consider projects in other locations?

Thus there are two aspects to the negotiation. First, if the minimum subsidy that the developer will accept is less than the maximum that the public agency is willing to provide, then there could be a negotiated subsidy per unit. Second, if it is feasible to modify the zoning or offer loan guarantees, it would be possible for the developer to undertake more profitable development in return for including more units for low-income residents. Example 12.4 works through a quantitative example of how the negotiations might proceed.

EXAMPLE 12.4   Public Incentives for Low-Income Housing

Public and private interests may come together to promote multiple objectives by allowing denser development in a suburban setting. Rezoning land for denser development can offer great opportunities for developers; requiring the new developments to serve social purposes, such as housing for families with low income, may be the grounds for a PPP.

Suppose that a developer is interested in constructing apartment houses in a suburban town where zoning currently allows only single-family housing on 1-acre lots. The developer has plans for constructing three buildings with 10 apartments each on a 5-acre site. The expected cost per unit is $150,000, and the developer plans to lease the units for $2,500/month. Operating expenses are expected to be $500/month per unit. The annual net income is therefore:

$$\text{Annual Net Income} = 10\,\text{units}(\$2,000/\text{month/unit})(12\,\text{months/yr})$$
$$= \$240,000\,\text{per yr}$$

Since the 10-unit building is expected to cost $1.5 million to construct, the expected return on investment (ROI) is expected to be $240,000/$1.5 million, or 16%. The developer's maximum acceptable rate of return (MARR) is 12%, so this is an attractive proposition. However, unless the zoning is changed, the 5-acre site can be used for only five single-family houses. Without the zoning change, the developer will have to sell the recently acquired site and seek development rights elsewhere.

The town is interested in creating housing that will be suitable for low- and middle-income families. The city planners thought it might be possible for the town to build low-income housing, which would be made available to town employees at a maximum rent of $1,000 per month. They found that the construction costs for a 5-unit building would be $160,000 per unit with monthly operating costs of $600 per unit. The net rent per month would therefore be just $400. The town could sell bonds with an interest rate of 4%, so that the annual interest cost per unit would be 4% ($160,000) = $6,400. The net rent of $400 per month or $4,800 per year would be insufficient to cover these interest payments. If the town went this route, it would have to include an additional $6,400 − $4,800 = $1,600 per unit in the town's budget, which it would prefer not to do. The town therefore approaches the developer about the possibility of allocating some of the units in the proposed apartments to low-income residents whose rent would be set at $1,000 per month.

The first question is whether there is some basis for a partnership. To answer this, we need to determine the maximum reduction in rent that the developer can accept while still earning an acceptable return on the project. With an MARR of 12%, the developer needs an annual return of $180,000 (12% of the $1.5 million investment), which is $60,000 less than the expected rent of $240,000 per year. Reducing the rent from the market rate of $2,500 per month to the desired rate of $1,000 per month would cause a loss of revenue of $1,500/month or $18,000/yr for each unit. Thus, even if it had to make three units available to town employees at the lower rent, the developer would still have an acceptable MARR:

$$\text{ROI with 3 low-income units} = (\$240,000 - 3 \times \$18,000)/\$1.5\,\text{million} = 12.4\%$$

If forced to decide between abandoning the project and accepting a project in which three units are reserved for low-income families paying lower rents, the developer will likely accept the deal. Of course, the developer is likely to say, "If you provide a subsidy of $1,500 per month per unit ($18,000 per year), you can reserve as many units as you like for low-income families."

The town will probably set $1,600 per year as the maximum subsidy they will consider, since they could build their own complex if they were willing to provide that level of subsidy. Thus they would be unwilling to provide anything close to the desired subsidy.

On the other hand, the town could perhaps offer something else. Suppose the state had approved legislation aimed at promoting the development of low-income housing by allowing the state to guarantee the interest on loans associated with constructing housing in which at least 25% of the units were reserved for qualifying low-income families. Under this legislation, the interest rate on the developer's loans would drop by 2% if the development qualified. If the developer had a loan of $1.5 million, a 2% reduction in the interest rate would be worth more than $30,000 per year. This would be equivalent to $10,000 per unit if three units were reserved for low-income families.

Is this enough to close the deal? Maybe, and maybe not. It depends on how much the developer wants to proceed and how aggressive the town is willing to be in considering the rezoning application. Conceivably, some residents in the town will prefer not to attract low-income families—and conceivably, others will be very supportive of initiatives that allow young families and public employees to live in the town. Another possible step would be to allow the developer to add a few more units to each building, thereby increasing net income and making the overall development more attractive.

12.5   COMPLEMENTARY STRENGTHS OF PUBLIC AND PRIVATE PARTNERS: TEMPE TOWN LAKE

The Tempe Town Lake is a 220-acre manmade lake built in the formerly dry portion of the Salt River that lies within the boundaries of Tempe, Arizona. The lake was initially opened to the public in 1999, and the main park areas were completed in early 2000. Private development of business, retail and residential space is essential to the financial success of the project, because the operations and maintenance costs, as well as a significant portion of the capital cost incurred by the City of Tempe, will be repaid by assessments levied on these developers. . . . The project's inducement of active recreation and rezoning for higher population density play a key role in its social value, economic value, and overall sustainability. . . . However, thirty to fifty years from now, the City of Tempe might begin to wish that it had considered water as a more scarce resource during its evaluation of the project. Still, even if the lake is drained, it would not be difficult to revert to a desert park scheme that would continue to bring business and visitors to the area for recreation and events.[5]

12.5.1   Project Description

The City of Tempe, Arizona, transformed an ugly dry riverbed into an urban oasis by creating a lake, surrounding it with walkways and parks, and rezoning surrounding areas to encourage higher-density development. The lake and the parks were completed by 2000, and substantial urban redevelopment commenced at the same time. The basic concept is illustrated in Figures 12.9 and 12.10. The parks attract large crowds for festivals, concerts, races, and general outdoor recreational activity. The riverbed previously served as an inadequate flood channel, and portions of it were used as a landfill site for industrial and residential waste. The project was funded by an innovative public-private partnership based on a scheme originally suggested in 1966 by a group of students working with Dean James Elmore of the Arizona State University's College of Architecture. More than three decades passed before a plan was agreed upon for this site, but land acquisition and construction finally began in 1996.

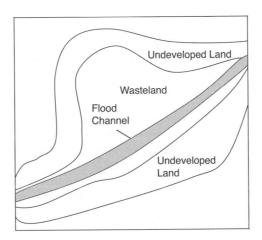

Figure 12.9 Schematic Illustrating Approach Used to Create Tempe Town Lake (not to scale, and not an accurate representation of the actual site). The city acquired a plot of land that included the flood channel, waste-strewn areas on either side of the channel, and underdeveloped land near the channel. The unsightly flood channel depressed land values and restricted recreational use of the area.

---

[5] This case study is largely based upon information in "The Tempe Town Lake: Oasis in an Urban Desert?" a term paper prepared for an MIT class (1.011 Project Evaluation) by Gwendolyn Johnson in May 2007. Johnson obtained her information from the City of Tempe's website, which has sections devoted to the history and management of the Tempe Town Lake (www.tempe.gov/lake).

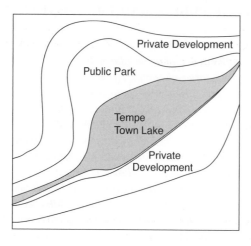

Figure 12.10 Schematic Illustrating Approach Used to Create Tempe Town Lake (not to scale). The city sold part of the site to developers, kept part of it for parkland, and created a 5-mile-long lake. The increase in property values enabled developers to agree to pay fees (on a long-term basis) that would cover the capital cost of cleaning the site and creating the lake plus the annual maintenance expenses.

The funding scheme was based on the following strategy:

- The City of Tempe acquired 840 acres of privately owned land and created the Tempe Town Lake Community Facilities District (CFD).
- The land was allocated to three major uses:
  ○ The lake would eventually be 5 miles long and cover 220 acres.
  ○ The city retained a portion of the land (400 acres) for parks.
  ○ The CFD sold a portion of the land (220 acres) to private developers who agreed to an operations and maintenance assessment (OAM) plus a lake assessment tax (LAT).
- The funds from the OAM would be used to pay for operations and maintenance for a period of 50 years.
- The funds from the LAT were designed to recover the capital costs of the project over a period of 25 years, assuming an interest rate of 5%.
- Some state and federal grants were received for the development.
- Tempe increased local sales and hotel taxes to help fund the project, since the project was expected to attract more visitors and tourists to the area.

The OAM assessment would ultimately be determined by allocating the actual costs of maintenance and operation to developers, based on their share of the developed property. At the outset, the OAM assessment was set at a much lower level, namely 20% of what each property owner's share would be if the development were completed.

12.5.2 Motivation

Tempe lies between Phoenix and Tucson. The Salt River was a frequent, major impediment to travel between these two cities during wet seasons, even though it disappeared during dry seasons. Major floods periodically made the river impassible and sometimes damaged bridges. Extensive flooding in the late 1980s destroyed key bridges and provided an impetus to do something to improve flood control in the area. Various state and local agencies realized that flood control, waste removal, and recreation would have to be considered together, so the decision to invest in flood control provided an opportunity to deal with the other two objectives.

Arizona State University's (ASU) ideas of creating a linear park along the Salt River had undergone various transformations since 1966. ASU had focused on the possibility for a greenbelt of parks and recreational facilities that would include Phoenix and the entire metropolitan area, not just Tempe. And there was even a proposal for a massive project to make Tempe an inland port.

In 1987, a county-wide initiative was proposed that would raise taxes to support a regional plan for flood control, waste removal, and recreation. When this plan was defeated at the polls, the City of

Tempe elected to proceed with a more limited plan using its own resources. Construction work on the lake began in 1997 and was completed in 1999; the parks were completed the following year. In 2003, the U.S. Army Corps of Engineers began to work with the city to help with cleaning up the Salt River and restoring habitat upstream from the lake.

12.5.3   Results

The cost of the project was approximately $60 million:

- Land acquisition: $11 million
- Planning: $5 million
- Construction: $45 million

A portion of the planning was covered by voter-approved taxes for the project; the rest of the planning and construction costs were covered by bonds with annual interest of 5%. The LAT was expected to raise $3 million per year once the private development was completed, and this tax would be more than enough to pay the interest on the bonds. Annual operating and maintenance costs were expected to be about $4 million per year, which would eventually be covered by the OAM assessment.

The public response to the Tempe Town Lake was resounding. The park attracted 100 major events per year and more than 2 million visitors. Large special events, numerous smaller events, and the opportunities for people to walk through and enjoy the park enhance the sense of community in Tempe.

The one discordant aspect of the Tempe Town Lake is its use of water, which could be viewed as extravagant for a desert region. Because of the dry environment, there is a great deal of evaporation from the lake plus a continual need for watering the extensive lawns within the parks. Together, the evaporation and irrigation use enough water to supply more than 5000 households, which amounts to perhaps 10% of the population of the greater Tempe area.

12.5.4   Lessons Learned

Tempe's experience shows how investment in public space can be financed through a combination of taxes and fees targeted to those who will most benefit from being close to the open space and those who will be attracted to the area because of the open space. Coordinating flood control, recreation, waste removal, and habitat restoration allowed tremendous improvements in what was previously a rather derelict portion of the city.

The decision to create a lake led to a continuing demand for water, which could become a serious issue for the region. Other options could have been considered that needed much less water, such as having a landscaped region surrounding a small creek or restoring the area to its original desert habitat.

The financing for the project has been successful and could serve as a template for similar investments in other cities. Figure 12.11 outlines the financing for Tempe Town Lake, and Table 12.2 summarizes the key elements of this project.

12.6   PUBLIC AND PRIVATE BENEFITS: THE SHEFFIELD FLYOVER, KANSAS CITY, MISSOURI[6]

The Sheffield Flyover increased the capacity and improved the performance of a major bottleneck in the rail network in and around Kansas City. At-grade crossing of high-density rail routes had not only led to train backups but also caused extensive delays to highway traffic when trains blocked

---

[6] This case study is based on material prepared by C. D. Martland for the National Cooperative Highway Research Program as an example of public-private partnerships. Additional information on this case study and other PPPs is available in Joseph Bryan, Glen Weisbrod, Carl Martland, and Wilbur Smimth Associates, Rail Freight Solutions to Roadway Congestion—Final Report and Guidebook, NCHRP Report 586, Transportation Research Board, (2007).

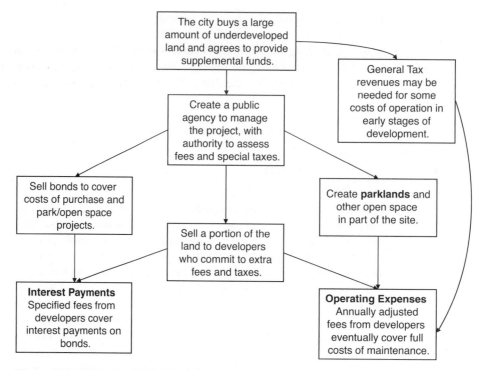

Figure 12.11 Financing for the Tempe Town Lake Project.

local streets. An innovative public-private partnership helped secure funding for and ensured the successful implementation of the flyover. This project illustrates how public agencies can work with the private sector to expand capacity and improve the performance of the local transportation system, with benefits to the region and the nation as well. Due to the success of the Sheffield Flyover, the railroads and public agencies decided to build a second major flyover in Kansas City to secure similar benefits.

Table 12.2   Key Elements of the Tempe Town Lake Project

Reason for PPP	Complementary strengths of public and private sectors: • Public responsibility for waterways, flood control, and parks • Private capabilities for real estate development
What was done	Transformed underdeveloped, unsightly land into parks and open space, thereby increasing value of the surrounding land and promoting development
Benefits	Economic: developing a significant portion of the region Social: creating open space and parkland that attracts thousands of visitors Environmental: cleaning a 5-mile section of the flood channel and creating parks
Financing	Value capture: The city purchased the entire parcel, kept portions for the lake and the parks, and sold the rest to developers. Because the public investments would increase the land value, developers were able to agree to higher taxes and fees that would cover the costs of creating and maintaining the parks and lake.
Major risks	• Development might not proceed rapidly enough to cover the costs of the project. The extra fees and taxes were therefore phased in over a period of years, and the city covered other costs. • Because of the growing demand for water in this desert region, keeping a lake might not be sustainable; however, the lake could be converted to a desert park.

12.6.1 Project Description

The project addressed a key bottleneck in the national rail system where the Burlington Northern Santa Fe (BNSF) main line crossed the Union Pacific (UP) and Kansas City Southern (KCS) main lines.[7] With 100 to 120 trains operating on the BNSF, 60 to 80 on the UP and KCS, and another 40 to 60 local trains operating in the area, this was described as the "third busiest railroad intersection in the country." Trains were inevitably delayed as dispatchers worked to route them through the bottleneck; the delayed trains blocked highway intersections for a mile or more. The resulting delays were especially difficult for trucks seeking to enter or exit a major industrial area hemmed in between the main lines.

By constructing a flyover, it was possible to eliminate rail and highway delays associated with train interference at the crossovers. The project covered nearly 3 route-miles almost entirely constructed on the Kansas City Terminal Railroad's right-of-way; it included a main bridge of 6740 feet and two other bridges of 890 and 150 feet. By double-tracking the flyover and keeping the existing tracks, it was possible to greatly increase the capacity of the intersection, improving flow of trains moving through Kansas City and also allowing better service to local rail customers. From the public's perspective, the most visible benefit was expected to be a reduction in delays at grade crossings. Transystems, a consulting firm involved in evaluating the project, estimated that 530 vehicle-hours would be saved daily for cars and trucks by elimination of grade crossings, based on the train volume, the average time that each train blocked a crossing, and the 4500 daily highway vehicle movements through the area. At $14/hour, this was estimated to amount to a savings of $1.85 million annually. In addition, with fewer trains and vehicles delayed in the area, emissions were expected to be sharply reduced.

Transystems did not provide details on the railway benefits, but indicated they would be approximately three times as great as the public benefits. This is borne out by a quick assessment of the benefits from reduced train delay. If 150–180 trains per day each saved 20 minutes in moving through this region (as estimated by Transystems), that would be a savings of more than 60 hours of train delay per day, or 20,000 per year. The cost per train-hour is commonly estimated to be on the order of $250/hour based on the hourly cost of equipment ownership plus the opportunity cost associated with the loads themselves. Hence the delay cost of an average 20-minute delay to these trains would exceed $5 million per year.

The project cost was $75 million. Raising the capital was a stumbling block for the railroads, even though they were willing to pay for the project on a continuing basis. Another problem was that construction would increase the assessed value of the property and therefore the property tax owed by the railroads. Various public agencies were interested in providing financial support, but there were barriers to using public funds. At one point, it appeared that a Federal Highway Administration (FHWA) loan would be approved to finance 25% of the project, based on the public's share of the project benefits. This loan possibility fell through when trucking interests objected to the use of highway trust money for rail projects. State agencies were interested, but were prohibited from investing in a private sector project.

The financing problem was resolved by creating a transportation corporation (T-CORP), a quasi-governmental entity that can be created under Missouri law and can receive highway funds. A T-CORP can issue 20-year, state tax exempt bonds to fund transport projects, and it receives real estate tax abatements. A T-CORP is represented jointly by the project owner and the Missouri Highway Department; the T-CORP owns the land and the project until the loans are paid off, at which point the land goes back to the previous owners. The net result for the Sheffield Flyover was that the T-CORP issued the bonds, the U.S. DOT provided a letter of credit, and the railroads agreed to repay the loans. In addition to benefiting from low interest rates, the corporation enjoyed a property tax abatement worth $1.4 million per year (estimated by Transystems as being nearly 20% of the annual amortization costs).

The project required a few other elements of cooperation. The project was supported by the Heartland Freight Coalition and the Greater Kansas City Chamber of Commerce, as well as the

---

[7] Transystems Corporation, "Kansas City Terminal Railway Flyover Project: A Public/Private Cooperative Success," Presentation to Financing Freight Transportation Improvements, Federal Highway Administration (FHWA) Conference, St. Louis, Missouri, April 29, 2001.

Missouri DOT, the FHWA, and the railroads. Some public land was needed for the flyover, and a land swap was arranged with the city. While the project was under way, work was done to modernize or coordinate 14 different utilities serving this industrial area. Also, a portion of one of the city streets had to be reconstructed and temporarily closed to enable completion of the flyover.

12.6.2   Motivation

Kansas City is the second-largest rail freight hub in the country after Chicago. It is served by four Class I railroads: Burlington Northern Santa Fe (BNSF), Union Pacific (UP), Norfolk Southern (NS), and Kansas City Southern (KCS), while the Gateway Western provides an independent route that reaches the CSX rail line in St. Louis. The Kansas City Terminal Railroad provides local switching services, and various short line and switching railroads serve the area. The metropolitan area has an intricate network of classification yards, industrial support yards, and through tracks. A major problem within the region is that major rail routes intersect in Kansas City, resulting in extensive delays to both trains and highway vehicles.

The Mid-America Regional Council documented the importance of rail to the region.[8] Rail handled just over half of the freight tonnage moving through Kansas City. Over 80% of the rail freight was moving through the area, and this traffic amounted to 150 million tons in 2000. Much of this traffic was intermodal (i.e., containers or trailers that are transferred by truck between customers and intermodal terminals and moved by rail between the intermodal terminals). The BNSF route from Los Angeles to Chicago, which handles 1.6 million containers and trailers annually, goes right through Kansas City. Another 23 million rail tons were received by Kansas City industries, while about 11 million tons were shipped out by Kansas City shippers. Rail's market share varies greatly with the type of movement. Rail accounted for approximately two-thirds of the freight moving into or though the region; truck accounted for all of the intra-regional freight and more than three-fourths of the outbound freight. The rail share versus truck was growing for through traffic, stable for traffic inbound to the region, and declining for outbound traffic.

During the 1990s, it became increasingly evident that various national trends in rail freight traffic were disrupting both rail and highway traffic in the city. Rationalization of the network was concentrating more traffic on fewer routes, leading to congestion and interference within the rail network as well as increasing delays to highway traffic. Trains waiting for authorization to proceed through an intersection often blocked automobiles and trucks at grade crossings, frequently for 20 minutes or longer. Mergers, traffic growth, and shifts in freight traffic patterns required greater capacity along key rail routes within the city, but the bottlenecks where key routes intersected threatened to limit growth of rail traffic.

The project therefore was seen to have both local and national significance. Grade crossings and local air quality were the obvious benefits for the local area. However, the movement of 1.6 million trailers and containers by train rather than by highway was recognized as much more than a local benefit since these shipments might otherwise be moving on the highways not just through Kansas City but also through many other cities across the country. Expanding the capacity of such an important rail hub was also of major significance for the national rail system. The 150 million tons of freight moving through the rail hub represented at least 7 million truck shipments, including the intermodal trailers and containers mentioned already. This is a good illustration of a network-level investment in which important improvements in system performance help retain existing customers and attract new customers.

---

[8] "Transportation Outlook 2030: Metropolitan Kansas City's Long-Range Transportation Plan," Mid-America Regional Council, October 2002.

12.6.3   Results

The project achieved its goals. Following the opening of the new facility in 2000, travel times for trains dropped from 40 to about 15 minutes.[9] This improvement in train efficiency translated directly into the hoped-for reduction in grade-crossing delays and air quality. The institutional structure also worked well enough to be expanded. In February 2002, BNSF announced that a second major flyover would be constructed to provide grade separation at the intersection of two of their main routes and improve access to Argentine Yard, their major freight facility in the region.[10] The "Argentine Flyover," which would cost about $60 million, was initiated using the same institutional arrangements as the Sheffield Flyover.

The project received broad recognition as an outstanding example of public-private cooperation. The Intermodal Advisory Task Force of the Chicago Area Transportation Study (CATS) identified this project as one of the best examples of "holistic" planning "involving major transportation industries, the political decision-makers, plus the industries (shippers and receivers, essentially) that stood to benefit."[11] Gerald Rawlings, Freeight Manager for CATS, noted the key roles played by the Chamber of Commerce and the Mid-America Regional Council, which funded preliminary freight studies and were able to focus interest on and achieve a consensus for the flyover and a few other critical projects.

12.6.4   Lessons Learned

Table 12.3 summarizes the key elements of this project. In this case, the train volumes were so high and the benefits so large that it was easy for local parties to agree that the benefits justified the costs of the project. At intersections of busy rail lines, trains back up and clearly block the local highway network. These local costs were easily identifiable and large enough to justify public participation, even though the national significance of the project is what motivated FHWA's interest. The benefits were equally clear to the railroads, as were the costs to operations if action were not taken.

This project provides various lessons for promoting public-private partnerships that seek to enhance a system in order to achieve both local and national benefits:

- The involvement and support of the local interests is essential.
- The willingness of the various partners to work together and to negotiate ways to share the costs is essential.
- Federal, state, and local cooperation can provide innovative financing mechanism and enable a complex project to be completed quickly.
- Environmental benefits may provide part of the story in support of the project, but the financing may need to be based on a clear understanding that the system improvements—both local and national—translate directly into enough cost savings to justify the project.
- The national scope of the project may add to the story and motivate federal involvement, but it may not directly affect the local assessment of the project. In other locations, where the local effects are not so evident, it may be necessary to make a stronger case for the indirect and national benefits in order to secure local support and a broader base of funding.
- Once a coalition is formed to identify, finance, and implement projects that fulfill clear needs, then that coalition can quickly move on to additional projects.

---

[9] Brian Cookson, "Railway Putting Flyover on Track," *Business Journal of Kansas City*, 5 November 2001 (from the 1 November 2001 print edition).

[10] BNSF News Release, "Second Flyover Bridge to Streamline Rail Traffic Through Kansas City." Kansas City, Kansas, February 15, 2002.

[11] Gerald Rawlings, comments posted on "National Dialogue on Freight" website (March 14, 2003). http://www.icfhosting.com/fhwa%5Cnfd_disc.nsf/CategoryAllThreadedweb/de3710dc6997c8fd85256bb50071a7ce?OpenDocument

Table 12.3   Key Elements of the Kansas City Flyover Project

Reason for PPP	Project was justified by the combination of public and private benefits, and it required innovative, cooperative relationships for implementation.
What was done	Rail lines and some public property were conveyed to a special public agency that constructed flyovers that eliminated some of the main at-grade rail-rail and rail-highway crossings in Kansas City.
Benefits	Financial: reducing the operating costs for the railroads Economic: reducing delays to highway traffic at grade crossings
Financing	Low-interest bonds and property tax abatement: The project was financed with low-interest, tax-free bonds that would be paid back by the railroads in proportion to their use of the facility. By transferring the land to the city, the railroads would not have to pay property tax.
Major risks	• Would construction costs be higher than predicted? • Would traffic volumes be high enough to produce enough revenue to cover the annualized costs of the project? This risk was thought to be minimal, since this region was the second-busiest rail hub in the country.

12.7   MAXIMIZE ABILITY TO UNDERTAKE PROJECTS: TORONTO'S HIGHWAY 407[12]

Toronto built a state-of-the-art highway and then leased it for 99 years to a private company for double what it cost to build. The consortium committed to extend the highway at no cost to the city. This example shows how one city was able to capture the value of its infrastructure while moving the burden for creating new infrastructure to the private sector.

In 1994, construction began for a 69-km, fully automated toll road that would create another through route in one of the most congested highway corridors in Canada. The Province of Ontario financed the C$1.5 billion project by selling taxable, general obligation bonds. The road opened in 1997, attracting 200,000 trips per day and annual revenues of C$70 million. Two years later, Ontario decided to privatize the road by leasing it to a private company that would agree to construct the planned 39-km extension, which was expected to cost C$0.5 billion.

The province asked for bids on the highway, assuming a 99-year lease. Since the value of the highway depended entirely upon the tolls that could be collected, certain restrictions were placed on tolls. The tolls were initially limited to 10 cents per mile for automobiles, with an increase to 13 cents per mile over 15 years plus adjustments for inflation. Tolls were two to three times higher for trucks. Beyond 15 years, there would be no limit on tolls so long as peak-hour traffic volumes exceeded 9000 vehicles per hour. No limit was placed on the potential ROI for the successful bidder.

The winning bid was submitted by a consortium that agreed to a purchase price of C$3.1 billion for a 99-year lease, which was slightly more than double the construction cost of the first section of the highway. The company was required to maintain nearly $400 million of working capital, and it was required to keep C$775 million equity in the project. These restrictions limited the amount the company could borrow, thereby limiting the risk of bankruptcy. The expected return on their investment was 11%.

12.7.1   Lessons Learned

This project demonstrates a way for governments to tap the enormous value of public infrastructure. Ontario had built the road as part of its ongoing program of highway construction. It built the road as a toll road on the assumption that the tolls would be sufficient to cover the interest on the construction bonds, along with the expenses of operating the highway. Ontario wanted to sell the road to a private

---

[12] This section is from the *Traffic Technology International Journal*, June–July 1999.

Figure 12.12 Downtown Toronto, September 2009. Due to a strong economy, rapid influx of immigrants, and opportunities to redevelop old industrial properties along the waterfront, construction is booming in Toronto. As a result, arterial streets and highways of Toronto carry some of the heaviest traffic volumes in Canada. Efforts to relieve congestion included the construction of Route 407, which was intended to divert through traffic away from the part of the city shown in this photograph. A close look reveals traffic stalled on the elevated highway at center right of the photo, indicating that congestion remains a concern. On the tracks shown in the foreground, Toronto operates a busy commuter rail service as another approach to relieving highway congestion.

company largely to move the debt associated with the road off the public accounts, thereby protecting or improving the province's credit rating. In fact, the deal did much more than take over the debt associated with the highway: it provided an extra C$1.5 billion to the treasury, and the winning bidder agreed to finance the next section of the road as well. In the United States, politicians in several states including Pennsylvania and New Jersey have considered similar projects, in which the rights to operate existing toll roads would be sold at auction. If the private operator is allowed to charge much higher tolls, then the toll roads could have a value far in excess of the original costs.

The key issues in this deal concern the level of tolls, future volumes of traffic, and the role of this highway in the regional transportation system. By charging tolls, the highway operator captures some of the general economic benefits of the highway. Conceivably, a private company could maximize its profit by charging very high tolls, thereby enabling this highway to operate with little or no congestion

Table 12.4 Key Elements of Toronto's Highway 407 Project

Reason for PPP	Public desire to capture value of its infrastructure by selling a long-term lease of a major asset
What was done	An existing toll road was leased for 99 years to the high bidder, who also agreed to extend the road.
Benefits	Financial: reducing debt in the province; anticipated long-term profits for the highway operator
Financing Major risks	Toll-based bonds: the project was financed by selling bonds backed by the tolls. • Would construction costs be higher than predicted for the new portion of the road? (This risk was borne by the private operator). • Would highway traffic volumes be high enough to produce enough toll revenue to cover annual interest payments on the bonds? (This risk was borne by the private operator; in some toll-road agreements, the public sector guarantees a minimum level of payments.) • Would the highway fulfill its intended role in the region? (This risk was borne by the public sector; the exact requirements regarding toll levels will determine the role played by this highway over the 99-year lease.)

while traffic that could easily go on this route would shift to other, much more congested highways. Selling parts of the system could make it more difficult to manage the overall highway network effectively.

12.8 PUBLIC INVESTMENT TO STIMULATE THE ECONOMY: PROVINCE OF NEWFOUNDLAND AND LABRADOR'S INVESTMENT IN OFFSHORE OIL EXPLORATION

In 1979, after 16 years of research and exploration, oil was discovered in the Hibernia oil field located off the coast of Newfoundland, Canada.[13] The high prices of oil at that time justified looking for oil offshore, even though the costs of offshore drilling are very high. Political disputes between the federal and provincial governments delayed exploitation of the oil field for more than a decade as both governments claimed jurisdiction over the area. Although the Supreme Court of Canada ruled in favor of the federal government in 1984, the Canada-Newfoundland Atlantic Accord Act of 1987 and the Canada-Newfoundland Atlantic Accord (Newfoundland) Act of 1990 created a framework for sharing the regulation of offshore drilling under the supervision of the Canada Newfoundland Offshore Petroleum Board.[14] With the political battles settled, construction finally began in 1991 on an enormous platform and related infrastructure for extracting oil and accommodating 185 workers. The structure was completed in 1997 and began production in 1998. Located in 80 m of water, the platform rises 244 m above the ocean floor. Since the oil field was located in a region known as "Iceberg Alley" (approximately 200 miles southeast of St John's), it was necessary to construct a 1.2-million-ton structure that could withstand a collision with a million-ton iceberg—an event expected to happen once in 500 years. The risk of such a collision meant that it was not possible to construct a much less massive, and much cheaper, floating platform.

The facility was designed to extract 150,000 barrels of crude oil per day or 54 million barrels per year, and it was believed that it would take 20 years to extract all of the recoverable oil. The facility was constructed so that it could then be decommissioned and moved to another site.

The project was undertaken by a consortium of five oil companies along with a new government agency, the Canada Hibernia Holding Corporation, that owned an 8.5% share in the project. The cost of the project was initially estimated at $5.8 billion plus another $2.1 billion for continuing exploration

Table 12.5 Key Elements of Newfoundland's Investment in the Hibernia Oil Field

Reason for PPP	Public officials in Newfoundland wanted to promote economic activity that provides jobs and income to a depressed province. Private companies needed additional capital to complete their project.
What was done	The province joined a consortium of oil companies in funding an offshore oil-drilling platform. By becoming a partner rather than just loaning money, the province anticipated sharing in profits that were expected to rise over the life of the project.
Benefits	Economic: providing more jobs and higher income for residents of Newfoundland Financial: realizing profits from the sale of oil, which would go to the province as well as the oil companies
Financing Major risks	The province purchased stock in the consortium. • Would the drilling platform be able to survive in "Iceberg Alley"? (This risk was mitigated by careful design of the structures.) • Would oil prices be sufficient to cover the annualized costs of the project? (Oil prices had been highly volatile, but the expected trend was upward).

---

[13] Much of the information in this section is based on a term paper prepared for my class on Project Evaluation by Michelle Vidal, David Rabago, and Armand deBardelaben ("Hibernia Offshore Oil Drilling Platform," May 13, 1997).

[14] G. E. Bridges & Associates, "Due Diligence Issues Report," British Columbia Offshore Oil and Gas Socio-Economic Issue Papers, Royal Road University, Victoria BC, Canada, May 2004.

of new drilling sites within the Hibernia oil field. The annual operating costs were estimated to be $325 million while annual revenues were expected to be $1 billion based on a market price of $20 per barrel. The net present value over the 20-year life was found to be $5.8 billion using a 12% discount rate.

Government participation in the project was designed to produce income for the Province of Newfoundland and Labrador, one of the poorest provinces of Canada. The construction and operation of the site plus additional energy-related jobs provided a major boost to provincial economy. There were 5000 construction jobs, and the growth in the energy sector boosted the region's economic activity by 3.1% during the first year of operation.

While oil prices fell below $20 per barrel soon after Hibernia began producing oil, prices subsequently rose well above $20/barrel and reached a peak of more than $140 per barrel in 2008.

12.9 SUMMARY

Principles of Public-Private Partnerships

The public and private sectors have differing types of objectives and complementary strengths in defining, financing, constructing, and managing projects. Thus, in many situations projects are likely to be better designed, constructed, or operated if the two sectors work together. The public sector is structured to allocate resources among competing public needs, whereas the private sector is structured to achieve financial objectives. The public sector has the ability to raise funds via taxation or by issuing bonds backed by specific revenue sources (e.g., tolls or user fees), but for financial or political reasons may have limits on the amount of money that can be raised for major infrastructure projects. The private sector has greater ability to attract investors for large projects that may be risky, but that have potential for high profits. The public sector has the power to impose and modify many of the laws and regulations that affect development, including zoning, other land use regulations, and environmental and social regulations. The public also has the power of eminent domain whereby a public agency may be able to force landowners to sell their land so that it can be used for a public purpose, such as a road, a dam, or a reservoir.

The public sector may have agencies devoted to certain types of infrastructure, such as highways and waterworks, but they may not have experienced employees capable of dealing with major new projects in important areas. Private companies with specialized skills will be able to apply those skills throughout the world. Private companies may also have greater experience and therefore greater efficiency in undertaking certain types of projects.

Creating a Framework for a Partnership

Negotiation will be necessary to ensure that all parties to the agreement believe that their costs will justify the costs and the risks they incur. The private sector will participate only if it perceives sufficient financial benefits. The public sector may in part be concerned with financial matters but must be sure that the projected benefits—whether they are financial or even economic benefits—are sufficient to justify whatever social and environmental costs and benefits may be incurred. The negotiations will consider design, financing, responsibility for construction management, responsibility for operations, and ownership. The risks include the usual risks associated with construction of and demand for a project. In addition, there will be a question as to the extent to which the public and the private sector share the risks of financial failure: to what extent, if any, will the public sector be responsible for paying off bonds? The public sector will be able to accept long-term social and environmental benefits as justification for their participation, but the private sector will require more immediate financial benefits.

Determining How Much to Invest

The analysis of a potential PPP will be approached by each party using the tools and techniques they would use for any other project. Each party will have to estimate the cash flows associated with the project, and the public agency will also have to estimate the economic, social, and environmental costs and benefits expected. The private partner will have to be satisfied that its NPV is likely to be positive or that its internal rate of return is likely to be acceptable, based on reasonable assumptions about the factors that will influence the outcomes of the project. Likewise, the public partner will have to be satisfied that the benefit-cost ratio is adequate, based on their perception of the possible outcomes of the project. Each partner will be able to determine the maximum amount that they would be willing to contribute to the project; if the combined amounts that they are willing to pay exceed the cost of the project, then it is feasible. The parties will naturally try to find an arrangement where their actual contribution is less than what they are willing to pay, a process that could take some time.

Types of Public-Private Partnerships

This chapter identifies and provides examples for four types of situations that are well suited to PPP:

1. *The project requires the complementary strengths of the public and private sector.* The City of Tempe, Arizona, acquired a large tract of land adjacent to a rather derelict seasonal streambed. By cleaning up the site and creating a lake and a large public park, Tempe was able to increase the value of the land while creating open space with multiple attractions for both residents and tourists. The project was funded by selling a portion of the land to developers, who agreed to pay extra assessments on their property tax that would be used to pay the costs of financing and maintaining the park.

2. *The project provides both public and private benefits, and only the combined benefits are large enough to justify the project.* In Kansas City, the intersection of three major railroad lines caused extensive delays on the railroads and the nearby grade crossings. The solution was to elevate one of the rail lines on a flyover, which eliminated many grade crossings and eliminated the conflicts among the railroads. To allow public funds to be used on the project, the railroads turned the route over to a newly formed public corporation. The corporation could sell tax-free bonds at a much lower interest rate than was available to the railroads. The railroads were willing to pay a per-car fee that would be used to help pay off the bonds. The public's share of the project expense was justified by the reduction in delays to highway traffic, which was viewed as an important social benefit, even though it had no direct financial impact.

3. *The public sector wishes to maximize its ability to undertake projects by gaining access to the financial markets.* The City of Toronto, Ontario, sold one of its toll roads to a private company for a price that exceeded the road's construction cost. The private company agreed to extend the toll road at its own expense. The cost of the new road and the purchase price were financed largely by bonds that could be sold because the tolls were expected to be large enough to cover the interest payments.

4. *The public sector decides to assist new industries.* The government of Newfoundland decided to invest in offshore oil drilling to help create what they hoped would be a much-needed boost to the region's economy.

ESSAY AND DISCUSSION QUESTIONS

12.1 What are the most important principles for establishing a PPP?

12.2 Consider a government agency that seeks bids for a PPP wherein the private partner will construct a major bridge, pay off its debt using tolls, and eventually turn over the bridge to the government agency. The government would like to guarantee that tolls will not be excessive, and the private partner would like to guarantee that it can in fact pay the interest on its bonds (and avoid bankruptcy). The government has provided a traffic study that shows expected traffic growth of 2% per year over the next 20 years, and it has indicated that bidders should consider the benefits of this traffic growth when deciding what toll levels they will include in their bid. What are the risks associated with this project? How are the risks shared between the government agency and the private company? What scenarios should a private company consider in formulating its bid? What scenarios should the government agency consider as it structures the bidding process and evaluates proposals?

12.3 Search online to find the current status of one of the case studies included in this project. Have any unforeseen problems been encountered? Are the projects still viewed as successful? Have significant modifications been necessary in the partnerships?

PROBLEMS

12.1 A state currently subsidizes ferry service to an island at a cost of $300,000 per year. To avoid the delays associated with a ferry, residents of the island have long lobbied for a bridge. A study has shown that a bridge would cost about $5 million to construct, incur annualized maintenance costs of about $500,000, and have an expected life of at least 70 years. The ferry service, which operates hourly, handles about 500 cars per day, and the average charge per vehicle is $4. If there were a ferry service, then the average time savings would be about 20 minutes per trip. Assume that the average value of time is $12 per vehicle, and assume that the number of trips would increase by 20% if the delays associated with the ferry were eliminated. Are the economic benefits related to the time savings high enough to justify the cost of building the bridge? Assume that the DOT uses a discount rate of 8% in its studies.

12.2 Consider the same bridge discussed in the previous problem. Suppose the DOT allowed a private operator to construct a bridge and pay for it with toll revenues. If the bridge were financed with corporate bonds paying 9% interest, what toll would the private operator have to charge to cover costs, assuming that daily traffic was 500 vehicles per day?

a. $2.50
b. $3.00
c. $3.50
d. $4.00
e. $4.50
f. $5.00

12.3 Consider the same bridge as in the previous two problems. If the bridge were constructed as a PPP, what would be the best role for each partner? How would you approach negotiations if you were working for the DOT? If you were working for the company? What are the key points to be negotiated?

12.4 A Turnpike Authority is in charge of 200 miles of limited access highway. Tolls on this road average 2 cents per mile for cars and 5 cents per mile for trucks. These tolls were initially designed to pay interest payments on construction bonds and to cover the continuing costs of maintenance and rehabilitation. These tolls are much lower than what is charged for using turnpikes on neighboring states (typically 8 cents per mile for cars and 20 cents per mile for trucks). However, according to the state law governing the Turnpike Authority, tolls can be raised only to cover the necessary costs of operating the road; toll revenues from the turnpike cannot be used to cover the costs of other state roads or to provide a revenue source for other state

activities. The average traffic volume on the turnpike is 150,000 vehicles per day, which consists of 120,000 cars traveling an average distance of 30 miles and 30,000 trucks traveling an average of 50 miles. The governor is considering a proposal that would lease the turnpike for a long-term period to a private company, which would be allowed to charge higher tolls. A company has offered to pay $1 billion to lease the turnpike for a 50-year period, assuming that it would immediately be able to increase tolls to the levels charged by neighboring states and subsequently increase the tolls to keep pace with inflation.

a. Assuming that the existing tolls are what is needed to cover the costs of maintaining the highway, what will be the annual net revenue (total tolls minus operating costs) for the private company?

b. What will the company's ROI be for its initial investment of $1 billion?

c. What would the increase in toll revenue be worth at the state's discount rate of 7%?

d. What would you advise the governor to do?

12.5 A city and a railroad would like to share the expense of constructing several bridges that would eliminate the congestion at rail–highway grade crossings while allowing trains to operate at higher speeds through the city. The proposed project would cost $9 million. Once completed, the project would reduce travel time by an average of 1 hour for the 10 daily trains operating on this line and eliminate delays averaging 6 minutes for 300 automobiles and 50 trucks per day. Assume that the average value of the time saved is $200 per train-hour, $10 per automobile-hour, and $20 per truck-hour.

a. How much should the railroad be willing to pay to help fund the project, assuming its MARR is 10%?

b. How much would the city be willing to pay toward the project, assuming that its MARR is 8%?

c. Is a PPP necessary? Why or why not?

d. If you were trying to facilitate a PPP, how would you recommend sharing the costs?

12.6 Sketch the cash flows associated with one of the case studies presented in this chapter. Show the cash flows from the perspective of both the public and private partners involved in the project:

a. Tempe Town Lake
b. Kansas City Flyovers
c. Toronto's Highway 401
d. Hibernia Oil Drilling Platform

12.7 Tempe Town Lake:

a. Using the information presented in the case, estimate the costs and benefits of the project to the public and to public agencies. Do you believe that this project provided sufficient public benefits to justify the public expenses?

b. Discuss how public participation made the project more attractive for developers.

c. Suppose other cities wanted to consider a project like this to reduce the risks associated with flooding while creating park space and encouraging economic development. Develop a spreadsheet that you could use to estimate the public and private costs and benefits as a function of the key variables affecting costs and benefits.

12.8 Kansas City Flyover:

a. Using the information presented in the case, estimate the costs and benefits of the project to the public and to public agencies. Do you believe that this project provided sufficient public benefits to justify the public expenses?

b. Using the information presented in the case, estimate the amount that the railroads would be required to pay per carload or per ton in order to pay 4% interest on T-CORP bonds.

c. Discuss how public participation made the project more attractive for the railroads.

d. Suppose other cities wanted to consider a project like this to alleviate the highway congestion related to rail–highway grade crossings and to promote more efficient rail operations. Develop a spreadsheet that you could use to estimate the public and private costs and benefits as a function of the key variables affecting costs and benefits.

CASE STUDY: POSITIVE TRAIN CONTROL

Various officials in the U.S. Department of Transportation have urged the railroad industry to adopt a system known as positive train control (PTC). To implement PTC, all trains must be equipped with digital communications, onboard computers, and onboard global positioning systems (GPS). Using data from the GPS system, the computer on the train can determine where it is and how fast it is going. Using the digital communications system, the train can receive authorization to proceed for a specified distance along the railroad from a central dispatching center. If the train fails to slow down as it approaches the end of the territory where it is authorized to operate, then the computer can set off a bell or other warning for the engineer; if the engineer does not respond, then the computer can cause the train to stop. Likewise, the computer can issue a warning—and if necessary, slow down the train—if the engineer fails to reduce train speed to stay within the speed limit (e.g., at a curve or near a location where track conditions require slower operation). PTC systems, had they been installed, could have prevented many fatal railroad collisions during the past 20 years, including several highly publicized accidents involving passenger trains that resulted in dozens of fatalities.

Numerous studies have addressed the feasibility of installing PTC for the U.S. rail system. Among the most significant findings:

1. The risks associated with passenger trains are an order of magnitude greater than the risks associated with freight trains, simply due to the much larger number of people at risk in the event of an accident.

2. The safety benefits, though real, are insufficient to justify the cost of the system, because (a) the rail system is already an extremely safe mode of transportation, and (b) the cost of the PTC system is very high. In fact, the safety benefits were found to be only enough to justify the maintenance costs of the system.

3. Substantial management benefits can be gained in addition to the safety benefits; one early study (completed in 1990) indicated that the railroads could justify their costs based on improvements in equipment utilization and improvements in service.

4. More recent studies noted that advances in communications technology meant that most of the managerial benefits could be obtained using cell phones at a quarter of the cost of installing the PTC system. (The safety benefits of PTC can be achieved only with a failsafe system that would require extremely reliable components; the management benefits do not require such a system, because a break in business communication would not result in any safety problems.)

5. The total cost of implementing the system nationwide is likely on the order of $5 billion.

Questions

Some advocates of PTC say that the freight railroads (which own and operate most of the U.S. rail network) should be forced to install PTC since the operating benefits would provide an adequate return on a $5 billion investment. Others advocate a partnership involving the freight railroads, Amtrak (the public entity that provides intercity rail passenger service in the United States), and the commuter railroads (these operate over only a small portion of the network, but they operate the routes with the highest volumes of passenger traffic). The freight railroad companies, Amtrak, and the commuter rail agencies would each pay for equipping their own locomotives (which would account for about half the total cost of PTC). The costs of establishing the communications links and control systems would be shared based on the route-miles owned and the number of trains or train-miles operated by each entity. With this arrangement, the freight railroads would pay on the order of 80% of the costs of the system since they own most of the locomotives and most of the network and operate most of the trains.

a. Do you think the railroads should be required to install PTC? Why or why not?

b. Do you think the proposed cost-sharing arrangement is equitable? Why or why not?

c. If you were negotiating a PPP on behalf of the rail industry, what cost-sharing arrangement would you advocate?

d. If you were negotiating on behalf of Amtrak or the commuter railroads, what would you advocate?

e. If you were trying to help the various parties reach an agreement for some implementation of PTC, what would you suggest?

Chapter 13

Dealing with Risks
and Uncertainties

Our fundamental need is not the elucidation of the mysterious, but an appreciation of the significance of the obvious.[1]

CHAPTER CONCEPTS

Section 13.1 Introduction
Section 13.2 Modeling Performance
    Using models to understand the interrelationship among the many factors that affect the success or failure of a project
    Deterministic analytical models and probabilistic simulation models
    Incorporating random variables into a spreadsheet model
Section 13.3 How Best to Improve System Safety: Probabilistic Risk Assessment
    Basic concepts of probabilistic risk assessment
    Estimating global risks associated with an infrastructure system
    The importance of perceived risks
    Factors that make people fear certain kinds of risks
    How improper responses to an accident or incident may lead to severe indirect consequences
Section 13.4 Performance-Based Technology Scanning
    Goals of performance-based technology scanning (PBTS)
    Modeling a technology in terms of its impact on performance
    Using PBTS to determine what types of new technologies are potentially most important for improving system performance and allocating research and development funds

13.1 INTRODUCTION

Various kinds of risks and uncertainties must be considered in evaluating infrastructure projects. These two terms are related but have somewhat different meanings and implications. Risks refer to things that can go wrong, possibly leading to the failure of the project. Common risks associated with any major project include the following:

- *Construction risks:* It may not be possible to construct the project as planned, on budget, within the original time schedule. Storms could cause extensive damage, and unexpected geotechnical problems could set back a project many months or years.

---

[1] John A. Droege, *Freight Terminals and Trains* (New York and London: McGraw-Hill, 1925), p.13.

- *Financial risks:* It may not be possible to obtain sufficient funding for a project; interest rates on bonds, mortgages, or loans may be much higher than anticipated; investors may demand a larger share of the company ownership or pull out of the project altogether; general economic conditions may decline sharply while the project is under way, possibly making it impossible to continue borrowing money to complete a project.
- *International risks:* If a project is undertaken in another country, exchange rates could change, thereby upsetting key assumptions about the value of cash flows. In some locations, there is a risk that local governments will change the regulations governing the project, attempt to take over the project, or fail to follow through on commitments made to support the project.
- *Infrastructure safety:* Avoiding infrastructure failure will be one of the fundamental goals of the engineering design process; the design, maintenance, and operation of infrastructure will affect the risks of accidents for operators, users, and abutters.
- *Demand risks:* The revenue projected for a project may fail to materialize due to a lack of demand or overly optimistic assumptions about pricing.
- *Operating costs:* The project might not work as well as planned; operations might be more costly, and service might not be as good as expected.

It is important to anticipate and to plan how to deal with all of these risks before starting to implement a project. Some of these risks, such as the risk of being unable to complete the project on time, can be reduced by allowing buffer time in the schedule and a significant amount for unexpected contingencies in the budget. Risks related to storms can be reduced by developing emergency plans and by taking precautions at the building site. Safety risks can be mitigated by testing materials, ensuring good design standards, and careful monitoring of all construction processes. Risks related to demand and operations can be handled by undertaking detailed studies of how the system will work and how potential users will respond to the system.

Risk and uncertainty are sometimes used interchangeably, but there are some differences worth mentioning. The term *risk* implies that something bad could happen, which might involve something physically bad (a building collapses) or something less tangible (mortgage markets dried up, and financing for the project could not be obtained at a reasonable interest rate). The term *uncertainty* does not have the connotation of something that is bad; instead, it refers to the inability to predict exactly what will happen. Interest rates could go up or down; demand for apartments could go up or down; the costs of gasoline or building supplies could rise rapidly, rise slowly, or oscillate.

EXAMPLE 13.1 Risk versus Uncertainty

A city and a company have entered a public-private partnership (PPP) to construct a new toll road. The city has allowed the company to charge a fixed toll during the first years of operation as well as have some ability to raise tolls every few years thereafter. The city also recognizes that demand will be lower in the first few years of operation, so it has agreed to provide a subsidy to supplement toll revenues during the first few years of operations. The city has agreed to guarantee the interest rate on the bonds issued by the company to pay for the construction costs. The company was created to construct and operate the toll road; its future depends upon the financial success of the project. What are the uncertainties and risks in this situation, and what are the key issues to negotiate?

The major uncertainties are as follows:

- The cost of construction, which will determine the annual interest to be paid on bonds.
- The time required for construction, which will determine the point at which toll revenues begin.
- The demand for the road is not known.

The major risks are as follows:

- If the company does not receive enough revenue from tolls and subsidies to cover the interest costs, it will have to declare bankruptcy and the city will have to take over the project.
- If the city has to pay too much for subsidies, then it will have to cut back on other projects or raise taxes, which will cause a political uproar.

- If tolls are set too high, and if the private company becomes very profitable, there could be a political backlash against the project, and the ability for the city to undertake future PPPs would be jeopardized.

The key issues to negotiate are as follows:

- *Initial levels for tolls:* If tolls are set too low, public subsidies will be too high; if set too high, public outrage could be a problem.
- *Escalation for tolls:* If tolls are allowed to rise sooner, rather than later, then the private company may be able to borrow additional funds to cover losses in the initial years. The contract could require the company to establish a fund or a line of credit that would be sufficient to cover a shortfall of toll revenues during the first few years.
- *Nature of the subsidy:* If a maximum is established, then the company bears more risk related to the uncertainty in demand; if no maximum is established, then the city bears all of the risk related to the uncertainty in demand.

If the city and the private company are careful in structuring the deal, then they should be able to reduce the risks that each of them face. If one party fails to recognize the major uncertainties, then it is likely to end up with more of the risks.

As described in Chapter 11, sensitivity analysis can be used to understand performance under various assumptions about the key factors related to project success. Sensitivity analysis is in effect a means of dealing with uncertainty. By looking at many possible combinations of factors, it is possible to determine if there is a risk that the project will fail, given the uncertainty in the assumptions related to supply and demand.

Scenario analysis is another method for dealing with risks and uncertainty. The basic idea of scenario analysis, as presented in Chapter 11, is that potential projects should be considered in the context of various possible visions of the future. In deciding whether to proceed with a project or how best to proceed with it, it is helpful to consider multiple futures and to consider the different risks that might be encountered.

This chapter presents several specialized methodologies that can be useful in dealing with some of the risks and uncertainties associated with infrastructure projects:

- Simulation, a very general technique that can be used when there is a great deal of uncertainty in key factors involved in evaluating a project.
- Probabilistic risk assessment, a technique that can be used in developing strategies to improve the safety of infrastructure systems.
- Performance-based technology scanning, a technique that can be used to estimate the ultimate success of investments in infrastructure by considering how changes in performance will result in changes in demand.

The emphasis in this chapter is on safety and demand, due to the importance of these two factors in infrastructure systems. Risk of failure and risks of accidents are major factors in the design and management of infrastructure systems, and it is important to have a systematic way to address these risks. Risk of financial failure because of inadequate demand is another concern in almost any large infrastructure project. Uncertainties in demand relate to the size and staging of projects and to the design of the project. Understanding how potential customers might respond to new infrastructure is therefore an essential aspect of project design.

Section 13.2 shows how models can be used to simulate the performance of a system, which can provide insight into the performance of a system, the likely effect of a project, or the amount of risk associated with a project. As discussed in Chapter 2, infrastructure-based systems can be measured in terms of many measures, including cost, service, capacity, and safety. Projects may be aimed at improving any or all of these aspects of performance, and it is helpful to structure engineering models of the system that can be used to investigate how the proposed project might improve performance and how the improvement in performance might affect demand or profitability of the system. It is also useful to have higher-level models that can be used to explore risks and uncertainty more thoroughly than is possible with simple sensitivity analysis.

Section 13.3 presents probabilistic risk assessment, which is a structured methodology for improving safety. This section delves into two aspects of risk assessment that are critical for many projects. First, what is the most effective way to reduce the risks that affect infrastructure safety? For example, what is the best way to reduce highway accidents or the best way to reduce the risks of floods? Second, how can the risks associated with the safety of a project be best communicated to the public? For example, how does an energy company deal with public perceptions of the risks of nuclear energy?

Many projects involve introducing new technologies or modifying an existing system in response to competitors' introduction of new technologies. Technology scanning refers to the search for new technologies that might be important to a particular industry. Section 13.4 introduces performance-based technology scanning (PBTS). This methodology provides a way to determine what kinds of technologies are most worth pursuing by focusing on two things:

1. The aspects of performance that might be affected by new technologies
2. The market response to changes in various aspects of performance

New technologies may reduce cost, improve operations, increase capacity, or enhance safety. Whether they are worth pursuing depends upon how potential customers or users will respond to these changes. The examples in this chapter concern the competitiveness of intercity rail versus air, bus, and automobile. While much public discussion calls for "high-speed rail," other aspects of performance may be just as critical, and different kinds of rail systems may be more effective.

Risks and uncertainties associated with construction are considered in Chapter 14, which provides an introduction to project and program management. It is beyond the scope of his text to address any of the more exotic strategies that can be used to minimize financial risks, such as hedging strategies that protect against changes in commodity prices, interest rates, or exchange rates.

When infrastructure projects are planned, especially for new types of infrastructure, there will be many uncertainties concerning operating performance, demand, profitability, environmental impacts, safety, and other social and economic impacts. Over time, experience reduces the uncertainty about these matters, and it will become evident where the system works well, where it has encountered problems, and where completely unexpected side effects have emerged. As better knowledge is gained, adjustments can be made to improve infrastructure performance, which is the topic of Chapter 15, "Toward More Sustainable Infrastructure."

13.2 MODELING PERFORMANCE: SIMULATION AND ANALYTICAL MODELS

Models can be used to address three major questions related to project evaluation:

1. How will system capabilities or performance change if the project is implemented?
2. If system capabilities or performance change, what will happen to demand?
3. How will uncertainties affect the success of the project?

The first question refers to the engineering factors that underlie performance. How does the system currently function? What are the key relationships? What are the key performance measures? What are the key factors that affect costs, capacity, or service quality? How will the proposed project affect these relationships, key factors, and performance measures? These are all engineering questions that can be addressed with models appropriate to the system. For example, putting a portion of a light-rail transit system into a tunnel can save travel time by eliminating delays at grade crossings. Modeling the operation of the system may show that the travel time savings would be 4 minutes, thus reducing the average time from 44 to 40 minutes. Another example might be an analysis of various designs for wind farms to determine which design is the most cost effective in producing electricity.

The second question concerns what might happen after the project is implemented—what happens next? So what if the average travel time on the transit system is reduced by 10%, or if the unit cost of producing electricity with wind turbines is reduced by 10%? People will still want to commute in their own cars, and the cost of the green electricity may well remain higher than the cost of electricity from efficient coal-fired plants. Thus the changes in performance may or may not result in any more people using the light-rail system or in any increase in demand for electricity produced by wind turbines. To determine the effect on demand, it is necessary to consider the factors affecting the demand for transit and the willingness of customers to choose to use green energy even though it is somewhat more costly.

The third question considers the interaction among the key factors that determine the success of the project, which include variables related to supply and demand as well as variables related to financing the project. Subsection 13.2.1 shows how to use random variables to understand how uncertainty might affect profits and the financial risks of a project. Subsection 13.2.2 provides a somewhat more complex simulation model that could be used to obtain a better understanding of the variability in scheduling the construction of a project.

13.2.1   Using a Simple Probabilistic Model to Study Uncertainties and Risks

In Chapter 12, sensitivity analysis was introduced as a way to find out which aspects of a project were most critical. By changing one variable at a time along a range of possible values, it is possible to calculate the effect on ROI, NPV, or any other measure of performance. By changing two or more variables, it is possible to see more complicated interactions. However, as emphasized in the discussion of sensitivity analysis, it takes some thought and considerable care to conduct an intelligent sensitivity analysis and to interpret and present the results of such an analysis. A good sensitivity analysis seeks to determine whether the success of the project is in doubt, given the uncertainty in the various assumptions that must be made. The conclusions of a sensitivity analysis might be stated as follows:

- As long as construction costs are no more than 20% of budget and net revenues are at least 60% of what we expect, the NPV of the project will be positive (a strong indication that this would be a good project).
- If construction costs are 10% over budget or if income is only 90% of what we anticipate, then we will be unable to cover the interest payments on our mortgage (a clear indication that this is a risky project).

A probabilistic analysis provides another approach to assessing the effects of uncertainty on a project. If a probabilistic model is created, then it can be run dozens or hundreds of times to determine what the most likely performance will be and the chance that the performance will be unacceptable.

To illustrate the probabilistic approach, let's consider the basic factors affecting the success of a project. Profit and return on investment (ROI) are important measures of financial performance:

$$\text{Profit} = \text{Revenue} - \text{Cost} \qquad \text{(Eq. 13.1)}$$

$$\text{ROI} = \text{Profit/Investment} \qquad \text{(Eq. 13.2)}$$

When evaluating a potential project, owners and investors naturally are concerned about making enough profit to achieve their minimum acceptable rate of return. If they are borrowing money for the project, then they want the profit from operations to be large enough to cover interest payments on the debt they have incurred while also providing an acceptable rate of return on their equity (i.e., what they themselves have invested).

$$\text{Investment} = \text{Equity} + \text{Debt} \qquad \text{(Eq. 13.3)}$$

$$\text{Profit} = \text{Revenue} - \text{Cost} - (\text{Interest rate})(\text{Debt}) \qquad \text{(Eq. 13.4)}$$

$$\text{ROI} = \text{Profit/Equity} \qquad \text{(Eq. 13.5)}$$

In any project, at the outset, each of the variables in these equations involves considerable uncertainty. The total investment is not known until construction is completed, and the amount of debt and the interest rate on the debt are not known until financing is arranged. Actual revenues and operating costs are uncertain for some time after project completion because it may take some years for the entire project to reach full operation.

All of the variables in Eqs. 13.1 to 13.5 can be treated as random variables. Each time a random variable is encountered in an analysis, a value for that variable is generated from a set of possible discrete values or a possible distribution of values. If a set of equations contains more than one random variable, then they could be independent or dependent variables. If they are independent, then the value selected for one variable does not affect the value selected for the other variable. If they are dependent, then the value selected for one variable influences the value selected for the other variable. Before proceeding to an example, let's see how to create random variables in a spreadsheet.

Creating Random Variables in a Spreadsheet

Random variables can be created in spreadsheets by using the function that generates random numbers. In Excel, for example, it is possible to use the function RAND() to generate a random number greater than or equal to 0 and less than 1.0. Whenever the spreadsheet is recalculated, a different random number is returned. This function can be used to create random variables that take on a value that is believed to be uniformly distributed between a minimum value A and a maximum value B:

$$\text{Random Value between A and B} = A + \text{RAND}()(B - A) \qquad \text{(Eq. 13.6)}$$

This type of random variable is used in Example 13.2 below.

A random variable can also be defined to take on discrete values. For example, suppose that a random variable could take one of two equally likely values, V_1 or V_2. This random variable can be created in a spreadsheet as follows. First, select a random number by using RAND(); then create a scheme like this:

$$\text{Discrete Random Variable} = V_1 \text{ if } 0 \le \text{RAND}() < 0.5 \qquad \text{(Eq. 13.7)}$$

$$\text{Discrete Random Variable} = V_2 \text{ if } 0.5 \le \text{RAND}() < 1 \qquad \text{(Eq. 13.8)}$$

If the values are not equally likely, then it is possible to adjust the ranges in Eqs. 13.7 and 13.8. For instance, if V_1 is three times as likely as V_2, then

$$\text{Discrete Random Variable} = V_1 \text{ if } 0 \le \text{RAND}() < 0.75 \qquad \text{(Eq. 13.9)}$$

$$\text{Discrete Random Variable} = V_2 \text{ if } 0.75 \le \text{RAND}() < 1 \qquad \text{(Eq. 13.10)}$$

This process can readily be extended to allow any number of discrete values, each with their own probability, by dividing the interval between 0 and 1 into increments such that the length of each increment equals the desired probability for each value. Thus we can now create a random variable R that takes the values of $V_1, V_2, \ldots, V_n$ with probabilities $P_1, P_2, \ldots, P_n$. An example of this type of analysis is given in subsection 13.2.2.

It is possible to create new random variables as functions of multiple random independent variables:

$$\text{Sum} = R_1 + R_2 \qquad \text{(Eq. 13.11)}$$

$$\text{Ratio} = R_1/R_2 \qquad \text{(Eq. 13.12)}$$

Thus, if investment, revenue, and cost are independent random variables, then profit and ROI will also be random variables.

It is also easy to create a random variable that is dependent upon the value of another random variable. This can be done by using statements like this in a spreadsheet:

$$R_2 = a + b R_1 \qquad \text{(Eq. 13.13)}$$

In this case, the value of R_2 depends upon the value of R_1. More complex equations can also be used to allow different calculations for R_2 depending upon the value of R_1. For example:

$$\text{If } R_1 < A, \text{ then } R_2 = 100 + R_1; \text{ but if } R \ge A, \text{ then } R_2 = 50 + R_1 \qquad \text{(Eq. 13.14)}$$

With some ingenuity, it is possible to design complex simulations using a spreadsheet. A time-based simulation can be structured by having values of variable in one time period depend upon the situation at the end of the previous time period. If all the variables, random numbers, and calculations are included in one (possibly very long) row of the spreadsheet, then copying that row over a range of rows will produce a simulation of how the system will change over time.

EXAMPLE 13.2   Will the ROI of a Proposed Project Be Acceptable?

A company is considering making an investment of $100 million to purchase the rights to operate an existing toll road. Toll revenues are expected to range from $20 to $25 million annually, while operating and maintenance expenses are expected to range from $10 to $15 million annually. The company plans to sell bonds to raise $80 million at 8% interest and to raise $20 million from equity investors who anticipate a 10% ROI. Assuming that revenue and operating costs are independent random variables, what is the likelihood that the ROI will exceed 10% in any given year? What is the likelihood that net revenues in any year will be insufficient to cover the interest costs? A quick glance at the minimum expected revenue and the maximum expected cost shows that the annual revenue could be as low $5 million, which would be insufficient to cover the $6.4 million interest on the bonds. Is this something that investors should worry about?

To answer this question, revenue and cost can be defined as independent random variables. Investment cost, total debt, and interest rate on debt are assumed to be $100 million, $80 million, and 8%. Profit will be calculated using Eq. 13.4:

$$\text{Profit} = \text{Revenue} - \text{Cost} - 8\%(\$80\,\text{million}) \tag{Eq. 13.15}$$

Table 13.1 shows results from a probabilistic model that was created in a spreadsheet. The table has 20 rows, each of which could be viewed as either a separate year over a 20-year period or 20 different random results for a single year. Investment, debt, and interest rates are assumed to be as given, but the spreadsheet has an area at the upper right where these factors can be modified. Each row of the main body of the spreadsheet has two random variables. The first is used to calculate a value for revenue, and the second is used

Table 13.1   A Spreadsheet That Uses Random Numbers to Explore Risks and Uncertainties Associated with a Project

Year	Random Number 1	Random Number 2	Revenue	Cost	Investment $100 million Debt $80 million Interest 8% Interest	Profit	ROI
1	0.887713	0.138114	24.44	10.69	6.40	7.35	36.7%
2	0.997132	0.077517	24.99	10.39	6.40	8.20	41.0%
3	0.651748	0.191994	23.26	10.96	6.40	5.90	29.5%
4	0.474722	0.464287	22.37	12.32	6.40	3.65	18.3%
5	0.922596	0.886441	24.61	14.43	6.40	3.78	18.9%
6	0.415191	0.818556	22.08	14.09	6.40	1.58	7.9%
7	0.739819	0.764117	23.70	13.82	6.40	3.48	17.4%
8	0.425825	0.106793	22.13	10.53	6.40	5.20	26.0%
9	0.771215	0.067184	23.86	10.34	6.40	7.12	35.6%
10	0.590406	0.437435	22.95	12.19	6.40	4.36	21.8%
11	0.607368	0.91192	23.04	14.56	6.40	2.08	10.4%
12	0.005216	0.406998	20.03	12.03	6.40	1.59	8.0%
13	0.382789	0.919088	21.91	14.60	6.40	0.92	4.6%
14	0.003209	0.711268	20.02	13.56	6.40	0.06	0.3%
15	0.479641	0.470464	22.40	12.35	6.40	3.65	18.2%
16	0.172015	0.24961	20.86	11.25	6.40	3.21	16.1%
17	0.648983	0.107807	23.24	10.54	6.40	6.31	31.5%
18	0.065481	0.587854	20.33	12.94	6.40	0.99	4.9%
19	0.930513	0.395585	24.65	11.98	6.40	6.27	31.4%
20	0.94164	0.431054	24.71	12.16	6.40	6.15	30.8%
Total			455.57	245.72	128.00	81.85	20.5%

to calculate a variable for cost, both using Eq. 13.6. Profit is calculated using Eq. 13.15, and ROI is calculated as profit/equity (Eq. 13.5). The average ROI is calculated as the average profit divided by the net investment of $20 million. According to this analysis, the financial aspects of the project are solid. There is very little chance that net revenues will be unable to cover costs, and the expected ROI is over 20%, well above what the investors have been promised. The annual values vary considerably, reflecting the uncertainty in the estimates of revenue and cost. However, the risk of failure is very low.

Once this spreadsheet was created, it was possible to conduct further analyses within a few minutes. For example:

- Running the analysis 10 times produced average ROI ranging from 16.8% to 23.2%; the median ROI was 18.1%.
- If interest rates were increased to 9%, the average ROI ranged from 8.7% to 18.4%; the median ROI in 11 trials was 14%. Increasing interest rates on the bonds naturally reduces ROI, but the project still can be expected to have better than a 10% return, which is what the investors are looking for.
- If there is a cost overrun, and the investors have to put up an additional $20 million, the ROI ranged from 6% to 11.4% with a median of 8.6%. Thus, controlling construction costs appears to be quite important.

This example shows how to create a simple spreadsheet that can be used to explore risk and uncertainty using random variables for the key factors in the analysis.

13.2.2   Using a Probabilistic Model in Project Scheduling

Consider a project that will be implemented in three sequential phases; that is, the second phase cannot begin until the first phase is finished, and the third phase cannot begin until the second phase is completed. How much time must be allowed for the complete implementation?

Based on past experience, the contractor anticipates the average time required for each phase to be as shown in Table 13.2. The contractor is pretty sure that each phase can be completed within a week of the expected time, and he is 90% certain that the work can be done within the maximum time. There is only a 10% chance that the work could be done in the minimum time. The times required for the tasks are independent; delays or early completions will not affect the time required in subsequent tasks. The expected time to complete the work is 23 weeks, but the contractor would like to add a buffer to be pretty certain to complete the project on time.

One possibility would be to add together the maximum times and then throw in an extra week just to be sure, for a total of 36 weeks. Will a 36-week schedule allow the contractor to be at least 90% certain of completing the project on time? Is this a reasonable schedule? With the information given, is it possible to recommend a shorter buffer?

The proposed 36 weeks appears to be a very safe schedule. To get to 36 weeks, one of the projects will have to take more than the so-called maximum time while the others take close to the maximum time. There is only a 10% chance that any of the phases will take more than the maximum time and a negligible chance that all three will be at or past the maximum $(0.1 \times 0.1 \times 0.1 = 0.001)$. Even if two of the phases are a week beyond the maximum, which is a 1 in 100 possibility, the third phase will have a fifty-fifty shot at being completed in the expected time, saving at least 2 weeks from the maximum time. While the 13-week buffer looks safe, it also seems to be a very large buffer; if we are leasing equipment or hiring people for the entire scheduled time, then we will end up with a lot of idle time and a lot of extra cost.

Can the given information be used to justify a shorter amount of time? Maybe, but we would have to make some assumptions about the probability distributions. In particular, we need to interpret the

Table 13.2   Estimated Time Required for Three Projects

Phase	Minimum Time	Expected Time	Maximum Time
I	4 weeks	5 weeks	7 weeks
II	7 weeks	9 weeks	12 weeks
III	6 weeks	9 weeks	16 weeks
Total	**17 weeks**	**23 weeks**	**35 weeks**

Table 13.3   Time Required for Phases I, II, and III of Prior Projects

Phase	2 Weeks Early	1 Week Early	Expected	1 Week Late	2 Weeks Late	3 Weeks Late	5 Weeks Late	7 Weeks Late
I	5%	10%	60%	10%	10%	5%		
II	10%	30%	30%	15%	5%	5%	5%	
III	5%	15%	25%	10%	15%	5%	15%	10%

contractor's remark that he was "pretty sure that each phase can be completed within a week of the maximum time." This suggests that the distributions may be reasonably well behaved (i.e., have a single mode and a declining tail). If we have studied statistics, we might wonder if the time required for the three processes could be treated as random variables that are normally distributed. The standard deviation of each distribution could be estimated as half the time between the expected and the maximum times, because approximately 90% of a normal distribution is less than the mean plus 2 standard deviations. Since the three phases are believed to be independent, the variance of the total time would equal the sum of the variances of the times required for each phase.

Using these assumptions, we can estimate how much time the contractor would have to allow to be 90% sure of completion within the allotted time. The standard deviations of phases I, II, and III would be 1, 1.5, and 3.5 weeks under these assumptions. The variances would be 1, 2.25, and 12.25 weeks^2. The total variance would therefore be 15.5 weeks^2 and the standard deviation of the sum would be the square root of that or 4 weeks. If we add two standard deviations to the expected time, we would get 23 weeks plus 8 weeks equals 31 weeks.

Now suppose the contractor supplies some additional information concerning the time required for Phases I, II, and III in previous projects (Table 13.3). With this information, we can construct a simple simulation model to estimate the combined trip time distribution. The trip time distribution will indicate how much time is needed to be 90% certain of being finished within the scheduled time.

The model can be constructed by treating the time in each phase as a random variable. Table 13.4 shows a simple model that was constructed in a spreadsheet. Each phase of the project is on a separate line. The right-hand portion of the table has the cumulative distribution of completion time, which is based on the distributions shown above in Table 13.3. The time is based on the value of the random variable generated for each phase. Each time the spreadsheet is recalculated, new random variables are generated as a number between 0 and 1.0. This random variable is then used to select the completion time. In the example shown in Table 3.4, the random variable generated for Phase I is 0.4206, which is used to select a point on the cumulative distribution of completion times for Phase I. Since 42.06% is larger than 15% but less than 75%, this random number corresponds to an on-time completion time. The logic in the model is, "If the random number for Phase I is between 15% and 75%, then the

Table 13.4   A Simple Model for Estimating the Distribution of the Total Time Required to Complete Three Independent Processes

Phase	Random Numbers	Time	Completion Time Relative to Expected Time							
			−2 Weeks Early	−1	0	1	2	3	5	7 Weeks Late
Phase I	0.4206	0	5%	15%	75%	85%	95%	100%	100%	100%
Phase II	0.5045	0	10%	40%	70%	85%	90%	95%	100%	100%
Phase III	0.0783	−1	5%	20%	45%	55%	70%	75%	90%	100%
Total		**−1**								

Table 13.5   Results of Two Sets of 100 Runs Each The table shows the distribution of completion times (i.e., the sum of the delays for the three phases of the project).

Trial	−4	−3	−2	−1	0	1	2	3	4	5	6	7	8	9+
A	1	2	6	13	13	19	17	9	6	1	3	2	3	5
B	1	2	7	11	27	18	12	7	4	6	0	1	4	0
Average	1	2	6.5	12	20	18.5	14.5	8	5	3.5	1.5	1.5	3.5	2.5

completion delay will be 0."[2] For Phase II, the random number is 0.5045, which also signifies on-time completion. For Phase III, the random number is 0.0783, which corresponds to completion 1 week early. Adding together the three times indicates that the completion of the entire project would be 1 week before the expected time—that is, 22 weeks.

Each time the spreadsheet is recalculated, the random numbers change, thus causing a new completion time to be selected for each phase; this leads to a new completion time for the project. By repeatedly generating the random numbers and keeping track of the results, it is possible to get an estimate of the distribution. Table 13.5 shows results for two sets of 100 runs each of this simple model.[3] The results are similar, but not identical, as would be expected. The average for the two runs, which is shown in the bottom row of Table 13.5, can be used as an estimate of the distribution of completion times. If the scheduled time were 5 weeks greater than the expected time of 23 weeks, then this analysis suggests that the chance of completion taking more than the schedule would only be 9%. In other words, if the schedule is 28 weeks, there would be a 91% chance of completing the project on time. By obtaining more information and conducting a more detailed probabilistic analysis, it has been possible to reduce the required buffer well below the initial estimate of 12 weeks (based on maximum times for each phase) or the estimate of 8 weeks (based on assumptions about the shape of the distributions of completion times for each phase).

The advantage of using the simulation approach is that it is straightforward to estimate the distribution of possible results for what could be rather complicated logical relationships. The speed and ease of using a spreadsheet makes sophisticated probabilistic analyses readily available.

13.3   HOW BEST TO IMPROVE SYSTEM SAFETY: PROBABILISTIC RISK ASSESSMENT

Probabilistic risk assessment was introduced in Chapter 2, where **risk** was defined as the product of two factors: the *probability of an accident* and the *expected consequences* if an accident occurs.[4] Global risks associated with a particular system can be estimated by summing over all types of accidents and all types of consequences. Investments to reduce risk can be compared by considering the ratio of the reduction in risk to the cost of achieving that reduction. Example 2.12 in Chapter 2 showed how to apply probabilistic risk assessment to the evaluation of strategies for reducing the risks associated with accidents at rail–highway grade crossings. That example demonstrated how a weighting scheme could be used to compare

---

[2] The logic is actually a nested IF statement that refers to the cumulative percentages and the delays shown in the table. The statement looks like this: =IF(B15<E15,E$14,IF(B15<F15,F$14,IF(B15<G15,G$14,IF(B15<H15,H$14,IF (B15<I15,I$14,IF(B15<J15,J$14,IF(B15<K15,K14,L$14))))))) where Cum1 to Cum7 refer to the numbers shown in Table 13.4 for the cumulative distribution, and Delay1 to Delay8 refer to the completion times. For Phase I, Cum1 is 5% and Delay1 is -2.

[3] Table 13.5 was created very quickly using paper and pencil along with the spreadsheet model. I first drew columns on a piece of paper with headings −4, −3, . . . , 9+ to indicate the completion time for that run relative to the expected completion time of 23 weeks. I then ran the spreadsheet model repeatedly, each time placing a hatch mark in the proper column. After making 100 runs, I added up the results. Of course the spreadsheet model could be enhanced to do all of this automatically, but my analysis was completed in a few minutes—far less time than it would have taken me to figure out how to do everything automatically within the model.

[4] Probabilistic risk assessment applies to studies related to safety (e.g., the likelihood of accidents that disrupt service and result in property damage, injuries, or fatalities). Other kinds of risk must also be considered, such as the risks associated with financial matters (currency exchange, interest rates) or demand (will demand meet initial projections?).

different types of consequences, such as fatalities, personal injuries, and property damage. Using such weights, it is possible to obtain a single metric for accident consequences.

Subsection 13.3.1 reviews the basic method, discusses the problem of placing a value on changes in expected fatalities, and provides an example of using probabilistic risk assessment to determine cost effectiveness of potential strategies for making automobiles less risky. The next three subsections address three additional aspects of risk assessment:

- Global risks (subsection 13.3.2)
- Perceived risks (subsection 13.3.3)
- Indirect consequences (subsection 13.3.4)

Subsection 13.3.5 shows how statistical analysis can be used to learn more about the causes of accidents. Understanding causality makes it possible to develop programs aimed at reducing the number of accidents, thereby reducing risks. Finally, subsection 13.3.6 presents an example to show how a city can use probabilistic risk assessment to enhance the safety of its residents. Fort Collins, Colorado took various measures aimed at reducing risks related to flooding, including restrictions on flood plain development and moving certain activities out of the flood plain.

13.3.1 Measuring Risks as the Expected Consequences of Potential Accidents

Clearly, individuals are quite willing to accept substantial risks in their everyday activities. Some people go skiing, a few go skydiving, kids ride down hills on their bikes, and many people like to ride their motorcycles. Despite the vast number of broken limbs associated with these activities, many people do seem quite willing to buy and use their skis, bikes, and motorcycles, even though many more prefer snowshoeing, walking, and jogging as less risky activities that are more clearly attached to the ground. In general, people may at times decide to avoid activities they view as too risky; at other times, they seek out activities for which the excitement, or the view, or the sense of achievement justifies whatever risks are encountered.

Collectively, people also are able to decide what kinds of risks are acceptable. In some cases societies not only condone, but promote activities that are well known to be risky. With more than 30,000 fatalities annually on the highways in the United States, it is clear that driving is risky. Most of us know of someone who was killed or severely injured in an automobile accident, and anyone who drives extensively is likely to be able to recount several close calls. Yet we do not limit access to highways; we do not post speeds of 30 mph to avoid high-impact collisions; we do not shut down the highways during snowstorms; and we do not require construction of automobiles that resemble tanks in their ability to survive collisions (Figure 13.1). We do not "do everything possible to save just one

Figure 13.1 Just Another Automobile Accident. Auto accidents are routine, but we continue to drive because the benefits of personal mobility outweigh whatever risks we perceive in driving. Quite possibly many of us refuse to believe that we ourselves are at risk, just as many of us believe that the cost of driving is free, so long as there is gasoline in the tank.

(Photo: Dublin, Ireland, September 2008)

human life." On the other hand, we do impose speed limits; we do have design standards for highways and vehicles; we do punish drunken drivers; and we do require seat belts. Some actions have been deemed worthwhile to pursue, and others have not.

Whether or not our public officials explicitly state the logic, it is useful for us to understand how analysis can help determine what risks are acceptable and what values can be used to determine the costs and benefits of changes in safety. The three key types of factors are as follows:

1. Probability of an accident (P)
2. Consequences of an accident (C_i)
3. Relative importance of the consequences (W_i)

Risk can be defined as the sum of the expected consequences of an accident:

$$\text{Risk} = P\Sigma(C_i * W_i) \qquad\qquad \text{(Eq. 13.16)}$$

If the weights (W_i) are expressed in monetary terms, then changes in risk can be used in benefit-cost analysis. Previous chapters have given two examples where specific values have been used in benefit-cost analysis for changes in expected fatalities. In the example about grade-crossing safety in Japan (Example 2.12 in Section 2.6), the Safety Research Lab of the East Japan Railroad used a weight of about $1 million per expected fatality, along with smaller weights for injuries and delays. In the discussion of the safety benefits of a new highway (Example 4.1 in Section 4.2), a value of $2.5 million was used as the benefit of reducing expected fatalities by one.

Given the importance of reducing risks and public concerns with any major accident involving infrastructure, it is worth spending some on what it really means to reduce the expected consequences of an accident. The direct medical costs associated with an accident can be estimated based on past history, but what about serious injuries or fatalities? There are many approaches to this issue. A strictly economic approach might consider the average lifetime earnings of an individual killed in an actual accident, but it is impossible to say who would have been involved in an accident that is prevented. A possible solution to this problem would be to consider the age and income potential of the people using the facility. For example, if the typical user is 35 years old and will earn $50,000 per year for 30 more years, then his or her future earnings would be more than $1 million. This approach is fraught with difficulties: "How can we put a value on a human life?" is a common refrain after any tragic accident, especially one involving children.

Another approach advocated by economists such as Ken Small would look not at the cost of a fatality but at the value of reducing the risks shared by all users. In trying to improve highway safety, this approach has the merit of focusing on a benefit that is shared by everyone using the highways, rather than directing attention to an unknown victim of an unspecified accident. With this approach, the basic question is, what would an individual pay to achieve a small reduction in risk? This question can be answered through detailed interviews with a cross section of users. Studies have found that people in the United States would be willing to pay a few dollars for a small reduction in the probability that they would die in a transportation accident. If individuals are willing to pay on the order of $2 to reduce the odds of a fatal accident by 1 in a million, then it is logical to say that society should be willing to pay on the order of $2 million to avoid a single fatality in an accident. Instead of putting a value on a human life, this approach puts a value on the reduction in risk that every traveler realizes.

Another approach would be to consider the damages awarded in lawsuits involving accidental death. And a much different approach might simply consider the need for a reasonable number to use in estimating benefits and costs for public investments and public policy. If an extremely high value is placed on expected fatalities, then it is possible to justify extreme measures to improve safety. If no consideration is given to safety, then it is be possible to construct facilities that are quite unsafe. Perhaps a society, through political and legal processes, will determine acceptable levels of risk and the amounts worth investing to reduce risk.

EXAMPLE 13.3   Which Automobile Safety Features Are Most Worth Pursuing?

If the probability of a fatal accident is 1 in 100 million miles, and if the average car is driven 10,000 miles per year, then the probability of a car being involved in a fatal accident would be 1 in 10 thousand per year. The risk of a fatality therefore is on the order of .0001 ($2.5 million) = $250. Consider three possible changes to reduce risk:

1. *Require seat belts:* $100 per car reduces the probability of a fatality by 50% (to .00005 per year); risk would now be $125/year.
2. *Airbags for driver and passenger:* $500 per car further reduces risk of fatality by 80% beyond what is possible with seat belts (to .0000125 per year); risk would now be $25/year.
3. *Automatic collision avoidance system:* $20,000 per car further reduces risk of fatality by 10% beyond what is possible with seat belts and airbags (to .0000113); risk would now be $22.50 per year.

To compare the risk reduction benefits to the costs of the systems, those costs must be converted to annual costs. With typical assumptions about financing automobile usage and financing, the annual costs of these three options would be approximately $12 for seat belts, $60 for airbags, and $2,400 for the collision avoidance system.[5] The reductions in annual risk per automobile would be $125 for the seat belts, $100 for the airbags, and $10 for the automatic collision avoidance system. The seat belt and airbags are clearly justifiable since the expected reduction in risk is far greater than the annual cost. The hypothetical collision avoidance system, which would cost an additional $2,400 per year, would provide an additional benefit of only $10 per year in risk reduction compared to a car equipped with seat belts and airbags. Therefore, it would be better to spend additional money not on automatic collision avoidance systems, but on some other means of reducing risk.

13.3.2   Global Risks

Global risk encompasses all the risks associated with the many types of accidents that could occur within a system. A company or agency should be interested in understanding global risk so that it can develop effective programs for mitigating overall risk. Without a clear understanding of overall risk, likely too much will be spent to avoid what in fact are minor risks while too little will be spent to avoid what are actually major risks.

Estimating global risk for an infrastructure-based system involves the following steps:

1. Identifying the types of accidents that might occur on each portion of the system
2. Estimating the probabilities of such accidents as a function of time or usage
3. Estimating the expected consequences of each type of accident
4. Calculating the risks associated with each type of accident by multiplying the probability of an accident by the expected consequences of an accident
5. Summing risks over all types of accidents

These steps can be carried out at various levels of detail. For a well-established system, estimates of accident probabilities and consequences can be based on past experience. For new systems, risk estimates will have to be derived from models, comparison with similar systems, or expert opinion.

Different types of risks will have to be considered:

- Risks related to construction
- Risks inherent in normal operations
- Risks related to maintenance and rehabilitation
- Risks related to deteriorating infrastructure
- Risks related to structural failure
- Risks related to natural disasters, such as earthquakes, floods, hurricanes, or tornadoes

Some risks are associated with potentially catastrophic accidents that could result in hundreds of fatalities, extensive disruptions in major cities, and extreme property damage. Examples of catastrophic

---

[5] The cost of each option was converted to an equivalent annual cost, assuming a 14-year life for the car and a discount rate of 8% for the owner.

Table 13.6 Risk Factors for Three Hypothetical Systems

	Major Canal	Skyscraper	Urban Highway
Risk of fatality per construction worker	0.05	0.001	0.001
Construction workers	20,000	1000	1000
Risk of fatality per user, normal operation	0.2 per million	0.02 per million	2 per million trips
Users per year	1 million	3 million	30 million
Daily usage	40 ships per day	3000 occupants	100,000 vehicles/day
Potential fatalities from structural failure	10	2000	100
Likelihood of failure in 100 years	0.01	0.01	0.01
Expected fatalities			
• Construction (total)	1000	1	1
• Operations (per year)	0.2/year or 20 in 100 years	0.06/year or 6 in 100 years	60/year or 6000 in 100 years
• Failure (life of project)	1 in 100 years	20 in 100 years	1 in 100 years

accidents would include the failure of a major bridge, collapse of a building, flooding of a major city, an explosion in a chemical plant, failure of a nuclear power plant, or a plane crash. Many aspects of system design, operating strategies, and inspection and maintenance policies are aimed at avoiding catastrophic accidents, and the probabilities of such accidents are thus likely to be very small. Nevertheless, because the potential consequences are very large, the risks associated with catastrophic accidents are likely to be an important component of global risk for any system.

Table 13.6 illustrates some major risk factors that might be associated with three hypothetical projects. Using these factors, it is readily possible to estimate the fatalities that could be anticipated over the 100-year lifetime of the projects. The table shows the expected fatalities associated with construction, normal operation, and structural failure. The first project is the construction of a major canal, assuming technology and medical knowledge available when the Panama Canal was constructed. The greatest risks are those associated with the construction process, since a very large workforce is necessary and experience suggests a death rate of 5%. The canal would involve some risk in operation and maintenance, but not that many people would be working at the canal on a daily basis, and the risk would be minimal in the event of a major structural failure. Thus, for the canal, the global risks are primarily associated with the construction, not with operation.

The second project in Table 13.6 is the construction of a skyscraper, using rough estimates of risks assuming only modest attention to safety during construction. Some risk is inherent in construction, since many workers will be working high off the ground. However, with nets and other safety procedures in place, few fatalities should occur during construction. Operation of a skyscraper should also be very safe, but it would surely be catastrophic if the building collapsed. The greatest concern for this project is therefore to ensure the structural integrity of the building. The third project is the construction of an urban highway that is expected to serve 100,000 vehicles per day. For this project, the risks in construction can be mitigated through use of proper safety procedures, but there will be a continuing risk associated with traffic accidents. The failure of a portion of the highway would surely attract national attention, but the nature of a highway is such that failure is unlikely to affect more than a few dozen vehicles. The greatest challenge is therefore to improve the safety of the highway.

Estimating global risks is an objective exercise that is limited only by the ability to envision the types of accidents that might be expected and the ability to estimate the consequences of those accidents. Once the risks are understood, it is possible to determine which risks are most serious and consider how best to mitigate those risks.

However, it would be a mistake to believe that risks can be dealt with objectively, without regard to the ways people perceive and respond to risk. As discussed in the next subsection, some risks are perceived to be much worse than others, so it is necessary to consider perceptions when assessing risks.

13.3.3   Perceived Risks

Estimating "perceived risks" requires more than assessing accident probabilities and expected consequences.[6] Adjustments are needed to reflect the perceived importance of certain types of accidents or consequences to various stakeholders, including users and the public. Quantifying perceived risks requires answers to questions such as "Who is at fault?" "Is it a catastrophic accident?" and "Is new technology involved?" In principle, these questions can be answered based on surveys, and researchers have created a framework for understanding how perceptions of risk vary with the circumstances related to particular types of accidents.[7]

One of the first studies of perceived risks found that there is a different trade-off between risks and benefits for activities undertaken voluntarily than for those undertaken involuntarily.[8] People will accept risks from voluntary activities (such as skiing) that are roughly 1000 times as great as they would tolerate from involuntary hazards that provide the same level of benefits.[9] This study concluded that the acceptability of risk from an activity is roughly proportional to the third power of the benefits for that activity.

Additional research on perceived risk identified other factors that affect perceptions of risk. Slovic, Fischhoff, and Lichtenstein (1979) compiled from their own and prior studies a list of nine characteristics of risk that might be important in risk ratings by the public:[10]

1. Whether the risk is assumed voluntarily or involuntarily
2. Whether the risk of death is immediate or delayed
3. Whether the risk is known to those exposed to it
4. Whether it is known by science
5. Whether the risk can be controlled by an individual's skill
6. Whether the risk is new and unfamiliar as opposed to old and familiar
7. Whether the fatalities are common or catastrophic accident
8. Whether the thought of the risk evokes a feeling of dread
9. Whether there is a risk of a fatal accident

Slovic and colleagues (1979) found strong intercorrelations among these nine characteristics and suggested that the risk perceptions could be explained in terms of two basic factors (Figure 13.2). The factor labeled "Dread Risk" is defined at its high end by a perceived lack of control, feelings of dread at

Figure 13.2 Reducing Various Characteristics Relevant to Risk Perceptions to Two Key Factors.

Source: adapted from P. Slovic, B. Fischhoff, and S. Lichtenstein, *Why Study Risk Perception*, Risk Analysis, Vol. 2, No. 2 (1982) p. 86.

---

[6] The material in this section is largely based on a report prepared for the East Japan Railway Company by C. D. Martland, J. M. Sussman, L. Guillaud, and J. Vanzo, "Risk Assessment: Improving Confidence in JR East", Final Report, Research Conducted for the East Japan Railways (Cambridge, MA: MIT Engineering Systems Division, March 2005).

[7] For an excellent overview of the fundamental research on risk perception, see Nancy Nighswonger Kraus, "Taxonomic Analysis of Perceived Risk: Modeling the Perceptions of Individuals and Representing Local Hazard Sets," Ph.D. Thesis, University of Pittsburgh, 1985.

[8] C. Starr, "Social Benefit versus Technological Risk," *Science* 165 (1969): 1232–38.

[9] Paul Slovic, "Perception of Risk," *Science* 236 (1987): 280–85.

[10] P. Slovic, B. Fischhoff, and S. Lichtenstein, *Perilous Progress: Managing the Hazards of Technology* (Boulder, CO: Westview, 1985).

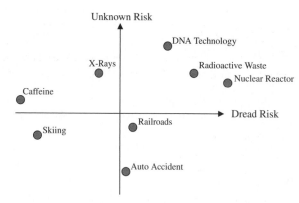

Figure 13.3 Location of Eight Hazards within the Two-Factor Space.

Source: adapted from P. Slovic, B. Fischhoff, and S. Lichtenstein, *Why Study Risk Perception*, Risk Analysis, Vol. 2, No. 2 (1982) p. 87.

the nature of an accident, and potential for a catastrophe. Factor 2, "Unknown Risk," is defined at its high end by hazards judged to be unobservable, unknown, new, and delayed in their manifestation of harm. Research has shown that laypeople's risk perceptions and attitudes are closely related to the position of hazards within this type of factor space.[11] Most important is the horizontal factor "Dread Risk." The more people perceived a hazard to be dreadful, the more they wanted to see its risks reduced, and the more they wanted to see strict regulation employed to achieve the desired reduction in risk.

Figure 13.3 shows how people viewed 8 hazards, including some related to daily activities, some related to new technology, and some related to infrastructure. The least dreadful risk was that associated with caffeine—we know it may not be good for us, but we still have to wake up in the morning! Likewise skiing might get you a broken leg, but that risk is not viewed as dreadful. The most dreadful risks for the people in this survey were those associated with nuclear reactors and radioactive wastes—which goes a long way toward explaining the difficulty of expanding the use of nuclear power in the United States and in many other countries. The other scale goes from known to unknown. The risks of automobile accidents are well known because they are so common; we continue to drive despite the risks because we often would rather be somewhere else—meeting with friends, going to work, taking a trip, or going shopping. We know the risks, and we accept them. The other end of the scale includes the risks associated with new or unusual technologies. The risks associated with nuclear reactors and nuclear waste are rather high on this scale as well: we don't know how bad the risks really are, but we know we don't like them. DNA technology, which relates to issues such as genetic alterations to increase agricultural yields, is the least known risk and therefore one of the most feared.

There are various strategies for dealing with perceived risks. One is to provide more and more safeguards so as to convince the public that the risks are well under control. The problem with this approach is that a rational response may do little to ease what in large part is likely to be an emotional problem. Moreover, the public may just not believe the technicians who claim that everything will work as planned. A second approach is to provide information concerning the plans that have been put in place to deal with incidents, however unlikely they may appear to be. By discussing the risks and how they are being handled, a company or an agency may be able to assuage public fears or at least provide an impression of competence. A third approach is to involve the public in the design and implementation of a comprehensive and credible risk management plan.

Another, more general approach, is to make sure that risks are put in the proper perspective. None of us live in a risk-free environment. We could be involved in a serious automobile accident; we could be hit by a car when crossing the street; we could be hit by lightning or slip on the ice or trip over a chair when rushing to answer the phone. Our house could be hit by a tornado, or we might be caught in a flood. We could be mugged, shot, or blown up in a terrorist act—which unfortunately is a real day-to-day risk in some countries. Hence, it may be reasonable to consider the extent to which the risks

---

[11] Slovic, "Perception of Risk."

associated with new activities increase our overall exposure to risk and the extent to which the risks associated with a new project relate to the overall risks experienced by society.

A study conducted for the East Japan Railway addressed a specific question concerning perceived risks: to what extent should the consequences of catastrophic accidents be weighted more heavily in developing a risk management program for the railroad? This was a practical question concerning the allocation of JR East's research budget and the amounts of money that should be spent on improving the protection at grade crossings in relation to the money spent on earthquake warning systems or installing collision avoidance systems for trains. The initial assessment of risks simply determined the expected fatalities per year from all kinds of accidents, which in effect weighted 100 fatalities in a train collision the same as 100 fatalities in 100 separate grade-crossing accidents. However, the review of the literature just cited indicated that the public—you and me—really are more fearful of catastrophic accidents, especially those related to new technologies. The conclusion, which is quoted in the accompanying box, was that JR East should indeed spend proportionately more to avoid catastrophic accidents.

Catastrophic, but infrequent accidents should be given more weight in strategic decisions than much more frequent, but relatively minor accidents. There are two main reasons for this conclusion: the possibility of indirect consequences and public perceptions of risk. The public or government agencies may over-react to a catastrophic accident by requiring very expensive and possibly irrelevant or ineffective investments to reduce future risks. Customers may lose confidence in the system and switch to other modes for extended periods of time. In addition, people perceive catastrophic accidents as being much more important than a large number of smaller, routine accidents that have the same total consequences.

Case studies of accidents involving transportation and other industries confirmed that indirect consequences of an accident can be many times higher than the direct consequences. Indirect consequences are more likely if an accident is viewed as a "warning signal"; i.e., an accident that may or may not be catastrophic, but does change the public's awareness of risks. In the emotional period following an accident, actions may be taken that, if considered at other times, would be recognized as unnecessary, too expensive, or misdirected. The best way to minimize indirect consequences seems to be to have a comprehensive risk management plan, periodic publication of information concerning safety and risk, and good public relations following an incident.

Source: C. D. Martland, J. M. Sussman, L. Guillaud, and J. Vanzo, "Risk Assessment: Improving Confidence in JR East," Final Report, Research Conducted for the East Japan Railways, MIT Engineering Systems Division, Cambridge, MA, March 2005.

13.3.4   Indirect Consequences of an Accident or Incident

An accident or incident may result in a greater awareness of risks, which then leads to changes such as enforcing proper operations of the system (e.g., enforcing speed limits on highways), improving the infrastructure (e.g., retrofitting buildings and bridges so they are better able to survive in an earthquake), or installing surveillance aimed at limiting deliberate attacks on the system (e.g., metal detectors at airports or procedures that limit public access to buildings). If these changes are indeed cost effective, then the expected performance of the system will actually improve relative to what was in place before the accident. Operating or equipment costs may indeed have gone up, but risks will have declined. What has changed is the understanding of the risks, which has resulted in rational decisions to reduce those risks. In this case, no indirect costs would be associated with the changes in operations, equipment, or regulation.

However, a rational approach may be difficult in the turmoil following a catastrophic accident. People—system managers, government officials, passengers, and the public—all want to do something to prevent a similar accident. In what may become a highly emotional environment, there may be pressure to make changes that do seem to address the immediate problem, but that would not satisfy a cost-effectiveness test. Three types of improper reactions might be encountered:

1. *Overreaction:* An accident or incident could result in changes that really are not related to, or that go far beyond, the specific causes of the accident or incident.
2. *Irrelevant reaction:* The proposed changes might in fact have little or no impact on risk.
3. *Ineffective or counterproductive reaction:* The proposed changes might increase risks or reduce the risk, but at only at a very high cost. Greater reductions in risk would have been gained by investing in other types of problems or other types of solutions.

Improper reaction leads to extra costs to the infrastructure operator and to society; such costs are not incurred unless there is a catastrophic accident. These costs are excessive because they are above what could be justified using the standard probabilistic risk assessment approach.

On the other hand, effective response to a disaster may give a company or agency credibility with the public, thereby allowing a more rational approach to risk management as well as project design and evaluation.

13.3.5   Using Accident Statistics to Determine Causality

Whenever an accident occurs (or an identifiable incident that could lead to an accident), there is an opportunity to collect data concerning the conditions at the time of the accident, the factors leading to the accident, and the consequences of the accident. This data, which can be used to learn more about the causes of accidents, will be helpful in devising policies and technologies to reduce the number or the severity of accidents.

Three types of analysis are feasible:

1. *Data analysis from thousands of accidents to understand the most important causal factors:* This approach can be used for studying automobile accidents because hundreds of thousands of accidents take place per year, and there are well-defined policies for filing accident reports.
2. *Intensive analysis of each major accident to identify causality:* This approach is used for accidents involving airlines, railways, and other transportation companies, where any accident can be catastrophic.
3. *Statistical analysis of a representative sample of accidents:* This approach can provide more accurate, in-depth documentation of the factors related to the accident in order to understand the sequence of events and factors that lead to an accident.

In the following paragraphs, each approach is discussed in more detail.

Using Data for a Very Large Number of Accidents

If you are involved in an automobile accident, you are required to fill out an accident report that describes what happened, when it happened, and who was involved. If the accident is serious, then police will conduct their own investigation and file their report. The data from such reports is entered into a database to allow statistical assessment of the causes of accidents. Data for hundreds or thousands of accidents can be used to determine how accident frequency varies by time of day, age of the driver, weather conditions, type and age of car, type and condition of road, and other factors. This data can be combined with other information about traffic volumes and the characteristics of licensed drivers to obtain estimates of accident rates, which could be measured as the number of accidents per million miles driven. This type of analysis shows that risks are higher for new drivers, very old drivers, drivers using cell phones, drunk drivers, drivers who speed, and drivers and passengers who do not wear seat belts.

Knowledge of these risks has led to public policies aimed at reducing the number or severity of accidents:

- *Policies aimed at new drivers:* requirements for driver training, restrictions on initial driver's license (limits related to time of day and number of passengers), and programs that increase public awareness of risks faced by new drivers
- *Policies aimed at very old drivers:* eye tests for renewal of licenses; programs that increase public awareness of risks faced by old drivers (which may help older drivers realize that they need to limit or stop their driving); research to develop new technologies to enhance visibility or to reduce reaction time of older drivers
- *Policies aimed at reducing drunk driving:* suspension or loss of license; public awareness programs; highly publicized extra enforcement on major holiday weekends
- *Policies aimed at reducing the severity of accidents:* requiring seat belts and airbags; construction standards for cars; requiring helmets for motorcyclists
- *Policies aimed at improving the design of highways:* limiting access to highways; minimizing curves; requiring wider lanes; eliminating dangerous intersections; improving sight distances by clearing vegetation; improving signage

Detailed Analysis of a Few Major Accidents

It is also possible to conduct detailed analysis of particular accidents or incidents in order to determine causality. Such investigations are undertaken by company and government officials for any dam failure, any incident at a nuclear power plant, airline crashes, rail collisions, and any other accident or incident that led to or could have led to a catastrophe. The results of the investigation may highlight problems with design, operations, maintenance, personnel training, weather, or other factors. A single accident or incident may result in new regulations, new designs for equipment, inspection of an entire fleet of airplanes, or new training practices.

Statistical Analysis of a Representative Sample of Accidents

The standard accident report submitted does not provide extensive information about the condition of the vehicle or the driver. It also may have incomplete, confusing, or inconsistent accounts of the accident. If the drivers and witnesses can be interviewed soon after the accident, it is possible to learn more about not just what happened, but why it happened. If the driver works for a transportation company, then the company may also have records indicating how long the driver had been on duty and how often the driver had been driving in the past several days.

The Federal Motor Carrier Safety Commission recently conducted a study of the causes of serious crashes involving large trucks.[12] The study was called the Large Truck Crash Causation Study (LTCCS). A serious crash was one that involved an injury or a fatality, and a large truck was defined as having a gross vehicle weight of more than 10,000 pounds. LTCCS investigated a representative sample of 963 crashes that occurred in 2001 to 2003 involving 1123 large trucks and 959 other vehicles; 73% of these crashes involved a large truck and at least one other vehicle; 77% of the large trucks were tractors pulling a single trailer, and 5% of the crashes involved hazardous materials.

The LTCCS was conducted by having teams visit the site of large truck crashes shortly after the event. The teams interviewed the drivers or the trucking companies involved in each crash, along with witnesses and passengers. From the trucking companies, they obtained the drivers' logbooks, which indicated where and when they had been working. They reviewed police reports, hospital records, and revisited the scene after assembling all of the relevant information.

---

[12] Analysis Division, Federal Motor Carrier Safety Administration, U.S. Department of Transportation, "The Large Truck Crash Causation Study," U.S. DOT Publication FMCSA-RRA-07-017, July 2007.

The LTCCS identified three types of variables related to causality:

1. *Critical event:* the event that made the collision unavoidable
2. *Critical reason:* the reason that led to the critical event, which could be driver error, vehicle failure, or a problem in the environment (weather or highway condition)
3. *Associated factors:* any factor related to the vehicles, people, or environment in the period leading up to the crash

The research team identified one critical event and one critical reason for each accident. The critical event could be assigned to the truck or to another vehicle involved in the accident. Three types of critical events were assigned to trucks:

1. Running out of the travel lane into another lane or off the road (this was the critical event for 32% of the accidents involving large trucks in which the critical event was assigned to the large truck)
2. Loss of control due to traveling too fast for the conditions, cargo shift, vehicle systems failure, poor road conditions, or other reasons (29%)
3. Colliding with the rear end of another vehicle in the truck's travel lane (22%)

In other words, when the truck was at fault, 83% of the accidents involved a truck that ran out of its lane, lost control, or ran into a vehicle in front of it. Once these events occurred, the crash was inevitable. But what were the critical reasons that caused these events? Only 3% related to the weather or to roads conditions, and only 10% related to some problem with the vehicle. If the critical event was assigned to the truck, then the vast majority of the critical reasons related to the truck driver:

- *Poor decision* (38%): driving too fast, following too closely, misjudging the speed of other vehicles, etc.
- *Failure to recognize a problem* (28%): inattention, distraction, or some other reason for failing to observe the situation
- *Nonperformance* (12%): falling asleep, heart attack or seizure, or other impairment
- *Poor performance* (9%): panicking, overcompensating, or exercising poor directional control

This logic could be pursued even further: What was the situation that led to the critical reason? Did the driver fail to recognize the problem because of poor visibility, illness, or lack of familiarity with the road? The LTCCS therefore identified a large number of potentially causal factors associated with each vehicle involved in the crash. The top 10 factors were brake problems, congestion, prescription drug use, traveling too fast, unfamiliarity with the roadway, roadway problems, required to stop (e.g., traffic signal), over-the-counter drug use, inadequate surveillance, and fatigue. To determine which factors were most related to the risks associated with large trucks, the study divided the trucks into two groups—those that were assigned the critical reason for the crash and those that were not. The percentage of crashes with the associated factor was then calculated for each group of trucks. The relative risk for each factor was determined by comparing the two percentages:

Relative Risk = % of at-fault trucks with this factor/% of not-at-fault trucks with this factor

(Eq. 13.17)

If the percentages were the same, then the relative risk would be 1.0, and that factor would not increase the risk of being in an accident. This was the case for 3 of the top 10 factors: congestion, use of prescription drugs, and being required to stop.

If a factor was more commonly associated with at-fault trucks than with other trucks involved in accidents, then that factor would have a higher relative risk. For example, approximately 30% of the drivers of the at-fault trucks were unfamiliar with the roadway, but only 15% of the drivers of the not-at-fault trucks. The relative risk was therefore 30%/15% = 2.0 if the driver was unfamiliar with the roadway. The highest relative risk was 56.3, for the "cargo shift" factor.[13]

---

[13] If the cargo shifts in a heavy truck, then the load may become unbalanced, which can cause a tractor-trailer combination to behave erratically or even overturn in a curve.

Table 13.7   Most Important Factors for At-Fault Trucks

Factor	Frequency (percentage of trucks with the factor)	Relative Risk	Frequency × Relative Risk (F × R)
Driver: made illegal maneuver	9%	26.4	2.4
Vehicle: cargo shift	4%	56.3	2.3
Driver: traveling too fast for conditions	23%	7.7	1.8
Driver: inattention	9%	17.1	1.5
Driver: inadequate surveillance	14%	9.3	1.3
Driver: following too close	5%	22.6	1.1
Driver: fatigue	3%	8.0	1.0
Driver: illness	3%	34.0	1.0
Vehicle: brake problems	29%	2.7	0.8

The study highlighted the factors with the highest relative risk and the highest frequency. Table 13.7 shows the factors that had the highest potential for causing accidents. The far right column shows the product of the relative risk and the frequency. None of the other factors had a combined score that was close to 0.8. In particular, substance abuse was not a major factor for truck drivers: only 2% of the truck drivers involved in crashes were using illegal drugs (with a relative risk of 1.8 and $F \times R = 0.036$), and only 1% were using alcohol (with relative risk of 5.3 and $F \times R = 0.05$).

The most important factors include some that are uncommon but have high relative risk, such as driver fatigue or a vehicle with a load that has shifted. The list also includes some fairly common factors with modest relative risk, such as driving too fast or a vehicle with brake problems.

The study also looked at the accidents involving a large truck and a passenger vehicle (car, SUV, or pickup truck). Some marked differences were noted in the factors that were most important for trucks and for passenger vehicles. Illness, drug use, alcohol use, and fatigue were two to five times more likely to be associated with the passenger car driver than with the truck driver. Following too closely and distractions outside of the vehicle were more likely to be associated with the truck driver than the passenger car driver.

This study demonstrates how a detailed analysis of accidents can provide insights into the factors that are most likely to contribute to risk. The factors will be different for different types of accidents, and therefore the strategies for reducing risks must also differ.

13.3.6   Mitigating Risks: Floodplain Management in Fort Collins, Colorado

As described in the previous section, probabilistic risk assessment is a technique that can be, and has been, used to determine how best to reduce risks. To show the value of this approach, consider the case of Fort Collins, Colorado, a city of more than 100,000 people that is located 60 miles north of Denver.[14] Six small streams, any of which can flood during heavy rains, flow through the region. One of the streams, Spring Creek, had flooded in 1902, 1904, 1938, 1949, and 1951. Nearby regions suffered from a major flood in 1976, and many floods have occurred in and near Fort Collins. In 1979, Fort Collins began participating in the National Flood Insurance Program, which lowers flood insurance premiums for policyholders if the local government adopts measures to limit damage from flooding and erosion. A dam failure in 1982 prompted Fort Collins to begin a comprehensive floodplain management program. The program included various activities and projects:

- Education
  - Increased public awareness of flood dangers and possible protection measures
  - Information for property buyers and sellers

---

[14] Mike Grimm, "Floodplain Management," *Civil Engineering* 68, no. 3 (March 1998): 62–64.

- Regulation of floodplain development and design criteria for structures located in the floodplain
- Acquisition of floodplain land
- Drainage system maintenance
- Creation of a database with local flood and erosion hazards

In 1990, Fort Collins entered the Community Rating System (CRS), established by the Federal Emergency Management Agency (FEMA) to encourage communities to manage floodplain risks. Part of the CRS is based on the location of critical facilities such as hospitals and nursing homes, police and fire stations, and storage facilities for highly volatile, flammable, explosive, or toxic materials. In 1995, Fort Collins changed its zoning to prohibit the location of critical facilities within the 500-year floodplain (i.e., the area that would be flooded by the worst flood of the past 500 years), and the city relocated a retirement home that it had recently acquired to a safer site outside of the floodplain. In 1996, FEMA recognized Fort Collins as one of the top 10 out of nearly 1000 communities participating in the CRS.

These preparations were made none too soon. On July 28, 1997, Fort Collins was subjected to torrential rains: over a 30-hour period, 13 inches of rain fell in the Spring Creek drainage basin; 8.4 inches fell in less than 5 hours, which was believed to be the heaviest recorded rainfall in the history of Colorado.

In preparing for such an event, the city had spent $5 million on various flood control measures for Spring Creek:

- Acquisition and removal of 41 structures from the most hazardous portions of the floodplain, including 30 mobile homes as well as the retirement home
- Channelization that effectively removed 45 other structures from the floodplain
- Storm drainage improvements
- Reinforcement of a railway embankment
- Preserving approximately one-third of the floodplain area in the city as open space

Despite the preparations, the deluge resulted in five fatalities and property damage estimated to be $9 million, and it shut down the city. However, the removal of so many structures from the floodplain was estimated to have avoided $5.6 million in further damages. FEMA estimated the damage as the value of the 41 structures that were removed from the floodplain, not including their contents. More important, without these measures, many of the nearly 100 people whose homes were relocated might well have perished in the flood.

This example shows how probabilistic risk assessment can be used in developing strategies for reducing the risks associated with civil infrastructure, in this case the risks to a city from floods. The strategies employed some projects that were aimed at protecting structures from damage, some projects that moved structures out of high-risk locations, and some government actions that prevented development in high-risk locations.

13.4 PERFORMANCE-BASED TECHNOLOGY SCANNING

A project may be completed on time, within budget, and with no serious accidents, yet still be unsuccessful. The project fails to operate as intended or, more likely, the actual demand for the project may be much less than expected. To minimize the risks associated with building projects that too few will use, it is essential to think about how people will respond to the project long before the design is finalized. The usefulness of the project and the ultimate demand for the project are based not on the technical difficulties of its construction or the personalities of the leading characters promoting the project, but on the actual changes in system performance made possible by the project. It is well to consider whether the proposed improvements in performance are really sufficient to attract enough users and supporters who are willing to pay enough to cover the costs of construction and operations. It is not necessary to know how to build a bridge in order to begin thinking about the potential benefits if the bridge were built. Likewise, it is not necessary to know how to make pure drinking water more widely available in developing countries in order to estimate the health benefits that could be achieved.

The process of deciding what kinds of projects might be most useful is similar to the process of deciding what technologies might be most useful. If new technologies are to be adopted, whether the technologies involve electronics, bioengineering, or stronger materials, a series of projects are needed to implement those technologies. Methodologies that have been used to determine which new technologies might be most useful turn out to be applicable in deciding the types of projects that might be most useful. This section therefore introduces concepts broadly referred to as **technology scanning**.

The goal of technology scanning is to identify and evaluate new and emerging technologies that are potentially important to an industry, its competitors, and its customers. Performance-based technology scanning (PBTS) is a methodology for identifying areas where new technologies can have greatest performance benefits over the next 20–30 years in terms of reducing costs, increasing market share, and achieving higher profitability.[15] Subsection 13.4.1 describes the steps involved in PB TS; subsection 13.4.2 presents an application of PBTS to intercity passenger transportation.

13.4.1 Overview of the Methodology

New technologies, however exciting, are germane only if (a) the technologies lead to significant improvements in system performance (speed, safety, cost, ease of use, etc.), and (b) those performance improvements are indeed important for users of a system. Effective technology scanning can help an industry identify new technological approaches, formulate a broader yet better focused research and development (R&D) program, and improve its investment strategies. A superficial technology-scanning program readily identifies exciting technologies, but it can be distracted and diverted into finding high-tech solutions for minor problems rather than seeking technological assistance in dealing with fundamental problems. A balanced technology-scanning program should consider how technology can help meet customer needs and overcome fundamental operating constraints.

Table 13.8 shows the range of possible technology-scanning activities as they might apply for companies or industries or agencies that manage or develop infrastructure systems. At the broadest level of technology scanning, there is a "General Search for Technologies." This search is necessarily somewhat unstructured since it is not initially clear what new and emerging technologies will be available or what relevance they will have for the industry. This search should involve people with varied backgrounds and different working contexts, so that the search is truly broad. The three intermediate

Table 13.8 Performance-Based Technology Scanning

General Search for Technologies
Conduct a broad review of new and emerging technologies that might be beneficial to the system.
Technology Mapping
Conduct structured investigations into the performance capabilities of the system and identify the points of leverage for technological developments related to cost, reliability, safety, or capacity of the system and any competing systems.
Systems Modeling
Develop and maintain a set of models that can be used to evaluate technological improvements as they affect specific aspects of systems performance.
Customer Requirements Analysis
Investigate the requirements of selected groups of customers or users and identify how new technologies might enable new ways of doing business; estimate the benefits to customers that will result from improvements in cost, speed, reliability, safety, or capacity.
Analysis of Specific Technologies
Examine specific technologies identified as having potential for improving system performance.

---

[15] C. D. Martland, "Performance-Based Technology Scanning Applied to Containerizable Freight Traffic," *Journal of the Transportation Research Forum* 57, no. 2 (Spring 2003): 119–34.

activities provide ways to narrow the technology scan from the "general search for technologies" to the "analysis of specific technologies."

Technology mapping is the most general of these activities, since it predicts the effects of hypothetical technological changes in order to find the most important technological constraints on system performance. This activity, for example, might consider the relative advantages and disadvantages of increasing capacity, reducing costs by enabling more efficient operations, increasing the life of major system components by introducing more effective maintenance or inspection technologies, and enhancing safety or security. Technology mapping begins with a base case that illustrates performance of a representative portion of the system. Next, high-level models predict performance for particular types of services as a function of technological capabilities, using inputs that capture the desired or anticipated results of deploying new technologies on each aspect of performance. Separate sets of models can be developed to predict the performance of typical types of systems at sufficient detail to provide realistic performance data and to capture the major competitive issues for different market segments.

Systems modeling is a more detailed activity. The objective here is to estimate the effects of particular technological improvements on system capabilities and performance. This is where very detailed engineering models are useful.

Customer requirements analysis can also be carried out at various levels of detail. There are several basic questions: How will a particular group of customers respond to potential changes in price, service, safety or capacity? What constraints, if any, limit the amount of services that will be purchased by these customers? How important are improvements in equipment design as opposed to improvements in trip times and reliability?

Finally, at the most detailed level of technology scanning, there is a need for analysis of specific technologies to demonstrate that a particular technology is indeed suited for the industry. Care is required in selecting technologies for this expensive stage of technology scanning.

The overall process is called performance-based technology scanning (PBTS). It includes consideration of the basic technologies but also investigates how technologies translate first into better technological performance and then into better system performance in terms of the competitive market environment. The best technologies will relieve constraints that currently limit competitiveness. Using PBTS, it is possible not only to find new technologies but also to identify gaps in an existing research program or to rearrange investment priorities in order to achieve more rapid implementation of the most effective technologies.

13.4.2   Applying PBTS to Intercity Passenger Transportation

This subsection applies PBTS to intercity passenger services, drawing upon research conducted for the UIC (Union Internationale de Chemins de Fer (UIC).[16] Because the research was conducted for a railway organization, the focus was on how to enhance the competitiveness of rail transportation; that perspective is taken throughout this subsection. The research ultimately related to the types of projects that would be desirable for attracting more intercity travelers to the railways of the world. Increasing the maximum train speed is often seen as the best way to improve the competitiveness of passenger rail service, but considering technological options using the PBTS framework suggests many other opportunities for improving rail competitiveness. Construction of high-speed rail corridors will be helpful in some locations, but other types of projects may prove to be just as useful and more cost effective.

The role of rail in intercity transportation varies widely around the world. In China, India, and Russia, where incomes are low and the rail network is extensive, rail has the largest market share for intercity travel. In much of Latin America, bus is the preferred mode even for distances greater than 1000 km. In the most developed countries, railways must compete with air for longer-distance travel

---

[16] C. D. Martland, Alex Lu, Dalong Shi, Nand Sharma, Vimal Kumar, and Joseph Sussman, *Performance-Based Technology Scanning for Rail Passenger Systems*. UIC/MIT-WP-2002-02. (Cambridge, MA: MIT, July 2002).

and with autos for shorter-distance travel. Still, when railways are able to offer service on the order of 100 mph along 200–300 mile corridors, they can capture more than half the non-auto market. Examples include Paris–London, Stockholm–Gothenburg, and Rome–Bologna. Where railways offer service in excess of 120 mph, they can dominate such markets (e.g., Paris–Brussels, Paris–Lyons, and Tokyo–Osaka). In the United States, passenger rail services are highly competitive only for the Northeast Corridor, and rail market share is very low elsewhere.

Competition for Intercity Passenger Services

Competition for intercity passenger services is based on cost, time, and quality of the available services. These elements can be modeled using the economic concept of utility, which was introduced in Chapter 3. Travelers' utility can be increased by reducing costs, increasing speed, or improving the quality of the passengers' experience. Travelers will choose the mode that allows them to reach their destination with the greatest utility. Technological changes can improve utility for potential customers and therefore increase market share, as demonstrated in this subsection.

Air, bus, and auto are the primary modes competing with rail for intercity passengers. For air, the key factors are the time required at the terminal and the number of stops, as well as the actual flight time. For rail-competitive trips (less than 1000 miles), flight time is at most several hours, often less than half the total trip time. Fares, access time, terminal processing, and time and hassle associated with connections are key elements affecting travelers' utility.

Bus is much simpler than air or rail because the terminal time and amenities are both minimal. The average trip time depends on highway conditions and the number of stops. Travel by bus allows opportunities for work, and seats in the best buses are at least as comfortable as those in coach class on most planes. Design of bus networks is extremely flexible and readily integrated with air or rail networks.

Auto travel is the most flexible, the most comfortable, and often appears the cheapest. Most people ignore depreciation and treat insurance and taxes as fixed or sunk costs, worrying only about out-of-pocket costs (fuel and tolls for personal automobiles, plus daily and mileage fees for rental cars). Auto competition varies greatly across the world in terms of availability, service, and cost. Where roads are poorly developed or extremely congested, auto is too slow for anything but short trips. Where auto ownership is high and highways well-developed, auto is a convenient, cheap option for traveling quite long distances. In Europe and Japan, out-of-pocket costs are high because of tolls and fuel taxes. In China, railways are losing mode share to autos and especially buses as the highway network is expanded.

The Utility of Time

Economists use the concept of utility as a means of understanding how people make economic decisions. People are assumed to make choices that maximize their utility, perhaps unconsciously, thus providing a basis for understanding and modeling the way people make choices. In principle, utility can encompass cost, travel time, comfort, and other factors. Surveys and statistical methodologies can be used to develop models of utility based on user choices or their stated preferences. Assuming that such models exist, we can compare the utility associated with using rail, air, or other modes. If utility is identical for two modes, then we would expect travelers to be indifferent to which mode they use. If utility varies with distance, there may be a break-even distance at which mode shares are equal. Models that estimate the percentage of travelers who will choose a particular mode are called travel demand models. These models are used extensively in evaluating transportation projects, since the predicted benefits of any transportation project will depend heavily upon the demand for the new facility or service.

Some general insights concerning utility have been gained from research on travel demand:

- Trip times and reliability are important factors in addition to out-of-pocket cost.
- Value of time is related to, but less than, the hourly wage and may depend upon mode or trip purpose.

- Time spent in different activities is valued differently; time spent moving in a vehicle is generally less onerous than time spent waiting in the terminal.
- Ease of access and ease of using the mode are important.
- Time of day, trip purpose, and service frequency affect choice of departure and arrival times.

These results document great variations in value of the time for different groups of people in various activities. A study of intermodal facilities for intercity rail, bus, and transit facilities suggested using one-third of the prevailing wage for the travel time from home to work, one-sixth of the prevailing wage for nonwork travel, and 200% of the prevailing wage for work-related travel.[17] Safety and security can also be included in utility analysis.

Studies of demand typically use rather general independent variables, often just separating trip time into in-vehicle and out-of-vehicle time. A study of high-speed rail conducted for the U.S. Federal Railroad Administration, for example, used trip time, fares, and frequency of service in developing demand models for various market segments.[18] That study addressed the potential markets for high-speed rail as an alternative to air and auto travel; it defined service to be total trip time, without attempting to distinguish the utility associated with different trip segments. However, passengers make much finer distinctions concerning utility than this, so it is necessary to make some assumption about passenger utility as part of PBTS.

The researchers in the UIC study assumed that time and comfort utilities can be expressed in monetary terms and compared directly to fares and other out-of-pocket costs. They then made assumptions concerning utilities for different segments of a trip in order to illustrate the relative importance of these segments and the opportunities for technological improvements. After all, it is clear to anyone who has ever traveled that the time spent in some portions of the trip is very onerous, while the time spent in other portions of the trip may be neutral or pleasant. Saving a few minutes in travel time by introducing faster trains may not be nearly as beneficial as using better information technology to save the same few minutes in terminal processing, or providing in-vehicle communications and entertainment to make travel time more productive or more enjoyable.

A Preliminary Model of Passenger Utility

To begin, consider how a business traveler with an average billable rate of $100/hour and a salary of $40 per hour might view the various segments of an air trip:

- *Drive to airport, including buffer time required because of access unreliability:* Unproductive time is valued at 50% of the average salary, or $20/hour
- *Process time:* Standing in lines, checking in, going through security, and boarding are not only unproductive, but uncomfortable and stressful, so this time is valued at $50/hour.
- *Extra time at the airport:* Conceivably useful for shopping, eating, or reading, but likely broken into segments too small to be productive; this time is valued as somewhat better than driving at $10/hour.
- *Time on the plane resting, eating (peanuts), waiting:* Similar to the time in the car, probably negative, but at something less than average salary; this time is valued at $20/hour.
- *Time on the plane having fun:* Time spent watching a movie, eating (a real meal), or reading a book may be indistinguishable from time spent at home, so some of the time could be considered neutral—that is, $0/hour.
- *Time on the plane working:* This could be billable time with a positive value of $100/hour.

---

[17] Alan Horowitz and Nick Thompson, *Evaluation of Intermodal Passenger Transfer Facilities.* Final Report to the U.S. Federal Highway Administration, DOT-T-95-02. (Washington, DC: U.S. DOT Technology Sharing Program, September 1994).

[18] Federal Railroad Administration, *High-Speed Ground Transportation for America* (Washington, DC: U.S. Department of Transportation, September 1997), 5–10.

Table 13.9 Calculating Out-of-Pocket Cost, by Various Modes

	Air Nonstop	Air via Hub	Train	Auto	Rental Car
Circuity	1	1.2	1.1	1.1	1.15
Distance one way	250	300	275	275	287.5
Days at destination	2	2	2	2	2
Reservations (hours)	0.25	0.25	0.25	0	0.1
Cost (one way)					
Fare—fixed	$100	$50	$25		
Fare per mile	$0.50	$0.40	$0.30		
Expenses per trip					$40
Expenses per mile				$0.30	$0.05
Expenses per day					$40
Access to destination	$20	$20	$10	$0	$0
Parking per day	$20	$20	$20	$20	$20
Total out-of-pocket cost	$289	$234	$162	$123	$178

This individual would presumably associate similar utilities with the corresponding segments of a trip by rail, bus, or automobile, although the duration of similar segments could be quite different for each mode. If we break the competing travel options into logical trip segments and use consistent values of time for each activity, then we can estimate the utility associated with the various options available for any trip.

Let's begin with a 250-mile trip, a distance long enough for rail to be competitive with auto and short enough to be competitive with air. Tables 13.9 to 13.11 give representative inputs for evaluating trip utility. Table 13.9 shows sample inputs for calculating out-of-pocket costs. Air is the most expensive ($289 one way), automobile is the least expensive ($123),[19] and rail is in the middle ($162). The table also shows the time required to make a reservation, which is not an out-of-pocket expense but will affect utility.

Table 13.10 shows the factors used to estimate total travel time, including access, terminals, and buffers sufficient to cover likely delays. Nonstop air is the fastest, requiring 5.25 hours; rail and auto are nearly an hour longer.

Table 13.11 shows hypothetical values of time that might be reasonable for a business traveler in the United States for the various activities specified in Table 13.9; the final row shows the value per hour for the extra time gained by using the fastest mode. Most likely, the extra time is a net benefit to travelers at something close to their average value of time. However, it could be more or less. For a business traveler, the extra time might be spent with the client, leading to a higher probability of having a successful meeting. Table 13.11 therefore shows that the extra time is worth $150/hour, 50% higher than the value of work time for our hypothetical business traveler. Other travelers might have completely different perspectives on the value of this extra time. For a student traveling home for the holidays, extra time on the train might be valuable time to finish an assignment, or it might mean missing the start of a great party. A vacation traveler might lose 2% of the daylight hours available on the beach during the vacation, or gain time to finish up work before relaxing on the beach.

---

[19] Some business travelers are reimbursed for using their own automobiles on company business. The travel allowance is likely to be based on the fully allocated costs of owning and operating a car, which is on the order of $0.50 per mile. For a 500-mile round trip, a business traveler might thus be reimbursed $250, which would be $100 more than the variable costs of gas, tolls, and wear and tear on the vehicle, which are estimated in Table 13.8 as $0.30/mile or $150 for the trip. When passenger service was canceled between Pittsburgh and Harrisburg in 2009, this factor was cited as a major reason for the lower-than-expected ridership: the rail service was competing with the private service operated by the potential passengers themselves and was subsidized by their employers in the form of mileage reimbursement for use of their cars!

Table 13.10   Calculating Total Trip Time, by Mode

	Air Nonstop	Air via Hub	Train	Auto	Rental Car
Time for trip					
Access to station	0.75	0.75	0.5		0.5
Buffer for access unreliability	0.25	0.25	0.2		
Process time	0.1	0.15	0.0		0.25
Queue time	0.25	0.35			
Available time in station	0.5	1.5	0.25		
Boarding time	0.2	0.4	0.2		0.2
Travel time—fixed	0.75	1.5	0.2		
Travel time per 100 miles	0.2	0.2	1.25	2.0	2.0
Total travel time in vehicle	1.25	2.1	3.64	5.5	5.75
Travel time—work	75.0%	75.0%	75.0%	0.0%	0.0%
Travel time—entertainment	0.0%	0.0%	0.0%	10.0%	10.0%
Travel time—rest & other	25.0%	25.0%	25.0%	90.0%	90.0%
Travel time—work	0.94	1.58	2.73	0.0	0.0
Travel time—entertainment	0.0	0.0	0.0	0.55	0.58
Travel time—rest & other	0.31	0.53	0.91	4.95	5.18
Exit time from vehicle	0.2	0.4	0.2	0.0	0.25
Exit time from station	0.25	0.25	0.1		
Access to destination	1.0	1.0	0.5	0.25	0.25
Buffer for access unreliability	0.5	0.5	0.5	0.25	0.25
Total time	5.25	7.65	6.09	6.0	7.45

With these detailed inputs concerning travel time and the value of time, we can estimate our traveler's utility for each mode (Table 13.12). Time is shown as a "disutility," so it has the same sign as cost; the mode with the lowest disutility is thus the preferred mode. The quality of time spent traveling is clearly important; ranking the available options in terms of their disutility gives much different results than ranking by either out-of-pocket costs or time. In particular, rail looks much better because it affords extra time for work and less for processing and access. Although rail takes an hour longer, its

Table 13.11   Hypothetical Value of Time, by Mode and Type of Activity

	Air Nonstop	Air via Hub	Train	Auto	Rental Car
Reservations	50	50	50	50	50
Time for trip					
Access to station	20	20	20	20	20
Buffer for access unreliability	20	20	20	20	20
Process time	50	50	50	20	50
Queue time	50	50	50	20	50
Available time in station	10	10	10	10	10
Boarding time	50	50	50	50	50
Travel time—work	−100	−100	−100	−100	−100
Travel time—entertainment	0	0	0	0	0
Travel time—rest & other	20	20	20	40	50
Exit time from vehicle	50	50	50	0	0
Exit time from station	50	50	50	50	50
Access to destination	50	50	50	50	50
Buffer for access unreliability	10	10	10	10	10
Extra travel time	150	150	150	150	150

Table 13.12 (Hypothetical) Disutility of Travel, by Mode

	Air Nonstop	Air via Hub	Train	Auto	Rental Car
Direct costs	$289	$234	$162	$123	$178
Reservations	$13	$13	$13	$0	$5
Travel time					
Access to station	$15	$15	$10	$0	$10
Buffer for access unreliability	$5	$5	$4	$0	$0
Process time	$5	$8	$0	$0	$13
Queue time	$13	$18	$0	$0	$0
Available time in station	$5	$15	$3	$0	$0
Boarding time	$10	$20	$10	$0	$10
Travel time—work	−$94	−$158	−$273	$0	$0
Travel time—entertainment	$0	$0	$0	$0	$0
Travel time—rest & other	$6	$11	$18	$198	$259
Exit time from vehicle	$10	$20	$10	$0	$0
Exit time from station	$13	$13	$5	$0	$0
Access to destination	$50	$50	$25	$13	$13
Buffer for access unreliability	$5	$5	$5	$3	$3
Extra travel time	$0	$360	$126	$113	$330
Total travel time disutility	$43	$381	−$58	$326	$636
Total disutility	$344	$627	$117	$448	$820

disutility is less than the disutility of flying. For someone who can work on the train, driving is not a good option. Renting a car, which looks good in terms of direct cost, is by far the worst choice. It takes time to rent the car, and it is usually impossible to work in the car, so the disutility of the time is quite high relative to train or plane.

This particular example emphasizes the importance of "work time" to the decision and shows that the cumulative benefits of lower terminal time, easier processing, and greater accessibility help rail relative to air travel (but hurt rail relative to driving your own car). It also suggests a framework for comparing technologies or projects. Any intercity market will have groups of travelers with diverse needs and values. Some people may be able to think effectively when driving, so they may look forward to having several quiet hours in a car. Vacation travelers are concerned with baggage-handling facilities, but day-trippers are not. Self-employed businessmen undoubtedly view time and costs of travel far more carefully than do corporate travelers, whose personal finances are unaffected by their travel choices. The value of terminal services depends upon the expectations of the customer. Hungry students devour fast food, as long as it is cheap and plentiful; wealthy couples en route to a resort prefer to pass an extra hour enjoying a fine meal; a "road warrior" might grab pizza and a beer and check e-mail. The next subsection considers how passengers in four market segments might respond to various changes in mode or trip characteristics.

13.4.3 Estimating Mode Share

Given the utilities (or disutilities) for each available mode, it is possible to estimate mode shares using what is called a logit model. The mode share for mode j is calculated as follows:

$$\text{Mode Share} = \left(e^{-\text{disutility mode j/scale factor}} \right) \Big/ \left(\sum e^{-\text{disutility mode k/scale factor}} \right) \qquad \text{(Eq. 13.18)}$$

This type of model is commonly used in travel demand studies. If the disutility of two modes is within 5 or 10%, each mode has a sizeable market share; if the disutility of one mode is much greater, then it has a minor share of the market. The scale factor was assumed to be 25% of the average disutility of the mode with the lowest disutility for each market segment. This factor determines how strongly mode shares vary with the relative costs.

The base case for the sensitivity analysis added three market segments to the example from the previous subsection: general business, vacation, and student. The latter three market segments have values of time that are 50%, 25%, and 10% of the values for the executive considered earlier. Each market segment was assumed to have an equal number of travelers.

13.4.4   Sensitivity Analysis

Six cases were investigated in addition to the base case (Table 13.13). The first two considered airline strategies:

- *Case 1—Discount Air Fares:* A new carrier enters the market, halving air fares but doubling processing times. Rail retains more than half the market, because the trip is too short for air speed to make much difference. Since business travelers expect to be productive, the rail option still looks good.
- *Case 2—Business Shuttles:* Major airlines introduce a service aimed at business travelers. Fares match the discount airlines, but processing, queuing, and wait times are halved. This service captures more than 90% of the business market. Vacationers also appreciate the time savings; more than half switch to air. Students, still searching for the best deal, divide fairly evenly among the two air modes, rail, and auto. Overall rail market share plummets to 10%.

The next three cases address possible rail responses to the business shuttle. Each helps retain market share, and the greatest benefits for this particular example come from improving access:

- *Case 3—Lower Rail Fares:* Railways respond to the shuttle by cutting fares by 20%. Executives don't even notice the change; the other groups increase their rail mode share to a quarter or a third. Overall, the rail share recovers to 24% of the market.
- *Case 4—High-Speed Rail:* Average rail operating speed is 150 mph rather than 80 mph. This is more successful than simply lowering fares, and rail is projected to gain 43% of the market. However, a major project would be needed to achieve such high speeds, and it is unclear if prices could remain unchanged.
- *Case 5—Easy Access:* The average speed is again 80 mph, but times are halved for rail processing, access, and reservations; also, better onboard seating and services increases the value of time by 20% for business travelers. The value of terminal and onboard entertainment time is increased for

Table 13.13   Sensitivity Analysis for Mode Share

	Air Nonstop	Air via Hub	Train	Auto	Rental Car
Base case	2%	1%	67%	29%	1%
Discount air fares	18%	3%	56%	22%	1%
Business shuttle	72%	9%	10%	9%	1%
Lower rail fares	58%	14%	24%	4%	0%
High speed rail	40%	12%	43%	4%	0%
Easy access	17%	8%	71%	3%	0%
Two travelers	2%	0%	54%	40%	4%
Easy rail & business shuttle					
125 miles	7%	4%	69%	20%	1%
250 miles	17%	8%	71%	3%	0%
375 miles	35%	12%	51%	1%	0%
500 miles	56%	16%	27%	0%	0%
625 miles	68%	19%	13%	0%	0%

everyone due to more entertainment, retail, and culinary opportunities in the stations and better food and services on the train. Executives are assumed to increase their working time from 70 to 80% of the trip time. The results are very strong for the railways, which become dominant in the first three markets and capture a third of the students.

Sometimes a group is traveling:

- *Case 6—Two Travelers:* Travelers share the cost of auto trips or cab rides. The dominant result is to make driving a very good option, and almost all air traffic and more than 20% of the rail traffic divert to auto. Rental cars also improve, increasing their share from 1 to 4% and becoming a good option for vacationers and students. Clearly, if a family is going on vacation with children, the automobile will look better for even longer distances. Likewise, if three or four people are traveling together on business, then renting a car may look better, particularly if they can conduct some business while driving.

Distance is obviously another key factor for sensitivity analysis, since rail works best for distances that are rather long for highway travel, yet rather short for airlines. "Easy Access vs. the Air Shuttle" was used as the base case. For the 125-mile trip, rail captured 69% and autos took 20% of the market. For the 250-mile trip, the highway modes essentially drop out; direct air flights capture 17% of the market. As distances increase to 625 miles, the rail share drops steadily while the air share grows. Air travel via a hub is increasingly attractive for the longer distances since the cost savings become large enough to justify the additional time.

13.4.5 Implications for Carriers and Terminal Operators

The implications of utility analysis are generally well understood. There is value in reducing travel time, minimizing process time, and increasing passenger comfort. There is value in providing a variety of ways for travelers to spend their time and their money. Carriers attempt to capture this value by offering premium services at higher prices. First-class and business-class travelers enjoy quicker check-in, comfortable and productive waiting areas, larger seats, and better food—and they are willing to pay a premium of $100–$200 per flight hour for these privileges. This premium is high compared to the coach fare, but not unrealistic when compared to executive salaries or consulting rates. Carriers also advertise their onboard services, including telephones, movies, games, magazines, and shopping opportunities.

Terminal operators may have been slower to understand the importance of time and utility, but they have certainly responded well over the past 10–20 years. New airports feature greatly enlarged shopping opportunities, food courts, fine restaurants, lounges, TVs, Internet access, ATMs, and other amenities that make waiting time more valuable to the traveler (and more profitable to the terminal owner). Government agencies and airlines are also concerned about airport access; they recognize the importance of time and comfort to the user as well as the costs of the infrastructure. Similar trends have affected some major train stations, which now offer varied retail and dining opportunities.

13.4.6 Implications for Technology Scanning

This example of performance-based technology scanning shows how markedly different technologies can be compared in terms of their potential effects on passengers' utility. The most striking comparisons are among the three generic responses to the business shuttle for the 250-mile trip (Cases 3–5). Lower fares could be interpreted as representing any of the many technologies that might reduce cost while leaving service and access unchanged. High-speed rail is of course a dominant theme in the evolution of rail technology, in rail R&D, and in proposals for rail investment. Easy access relates to entirely different technologies, including terminal processing and terminal access. For this example, access is somewhat more important than train speed and much more important than cost reduction. In general, saving time in access and processing or allowing more productive use of time may be more effective—for the customer—than saving time by running faster.

The rail industry and public agencies are well aware of the potential for high-speed rail systems to attract traffic from congested airports and highways, and extensive R&D and investment programs are in place to advance such systems. In the United States, the "Next Generation High-Speed Rail Technology Demonstration Program" was funded at more than $25 million annually in fiscal 2001 and 2002, exceeding the rest of the Federal Railroad Administration (FRA) budget for both passenger and freight R&D.[20] However, as demonstrated in this section, higher speed is not the only way to reduce travel time or enhance travelers' utility, and quite different kinds of technologies and projects may be equally effective in enhancing rail competitiveness.

13.5 SUMMARY

This chapter presents several specialized methodologies that can be useful in project evaluation. The basic concepts have already been introduced in previous chapters, but this chapter provides more details on each methodology and more examples of how they can be used to assist in project evaluation.

Modeling Performance: Simulation and Analytical Models

Models can be very useful in providing insight into the performance of a system, the likely effect of a project, or the amount of risk associated with a project. Models will be needed to explore three key questions:

1. How will system capabilities or performance change if the project is implemented?
2. If system capabilities or performance change, what will happen to demand?
3. How will uncertainties affect the success of the project?

Models can be simple or complex, depending upon the stage of the analysis and the nature of the issues being investigated. Deterministic or probabilistic models can be developed, and even quite simple models can be helpful in determining the key factors affecting the success or failure of a project. By treating key factors, such as project cost or demand, as random variables, it is possible to see how the expected variations in these factors might interact and to estimate the likelihood that a project will or will not succeed.

How Best to Improve System Safety: Probabilistic Risk Assessment

The probabilistic model is a structured methodology for understanding risk, perceptions of risk, and how best to allocate resources for reducing risk. Any infrastructure-based system can experience accidents resulting from poor design, structural failure, bad weather, earthquakes, mismanagement, human error in operations, or other causes. The risk associated with any type of accident is defined to be the probability of an accident multiplied by the expected consequences of an accident. An accident has many possible consequences, including property damage, minor injuries, serious injuries, and fatalities. If weights are applied to each type of consequence, it is possible to come up with a single measure of risk. The weights may be stated in monetary terms, in which case the weight can be interpreted as the value of a unit reduction in each type of risk.

Research into human behavior has shown that people are more concerned with risks associated with unknown factors (e.g., introduction of genetically modified crops) or catastrophic accidents. People apparently believe that more care should be taken to reduce the risks associated with potentially catastrophic accidents, such as a meltdown at a nuclear power plant or a chemical explosion, than needs to be taken for well known, but non-catastrophic accidents such as those occurring on ski slopes or highways.

---

[20] Federal Railroad Administration, *Five-Year Strategic Plan for Railroad Research, Development, and Demonstration* (Washington, DC: U.S. Department of Transportation, 2002), chap. 8.

Risks can be summed for different types of accidents and different locations. Global risk of a system is the summation of the risks of all types of accidents over all locations within the system. Projects may increase some risks while reducing other types of risk. The increases or decreases in risk can be treated as costs or benefits of a project.

Public perceptions of risks are likely to differ from what engineers and risk experts calculate to be the risks. A highly publicized catastrophic accident—or even an incident that could have been (but was in fact not) catastrophic—may cause public uproar and outrage. In the immediate aftermath of such an accident or incident, there could be extreme pressure for public action to ensure that such an accident "never happens again." Such a response could be an overreaction that goes beyond the specific cause of the actual accident, an irrelevant reaction that has little or no impact on risk, or an ineffective reaction that may reduce risk, but only at an excessive cost. Infrastructure managers can reduce the likelihood of such improper responses by understanding the risks associated with their system, adopting and publicizing a risk management program, and responding quickly and effectively to any accidents or incidents that may occur.

Studies can be undertaken to increase the understanding of risks and thereby guide the selection of projects that will be most cost effective in reducing risks. For commonly occurring accidents, such as highway accidents, it is possible to assemble a database with basic information on every significant accident. The database can then be used to support statistical analysis regarding accident causes or severity. For accidents that are rare and potentially catastrophic, such as those involving a power plant or a plane crash, every accident should be studied in great detail to determine whether any previously unknown risks need to be addressed. A special study can be conducted to identify the critical event that led to each accident, the critical cause that precipitated the critical event, and the factors most likely to be associated with the serious accidents. By understanding the causal chain that led to the accident and by isolating the most important associated factors, investigators can formulate a strategy for reducing the risks of this type of accident, including recommendations for changes in design of new facilities or projects that will correct deficiencies in existing facilities.

Performance-Based Technology Scanning

Many projects involve the introduction of new technologies or the modification of an existing system in response to competitors' introduction of new technologies. Technology scanning refers to the search for new technologies that might be important to a particular industry. Performance-based technology scanning (PBTS) is a structured approach to technology scanning that focuses on the way the technology might affect system performance rather than on the details of the technology. Through this methodology, a technology can be represented as an option that has particular cost characteristics (e.g., investment cost, operating cost, lead time required for implementation) and performance impacts (e.g., reduction in operating cost, increase in capacity, or improvement in level of service). It is even possible to consider a range of hypothetical technologies with various cost and performance characteristics. Whether the technologies are useful will depend on a market analysis. Will the changes in performance attract new demand? How will the new services be priced? How will competitors respond? Answering these questions may require a great deal of analysis, but not necessarily any great detail concerning the technology. Methodologies developed for PBTS are readily useful for evaluating potential projects.

An example shows how PBTS was used to examine technologies that might be important for improving rail passenger service. The analysis used a simple model of travelers' utility functions to compare trips made by rail to the corresponding trips made by auto or air. The analysis showed that train speed was only one factor, and perhaps a relatively minor factor, influencing travelers' decisions. Total door-to-door trip time, the quality of time spent in each portion of the trip, and the opportunity to use the time for enjoyable or profitable activities were all important factors. This study showed that the most lauded high-speed rail technologies may not be the most effective ways to increase rail ridership. Comfortable trains operating over a dense network at reasonably frequent intervals can compete effectively with both air and auto for trips of 100 to 500 miles.

ESSAY AND DISCUSSION QUESTIONS

13.1 *We should do everything possible to prevent this from ever happening again!*: Statements like this are common whenever a tragic accident or incident results in a fatality. Of course, we seldom "do everything possible" to improve safety. As a result, tens of thousands die annually in car accidents, and thousands of others die as a result of contaminated water, cancers resulting from toxic chemicals, and other infrastructure-related problems. And we have recently seen headlines regarding fatalities in mining accidents, not to mention the daily carnage from terrorist activities. Should we be doing "everything possible?" If not, what should we be doing?

13.2 Catastrophic accidents shape the design, restrict the operations, and hinder the expansion of infrastructure-based systems. Discuss how catastrophes have influenced the following:

a. Bridges
b. Nuclear power plants
c. Highways
d. Skyscrapers
e. Aviation
f. Dams

13.3 Many infrastructure projects are justified by time savings that will be gained by many people. What is your value of time in the following situations?

a. You are trying to decide whether to save 10 minutes by taking a cab to school rather than waiting for the bus. How much extra would you be willing to pay if you are

i. Heading to the library to get started on a research paper
ii. Afraid of arriving 5 minutes late for class—and you have a quiz today

b. You have a choice of paying for your ticket on the train or standing in line to pay for the ticket. If you buy the ticket on the train, the cost will be $3 rather than $2. How long would you be willing to stand in line to get the cheaper ticket?

c. You are headed home from school for the weekend. The bus is cheaper, but it takes an hour more than the train. How much extra are you willing to pay for taking the train?

13.4 Research indicates that people fear some types of accidents more than others. Should safety officials and infrastructure executives take perceived risks into account, or should they try to minimize global risks of their systems?

13.5 Rapid technological advances are being made in telecommunications, materials, computing, and remote sensing. Consider how technological advances in these or other areas might benefit a particular type of infrastructure-based system (e.g., transportation, water supply, office buildings, sewerage treatment, electric power generation). For this type of infrastructure:

a. What are the major categories of users of this infrastructure?
b. What are the key areas of performance for each class of user?
c. Which technologies are most likely to affect the key areas of performance?

PROBLEMS

13.1 Create a spreadsheet model that will allow you to conduct simple simulations similar to the one presented in subsection 13.2.1. The spreadsheet will allow you to examine the variability in a mathematical function that includes three random variables (e.g., how will profits for a new apartment building vary given the possible variations in construction cost, rental rates, and interest rates?). The general format of the spreadsheet should be as follows:

• *Heading:* your name, date, and title for the spreadsheet
• *Variable A:* an area where you enter parameters for the first random variable and allow space to obtain 20 random values for this variable
• *Variable B:* an area where you enter parameters for the second random variable and allow space to obtain 20 random values for this variable.
• *Variable C:* an area where you enter parameters for the third random variable and allow space to obtain 20 random values for this variable
• *Other variables:* an area where you enter values for any other variables needed to calculate the mathematical function you are interested in
• *Value of the function:* an area where you calculate the value of the function given (a) the values of the "other variable" and (b) the values of each set of random variables

Structure the areas for variables A, B, and C as shown in the following table. The user will enter the following information for each variable:

• *Value 1 to Value 6* (row 3): These are six possible values for the random variable.
• *Cumulative probabilities P1 to P6* (row 4): This is the cumulative distribution for the random variable, which is an input to the model. The value is Value1 if the random variable generated is less than or equal to P1; the value is Value 2 if the random variable generated is greater than P1 and less than or equal to P2; and so forth.
• *A1* (row 5): This is the first value generated for A1. Cell B5 is a random number (the spreadsheet function for a random number will generate a number between 0 and 1). Cells D5 to I5 are logical expressions that return "1" if the random number is within the proper range and "0" otherwise.
• *Value for A1* (cell C5): This is the value for variable A given the random number in cell B5. For example, if cell D5 = 1, then Value = Value 1.
• *A2 to A20* (rows 24 to 27): These are 19 more values for variable A based on 19 additional random numbers generated in cells B6 to B27. If you structure row 5 carefully, then you will be able to copy that row into rows 6 to 27.

- *Mean and standard deviation* (cells C25 and C26): You can use spreadsheet functions to get the mean and the standard deviation for variable A based on the values in column C.

Problem 1, Table 1: Basic Format for a Spreadsheet for Simple Probabilistic Analysis: Estimating the Variability of Variable A

	A	B	C	D	E	F	G	H	I
1	Name, date, title, etc.								
2									
3	Variable A	Random number	Value	Value 1	Value 2	Value 3	Value 4	Value 5	Value 6
4	Cum. prob.			P1	P2	P3	P4	P5	P6
5	A1								
6	A2								
7	A3								
...	....								
24	A20								
25	Average								
26	Std. dev.								

Once the area set up for variable A is completed, you may copy that to areas for variables B and C.

What you now have are 20 independent sets of values for variables A, B, and C. You can now create a section where you calculate 20 values for a mathematical expression involving A, B, C, and the other variables. For a simple expression, you could create an area of your spreadsheet that looks like this, assuming there is just one other variable D. The values for Ai, Bi, and Ci would be copied from the previous sections of your spreadsheet; the value for D is the same in each case.

Problem 1, Table 1: Basic Format for a Spreadsheet for Simple Probabilistic Analysis: Estimating the Variability of a Function of Several Variables

	A	B	C	D	E	F
1	Trial	Variable A	Variable B	Variable C	Variable D	Function
2	1	A1	B1	C1	D	f(A1,B1,C1,D)
3	2	A2	B2	C2	D	f(A2.B2,C2,D)
4	3	A3	B3	C3	D	F(A3.B3.C3,D)
...	...					
21	20	A20	B20	C20	D	f(A20,B20, C20, D)
22						
23	Mean					
24	Std. dev.					

13.2 A project has three phases. The project manager has a preliminary budget of $100,000 for each phase, but he is concerned that it may not include sufficient reserves. Based on prior, similar projects, he estimates that the chances of coming in under or over budget are as shown in the following table. What is the expected distribution of total cost? What should the budget be for the project to have at least an 80% chance of being completed without an overrun?

NOTE: Use the spreadsheet that you developed in problem 1 to solve this problem. The cost in Phases I, II, and III will be the variables A, B, and C; the function you are interested in is A + B + C.

Problem 2: Costs Expected for Phases I, II, and III of a Proposed Project

Phase	−$10,000	−$5,000	Preliminary Budget	+$5,000	+$10,000	+$20,000
I	5%	10%	60%	10%	10%	5%
II	10%	15%	50%	15%	5%	5%
III	15%	20%	40%	20%	5%	

13.3 A company is considering whether to go ahead with a project. They expect their investment cost to be about $10 million, annual revenues to be about $2 million, and annual operating costs to be about $0.5 million. Hence, they expect their annual profit to be about $1.5 million and their annual return on investment to be about 15%. However, there is uncertainty in all of these numbers, as suggested in the following table. How likely are they to achieve at least a 15% ROI?

NOTE: Again, use the spreadsheet that you developed for problem 1.

Problem 3: Expected Variability in Investment, Revenue, and Operating Cost

	−20%	−10%	Expected	+10%	+20%	+30%
Investment cost	0%	10%	75%	10%	5%	0%
Revenue	10%	15%	50%	15%	5%	5%
Operating cost	10%	20%	40%	15%	15%	

13.4 Research suggests that the public weighs catastrophic accidents more heavily than routine accidents. For some types of transportation accidents, the weighting factor appears to increase with the cube root of the number of fatalities.

a. If this is so, then a single accident with 350 fatalities would be equivalent to how many accidents with one fatality each?
b. Suppose that the probability of the catastrophic accident is believed to be 1 in 100 years, although as yet no such accidents have occurred. Over the past 10 years, minor accidents occurred at a rate of 10 per year. What is the expected number of fatalities per year?
c. Suppose you have a safety research budget of $1 million that you can allocate as you see fit between reductions in the risks of catastrophic or minor accidents. What is your budget for work on catastrophic accidents, assuming you allocate the funds based on (i) causes of actual accidents during the past 10 years, (ii) expected fatalities over the next 10 years, and (iii) perceived risks over the next 10 years?

Chapter 14

Managing Projects and Programs

A journey of a thousand miles begins with a single step.

Chinese proverb

CHAPTER CONCEPTS

14.1 INTRODUCTION

Project evaluation leads to selection of particular projects, which must then be implemented. Project management encompasses the detailed planning, scheduling, monitoring, and adjusting of the many processes involved in implementing a project. A series of related projects can be managed as a program; public policy often dictates the criteria to be used in implementing projects as part of a program. Once a program has been approved and funded, many projects can be selected and implemented based on specified evaluation criteria.

14.2 STAGES IN PROJECT MANAGEMENT

Let's assume that a project has been defined, refined, and approved, so that it is now time to proceed with financing, acquisition of land, assembly of resources, and construction. This is the time for project management. The project manager's main job is to transform the ideas for the project into reality while meeting expectations regarding cost, quality, timeliness, safety, and security:

> *Successful organizations create projects that produce desired results in established time frames with assigned resources.*[1]

Project management can be viewed as having five stages:

1. Preparing to implement the project
2. Managing construction
3. Managing the transition from construction to operation
4. Completing the project
5. Assessing or auditing the project

Project management can be viewed narrowly as just the second of these stages (i.e. the processes involved in managing construction. However, given the scope of this textbook, it is natural to think of project management as encompassing all of the activities involved in implementing a project, including a post-audit that could take place several years after the construction is completed. The remaining portions of this section provide an overview of the key aspects of each of these five stages of project management.

14.2.1 Preparing to Implement the Project

The first phase of the project begins when there is general agreement as to the basic design and purpose of the project and ends when with the start of construction. This phase entails the following tasks:

- *Financing*: Ensure that final plans for financing the project are in place before construction begins. As a minimum, cash must be available, or a means of raising cash as needed to pay for resources and construction. A plan for making interest payments or mortgage payments during the construction period must also be in place, plus a plan for what to do when the project is complete (e.g., refinance the loans, sell bonds, sell the project, turn over the project to a government authority after a period of operating the project).
- *Acquisition of land*: Take the final steps to acquire land, including the construction site as well as any temporary rights-of-way or staging areas that may be needed.
- *Public approvals*: Obtain zoning variances, construction permits, environmental permits, and any other public approvals that are needed to begin construction.

---

[1] S. E. Portny, S. J. Mantel, J. R. Meredith, S. M. Shafer, M. M. Sutton, and B. E. Kramer, *Project Management: Planning, Scheduling and Controlling Projects* (Hoboken, New Jersey: Wiley, 2008), 2.

- *Construction strategy*: Develop a general strategy for constructing the project, which may include deciding where to start and how quickly to proceed, identifying the supply strategy, determining how quickly to proceed, and breaking the construction into multiple phases. If necessary, complete the final engineering designs for the project.
- *Marketing*: Secure tenants for any rentable space, establish customers for any services to be made available, and explain the purposes of the project and the plans for construction to abutters and to the general public.
- *Transition planning*: Develop a strategy for the transition from construction to operation.

Financing, marketing, and transition planning will continue throughout the construction of the project. What can go wrong:

- Changes in the world credit markets can make it impossible to obtain financing at reasonable rates, thus causing projects to be canceled or delayed.
- Inability to acquire the necessary land can cause a project to be restructured to fit into a smaller piece of land (a very rich resident of New York City refused to sell his land, so the design for Rockefeller Center had to be revised; many highway projects have been rerouted to avoid taking land belonging to politically powerful individuals).
- A major client may back out of a deal to lease office space, so that it becomes impossible to get a construction loan for the building.

14.2.2 Managing Construction

The second phase of project management concerns management of the construction process. This phase is often considered the core of project management, and many management techniques have been developed to assist managers during this phase of the work. The key subtasks include the following:

- *Preparing a network diagram*: Identify the activities that must be undertaken to complete the project, along with the milestones that can be used to monitor progress during construction. The network diagram indicates how activities relate to one another by showing which activities can be started immediately and showing how other activities can be started only after the completion of prior tasks.
- *Scheduling*: Prepare a schedule for the activities, taking into account the relationships among activities as identified in the network diagram and the time and expense associated with each activity. The schedule will need to consider the possibility of delays, the availability of resources, and options for changing the time and expense associated with activities. For example, it may be possible to use more people to complete a task sooner, but it may be cheaper to use fewer people who will be more fully utilized over most of the construction period. To be feasible, the schedule will have to structure the sequence of activities so that they can be completed on time using the available resources.
- *Budgeting*: Convert the rough cost estimates used previously to justify the project into actual budgets that are consistent with the final plans and work schedule for the project.
- *Risk management*: Prepare plans for eliminating or reducing the risks associated with the construction process itself. Also prepare plans for securing the construction site and preventing incidents involving abutters or trespassers.
- *Mitigating social and environmental impacts*: Prepare plans to mitigate the impacts of construction on the neighborhood; this may involve working with local officials to change traffic flows around the site, placing barriers to prevent contamination of wetlands or bodies of water, limiting hours of work to avoid complaints related to noise, or taking precautions to minimize release of toxic substances.

What can go wrong:

- Budgets and schedules may reflect hopes or experience rather than the realities of the actual project. Most projects fail to meet either their budgets or their initial timetables.

Figure 14.1 Separation of Storm Sewers in Cambridge, Massachusetts. The original sewer system in Cambridge and many other cities channeled rain runoff into sewerage treatment plants. During storms the runoff exceeded system capacity, and mixed sewerage and rainwater overflowed into low-lying areas or was entirely diverted into the Charles River or Boston Harbor. Separating the rain runoff keeps the peak load from clogging the waste treatment facilities and stops raw sewerage from spilling into waterways. The new system required digging up Massachusetts Avenue to install new pipes. This is a major arterial street connecting Boston and Cambridge, so project managers were very concerned about disrupting traffic.

- The work may be much more dangerous than anticipated. Tens of thousands of workers and their relatives died from disease during initial attempts to construct the Panama Canal.
- The construction process might prove to be unworkable. Initial attempts to lay the North Atlantic cable failed because workers could not splice pieces of cable that were transported and set down from multiple ships. Success was possible only when the largest ship in the world (the *Great Eastern*) was used to carry and distribute the entire length of cable.
- A project might not be sustainable socially. Public protests against the disruption of cities stopped construction of superhighways through portions of Boston and San Francisco in the early 1970s. Twenty years earlier, public outrage at the ugliness of the elevated Central Artery forced Boston to put much of that superhighway into a tunnel.

14.2.3 Managing the Transition from Construction to Operation

From the construction manager's perspective, the project ends when the structures are built; but in reality, the project is just beginning then. Whatever the goals of the project, they can be achieved only once construction is completed and the transition to operations succeeds. Marketing continues to be a major concern, especially for construction of buildings that require tenants who will pay enough rent to pay off the construction costs. Once tenants are found, they need some time to prepare their space for occupancy, and they may require some minor modification in their accommodations. Depending upon the project, the transition from construction to operation may be more or less complex. Transportation projects require a detailed procedure for opening new terminals, roads, subway lines, or any other facility: someone has to make sure that signs, signals, and management systems are in place to manage the system; that drivers, vehicles, and maintenance facilities are ready to operate; and that the public knows how to use the system. Some systems, such as new computer facilities, may require extensive testing to ensure that all hardware and software is operational.

What can go wrong:

- The marketing plan may be upset by national or international economic conditions. The Empire State Building, which opened in the middle of the Depression, suffered from low occupancy for many years.
- The design may fail. The Tacoma Narrows Bridge, a suspension bridge whose spectacular early demise was filmed, collapsed due to unexpected dynamic forces caused by winds acting on what proved to be too flimsy a structure.

- Accidents may occur. On Memorial Day, a week after the Brooklyn Bridge was first opened to the public, thousands of people took the opportunity to walk across the bridge. Somehow, word spread that the bridge was collapsing; it wasn't, but 12 people were crushed in the ensuing stampede.[2]

14.2.4   Completing the Project

The project is likely to have a clear-cut end date, at least from the perspective of the project manager. There may be a date when the facility is turned over to the owner for normal operation or when the facility is opened to the public. Or there may be a test period of months or years, during which the contractor is required to fix any mistakes or problems that are encountered during operation. For many projects, the key event may be obtaining permanent financing that enables the owners to pay off their construction loans or securing a stable funding source for operations and maintenance. For owners and the public, it may well take many years to judge whether a project was successful, and the apparent success or failure of a project may well change over time.

What can go wrong:

- A building may fail to attract tenants, or tenants may renege on their leases.
- The Hancock Building, a 50-story office tower in Boston, had sides covered entirely by reflective windows, which created a stunning visual effect when the building mirrored the city skyline and the passing clouds. Unfortunately, the windows began to fall out. All of the windows eventually had to be replaced at a cost of many millions as well as several years of legal wrangling to determine who would pay for them.
- A failure in operations can occur. The automated baggage control system at the Denver International Airport did not work properly when the airport was opened; this operations failure resulted in extreme delays for travelers waiting for their luggage, very bad publicity for the airport, and considerable expense to correct the problem.

14.2.5   Assessing or Auditing the Project

The final stage of project management may actually begin during construction, and it may continue for years after construction is completed. A project audit can answer three basic sets of questions:

1. *Evaluation of the construction process:* How good were the assumptions regarding cost, time, schedules, safety, and security? Was the construction strategy effective? How could the project have been constructed more quickly, at lower cost, or more safely? How does experience with this project affect project management assumptions that will be used in future projects?
2. *Evaluation of the construction design:* How well did the construction materials and design work? Should different materials or designs be used in the future?
3. *Evaluation of the project with respect to its goals:* How well did this project meet the original financial, social, or environmental goals?

The first set of questions can be answered in the midst of construction, since the cost and time required for each activity can be compared to the plans. The second and third sets of questions may take much longer, although there will certainly be some early indications as to the success or failure of a project. How did initial demand for the facility compare to projections? Was the space rented, or could the output of the facility be sold at a profitable price? Matters related to social and environmental impacts may take much longer to resolve because understanding of the problems may not be noticed for decades.

---

[2] David McCullough, *The Great Bridge: The Epic Story of the Building of the Brooklyn Bridge* (New York: Simon & Schuster, 1972), 544.

Table 14.1 Stages of Project Management

---

1 **Preparing to Implement the Project**
 - Financing
 - Acquisition of land
 - Public approvals
 - Construction strategy
 - Marketing and transition
2 **Construction Management**
 - Network of activities and events
 - Schedule
 - Budget
 - Safety
 - Security
 - Social and environmental impacts
3 **Transition Management**
 - Marketing
 - Final setup of facilities
 - Operating strategy
 - Opening day
4 **Project Completion**
 - Finishing touches and corrections
 - Finances: sell, refinance, turn over to operating management
5 **Project Assessment or Audit**
 - Quality of construction—was project constructed according to the plans?
 - Construction goals—did construction process meet goals set for cost, time, safety, and security?
 - Goals of project—did project achieve desired results?

What can go wrong:

- Materials used may turn out to be health hazards. Asbestos was widely used for insulation before it was known that exposure to asbestos caused lung cancer and other respiratory problems, especially for workers.
- Fatal accidents may occur that could have been prevented (e.g., by using nets to protect workers who fall from bridges or from the rising superstructure of a skyscraper).
- Environmental effects may turn out to be extreme. Reliance on the automobile leads to massive problems related to safety, emissions, and land use.
- The design may turn out to be inappropriate for the ultimate users. Public housing projects in St. Louis had to be destroyed after less than 20 years because their design (long corridors in tall, sterile buildings) resulted in unsafe, unattractive, and unhealthy buildings.

Table 14.1 summarizes the stages of project management and the subtasks within each stage.

14.3 PROJECT MANAGEMENT TECHNIQUES

Various techniques are commonly used to manage the construction phase of a project, including the following:

- Statement of work
- Work breakdown structure
- Network diagrams
- Schedules
- Budgets

These techniques are discussed in the following subsections.

14.3.1 Statement of Work

The statement of work specifies what is to be done, using what resources, within what timetable. This document can be part of the contract between the owner and the builder, or it can be an assignment to a project manager who will manage a project undertaken by his company. The statement of work is often part of a contractual agreement or an organizational understanding that includes procedures for dealing with the changes likely to become necessary or desirable as the project is being implemented. The client may wish to add to the scope of the project, thus increasing both the time and cost; the contractor may discover that costs are higher than expected, or that progress is slower than expected, so that some modifications will be needed to complete the project. Trade-offs are likely to be made among opportunities for speeding up the work (allowing overtime or working an extra shift), reducing costs (using less expensive, but less productive equipment), or changing the project scope (reducing the size of the project or deferring portions of it). The project manager will work with the client over the course of the construction to decide whether and how to change the statement of work, the budget, and the timetable while still meeting the objectives of the project.

Before agreeing to the statement of work, the contractor will need some assurance that the project can actually be completed. This assurance could come from experience ("we've done it before, so we can do it again"), self-confidence ("we've done similar things before, and this will be no different"), or careful assessment of the costs of various options for completing the project ("we've studied this from all sides, we have a good plan plus two backup plans, and we're sure we can do the job right within the budget and time frame that we have proposed").

Table 14.2 Coming Up with an Initial Estimate for the Budget and Timetable

Basis for Budget and Timetable	Examples
Past experience	*Construction of another McDonald's, Home Depot, or Wal-Mart*: The job will be similar to dozens or hundreds of previous jobs.
	Construction of a suburban office building: Thousands have been built, and construction costs are published annually for typical designs. These include unit costs for the individual elements required (e.g., doors, windows, roofing, walls, and flooring).
	Construction of a railroad or a highway in rural areas: Methods, machinery, and materials are well-developed, and unit costs and productivity are well known.
Self-confidence	*Justified*: Construction of the Eiffel Tower was undertaken because Eiffel had gained tremendous experience in using the same materials and methods to construct bridges.
	Partially justified: Design and construction of the Brooklyn Bridge were based on smaller suspension bridges built in Cincinnati and Buffalo. But the project ran into unexpected difficulties as workmen suffered from or were incapacitated by the "bends" when they ascended too quickly from their work deep under the river, where they were excavating the foundations for the towers.
	Unjustified: French efforts to construct the Panama Canal were based on experience in constructing the Suez Canal, but the problems in the wet jungles and mountains of Panama were far different from the dry sand dunes of the Middle East. Not only were the technical requirements much greater, but malaria and other diseases took a terrible toll on the workforce.
Careful planning	*Construction of dams*: Careful planning is necessary to ensure that the dam will be strong enough to survive.
	Space stations: Every aspect of every activity must be thoroughly investigated to ensure that the materials can be sent into orbit and then assembled by a few astronauts working under extreme conditions.

Table 14.3   Work Breakdown Structure for Constructing a New Rail Line

Prepare the Route
- Survey the route (identify the exact route, taking into account curvature and grades).
- Perform cut and fill operations to create a flat surface for the route.
- Build drainage ditches and culverts as needed.

Bridges
- Select locations for bridges.
- Design bridges.
- Construct bridges.

Connections to Other Lines
- Install interlockings where the line crosses other lines.
- Install turnouts where the line connects to other lines.

Install Track
- Install ballast.
- Install ties.
- Install rail.
  - Attach tie plates to ties.
  - Place rail on tie plates.
  - Attach rail to tie plates.
  - Install angle bars (connect rails to one another).

Install Signals
- Install turnouts and crossings (for safe movement through intersections).
- Install block signals (to ensure safe train spacing).

14.3.2   Work Breakdown Structure

Given the work statement and the general strategy for constructing the project, the next step is to identify all activities that must be undertaken to complete the project. The activities can be organized within a hierarchical framework to create what is called the **work breakdown structure** (**WBS**). The WBS is very helpful in managing the project because it breaks down what may be a complex project into what could become a large number of manageable activities.

A competent engineer or manager will be able to create a WBS by identifying and categorizing all the steps that will be required, based on some combination of experience, knowledge of the task at hand, and common sense. Portions of the WBS for a complex project may be developed by different people or groups of people who have the necessary expertise to identify the activities.

The WBS can be shown as a table or as an organization chart. For example, Table 14.3 shows a WBS for constructing a new rail line, while Figure 14.2 shows much of the same information as an

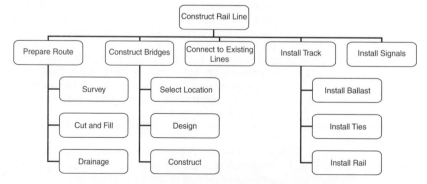

Figure 14.2 Work Breakdown Structure for Constructing a New Rail Line.

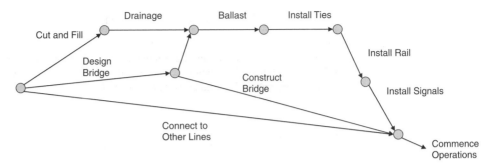

Figure 14.3 Network Diagram for Constructing a New Rail Line.

organization chart. The major activities are preparing the route, constructing bridges, making the connections to existing lines, laying the track and installing the signals. Each of these activities has multiple sub-activities, and it is possible to go to a very fine level of detail.

14.3.3   Network Diagrams

Once the activities have been identified, it will be possible to create a **network diagram** that shows how the activities are related. Some activities could begin at the outset of the project, while other activities must wait until certain prerequisites are completed. For any activity, it will be possible to identify a) what other activities must be completed before this one can begin and b) what other activities cannot be started until this activity is completed. These interdependencies among activities may reflect logic and physics (i.e., the rails cannot be installed before the ties are put in place and the roof cannot be put on a building until the superstructure has been erected) or they may reflect constraints imposed by regulations, standard operating procedures, contractors' preferences, or the choice of technology.

Figure 14.3 shows a network diagram for some of the tasks associated with constructing a new rail line. The diagram begins at the left, at a point where the route has been selected. At that point, work can begin on three tasks: the cut & fill operation that will create the rail route (Figure 14.4), design for the bridges that are needed, and creating the intersection where the new line will connect with existing lines. Once the route is laid out and the bridge has been designed it will be possible to begin to install the track, beginning with ballast. (The unnamed link between completion of bridge design and the beginning of ballast indicates that it is necessary to know exactly how and where the track on the bridge will meet the track on the approaches before determining how much ballast will be needed on the approaches to the bridges.)

Figure 14.4 Preparing the Route for a High-Speed Rail Line across Saudi Arabia (August 2005). Saudi Arabia has recently begun construction of a high-speed rail system to serve many of its major cities, ports, and industrial centers. No settlements exist in most of the region crossed by the rail network, but it is difficult to create a reasonably level route through a hot, dry landscape with constantly shifting sand dunes.

(Photo: Vinay Mudholkar)

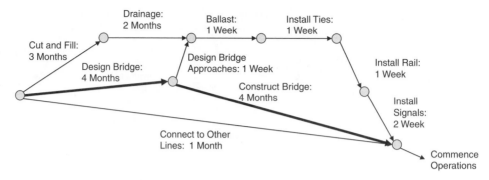

Figure 14.5 Critical Path for Constructing a New Rail Line.

14.3.4   Schedules

The network diagram only shows how the various activities must be sequenced. To determine the cost of and the time required for the project, it is necessary to consider the time required for each activity and the resources that are available to the project. The actual work time to complete an activity depends upon the quantity and productivity of the resources that are employed in that activity, as well as the nature of the work to be done. The elapsed time to complete the activity will also depend upon the work schedule and the availability of resources. It may be possible to complete the activity within a shorter time by assigning more people to the task, by working longer hours, or by given the people better tools.

The time required to complete each activity can be added to the network diagram, as shown in Figure 14.5. Once the time is added, it is possible to find the time required to follow any path through the network. For example, following the path along the top of the diagram indicates that it will take 6.25 months to complete the activities related to the construction of the rail line: 5 months to prepare the route (cut & fill and drainage operations) plus another 5 weeks to install the track (ballast, ties, rail and signals). Following the solid line through the middle of the chart shows that it will take 8 months to design and construct the bridge. The final activity, connecting the new line to the existing lines will take a month. The **critical path** is the path that takes longest amount of time; this is critical in the sense that any delays to activities on this path will delay the overall project. The other activities all have some **slack time**, because they can be delayed without causing any delay in the completion of the project. The slack time for the cut & fill operation is 1.75 months, because this operation could be delayed by that amount and it would still be possible to complete the subsequent operations without delaying the start of operations. The **latest start date** for the cut & fill activity is therefore 1.75 months after the start of the project.

Now consider the ballast operation, which cannot begin until the drainage activity is completed and the bridge approach has been designed. The **earliest finish date** for the drainage work will be at the end of five months, as cut & fill will take 3 months and drainage will take another 2 months. Likewise, the earliest finish date for designing the bridge approaches will be 4.25 months, which is the sum of the time needed for designing the bridge plus the time for designing the bridge approaches. The **earliest start date** for the ballast operation will be the latest of the earliest finish date for the activities that must precede this activity, which in this case will be the minimum of 5 months required for the completion of the drainage work. It is possible to proceed though this diagram and calculate the earliest and latest start dates and the earliest and latest finish date for each activity. These dates will be useful for the project manager who is trying to complete the project on time; recognizing which activities are on the critical path will help the manager focus on the areas where delays are most critical. Identifying the activities with the greatest slack–i.e. the greatest difference between the earliest and latest start dates– will provide insight as to which activities can be delayed if multiple activities will be using the same resources. For example, the same people who are designing the bridge and the bridge approaches are

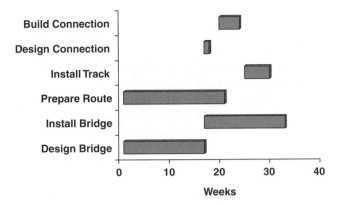

Figure 14.6 Task Schedule for Construction of New Rail Line.

likely to be involved in designing the connections where the new route meets existing lines. As is evident from the diagram, planning for that connection can easily be delayed until after the bridge design is completed, as work on the connection activity can be deferred to the end of month 7 without delaying the project.

Once the relationships among the tasks and the duration of the tasks are understood, a plan can be developed for proceeding with the work. Figure 14.6 shows how the work might be scheduled. The bridge work is on the critical path, so the bridge design must begin immediately and the construction is scheduled to start as soon as the design is completed. There is some slack in the other tasks, so they do not have to start immediately. Nevertheless, the cut & fill and the drainage tasks, which are shown as a single task ("prepare route") are scheduled to start immediately in order to allow some additional time in case bad weather or unexpected excavating difficulties are encountered. The installation is planned to begin with the design and construction of the connections to the existing lines and then to proceed with the ballast, tie, rail, and signal installations (shown in the chart as "install track"). These tasks are scheduled sequentially rather than simultaneously because many of the same people will be involved in making the connections to the existing lines and in installing the track. This type of chart is called a Gantt chart, after Henry L. Gantt, an early practitioner of scientific management who developed this scheduling aid in 1917. Many variations of this chart are possible, including highlighting of the tasks on the critical path or including the earliest and latest start dates.

Computer programs have been developed to assist project managers in planning and monitoring progress on projects. Once the activities and their relationships with one another have been defined, it is possible to create a network diagram. Once estimates of the time required for each activity have been added, it is possible to find the critical path and to define the slack associated with each activity. The managers can then adjust the start dates of various activities in order to balance workloads or to provide buffers that will allow the schedule to continue despite minor delays.

The network diagram and the critical path are meaningful only after considerable thought has been given to the strategy for implementing the project. There are likely to be many possible strategies, some of them better at minimizing time while others are better at minimizing cost or reducing risks. In the railroad example given earlier, senior management may insist on completing the work within 6 months, which would seem to be impossible if we really believed the logical relationships and the time estimates as shown in Figure 14.3. However, the project manager could confer with the design group and the engineers responsible for supervising the construction. The challenge would be to reduce the time required for the activities on the critical path. It might be possible to use a standard bridge design that would eliminate most of the time estimated for bridge design. If that is not possible, perhaps the size of the construction crew could be increased to allow the bridge to be installed in 2 months rather than 3 months. Once a plan is devised to install the bridge within the desired 6 months, then the upper row of route preparation and track installation would become the critical path, with an expected time of 6.25 months. Squeezing 1 week out of this sequence of activities would probably be quite straightforward. One option is to start the drainage work as soon as a mile or two of the cut and fill operation were completed; another option would be to work on cut and fill and drainage simultaneously, if additional people and equipment were made available. To save time in installing track, one option would be to work a couple of weekends; another would be to have the tie and rail gangs follow closely behind the ballast gang so that these tasks could be overlapped. Choosing the best option would depend upon the availability of people and equipment and the costs associated with the various strategies.

Once the schedule is settled, it will be possible to identify who is working where and on what activities over the course of the project. This can be expressed as a table showing the workforce associated with each activity. The schedule may need adjusting to balance the workload to meet the available resources or to reduce costs. For example, it may be better to have 50 people working for 4 months than to have 100 people working for 2 months. For specialized workers and supervisors, the workload will have to reflect the time that they actually have available to work on the project.

The next section shows how a chart can be created that delineates responsibilities for completing a project. Section 14.3.6 then provides a detailed example of how it is possible to estimate the time and costs associated with a particular activity.

14.3.5 Linear Responsibility Charts and Estimating Resource Requirements

To complete a project, a group of people must work together to complete all of the discrete tasks that make up the project. An important part of project management is ensuring that everyone knows what they are supposed to be doing and that every task is assigned to people capable of completing it. Responsibilities of individuals can be laid out on a **linear responsibility chart** that indicates the group or individuals involved in each task.

The steps involved in creating a linear responsibility chart are as follows:

1. Develop a complete list of all the activities that must be carried out to complete the project.
2. Identify all of the people who are involved in the project.
3. Draft a linear responsibility chart that indicates who is playing what role in each project.
4. Review the responsibilities with all of the key people involved in the project, and revise the chart as necessary.
5. Repeat the steps 3 and 4 until everyone understands and agrees to the chart.
6. Review and update the chart as needed to clarify roles and responsibilities or to overcome organizational problems that may be encountered.

Making sure that everyone involved understands their role and their responsibilities can go a long way to avoiding misunderstanding and conflicts as the project proceeds. If projects are complex, then it is useful to create a hierarchy of linear responsibility charts, each one dealing with a well-defined portion of the project. The project manager will need to work with people who have sufficient expertise to judge how best to organize the various technical activities that must be undertaken as part of a large project.

EXAMPLE 14.1 A Linear Responsibility Chart for the Railroad Project

Table 14.4 summarizes how a railroad might assign responsibility within the engineering department for the construction of the new rail line described in previous sections. This summary chart shows only the individuals who have the ultimate authority to approve each task or the primary responsibility or the secondary responsibility for carrying out each task. The project manager must approve all aspects of the design and implementation of the project. There are two assistant chief engineers, one in charge of design activities and one in charge of all field activities. The construction will be undertaken by three work gangs; one will prepare the route, another will make the connection to the existing line and install the new track,

Table 14.4 Linear Responsibility Chart for Construction of a New Rail Line. (A—approval authority; P—primary responsibility; S—secondary responsibility)

Task	Project Manager	Assistant Chief Engineer—Design		Leader, Route Preparation Gang	Leader, Track Gang	Leader, Bridge Gang
		Design	Field			
Design connection	A	P				
Build connection	A		P		S	
Install track	A		P		S	
Prepare route	A		P	S		
Design bridge	A	P				
Install bridge	A		P			S
Logistics	P	S	S			

and a third will install the bridge. The project manager has primary responsibility for managing the logistics involved in getting people and material to the worksite as needed to complete the job; this task requires coordination with suppliers and the railroad's operating department.

More detailed charts could be developed for each of the tasks shown in Table 14.4. For example, installing track requires acquiring, transporting, and installing the ballast, ties, rail, and other materials that make up the track structure. Assembling a work gang is a complex task; it requires scheduling of various expensive machines, assigning individuals to the gang, and arranging transportation and living arrangements for the gang while they are at the site. Each task in the table could be exploded into a more detailed linear responsibility chart, as suggested by Table 14.5. The project manager works with the chief engineer's office to arrange for the purchase and transport of materials and the assembly of the track gang. The actual installation proceeds under the direct supervision of the leader of the track gang, working with the leaders of separate groups that install each of the track components. The next example shows how project management must get into the details of the activities that must be performed: How long will they take? How many people will be needed? What is the exact sequence of how things must be done?

Table 14.5 Linear Responsibility Chart for Installing Track (A—approval authority; P—primary responsibility; S—secondary responsibility)

Task	Project Manager	Leader, Track Gang	Office of the Assistant Chief Engineer—Field		Leader, Ballast	Leader, Ties	Leader, Rail
			Purchasing	Personnel			
Purchase materials	A		P				
Transport materials to site	A		P	S			
Lease equipment	A		P				
Assemble track gang	A	S		P	S	S	S
Install ballast		P			S		
Install ties		P				S	
Install rail		P					S

EXAMPLE 14.2 Estimating Resource Requirements

The project manager, the leader of the track gang, and the person in charge of rail installation must figure out exactly how they will install the rail. They will want to figure out how many people are needed and how long it will take to install new rail on 10 miles of new track. Rail installation can begin once ballast and ties are installed and rail, tie plates, and fasteners have been placed alongside the right-of-way. The size of the crew and the time required depend upon the technology used. Today, the engineering department would be able to use experience to determine that its standard rail gang, with modern equipment, could lay rail at the rate of about 1 mile per day on tangent track and a half mile a day on curves. A hundred years ago, this

Table 14.6   Labor-Intensive Approach to Laying Rail

Task Number	Task	Gang Size	Time Required
1	Position tie plate and spikes.	2 men	5 minutes
2	Spike tie plates.	4 men	5 minutes
3	Place rail.	50 men	5 minutes
4	Spike rail.	4 men	5 minutes
5	Attach joint bars.	2 men	15 minutes
Total		62 men	35 minutes

work might have been undertaken by a crew of men using hand tools, and the rate of progress would depend upon the number of people working and the level of coordination.

To illustrate the process inherent in estimating resource requirements, let's first consider the work of laying rail as it was carried out during the nineteenth century, when the vast majority of the North American rail system was first built. The tasks and the time required for each task were something like this:

- Place tie plates on the ties and drop 4 spikes by each tie plate (a tie plate is a steel plate with holes that is attached to the tie by steel spikes; two men could place tie plates on a dozen ties in 5 minutes and leave spikes next to each tie plate).
- Hammer two spikes through each tie plate to attach it to the tie (four men could hammer two spikes through three ties in 5 minutes).
- Lift a 30-foot rail and place it on the tie plates (this rail would weigh about 1.5 tons, so it could be lifted by a team of 50 men and put in position in another 5 minutes).
- Hammer two more spikes through holes in the tie plate to attach the rail to the tie plate and further anchor the tie plate to the tie (four men could work simultaneously and attach the 30-foot rail to a dozen ties in another 5 minutes).
- Attach joint bars to link this rail with the previous rail (two men could complete this task in 15 minutes).
- Repeat all five tasks for the other side of the rail.

Table 14.6 summarizes the tasks and the time required.

How fast could this team lay rail? The work must be conducted in sequence, so it might appear to take 62 men a total of 35 minutes per rail, or 70 minutes for two rails. Actually, only the first rail took that much time. After that, the men laying the tie plates and the men attaching the tie plates to the ties could move ahead to prepare for the next rail. Likewise, each group moved on to the next rail when they completed their task. Table 14.7 shows how the work might be proceeding if the 62 men

Table 14.7   The 62-Man Gang Spreads Out along the Track, Working on All Tasks Simultaneously

Time Interval (minutes)	1st Rail	2nd Rail	3rd Rail	4th Rail	5th Rail	6th Rail	7th Rail
0–5	1						
5–10	2						
10–15	3	1					
15–20	4	2					
20–25	5	3	1				
25–30	5	4	2				
30–35	5		3				
35–40		5	4	1			
40–45		5		2			
45–50		5		3	1		
50–55			5	4	2		
55–60			5		3	1	
60–65			5		4	2	
65–70				5		3	
70–75				5		4	1
75–80				5			2
80–85					5		3
85–90					5		4
90–95					5		

Table 14.8   The 62-Man Gang Spreads Out along the Track, Working on All Tasks Simultaneously

Time Interval (minutes)	1st Rail	2nd Rail	3rd Rail	4th Rail	5th Rail	6th Rail	7th Rail
Early work	1, 2	1, 2	1, 2	1, 2	1, 2	1, 2	1, 2
0–5	3						
5–10	4	3					
10–15	5a	4	3				
15–20	5a	5b	4	3			
20–25	5a	5b		4	3		
25–30		5b	5a		4	3	
30–35			5a	5b		4	3
35–40			5a	5b			4
40–45				5b	5a		
45–50					5a	5b	
50–55					5a	5b	
55–60						5b	5a
60–65							5b
65–70							5c

just kept moving on down the railbed, working on their own task (numbers 1–5 in the table). Within a half hour, all of the men are working, but the groups assigned to each task are working in or moving to different locations. It now appears that a rail can be completed every 15 minutes once everyone is fully occupied. If the gang continued like this for an entire 9-hour day, with time off for lunch and a couple of 10-minute breaks, it would lay 32 rails.

The gang boss might feel proud at his men's progress, reporting to the general manager that they laid 16 rails in a single day, for a total of 480 feet of track. However, the boss probably exploded—because it would take 11 such days to complete a mile of track and 110 days to complete 10 miles of track—and by then it would be winter and the gang would be unable to work. The humbled gang boss would go back to his tent to work out a better scheme for the next day. If he were looking at Table 14.7, he would realize that one problem was that not enough men were working on the joint bars (task 5 in the tables), since these are the only workers engaged continuously. Another problem is that most of the gang was sitting idle for the first 10–20 minutes of the day. He might therefore make these changes to his plan:

1. Have a few men come in early to position all of the tie plates and spikes for the day's work.
2. Have a few men begin spiking the tie plates before the main portion of the gang arrives.
3. Have more men work on the joint bars, which is what was holding up progress.
4. He would personally monitor the progress of the group to make sure they weren't wasting time between rails.

The resulting productivity on day 2 was much better, as shown in Table 14.8.

With this new plan, it was possible to finish 6 rails in the first hour and 48 rails during the second day, for an improvement of 50% over the first day's effort. Moreover, when the gang boss reviewed this day's effort, he realized that the joint-bar task was still holding up progress, so he assigned a third team for the next day's work, continuing with the early work in placing tie plates and attaching them to the rails. With this revision, he was able to lay 80 rails in a single day, which is a rate of 1200 feet of road per day, or approximately 1 mile per 5-day week, allowing some extra time for delays related to bad weather.

These tasks required strength and effort, but they were quickly learned. If another 60 men could be diverted to laying rail, then they could work from the other end of the track and the whole job might be completed in half the time. And if someone could figure out how to rig up some pulleys, then it wouldn't take 50 men to move the rail, and smaller gangs could do as much or more work in a day.

Now consider the modern process, which relies heavily upon machinery and which combines replacement of rail with replacement of ties and upgrading the ballast and subgrade. A special train, known as the P811, is able to replace ties, move the old rail to the side of the track, and place the new rail onto the new ties. If concrete ties are used, then the tie plates can be attached in the factory and the rail is attached to the ties using a special clamp rather than spikes. It takes less time for one man to

Figure 14.7 Installing Track Is Now a Capital-Intensive Process Requiring Special Equipment for Each Task. At this site on a Burlington Northern Santa Fe line in Nebraska, new concrete ties and heavy rail have been already installed, and extra ballast has been dumped along the track between the rails. This machine is spreading ballast between the ties to help hold the rail in place.

hammer the fastener into place than it used to take to hammer in four spikes. This impressive piece of equipment can lay a mile of track in an hour, so it would be able to relay 10 miles of track in a couple of days, assuming that it had at least 8 hours per day to do the job. Additional types of equipment are used to finish the job (Figure 14.7).

Many infrastructure projects involve the kind of repetitive work described in Example 14.2. As the gang leaders and workers gain experience, they can do the job better and faster. Because companies understand the process better, they can design machinery that will help provide further improvements in productivity. Raising the steel structure of a skyscraper is a task that can be done very efficiently once all the materials and workers have been assembled and had a chance to work out the details of how best to proceed. Another example is described in Box 14.1: Henry Flagler's ill-fated attempt to build a profitable rail line across the Florida Keys to fill in what he viewed as the missing link in the best route between the United States and Cuba.

Henry Flagler became the second-richest man in the world because of his expertise in negotiating rail rates for John D. Rockefeller and Standard Oil. His interest in railroads continued after he left the oil business, and he financed the construction of the Florida East Coast Railroad. After linking Jacksonville to Miami (then merely a small village serving as watering stop for ocean shipping), he decided to extend the railroad all the way to Key West to capture what he expected to be lucrative trade with Cuba and a valuable way station for ships headed to the Panama Canal. While he completely misunderstood the economics of ocean shipping (the point is not to minimize the distance shipped on water, but to minimize the distance transported on land), Flagler was able to complete the railroad all the way to Key West, including nearly 40 miles of bridges over the open ocean. The railroad construction involved seemingly endless repetition of construction tasks, supported by an intricate logistics network. The description of this effort paints the picture of a well-managed construction project.

Box 14.1   The Railroad That Crossed an Ocean

Of the more than one hundred miles that remained from Jewfish Creek to Key West, about half were over open water. Some of the spans were a few hundred yards. Others stretched for miles. Still, to a pragmatist like Flagler, the route seemed possible. When questioned how he would cross these mammoth stretches of open water, Flagler replied, "It is perfectly simple. All you have to do is to build one concrete arch, and then another, and pretty soon you will find yourself in Key West." (p. 88)

One portion of the route required a bridge that would cross the 9-mile gap between Knight's Key and Pigeon's Key. The bridge had four sections:

The first three segments were to be constructed of steel deck girders laid atop a series of 546 support piers, while the Pacet Channel segment was to be built in the spandrel arch style—210 of them. . . .

All of the heavy construction equipment used at the time was powered by steam engines . . . which consumed great quantities of coal and fresh water, all of which had to be transported over great distances to the work site, along with most of the materials used in the construction. Each one of the bridge piers required enough sand, gravel, cement, lumber and steel to fill a single five-masted schooner. Most of the cement used for construction above the tide line was of domestic origin, much of it came from New York. The cement used for submarine applications was brought from Alsace. Steel girders and track segments were fabricated in Pittsburgh, pine lumber came from Georgia and Florida, hardwoods from sources in the Midwest. . . .

And yet the very size and scope of such a task produces its own inescapable inertia. As any child knows, "I think I can" soon becomes "I know I can," and Flagler's crews were soon working at a record pace. . . . If no problems intervened, crews could normally complete as many as four support piers a week, and ironworkers could join as many spans in the same time frame. (pp. 159–161)

Hurricanes wiped out the track several times during the construction, and they eventually doomed the railroad; in the 1930s, the structures originally created for the railroad were used for the Overseas Highway to Key West. Nevertheless, we cannot help admiring the monumental efforts and teamwork that enabled a railroad to be built over an ocean.

Les Standiford, *The Last Train to Paradise: Henry Flagler and the Spectacular Rise and Fall of the Railroad That Crossed an Ocean* (New York: Crown Publishers, 2003).

14.3.6 Cost Estimation and Budgets

Costs can be estimated in various ways and at various levels of detail. In the early stages of project evaluation, an order of magnitude estimate may be all that is needed to provide some assessment of project feasibility. A cost estimate that is within 30–40% will probably be more accurate than the estimates of possible benefits and the costs of financing. Rough estimates of costs are suitable for winnowing a large set of options into a few that will be studied in greater detail. Also, at that stage of project evaluation, planners and engineers are more likely to make mistakes by overlooking costs than by misestimating them (e.g., by considering only the costs of construction and forgetting about the costs of operation or the externalities that might be associated with a project). Once evaluation has reached the stage of considering only one or two basic concepts, more accurate estimates are desirable, but estimates within 10–15% are still suitable. Eventually, once a particular project has been defined, it will be necessary to develop a budget for the overall project and for each element of the project. These budgets ideally should be accurate within a few percent.

The objective in estimating costs is to provide a suitable estimate at a reasonable cost. It makes no sense to develop a detailed budget for dozens of options that are being considered, when a rule of thumb can be used to get a feel for cost. For example, contractors will be able to use their own experience and industry publications to estimate the cost per square foot for constructing a typical office building, and a highway department will be able to estimate the cost per mile for adding a lane to an existing rural highway. However, when it comes time to actually construct the office building or add that lane to the highway, a budget based on the actual site conditions and the final design is essential.

The budget follows directly from the work schedule, which indicates the resources required for each task and the duration of each task. If the work schedule and the project design are accurate, then the budget can be created. This is done by adding up the materials needed, the machines and vehicles used, and the labor force required and then applying the appropriate unit costs to each item.

EXAMPLE 14.3   Creating Rules of Thumb for Costing

How can a contractor estimate the cost per square foot for constructing office buildings? Such an estimate could be based on a careful assessment of all of the costs that go into constructing a building:

Step 1: Obtain unit costs for all the elements and activities required to construct a particular type of structure, including such things as the following:
- Excavation
- Foundation
- Steel structure
- Walls
- Floors
- Windows
- Doors
- Roofing

Step 2: Design a model structure (e.g., an 8-story office building with 12-foot stories and 100,000 square feet of floor space).

Step 3: Count the total number of units required for constructing this structure.

Step 4: Multiply the cost per unit by the total number of units to obtain the total cost of the structure.

Step 5: Divide the total cost by 100,000 sq ft to get the estimated cost per square foot

If the model was carefully chosen to be representative of the types of buildings that the contractor hopes to build, then the estimated cost per square foot could be a useful rule of thumb to use in preliminary discussions of possible projects.

Some companies specialize in developing and publishing information that is useful in cost estimation. For example, RS Means publishes unit costs and building costs for a variety of standard building types. Their publication provides detailed information concerning unit costs (Table 14.9) and the average costs of buildings (Table 14.10). According to these estimates, a standard 5- to 10-story office building similar to what is seen in many an office park by an interstate highway would cost $85 to $95 per square foot, depending upon the size of the building. Thus, a steel frame building with 100,000 sq ft of space would cost $85.85 per sq ft (SF in the tables), or $8.6 million.

Table 14.9   Estimates of Unit Costs

Task	Unit	Unit Cost	Cost/SF of Building
1-1-2 Excavation and preparation for slab	SF of ground	$ 1.10	$0.14
1-2-1 Exterior walls, precast concrete panels (80%)	SF of wall	$ 17.59	$6.08
1-2-2 Windows (20%)	Each	$375	$2.16
1-3-6 Floor finishes—60% carpet; 40% tile	SF of floor	$ 5.02	$5.02
1-4-1 Roof covering—tar and gravel with flashings	SF of roof	$ 2.32	$0.29

Source: RSMeans Square Foot Costs 2000. (Kingston, MA: R.S. Means Company, 2000).

Table 14.10   Estimates of the Cost per Square Foot of Floor Area for 5- to 10-Story Office Buildings

Exterior Wall		50,000	60,000	70,000	80,000	90,000	100,000
	Square Feet Area	50,000	60,000	70,000	80,000	90,000	100,000
	Linear Feet Perimeter	328	370	378	410	441	450
Precast Concrete Panel	**Steel Frame**	$92.45	$92.75	$88.20	$87.10	$86.20	$85.85
	Reinforced Concrete Frame	$92.85	$92.15	$89.55	$88.45	$87.55	$86.20

Source: RSMeans Square Foot Costs 2000. (Kingston, MA: R.S. Means Company, 2000).

Experience can also be used to develop rough cost estimates. Three factors need to be taken into consideration in projecting future costs based on experience:

1. *Inflation*: Costs need to be adjusted for inflation before making any comparisons.
2. *Economies of scale*: The size of a project may affect the expected cost of the project, so an adjustment will be needed that takes into account the size of past projects as well as the size of the future project.

3. *Experience*: As a task is repeated, individuals and companies learn how to do it better, so it is desirable to make an adjustment that allows for some efficiencies to be gained as more tasks or projects are completed.

Adjusting for Inflation

If there is an appropriate cost index, then it will be possible to express all costs in constant dollar terms using that index:

$$\text{Current \$ Cost} = (\text{Actual Cost Year t})(\text{Current cost index/cost index year t}) \qquad \text{(Eq. 14.1)}$$

Costs can be projected into the future using estimates of inflation:

$$\text{Future \$ Cost} = (\text{Current \$ Cost})(1 + \text{Expected Inflation}) \qquad \text{(Eq. 14.2)}$$

Various industries publish cost indexes for construction and operations. If no cost indexes are suitable for a particular project, then the producer price index can be used to adjust all costs to the current year.

EXAMPLE 14.4   Adjusting Costs for Inflation

A series of construction projects were completed between 1981 and 2006; total costs ranged from $70,000 in 1981 to $200,000 in 2006. Costs rose steadily over this period as indicated by the cost index, which rose from 100 in 1981 to 278.3 in 2006. Costs were expected to rise much more slowly (an additional 3%) through 2010 because of the worldwide recession and the decline from peak energy prices. After adjusting for inflation, what is the expected cost of constructing a similar project in 2010?

Table 14.11 shows the initial project cost, the cost index, the cost expressed in 2006 dollars (using Eq. 14.1) and the cost projected to 2009 dollars (using Eq. 14.2). The average cost of these projects, expressed 2010 dollars, is $218,000.

Table 14.11   Adjusting Costs for Inflation

Year of Project	Actual Cost	Cost Index (1981 = 100)	Current $ Cost (2009 dollars)	Future $ Cost (2010 dollars)
1981	$70 thousand	88.3	$221 thousand	$227 thousand
1985	$80	116.6	$191	$197
1997	$130	167.6	$216	$222
2001	$160	195.0	$228	$235
2004	$170	219.9	$215	$222
2009	$200	278.3	$200	$206
2010		278.3 (1.03)	$212	$218

Adjusting for the Size and Type of Project

If there is sufficient experience—or sufficient knowledge of the construction costs—it is possible to adjust average costs for an increase or decrease in size. One approach is to use an algebraic cost function that includes size as one of the parameters (as amply demonstrated in earlier chapters and assignments). If models are available, and if there is sufficient time to conduct an analysis, it should be possible to estimate the costs as a function of size, quality, or any other parameter of design or construction. In many situations, such as informal discussions over lunch or preliminary meetings with a potential client, it is not possible to refer to a model or take the time to do a complicated analysis. Another approach is to use aggregate data regarding past projects to come up with a relationship between project size and project cost that can readily be used in thinking about future projects (see Example 14.5).

A third approach is to use an estimated relationship between project size and project cost. Such a relationship could be based on analysis of past projects, modeling related to future projects, or expert

Table 14.12 Cost of Project Estimated Using Power Sizing

b	50	75	100	150	200	500	1000
			Size of Project (base case = 100)				
0.6	66	84	100	128	152	263	398
0.8	57	79	100	138	174	362	631
1.0	50	75	100	150	200	500	1000
1.2	44	71	100	163	230	690	1585
1.4	38	67	100	176	264	952	2512

judgment. In a technique known as **power sizing**, costs are assumed to be proportional to the size of the project raised to a power:

$$\text{Cost A}/\text{Cost B} = (\text{Size A}/\text{Size B})^b \qquad \text{(Eq. 14.3)}$$

Power sizing is a quick means of estimating how cost would vary with the size of the project. If b is equal to 1, then cost is directly proportional to size. If b is less than 1, then costs rise more slowly than size (i.e., there are economies of scale in construction). If b is greater than 1, then there are diseconomies of scale. Table 14.12 shows how cost would vary with size for several values of b. In each case, costs are stated relative to a base case with size 100 and cost 100. The exponent b could be based on statistical analysis, physics, standards, or expert opinion.

EXAMPLE 14.5 Using Past Data to Estimate Cost as a Function of Size and Quality

Suppose some developers have experience in constructing apartment buildings in a major city. They believe that quality and size determine costs, as is evident in Figure 14.8, which illustrates costs for completed projects. The total cost of past projects varied from $20 million for a "Budget" complex with 10 units to nearly $80 million for a "Premium" complex with 75 units. The unit price was just over $500,000 for the largest of the budget projects, but more than $4 million for the smallest of the premium units. The developers would like to have a simple algebraic model for use in estimating costs of similar projects.

The developers decide to use the data in this figure, plus their own judgment, to develop a cost function. The structure of the cost function is affected by three observations. First, the data seem to indicate that the total cost is a linear function of the size of the complex, with a fixed cost of about $20 million for the standard apartments. Second, the developers believe that quality should be reflected as a multiplier (i.e., the total cost would be substantially higher for premium units and substantially lower for the lower-quality units). Third, the developers know that there were some differences among the past projects, in part reflecting location and in part reflecting the idiosyncrasies of the individual projects; they therefore are happy with a simple, approximate, easily remembered relationship. The first two observations determine the functional form for the cost equation:

$$\text{Total Cost} = [\text{Fixed Cost} + (\text{Cost/Unit})(\text{Number of Units})](\text{Quality Factor}) \qquad \text{(Eq. 14.4)}$$

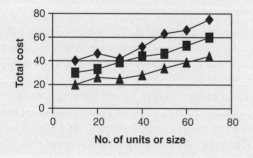

Figure 14.8 Total Cost as a Function of Size for a Hypothetical Project (in millions of 2009 dollars).

The third consideration means that it is not necessary to seek a precise answer. By entering the data into a spreadsheet and trying various values for the three variables, it is evident that the cost function, in round numbers, is something like this:

$$\text{Total Cost} = [\$20\,\text{million} + (\$0.5\,\text{million/unit})(\text{units})]\,\text{Quality Factor} \qquad \text{(Eq. 14.5)}$$

The quality factor is estimated as 140% for premium, 100% for standard, and 80% for the lowest-quality apartments. (This level of precision is more than enough for the developers, who understand that the purpose of this model is simply to allow them to make a ballpark estimate of the costs of any similar project.) The results are shown in Figure 14.9.

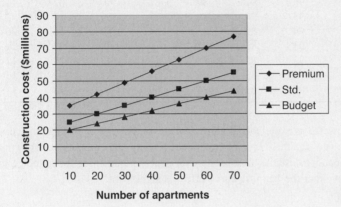

Figure 14.9 A Cost Function Based on the Developers' Experience.

Adjusting for Experience: The Learning Curve

Major projects are not generally mass-produced, so it is seldom possible to start with the most efficient construction approach. However, as illustrated earlier in Example 14.2, the people involved in a repetitive construction process develop greater skill at their tasks, managers figure out how to organize teams more effectively and improve coordination among teams, and support staff are likely to find better ways to deliver materials from suppliers to the site. As a result, repetitive tasks are likely to become more efficient; thus, if similar projects are undertaken, then the overall costs of those projects are likely to decline.

A common assumption is that costs will decline by a constant factor whenever the number of repetitions doubles. If the first project cost is K, and if the constant factor is s, then the cost would decline by a factor of s every time that experience is doubled:

- The second project would cost K × s.
- The fourth project would cost (K × s) × s.
- The eighth project would cost [(K × s) × s] × s.

Table 14.13 shows how costs would decline with various values of s. The terminology is used as follows: s is called the slope of the learning curve; if s is 90%, then there is a 90% learning curve. With a learning curve of 99%, costs will decline very little with experience; even after a thousand repetitions, costs will decline by only 10%. At the other extreme, with a learning curve of 75%, costs drop by nearly half after just 4 repetitions and by more than 90% after 500 repetitions. Common sense should indicate whether the learning curve is closer to one of these extremes or to something in between. With a little experience, project managers will gain some notion of what they can expect to gain from the learning curve effect for a particular type of activity or project. For example, suppose a company finds that it has managed to reduce costs by 15% after completing eight similar projects. By looking at the row for eight repetitions in Table 14.13, it is clear that a 95% learning curve would be appropriate.

Once a company has some feeling for the learning curve associated with a particular type of activity or project, it can estimate the cost of future activities. The learning curve is most important in two types of situations. First, if there is little or no experience with the proposed activities or projects, then some

Table 14.13 Effect of Learning Curve on the Cost of a Project (Base cost = 100; the table shows the estimated cost per project after the specified number of repetitions.)

Repetitions	Doublings	99%	98%	97%	95%	90%	85%	80%	75%
1	0	100	100	100	100	100	100	100	100
2	1	99	98	97	95	90	85	80	75
4	2	98	96	94	90	81	72	64	56
8	3	97	94	91	86	73	61	51	42
16	4	96	92	89	81	66	52	41	32
32	5	95	90	86	77	59	44	33	24
64	6	94	89	83	74	53	38	26	18
128	7	93	87	81	70	48	32	21	13
256	8	92	85	78	66	43	27	17	10
512	9	91	83	76	63	39	23	13	8
1024	10	90	82	74	60	35	20	11	6

improvement can be expected as experience is gained, whatever the slope of the learning curve. If only a few activities are involved, and there is some experience already (e.g., more than 10–20 repetitions), then the best bet is probably to use the most recent experience in budgeting for future costs—unless there is reason to believe that the learning curve effect is significant.

EXAMPLE 14.6 Using a Learning Curve to Estimate the Cost of New Projects

Suppose that a company is bidding on a project to strengthen eight similar highway bridges to reduce the risks from earthquakes. Its project developers have never done projects exactly like this, but their experience with other kinds of bridges suggests that they can expect a 90% learning curve related to any kind of major rehabilitation of bridges. They have estimated the cost of strengthening the first bridge to be $1 million. What should they expect for the cost of strengthening all eight bridges? How much benefit can they expect from the learning curve effect?

With a 90% learning curve, the cost of the second bridge will be $0.9 million, the cost of the fourth bridge will be $0.81 million, and the cost of the eighth bridge will be $0.73 million. The cost of the first bridge plus these three bridges is therefore estimated to be $3.44 million. The costs of the other four bridges can be estimated by interpolation:

- Bridge 3 will be between $0.9 million and $0.81 million; say, $0.85 million.
- Bridges 5, 6, and 7 will be between $0.81 and $0.73 million; say, an average of $0.77 million for a total of about $2.31 million.

Thus the total for the eight bridges is about $6.6 million, as compared to a cost of $8 million without a learning curve effect.

EXAMPLE 14.7 Using a Learning Curve and Experience to Estimate the Costs of Ongoing Activities

Suppose a company is in the business of erecting windmills for major power companies. It does not purchase the structures; it merely erects them. Over the past five years, the firm has erected just over 500 structures, and it has been able to reduce its costs from $50,000 per unit to $40,000. The firm is bidding on a contract to install 500 more units. Assuming that these installations are similar to what it has already done, how much should the company budget for the task?

The chief engineer argues that costs should be budgeted at $40,000: "We've installed a load of these units, our teams are very experienced, and we've already squeezed all of the waste we can find out of our construction process!"

The executive preparing the proposal is unconvinced. He wants to use an average cost of $32,000: "We cut costs by 20% on the first 500 installations, so why can't we cut costs another 20% on the next 500? If costs really are lower, we can lower our bid and have a better chance of getting the contract."

But if a low bid is made and costs cannot be lowered, then the company will be losing money. What is reasonable for average costs, based on the concepts embedded in learning curves? Should the chief engineer compromise with the marketing guys and allow the proposal to be based on a cost of $36,000 per unit?

Answer: This company has reduced costs by 20% after erecting 500 windmills. In looking across the row for 512 repetitions in Table 14.13, it appears that the slope of the learning curve is between 98% and 97% (i.e., about 97.5%). Once the company completes an additional 500 units, it can therefore expect only another 2.5% reduction in cost. Thus the cost of the last unit can be expected to be $(0.975)(\$40{,}000) = \$39{,}000$. This means the company can go slightly below $40,000 in budgeting the average cost per unit, perhaps to $39,500; it certainly cannot expect costs to drop to $32,000 or even $38,000.

Combining All Three Adjustments

All of the adjustment described in this section can be used in estimating the cost of a proposed project. The first step is to adjust all experience for inflation, so that all costs are expressed in comparable current dollars. The next step is to allow for inflation, since any proposed project will be completed sometime in the future. The third step is to adjust for the size and quality of the project. The final step is to allow for any significant learning curve effects that might be anticipated.

A good cost estimate is helpful in bidding on a project. With a good estimate of your own costs, you can specify an additional amount as a buffer for contingencies and then add what you believe to be a suitable profit margin. The better your understanding of costs, the smaller the buffer you will need, and the more aggressive you can be in going after a contract. If you do not have a good handle on your costs, then you are in danger of submitting a poor bid. Winning with a bid that is too low will make your company lose money on the project, but if you pad your bid too much, you may not win the contract.

Including Overhead Expenses

The cost of a project may include some share of the company's overhead expenses. Overhead includes items such as the salaries associated with administration and marketing, lease payments on building, utilities, libraries, and property taxes. Overhead expenses must somehow be covered if a private company is to remain profitable, even though it may be difficult to assign these expenses to any project. Public agencies and nonprofit organizations also have substantial overhead expenses to cover.

Economists focus on the variable costs associated with a project and ignore overhead. The logic for this approach is that any project that makes a positive contribution to overhead and profit is worth pursuing. Even if a project makes only a small contribution beyond covering variable costs, it will reduce the amount of overhead that must be covered by other projects or other parts of a company's operations.

Companies do not always adopt procedures recommended by economists. Many companies require projects to cover what is believed to be their fair share of overhead expenses. The overhead might be applied to each project or to each activity as a percentage of the estimated cost for each project or activity. The required percentage, called the **overhead rate**, is based on the relative proportion of total expenses viewed as overhead:

$$\text{Overhead Rate} = (100\%)(\text{Overhead Expenses})/(\text{Other Expenses}) \qquad \text{(Eq. 14.6)}$$

For example, if 20% of a company's total expenses are related to overhead items, then other expenses account for 80% of the total. Thus the average overhead rate has to be 25%. The company might therefore add 25% to the cost estimate for each project when submitting a bid.

The way that a company charges overhead affects the bids it makes for constructing a project. High overhead rates may limit a company's ability to bid on certain projects, and poorly constructed procedures for establishing overhead rates may make certain kinds of projects difficult or impossible to pursue. In a large company, accounting rules for allocating overhead can greatly affect costs as perceived by those who bid on projects, and changes in those rules could have unintended consequences with respect to the motivation of employees.

EXAMPLE 14.8   Irrational Allocations of Overhead Can Distort Incentives

A major research university charges overhead as a percentage of the direct expenses associated with a research project. Typical overhead rates are 60–65%. The overhead expenses cover a portion of the university's administration expense, much of the expenses associated with libraries, all of the expenses of research labs, and a large portion of the expenses associated with other buildings (including the construction of new buildings). Direct research expenses include summer salaries for professors, salaries for research and administrative staff as well as graduate research assistants, vacation and other benefits for employees (expressed as a percentage of salaries charged to the project), transportation, communications, and supplies. Conspicuously omitted from the direct expenses are the salaries of professors who supervise the research. This omission reflects a policy change made during the 1990s to "harden" faculty salaries (i.e., to treat faculty salaries during the spring and fall terms as academic expenses rather than requiring faculty to cover a portion of their salaries by securing research funding). This policy change was introduced in part to bring this university into line with other research universities that competed for the same federal research funds.

Consider the effect of this change on the cost of a proposal as it would be submitted by a professor or someone on the research staff before or after the change (Table 14.14). In both cases, the project supervisor expects to allocate 2 months of effort to the project in supervising two graduate research assistants, writing reports and papers, and coordinating the research with the client. Before the changes, the proposal would require a budget of $123,200, whether it was submitted by a professor or a research associate, assuming that they both had the same salary. After the change, the proposal is reduced to $89,600 if submitted by the professor, for a savings of $32,600; but the cost remains at $123, 200 for the research associate. The change in accounting, which has no affect at all on the expected levels of effort that will be devoted to the research project, has the effect of increasing the cost of the project to a prospective research sponsor by one-third if it were supervised by a research associate rather than a faculty member. The faculty member also has the option of adding a third grad student to the project for 7.5 months while keeping the budget at $123,200. To the extent that research staff and faculty submit proposals to the same potential sponsors, the change in accounting basically makes it impossible for the research staff to compete. This change in accounting has made it easier for professors to fund graduate students and easier for research associates to contemplate early retirement (which may or may not have been the intent). As we have already seen in Chapter 10, the rules of the game are important.

Table 14.14   Effect of Accounting Changes on a Proposal Submitted by a Professor

Proposed Budget Items		Expenses, before Change in Overhead Rules	Expenses, after Change in Overhead Rules
Professor (2 months)		$ 15,000	0
Grad students (two students, 18 months total)		$ 36,000	$36,000
Administrative assistant (1 month)		$ 4,000	$ 4,000
Total Salary		$ 55,000	$40,000
Benefits	40% of total salary	$ 22,000	$16,000
Total Direct Expense		$ 77,000	$56,000
Overhead	60% of direct expense	$ 46,200	$33,600
Total Budget		$123,200	$89,600

Creating a Budget for a Project

A budget is much more precise and much more detailed than a cost estimate. As indicated at the start of this section, an order of magnitude approximation may be fine for a preliminary cost estimate, and a cost estimate that is within a few percent may be suitable for bidding on a project. However, the budget specifies exactly how much should be spent on the project, and it serves as an important management tool. After all, the key goals for any project are to complete the work as planned, safely, on time, and within budget.

The budget for a project should be based on a clear understanding of (1) what must be done as laid out in the work breakdown structure (WBS); and (2) who will do the work, as laid out in the linear responsibility chart. A good cost estimate can provide a target for the total budget for a project, but the costs must be broken down into line items for specific expenses related to such things as personnel, equipment, supplies, communications, transportation, and subcontractors.

An accounting system can be set up to keep track of all of the expenses incurred by a project. Each expense item can be tied to a particular part of the project by defining expense categories for the project and establishing rules for assigning expenses to the proper category.

A budget is a more effective management tool if it is broken down along the same lines as the WBS. Periodic budget reports can then be used to follow progress on each task associated with the project.

14.4   ORGANIZATIONAL STRUCTURE

Several steps are always necessary in the implementation of a major project, as indicated previously in this chapter; but the steps need not be carried out by a single party. Broadly speaking, the three stages are design, build, and operate. These three steps easily can be carried out sequentially by different firms. For example, a public agency could use the following procedures to implement a project, such a new bridge, public housing, or a school:

1. Design competition; request submission of conceptual designs.
2. Select best concept, and contract with winner to prepare final design.
3. Open competitive bidding for the construction job.
4. Select construction contractor.
5. Supervise contactor as construction proceeds.
6. Upon completion of facility, the public agency begins operations.

This process allows numerous points for public review and for competitive inputs. The advantage of this approach is that the agency can select firms based on their specific competencies; but the process could well be delayed by having multiple phases of bidding, review, and contract selection. By separating design, construction, and operation, the agency may have created a process that causes contractors to lose sight of the ultimate goal, which is a successful project. Designers may focus on appearance or novelty rather than ease of construction or suitability for long-term operation. Contractors may focus on minimizing construction cost rather than minimizing the life-cycle cost, including the ongoing costs of maintenance and operations. The facility as designed and constructed may prove to be difficult to maintain and cumbersome to operate.

Other approaches to contracting can be used to promote a broader perspective in design and construction. The same contractor could be charged with designing and building a facility, which should encourage designers to pay more careful attention to the details and potential difficulties of construction. This contractual structure is known as **design/build** (**DB**). This approach still doesn't address the potential problems that may result with respect to operating and maintaining the facility. To handle those issues, the contractor could be required to design, build, and operate the facility. This arrangement presumably gives the designer powerful incentive to ensure that it is feasible to operate efficiently, that the durability of components and the ease of maintenance are considered in the design, and that the construction is done carefully to avoid defects that will eventually disrupt operations. This approach is called **design/build/operate** (**DBO**). This approach may require the contractor to operate the facility for a period of time and then turn it over to the appropriate government agency; it is then referred to as **design/build/operate/transfer** (**DBOT**).

The justification for the DBOT approach may be that the government agency lacks the expertise to run the facility, so that having a private entity operate the facility for 10 or 20 years will provide time for local managers and engineers to gain the expertise necessary to continue operations. Another justification might be that the government agency lacks the capital to pay for the project and would prefer the private company to finance the project. After the project is successfully up and running, it reverts to the government agency, which can then enjoy the benefits of the project. A third, somewhat Machiavellian reason for a government agency to use this approach is to have a private company that is isolated from political interference establish tolls or fees. When the facility reverts to the public, the high fees are already in place and the public agency has a financially secure operation.

Design standards, construction quality, maintenance, and operating practices need to be carefully considered in the DBOT option because the contractor may find a way to cut costs in any of these areas and thus transfer a second-rate, worn-out, or unsafe facility to the government agency. The DBOT contract should include provisions addressing construction and design standards, maintenance practices, and the quality of the facility when it is transferred. For example, some transit systems have been built using DBOT. These systems need to be concerned with the capacity and quality of the track, stations, and equipment when the government agency takes over operation.

14.5 MANAGING VERY LARGE PROJECTS

Very large projects can be broken up into smaller pieces, each of which can be managed as a separate project in terms of design, construction, and perhaps even operations. Besides the normal project management techniques covered earlier in this chapter, very large projects need some additional management techniques:

- *Overall project coordination:* Some entity has to be prepared to coordinate the interactions between and overlapping aspects of the various smaller pieces. This is partly contract supervision, partly arbitration and negotiation, and partly ensuring the timely flow of information among the parties that need to cooperate.
- *Political coordination:* By definition, very large projects are much more important in terms of their costs, benefits, externalities, and impacts on people and the environment. As seen in the following section, managing the politics can be a major headache for the project managers; but if the politics are not well-handled, there will be no project.
- *Public relations:* A big project is a large target for criticism from the time the first concepts are made public through the preliminary phases of design and environmental evaluation, during construction, and during operation. Hence, public relations are needed.
- *A capable, diverse management team:* Very large projects are almost certainly going to be unique, at least to the region. Even if managers have had experience on other large projects, they must expect to meet new challenges and unforeseen problems. It is therefore essential that the project managers create a work environment that brings together highly competent people, with a range of backgrounds and experience, to deal with whatever issues arise.

Very large projects often span decades. Based on his three decades of involvement in major transportation projects in Boston and elsewhere, Fred Salvucci identifies six phases of a very large project (Table 14.15). During the first phase, which he calls prehistory, the symptoms of major problems are evident; people are trying to understand the underlying problems and identify some potential solutions. Traffic congestion, a declining urban core, and rapidly increasing automobile ownership certainly seem like they may be related; but is the answer more roads or better transit, or some combination of tolls and fuel taxes aimed at reducing peak loads, or a major effort to rehabilitate the downtown retail and commercial districts? Phase two begins after one or more possible physical solutions have been identified, and it is possible to investigate the potential costs and benefits of these solutions. This phase seeks to find a concept for the project that will deal with the fundamental problems; that will be supported by the various actors and institutions within the region; that can be financed; and that can be constructed, operated, and maintained. During this phase, the concept, which is initially very sparse and quite fluid, begins to crystallize into a reasonably well-defined possibility. Internal debates within the public agencies and companies involved, as well as public interaction, lead to refinements and alternatives to the plan. The draft environmental impact statement is produced during this phase, which allows an in-depth review and response to the major social and environmental issues. This phase ends when the concept has been

Table 14.15   Six Phases of Very Large Projects

1	Prehistory	Serious, apparently unrelated problems generate public concerns; search for problem definition and potential solutions.
2	Project conceptualization, environmental impact assessment, and plan for finance and procurement	Define one or more physical solutions; analyze solutions; develop a project concept to evaluate and refine through internal debate and interaction with the public; consider alternatives and modifications to the concept; end up with a trustee for the project goals, a plan for financing the project, and a strategy for obtaining and managing the people and resources needed to construct the project.
3	Procurement	Create management team that will procure resources for design and construction, arrange for eventual start of operations, and deal with land use adjustments and socioeconomic considerations.
4	Final design	Transform the conceptual design into detailed designs that will form the basis for construction contracts.
5	Construction	Manage and oversee the construction process.
6	Operation	Operate and maintain the project; consider land use changes and socioeconomic impacts.

Source: Frederick Salvucci, lecture to 1.011 Project Evaluation, 2002.

approved, and a structure has been put in place to secure the funding and advance through design to construction and implementation. The final four phases—procurement, design, construction, and operation—are similar to what is needed for smaller projects, but the problems of coordination and control are magnified. That is why Salvucci emphasizes the need for a trustee for project goals; this might be a project leader with a small staff whose job is to ensure that the objectives of the project are not lost in the details and complexities of actually building the project. Section 14.7 describes the first two phases of a project that Salvucci managed for Massachusetts—namely, the depression of the interstate highway that slices through downtown Boston and the extension of the Massachusetts Turnpike under Boston Harbor to link up to Logan Airport and the region's major highways extending to the north and northeast.

Salvucci also offers some interesting ideas about how to think about very large projects. First, he says, you need to be concerned with doing the right job; only after you are sure that you are working on the right job do you shift to thinking about doing the job right. Second, you have to recognize that a complex project takes a long time to be conceptualized; ideas about the problems and potential solutions will change over time, as will the opportunities and possibilities for change. Third, you need to be flexible in your thinking, and you need to be willing to make some working hypotheses and use screening analyses to advance the study. In other words, you can try out different sets of assumptions to see what really makes a difference in the evaluation of various concepts. Fourth, interaction with and participation of a large number of people are essential, because both knowledge and power are distributed. Fifth, the environmental impact statement, properly structured, is an essential step because it involves the public, it highlights issues to be addressed, and it can help structure a project that is more broadly conceived and more beneficial than initially anticipated. A large public project aimed at transportation, for example, may offer opportunities to deal with environmental issues such as drainage or the improvement of parks.

14.6   THE BIG DIG: MANAGING A VERY LARGE, POLITICALLY CHARGED PROJECT

Very large projects require much more than a well-conceived design and an efficient construction strategy, and the most difficult aspects of project management may be in the early stages of the project. Boston's so-called Big Dig provides an excellent example of the tortuous process that may have to be followed when undertaking major infrastructure projects in a prosperous metropolitan area.[3] The Big Dig involved two mammoth projects:

1. Depressing the Central Artery: replacing an elevated highway that was built through the city in the 1950s with a wider, depressed highway in the same corridor.
2. Building a third tunnel under Boston Harbor that would connect the Mass Turnpike to Logan Airport and the major highway routes to the northeast of Boston.

The following were among the many benefits anticipated for this project:

- Relief of highway congestion within the core of the metropolitan area
- Vastly improved access to Logan Airport
- Elimination of the elevated Central Artery, an ugly structure that served as a visual and often physical barrier between the downtown and both the waterfront and the North End (a thriving residential community)
- Creation of 27 acres of open space along the corridor previously occupied by the elevated highway
- Enhancement of land values along the open space

These two projects were first recommended as part of an extensive review of transportation within the Boston metropolitan region that was undertaken by Governor Frank Sargent in 1970. The main focus of that review, however, was not increasing highway construction but stopping it. As a result of the study, Governor Sargent stopped two major projects that were part of the Interstate Highway System:

1. *The Inner Belt:* This project would have been a ring road slicing through the inner city and disrupting many neighborhoods in Boston and adjacent cities, including Cambridge and Somerville.
2. *The Southwest Expressway:* This project would extend I-95 (from Providence, RI) straight through Dedham and several Boston neighborhoods to link up with I-93 and the Inner Belt in downtown Boston. (The route that had already been acquired and cleared for this highway was eventually converted into a grade-separated trench that carried a multi-track rail line plus the Orange line, a heavy-rail transit line that was relocated from a nearby elevated structure).

Much of the allure of the project—and the key to its funding—was the availability of 90% federal funding for projects that were part of the Interstate Highway Program. Thus, if the project were approved as part of that program, then Massachusetts could recover 90% of the cost of the highway construction from the federal government. Despite this attraction, there was no interest in highways in Massachusetts in the early 1970s. Following the highly publicized Boston Transportation Planning Review, Governor Sargent stopped construction of interstate projects that had been approved for the region, and the focus shifted to investment in transit. After winning that fight, why go back to building highways? For several years, political and public support for the project was very low in both Boston and Massachusetts.

In 1974, Michael Dukakis defeated Governor Sargent, and Fred Salvucci became the secretary of transportation. Dukakis and Salvucci continued the investment in transit that the previous adminis-tration had begun. By the late 1970s, Massachusetts had completed or started major upgrades to the transit system, funded in part by the use of highway money freed up by the cancellation of the Inner Belt and the Southwest Expressway. Many of the subway stations had been upgraded, the Red Line was to be extended both to the north and to the south, and the Orange Line was to be relocated and taken off the decaying elevated structure that blighted neighborhoods along several miles of one of the city's main arterial streets. As someone who had advocated depressing the artery and building the

---

[3] Thomas P. Hughes, *Rescuing Prometheus: Four Monumental Projects That Changed the Modern World* (New York: Vintage Books, 2000).

tunnel when he worked for the City of Boston, Salvucci was now in a position to develop and to sell his ideas, first to the governor and then to the politicians and people of Massachusetts.

Salvucci sold the Central Artery/Tunnel (CA/T) project as a multifaceted approach to improving the quality of life in metropolitan Boston. As a highway project, it would double the capacity of the major north-south and east-west arteries through Boston, and the greatest relief of congestion would occur in the core of the region. Since the core of the region included Boston's greatest assets, keeping the core fluid would be a boon for the region's residents, for tourists, and for businesses. Depressing the artery would give the city access to its waterfront, and the city would enjoy "acres and acres of open space." Moreover, Salvucci viewed the highway construction as a way to advance transit and environmental matters: to get the highway, it would be necessary to mitigate the highway's impacts (e.g., by limiting downtown parking and providing additional transit investment). Salvucci also emphasized that these projects could be completed without relocating any residents and by relocating only a few businesses. Unlike the 1950s project that created the elevated structure, depression of the artery would bring the city together, improve aesthetics, and improve the functioning of the city.

Although political and business leaders accepted the ideas of the project, it still faced major obstacles. First, if the tunnel was built, it presumably would come up somewhere in East Boston; that plan was politically infeasible because of local opposition from residents. Second, the project was expected to cost more than $5 billion, which was viewed as a tremendous amount of money in the mid-1970s. While the federal aid made the cost much less of an issue in Massachusetts, it caused a political firestorm in Congress, and the Reagan administration viewed the project as completely unnecessary. Because the Speaker of the House of Representatives was from Massachusetts, it was possible to conceive of eventually gaining congressional support; but Tip O'Neill Jr. said he would support the project only if his constituents did. Residents in East Boston did not cease opposing the project until the Massachusetts Port Authority (Massport) allowed the tunnel to come up on airport property—and that did not happen until 1983, when Governor Dukakis's appointees finally controlled Massport. After another year or two of political battles, the project gained congressional approval. Table 14.16 summarizes the background for the Central Artery/Tunnel project.

On the assumption that congressional approval would finally be garnered, Massachusetts began planning for the project. Salvucci, as head of the transportation department, led both the political battles and the planning process. The Department of Public Works was designated as lead agency for the project. At the urging of DPW, two firms (Bechtel and Parsons/Brinckerhoff) with extensive international experience in construction design and management established a joint venture in Boston. The new firm, commonly referred to as the Joint Venture, began work on conceptual design in 1985 before final approval of the project was gained from Congress. The conceptual design was the basis for the draft final environmental impact statement, which was completed in 1990 and subjected to intense debate within the state for the next year; it was finally accepted by Secretary of Environmental Affairs John DeVillars on January 2, 1991—the last day of the Dukakis administration.

In the 1980s, we were fighting for our fair share, as we had had no federal funds for urban highways. Massachusetts built and paid for the Mass Pike, the Tobin Bridge, and the two harbor tunnels. We felt we deserved full 90% funding on the Big Dig, because we hadn't gotten any funding for these earlier projects. One federal response was "this is beautification—go to Housing and Urban Development." FHWA was furious with Massachusetts for killing other highway projects—so they needed careful treatment.

We engaged Bechtel and Parsons Brinckerhoff in part to gain credibility that we could do it. And Republicans liked public/private partnerships.

Fred Salvucci, CTS Presentation, MIT, November 3, 2000

Table 14.16 Background to Boston's Central Artery/Tunnel Project

Year	Action	Key Figures
1970	Both candidates for governor oppose Inner Belt and other interstate highways planned for metropolitan Boston.	Frank Sargent (R) defeated Kevin White (D), Mayor of Boston (D)
1970–72	Boston Transportation Planning Review—more than 50 people participate in review of all major transportation projects then planned for the region.	Alan Altshuler, MIT Professor who became MA Secretary of Transportation and Construction
1972	Governor Sargent cancels the construction of the Inner Belt and Southwest Expressway and proposes more investment in transit, depressing the Central Artery (CA/T) and constructing a third harbor tunnel to be used only for commercial vehicles and buses.	Governor Sargent Alan Altshuler
1974	Michael Dukakis becomes governor, supports transit investment; appoints Fred Salvucci Secretary of Transportation and Construction. Salvucci convinces Dukakis and others to support construction of the CA/T, taking care to avoid disrupting neighborhoods.	Governor Dukakis Fred Salvucci
1974–83	East Boston residents oppose tunnel because it would emerge within and disrupt their neighborhood. As long as the residents were opposed, their powerful congressman would not support the project.	Tip O'Neill Jr., Speaker of the U.S. House of Representatives
1983	The Massport Authority, bolstered by Dukakis appointees, approves having the third harbor tunnel emerge on Massport property, which eliminates the key local political problem.	Massport Board Tip O'Neill, Jr. Gov. Dukakis Fred Salvucci
1983–85	Draft environmental impact statement (EIS) prepared, approved by state and by EPA.	MA Department of Public Works U.S. EPA
1987	After years of political wrangling and analytical disputes involving the Federal Highway Administration, a bill is passed authorizing the construction of the CA/T as part of the Interstate Highway Program.	Fred Salvucci, Tip O'Neill, Jr. James Howard (D–MA) Chair of House Public Works Committee; Ray Barnhart (R–TX) head of FHWA

Source: Thomas P. Hughes, *Rescuing Prometheus: Four Monumental Projects That Changed the Modern World* (New York: Vintage Books, 2000), 202–210.

The single most controversial aspect of the whole project was left unresolved, however. DeVillars, along with many outspoken critics, believed that a better design was needed for linking the depressed artery in Boston to I-93 and the other major highways on the other side of the Charles River. The design at that time, known as Scheme Z because it was the 26th that had been considered, called for a complex set of circuitous bridges and ramps to provide the necessary connections between the major highways. Some traffic moving between the west and the northeast would actually have to cross the river three times in essentially the same location. The seemingly weird routing was necessary to allow traffic to merge safely and to avoid the situation on the existing bridge, which required large numbers of drivers to enter the highway at speed from the left lane and, just a short distance further on, to exit from the right lane. Clearly, this was a dangerous and accident-prone stretch of highway. Also, with a bridge, traffic would have to emerge from the tunnel, climb up and over the river, then descend to another depressed road, and then climb up over another nearby bridge. Many people preferred a tunnel to a bridge for aesthetic reasons, since the structure in Scheme Z was quite massive.

> *Boston's assets are the keys to the vitality of New England, Boston is the financial capital of the region; it has the Celtics, the Bruins, and the Red Sox; it has the Freedom Trail, the Museum of Fine Arts, the Science Museum, the Children's Museum. Boston is the largest city in New England and it is the capital of Massachusetts. It has New England's busiest airport and seaport. It has some of the best medical and educational facilities in the world. And what is unique in the United States, all of these assets are in a two-square-mile area! Boston is a very rail-oriented city. But, if you get to gridlock, then you've lost everything.*
>
> *The Big Dig is the largest highway project in the U.S., but it began with a transit agenda. It was funded in large part by the Interstate Highway System, which has a history of destroying cities, but the original concept was to help the city and the economy. Instead of shedding costs onto third parties, we sought mitigation and pareto optimal approaches (in which the winners compensate the losers). Our expectation is to accept the costs of the project, not to push them on to residents and abutters.*
>
> *The project had an unusually long gestation period, or so it seemed. But, like all major projects, conceptualization, design and construction all take a long time. Dealing with continuity is always a problem, since mayors and governors change every four years.*
>
> Fred Salvucci, CTS Presentation, MIT, November 3, 2000

When the Weld administration took office, the secretary of transportation called for a complete review of the crossing, rather than simply modifying Scheme Z. The debate over the crossing continued for most of 1991 as both the review team and the Joint Venture considered options. Finally, in November, Secretary Kerasiotes chose a new design for the crossing. The new design and what amounted to a 2-year delay added $1.3 billion to the cost of the project, which was at that time projected to be more than $7 billion. To probably everyone's surprise, the final design was marvelous. As David Nyhan put it: "The most spectacularly beautiful piece of urban architecture will be the cable-stayed bridge. . . . This is, for my money, the signature symbol of the largest civic works project in the nation, an artifact of phenomenal grace and power."[4]

> *In 1991, the cost estimate was $6 billion, and the project cost might have actually been $6 to $8 billion. Bush, Dukakis and Weld all looked at $6 billion, but now we're headed toward $14 billion, because bad things happened. The schedules changed, the environmental impact statement was reopened. Reopening the Charles River crossing added delay. Another factor, which I consider more important, was that many things were reneged upon: the project was supposed to employ local youths for construction jobs, but this was defunded in the name of economy. Many transit commitments haven't been put into place. . . .*
>
> *The project has brought $5 billion more into the region than Massachusetts paid into the highway trust fund. These are jobs in Boston, with a multiplier effect. The extra $8 billion—the opportunity cost—is what hurts. What projects were deferred? How serious was this cost to the region.*

---

[4] David Nyhan, "Halfway through the Big Dig, There's Light at the End of the Tunnel," *Boston Globe*, 29 January 1999.

> *Question: You use a political argument to justify the project; you use economics only when it suits you.*
>
> *Answer: I disagree. This was part of the last deal; no one else had to do benefit/cost analysis on urban interstates. I'm not ashamed of this project in this era of dot.com millionaires and trophy homes.*
>
> Fred Salvucci, CTS Presentation, MIT, November 3, 2000

Despite the uncertainty over the crossing of the Charles River, the Joint Venture began work on the preliminary design. The entire project was broken into 56 distinct design packages, which dealt with what would become more than 100 individual construction contracts. When the preliminary design was completed for one of these packages, DPW invited design firms to bid for the task of creating the final design. When the final designs were completed, construction firms were invited to bid on the construction packages related to that portion of the project. Breaking the very complex project into dozens of smaller projects allowed the tasks related to each contract to be simplified and limited in scope, although the interactions among the various design and construction firms were increased. With this structure, it was possible for preliminary design, final design, and construction to be proceeding for many parts of the project, even before the river-crossing issues were resolved. Table 14.17

Table 14.17   Designing Boston's Central Artery Project

Year	Action	Key Figures
1985–86	Basic organization established for the project: Department of Public Works (DPW) would have overall responsibility for the project; a joint venture firm would be created by Bechtel and Parsons/Brinckerhoff to manage the project.	Fred Salvucci DPW, MBTA
1985–90	Conceptual design	Fred Salvucci, DPW, Joint Venture
1990–91	Draft final EIS prepared, subjected to intense debate in MA, submitted and approved by FHWA; the Final EIS included numerous requirements for mitigation.	
1990–93	Preliminary design completed. Project divided into 56 pieces for design and 132 for construction. Final design firms bid on packages when they are advertised. Preliminary construction designs given to potential contractors to help them prepare construction bids.	Joint Venture Design firms
1990–97	Final design proceeds, and bids for construction solicited. Mass Highway selects winning bids based on recommendations from Joint Venture.	DPW Mass Highway Design firms Construction firms
1991	Dukakis administration is nearly ready to proceed with design and construction by the time it gives way to Governor Weld's republican administration. However, heated public opposition to the proposed design (Scheme Z) for crossing the Charles River continues to be a major obstacle for proceeding.	
1991	Weld administration establishes 42-member Bridge Design Review Committee in January; Joint Venture team also identifies alternatives. New transport secretary reviews options and selects the design in November.	Richard Taylor and James Kerasiotes, the first two secretaries of transportation in the Weld administration
1992	Scope of the CA/T project reasonably well-defined.	
1997	Final design essentially completed.	

Source: Thomas P. Hughes, *Rescuing Prometheus*, 210–243.

summarizes the key actions and actors involved in the design of the Central Artery/Tunnel project, a process that stretched from 1985 to 1997.

14.7 MANAGING PROGRAMS

A program can be created to promote, implement, and monitor the performance of a set of related projects. A program has several components:

- *Policy:* A policy statement identifies the objectives of the program and the types of projects that can be pursued as part of the program.
- *Selection criteria:* These are used for establishes whether a proposed project can be undertaken as part of the program; there may be criteria to determine which projects qualify plus a selection process to determine which projects will be approved.
- *Funding:* A source of funding is needed to pay for or subsidize the projects selected.
- *Construction standards:* Standards are developed for constructing projects that are funded as part of the program; these include requirements for mitigating social and environmental impacts.
- *Program management process:* This process addresses the following questions.
  - Project audits—was the money spent properly on projects, and were the projects implemented as planned?
  - Policy review—do the projects succeed in meeting program objectives?
  - Review of externalities—what are the positive and negative social, economic, and environmental consequences of the projects?
  - Revision—is there a process for recommending changes to any aspect of the program?

A program provides an effective means of managing what may be a very large number of related projects. By establishing selection criteria and construction standards that can be applied to all of these projects, a program simplifies the evaluation of projects and allows for more rapid implementation. With a program, it is not necessary to debate every single aspect of every potential project; instead, the debate (public or private) required to gain approval for a program is presumed to determine that projects selected and constructed according to the program's criteria are indeed worth pursuing.

Examples of public programs include the Interstate Highway System, public housing programs, and programs to promote the use of renewable energy. Examples of private programs include national retailers' plans for expanding their stores (e.g., Wal-Mart or Home Depot) or a railroad's decision to replace or upgrade bridges on its main routes so they can carry heavier freight cars.

Section 14.8 reviews the politics leading to creation of the Interstate Highway System. This program arose only after decades of dispute about what kind of system should be built, how it should be financed, and to what extent (if any) investments in highways should be linked to other development objectives. The program, as it was eventually implemented, took a narrow perspective of its role; it allowed highways to crash through cities while providing a bonanza for highway construction companies, the trucking industry, and the traveling public. As noted in Section 14.8, other possibilities were considered. Highways could have been built without so much disruption to cities, and other ways of financing and managing highways were (and still are) possible.

14.8 THE INTERSTATE HIGHWAY SYSTEM

> *It was a highway building program, pure and simple.*
> Mark H. Rose, *Interstate: Express Highway Politics, 1941–1956*, p. 98

The invention of the automobile and the truck at the beginning of the twentieth century created a demand for smoother, more durable roads.[5] In the United States, vast sums were spent on highways in the two decades following World War I. Federal, state, and local governments spent $35 billion between 1921 and 1939, building 418,000 miles of new roads and tripling the extent of paved roads, going from less than 0.4 million miles to 1.3 million miles.[6] It was clear to Congress and the public that a coordinated plan was needed for another major highway construction effort in order to keep pace with the rapid growth of auto ownership and the trucking industry. Moreover, a well-designed, limited-access highway would increase the efficiency of intercity transportation, basically cutting travel times in half while allowing for larger vehicles carrying heavier loads.

During World War II, war demands forced the government to curtail highway use by rationing gasoline and rubber (and therefore automobile tires). These wartime restrictions merely deferred the need for developing a better highway system. With the wealth and prosperity following WWII, the increase in motorization was extremely rapid. Since people had cars, they needed paved roads.

The wartime experience had also made it clear that a good road system was essential to national defense in terms of providing access to ports, military facilities, and factories. U.S. generals, including Eisenhower, had seen the autobahn in Germany, and they were aware of the critical role for good roads in moving both people and freight. At the height of the Cold War, politicians were concerned about the time it would take to evacuate a city in anticipation of a nuclear attack—thus providing another defense justification for a highway program.

14.8.1 Issues in Designing a National Highway System

Ideas for a national system of limited-access highways emerged during the 1930s. Road construction was seen by some as a means to redesign and reinvigorate cities. President Hoover thought that building roads could stimulate growth and help end the Depression. Thomas McDougal, of the U.S. Department of Agriculture, advanced plans for a "30,000 mile national expressway system, one aimed at speeding rural and urban traffic, eliminating urban and rural decay, and creating useful jobs." Thomas McDonald, chief, U.S. Bureau of Roads under Roosevelt, envisioned a 40,000-mile interstate system; he even had plans for major expressways into cities as well as inner and outer beltways.

Several sets of issues delayed federal action on these ideas:

- Transportation versus urban design
- Funding
- Design standards
- Capturing the economic value of new highways
- Allocating costs to different users

The first set of issues, **transportation versus urban design**, involved the **social impact** of highways. This question was not something belatedly discovered in Boston and San Francisco during the late 1960s; it was debated in Congress during the 1940s. Would the highway system be designed to serve interests of the major highway users (i.e., the trucking industry and its customers), or would it also be viewed as a major strategy for changing, reinvigorating, or segregating major cities? If roads became more expensive because of the routes selected or other social issues, why should highway users pay the extra costs? Roads were indeed viewed as barriers—and these barriers could promote

---

[5] This section is based on the following books about the development and impacts of highway construction in the United States: Mark H. Rose, *Interstate: Express Highway Politics, 1941–1956* (Lawrence: The Regents Press of Kansas, 1979); Howard Kunstler, *The Geography of Nowhere: The Rise and Decline of America's Man-Made Landscape* (New York: Free Press); Tom Lewis, *Divided Highways: Building the Interstate Highways, Transforming American Life* (New York: Penguin, 1997); Clay McShane, *Down the Asphalt Path: The Automobile and the American City* (New York: Columbia University Press, 1994).

[6] Rose, *Interstate*, 4. The political information perspective presented in this section is based on this excellent history of the personalities and conflicts that led to the creation of the Interstate Highway System.

neighborhood cohesion for some neighborhoods while dividing others. To what extent should the disruption of neighborhoods and the social costs of moving people and businesses affect the routing of highways? Should highways be designed in part to improve urban design or mitigate the disruption of neighborhoods, or should they simply be designed to minimize construction cost or costs to users?

The second set of issues related to the **funding** for highways. For intercity highways, the basic question was whether to fund the roads with tolls, with targeted taxes and user fees, or with general revenues. The Pennsylvania Turnpike was constructed before World War II, and by the early 1950s most of the other states in the Northeast had constructed turnpikes. Boston and other cities had also constructed major highways into and through their downtowns. Why should the cities and states that had already spent money on good roads be forced to pay for roads in other areas? More generally, there were questions about using funds gained by taxation and fees in rich states to fund infrastructure in poor states. Further, each state was dealing with conflicts among those concerned with urban issues (congestion and commuting) and rural issues (intercity connectivity and access to markets for their agricultural and other goods), and there were conflicts among those concerned with automobiles and those concerned with trucking. And perhaps most important of all, what share of the public funding should come from the federal government and what share from the states? How much could the federal government afford, especially as the Korean War sapped resources in the early 1950s?

The third set of issues is related to **design standards**. Should standards be determined at the state or the federal level? What procedures would be necessary to ensure that roads were in fact constructed to the design standards—or, to be blunt, how could the government ensure honest construction? Should design standards be the same in urban and rural areas? Was a four-lane, limited-access highway required everywhere? Could design standards for exit ramps be modified to allow the roads to fit into the existing land use?

The fourth set of issues related to who would benefit from the **economic value** created by the public investment. Most important, who would capture the increases in land value? When the government gave land grants to the railroads during the nineteenth century, the railroads were able to use the expected increase in land value to fund railroad construction. The federal and state governments that made the land grants did not give all of the land to the railroads, so the value of the public lands increased as well. Could some similar scheme be used to finance a highway system? The land values at interchanges would certainly be expected to increase; could some of those increases be used to fund the cost of the system?

A fifth set of issues arose when discussing how to create tolls, user fees, speed limits, and weight limits for **different classes of traffic**. There were engineering issues: how do you determine the costs incurred by different classes of users? And, how do you determine the design standards for the highway—what will be the maximum height, length, width and weight for trucks? There were operating issues: do you require all vehicles to maintain a minimum speed close to the speed limit,—or do you allow very heavy trucks to creep uphill as they did on many local and state roads? What is the speed limit, and how will the speed limits affect the design of the highway and the exit/entrance ramps?

Figure 14.10 Large Trucks Dominate Some Portions of the Rural Interstates. I-70 is among the interstate sections most heavily traveled by large trucks, which typically account for 20–40% of the traffic between cities such as Columbus, Indianapolis, and St. Louis. (This photo shows a fleet of trucks approaching Indianapolis on November 20, 2007.)

14.8.2   Creating the Interstate Highway System

In November 1944, the U.S. House of Representatives passed an amendment forbidding the condemnation of "rights-of-way larger than needed for traffic alone. No aesthetic or urban renewal considerations were going to trouble road builders."[7] A month later, this narrow view of highway spending was incorporated in the Federal Aid Highway Act of 1944, which created the "National System of Interstate Highways." This act stated that roads were to be built for traffic (i.e., not for other social purposes, such as urban renewal). The bill provided no special funding for an interstate system, but it did start planning for the system. By August 1947, federal and state officials had identified 37,000 miles for the Interstate Highway System.

Now the question was funding. Toll roads were being built, and they were successful, but businesses and truckers preferred freeways. There was support for dedicating certain fuel taxes and vehicle fees to a highway trust fund, but some feared that Congress would simply divert this money to other types of transportation or other sectors of the economy. "Diversion of funds" was a major concern to road proponents, who fought hard throughout the 1940s and 1950s to prevent any diversion.

The financial question was whether it would be feasible to sell bonds to raise money to build the system and then pay the interest on the bonds using proceeds from taxes and user fees. How much would the federal government be able to pay, and how much should it pay? Initial schemes suggested that in every state, the federal government and state government would each pay half of the costs. Later ideas increased the federal role to 75% and eventually to 90%. The higher percentages were possible because enough money would be coming in from the fees to cover the costs of construction, and there was tremendous public support for roads.

Congress finally passed the Federal Aid Highway Act of 1956, which created the program for financing and constructing the National System of Interstate and Defense Highways. The act allocated federal money to states based on their population, land, and road mileage (as was commonly done for other federal highway systems). The act provided 90% federal funding for construction of the Interstate Highway System, but none for maintenance. The money would come from fuel taxes and fees that were dedicated to a Highway Trust Fund. The act retained the basic structure of the system contemplated in the 1946 act—that is, highways went through cities and had accompanying ring roads. The act established standards for design and construction, calling for limited-access divided highways with at least two lanes in each direction and specifying lanes widths and bridge clearances for the entire system. But it ignored the social issues that had been debated for two decades, and it failed to anticipate the problems that overreliance on the automobile has contributed to the twenty-first-century problems of congestion, pollution, safety, nonsustainable use of fossil fuels, and climate change.

14.9   SUMMARY

Stages in Project Management

Project management involves five major stages. The challenges and potential problems are different in each stage:

1. *Developing a strategy for implementing the project:* Before it is possible to start construction, it is necessary to arrange financing, acquire the land, obtain the necessary public approvals, develop a construction strategy, and prepare a marketing strategy.

2. *Managing construction:* This stage is the core of project management. It involves the planning and control necessary for managing the construction, taking into account the need for safety, security, and the desire to minimize social and environmental disruption during construction.

---

[7] *Ibid.*, 25.

3. *Managing the transition from construction to operation:* At some point, this stage is reached. The control of the new facilities must shift from those responsible for construction to those responsible for management, and the facilities must be tested and eventually opened for business.

4. *Completing the project:* This stage involves two considerations; namely, putting the finishing touches on the structures and restructuring the finances. A large project likely has a lingering set of problems to correct, and it may be some years before the construction is finally completed. A project that was funded with a variety of construction loans, government subsidies, or other monies, can, upon completion, be financially restructured. Perhaps a more favorable loan rate can be achieved, or perhaps user fees can now be used to cover operating costs, so that public subsidies can be eliminated or diminished. The owners of a private project may decide to sell the completed project and move on to a new project.

5. *Assessing or auditing the project:* This stage involves three main issues. First, was the project completed on time and on schedule; and if not, then why not? Second, and perhaps more important, when the project was completed, did it actually function as expected? Third, and probably most important, did it actually serve the needs it was expected to serve without creating extensive, unexpected problems?

Project Management Techniques

The chapter introduced various methods that can be used to manage a construction project.

- *Statement of work:* This statement defines exactly what is to be done, the resources available, and the project timetable. It will be part of the construction contract, and the contractors and the owner may periodically have to adjust the statement of work in order to stay on time or on budget.
- *Work breakdown structure (WBS):* The key to managing the construction is to break down the work into logical steps and individual work elements.
- *Network diagrams:* The logical relationship between work elements can be expressed in a network diagram. Some work elements can begin immediately, but most work elements can begin only after various prerequisite elements have been completed.
- *Schedules and the critical path:* The time required for each work element can be estimated based on experience or expert advice. By adding the time required for each work element to the network diagram, project managers can determine how long it will take to complete any sequence of work elements. For each work element, it is possible to determine the earliest start date, based on the time required to complete any prerequisite work elements. By progressing in a logical fashion through the network diagram, managers can determine the critical path for the project, which is the sequence of activities expected to take the longest time to complete. Items on the critical path must be started as soon as possible and finished within the allotted time, or the entire project will be delayed. Other work elements can be delayed without delaying the project. The amount of slack time can be determined by calculating the latest start date and comparing that to the earliest start date.
- *Resource requirements:* The resources required for each work element can be based on prior experience or expert advice. The work schedule dictates when the resources are needed, and it is often desirable to adjust the schedule to reduce the peak demand for the labor force or for other resources.
- *Cost estimation:* The costs of a project can be estimated based on experience, expert judgment, or models of future experience. Experience is especially useful in predicting future costs, provided adjustments are made for inflation, the size and quality of projects, and the potential for becoming more efficient in implementing projects. Useful techniques include using cost indexes to adjust for inflation, power sizing to adjust for the size of projects, and learning curves to account for the benefits of experience. It is important to be efficient, organized, consistent, and careful in making cost estimates.

- *Budgets:* Budgets may initially be based on rules of thumb founded on prior experience, but as the construction design and work schedule are finalized, project managers can determine exactly what resources are needed and for how long. Cost estimates and preliminary budgets need only be as accurate as the other elements of the evaluation, such as financing costs and marketing expectations. Detailed budgets eventually are needed, and they should be tied to the work breakdown structure to facilitate management.

Organizational Structure

A company or an agency may choose to undertake all aspects of the project, from design through construction to operation. This approach has the advantage that the entity responsible for designing the project also has to build and operate the project; thus, there are clear incentives to ensure that the design can be constructed efficiently and that the facilities are constructed in a fashion that supports operations for the long term. Another approach is needed if the company or agency lacks the expertise for designing, constructing, or operating the facility.

One alternative approach is to hold a design competition, select a design, and then seek bids for construction and subsequently for operation. This process allows review at each stage, but it also tends to separate the designer's interests from those of the contractor and the operator. Design/build (D/B) is an approach whereby one entity designs and builds the structure; design/build/operate (DBO) has the same entity also operating the facility under contract to the owner; design/build/operate/transfer (DBOT) has the entity ultimately turn over the project to the owner after a period of what may be a few years or very many years. These approaches all have the merit of bringing common goals to the designer and contractor, and DBO and DBOT also require the same people to consider how to operate the facility. The DBOT option allows the ultimate return of the facility to the owner, who may then reap the ongoing benefits (or suffer the ongoing costs) of the project.

Managing Very Large Projects

Coordination is essential for a very large project. A standard approach is to break the project into multiple pieces, each having its own construction contract. This allows multiple firms to participate in the construction, which may allow greater flexibility in how to structure the construction process. It is essential to have a single entity with overall responsibility for managing and coordinating the project.

The Big Dig: Managing a Very Large, Politically Charged Project

The chapter included two case studies. Boston's "Big Dig" provides insights into the kinds of political, organizational, and financial problems that can affect very large projects. Since large projects require public approval for many factors, including funding and land use, political support at the local, state, and federal levels will be necessary. Projects such as the Big Dig require many years for design and construction, and a robust organizational structure is required to maintain continuity and to facilitate coordination among the many privates companies, government agencies, and citizens' group that are likely to be involved with or concerned about the project. Funding, a concern with any project, becomes an over-riding concern for very large projects simply because of the magnitude of such efforts. Changes in design, inflation, and unexpected events can add billions of dollars to a large project and delay or imperil successful completion.

Managing Programs

A program can be created to promote, implement, and monitor the performance of a set of related projects. A program includes a policy statement indicating the types of projects that can be pursued, along with criteria for evaluating particular proposals. The program may include funding or tax incentives or subsidies for qualifying projects as well as standards for construction and operation.

Program management will include audits for individual projects; periodic review of the policy statement; and a process for revising the program in response to changes in technology, needs, perceptions of the success or failure of the program, or changes in the ability to finance the program. For a program, it is not necessary to debate every single aspect of every potential project; instead, the debate (public or private) necessary to gain approval for a program is presumed to determine that projects selected and constructed according to the program's criteria are indeed worth pursuing.

The Interstate Highway System

The Interstate Highway System is the second case study in this chapter. The Interstate System is an example of a program established to meet the readily apparent need for better highways. Debates over how to proceed dragged on for years because of disagreements about the objectives of the program and the financing mechanisms. Once Congress approved the program, the funding sources set up for the program were sufficient to allow rapid creation of a national system of limited-access highways. Fifty years later, with congestion, renewable energy, and climate change as major public issues, the nature and role of the Interstate Highway System is being subjected to renewed debate.

ESSAY AND DISCUSSION TOPICS

14.1 What is the difference between a project and a program?

14.2 The Interstate Highway System was funded by fuel taxes and other fees that were paid into the Highway Trust Fund. In recent years, some of the funds from the HTF have been used to fund transit projects. For example, funds that had been authorized for the Southwest Expressway and the Inner Belt in Boston were made available for projects including the extension and relocation of subway lines. Today, a portion of the HTF is dedicated to transit projects. Do you think the HTF should be used for transit projects? Why or why not?

14.3 The debate that led to the Federal Aid Highway Act of 1956 ended up with legislation that funded the construction of highways—not the construction of highways as part of a larger program that could have been designed to help resolve some of the many problems related to urban land use, decaying neighborhoods, and declining downtowns. Do you think that the decision to focus solely on highways was the correct decision? If so, do you think it was fair to build highways straight through neighborhoods so as to minimize construction cost and driver convenience, while forcing residents and businesses to relocate? If not, how do you justify using highway user fees to tackle what appear to be urban development problems?

PROBLEMS

14.1 Define the following terms, and indicate why they are important for project management:

a. Work breakdown structure
b. Linear responsibility chart
c. Gantt chart
d. Network diagram
e. Earliest and latest start date
f. Earliest and latest finish date
g. Critical path
h. Schedule slack

14.2 Create a work breakdown structure for one of the case studies or major assignments included in this text:

a. The Panama Canal Case study
  i. The Panama Railroad
  ii. The French effort to construct the Panama Canal
  iii. The U.S. effort to construct the Panama Canal
b. The Middlesex Canal
c. Skyscraper

14.3 If you are doing a case study of a particular project as a term project:

a. Create a work breakdown structure for this project.
b. Structure a network diagram for the project.
c. Prepare a linear responsibility chart and a Gantt chart to illustrate how the major actors scheduled and managed the activities that were part of the project.
d. List the activities included on the critical path for that project.

14.4 Consider the network diagram shown below.

a. What is the earliest start time for task A4?
b. What is the latest start time for task B4?
c. What is the earliest finish time for task B3?
d. What is the latest finish time for task A5?
e. What tasks are on the critical path? How long will this project take if the tasks on the critical path are completed on time?

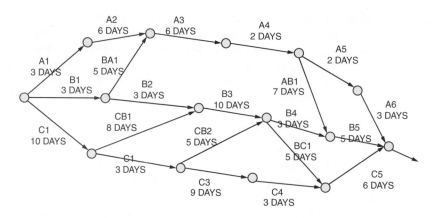

Problem 14.4 Task Relationships and Task Times for a Hypothetical Project

14.5 The Joint Venture that was responsible for managing Boston's Big Dig separated the design of that set of projects into 56 packages covering more than 100 separate construction contracts. Suppose you were managing the construction of a canal like the one considered in the Canal Assignment in Chapter 2:

a. How might you structure the design activities? Why?
b. How might you structure the construction activities? Why?

14.6 *Using a Cost Index*: The cost of a 10,000-square-foot apartment building was $400,000 in 1990, when a cost index for apartment buildings was at 180. The cost-capacity factor for this type of building is estimated to be 0.8. Estimate the current cost for a 15,000-square-foot apartment building, assuming that the cost index now stands at 240.

14.7 *Using a Learning Curve*: You estimate that your construction requires 100 hours to do a certain task for the first time. Based on a 95% learning curve, how much time will be required to do this task for the fifth time? What is the average time to do this task for the first five times?

14.8 *Preparing a Cost Estimate*: You are bidding for a 1000-square-meter warehouse construction job. You estimate that a crew of eight people will require thirty 10-hour working days to complete the job. Labor costs are $10/hour, material costs are $50/square meter, finishing costs are $40,000, and overhead costs are $60,000. What should be your bid per square meter, assuming that you want 15% profit?

Chapter 15

Toward More Sustainable Infrastructure

But it came to me then, I am sure, for the first time how promiscuous, how higgledy-piggledy was the whole of that jumble of mines and homes, collieries and pot-bands, railway yards, canals, schools, forges and blast furnaces, churches, chapels, allotment hovels, a vast irregular agglomeration of ugly smoking accidents in which men lived as happy as frogs in a dustbin. Each thing jostled and damaged the other things about it; each ignored the other things about it, the smoke of the furnace defiled the pot-bank clay, the clatter of the railway deafened the worshippers in church, the public house thrust corruption at the school doors, the dismal homes squeezed miserably amidst the monstrosities of industrialism, with an effect of groping imbecility. Humanity choked amidst its products, and all its energy went in increasing its disorder, like a blind stricken thing that struggles and sinks in a morass.

H. G. Wells, *In the Days of the Comet* (1906)

CHAPTER CONCEPTS

Section 15.1 Introduction

Section 15.2 Stages in the Evolution of Infrastructure-Based Systems
    Characteristics of eight stages of evolution
    Examples of issues and activities

Section 15.3 Skyscrapers and Building Booms
    How technology made skyscrapers feasible
    How real estate economics made skyscrapers inevitable
    How technology helped make skyscrapers cheaper—and more vulnerable

Section 15.4 Case Study: Evolution of the U.S. Rail System
    Case study of the evolution of an infrastructure-based system
    Roles of technology, management, and regulations
    Changing role of railroads as part of the national transportation system

Section 15.5 Twenty-First-Century Challenges: Sustainable Infrastructure
    Need for improved infrastructure performance
    Examples of more sustainable infrastructure

15.6 Upgrading Infrastructure: A Costing Framework
    Estimating the life of components
    Reducing the life-cycle costs of components

15.1   INTRODUCTION

Over 100 years ago, H. G. Wells—following in the tradition of Dickens, Jules Verne, and other great nineteenth-century novelists—railed against the pollution, ugliness, immorality, and disorder of city life in the industrial society. In one scathing paragraph, he raises vivid images of many of the disorders that have plagued cities for thousands of years and persist to this day: pollution, noise, corruption, thoughtless development, and horrid environments for children. Today, other ills and other fears could be added to the list: climate change, overuse of fossil fuels, looming crises involving the water supply, fears of lethal flu pandemics, terrorism, and even collapse of civilized life in more than one country. What is to be done?

What is to be done is what has always been done: recognize the problems, understand the causes of the problems, and try to do something to deal with the problems. The worst excesses of the industrial revolution that Dickens depicted so well have been eradicated by regulation of industry, better energy technology, and social services for the poor. The "dismal homes squeezed miserably amidst the monstrosities of industrialization" have, at least in the developed countries, been succeeded by urban regions that have evolved into places more suitable for people. Humanity struggled mightily during the twentieth century with its world wars, political upheavals, and innumerable lesser wars and disasters. But for the most part, we have not yet sunk into the morass, and we seek to improve the world for our children and grandchildren.

Over the past 200 years, investments in infrastructure have made astounding differences in the quality of life for most people in most areas of the world. First railways and then automobiles, buses, and airplanes brought mobility to unprecedented levels. With cheap transportation and warehousing, it is possible to ensure that food and other supplies are available throughout the world; droughts and natural disasters still cause hardship, but relief efforts can at least limit the amount of ongoing loss of life and despair that once followed such calamities. Construction of reservoirs, aqueducts, and water treatment facilities eliminated the ravages of cholera and other diseases and allowed cities to grow without destroying the health of their citizens. The telegraph, then the telephone, and now the Internet have made instantaneous communications a reality and global perspective a necessity. Innovations in design and materials have reduced the costs of construction, allowed construction of skyscrapers, and made possible the creation of great cities. Better regulations have helped to reduce air quality and water quality, and public and private initiatives have created hundreds of parks and ecological reserves.

We have made a great deal of progress. But much more remains to be done. People need to continue working to create a more sustainable world, and much of this effort must be allocated toward infrastructure-based systems.

The theme of this chapter is stated in its title: "Toward More Sustainable Infrastructure." This theme reflects several important ideas:

- Since infrastructure systems last for decades, they cannot be changed overnight.
- Over a period of decades, society's perception of infrastructure needs and performance can undergo major changes.
- It is essential for infrastructure systems to evolve toward systems that can be more easily maintained, that are better suited to society's needs, and that are less intrusive to the environment.

Infrastructure evolves through a series of incremental steps, including specific projects and programs, incorporation of new technologies, development of better management strategies, use of new types of materials and energy sources, and new regulations concerning the use and design of infrastructure. It is important to take a long-term perspective in thinking about moving toward more sustainable infrastructure. While progress may be slow, progress is certainly possible. Understanding how infrastructure systems evolve can be helpful in looking at any proposed project. The historical context will always be important, and knowing about the different stages in infrastructure evolution can be helpful in determining whether a particular proposal is meaningful step in the right direction or merely an ill-fated attempt to keep an obsolete system functioning for a few more years.

Very few infrastructure projects are either completely new or completely independent. Most projects are implemented using proven technologies and construction methods, and the completed project fits into

Table 15.1 Stages in the Evolution of Infrastructure-Based Systems

I	Technological Experimentation and Demonstration
II	Widespread, Uncoordinated Implementation
III	Development of Systems
IV	Consolidation and Rationalization
V	Technological and Institutional Advancement
VI	Responding to Competition
VII	Mitigating Social and Environmental Impacts
VIII	Retrenching and Obsolescence

an existing network or an existing pattern of similar facilities serving much the same purpose. Project evaluation may often boil down to rather simple ideas: rents are high, so this is a good time to build office buildings; roads are congested, so let's build more roads; demand for electricity is growing, so let's build more power plants; demand for water is growing, so let's build more reservoirs; the number of children is increasing, so let's build more schools. In all of these cases, the needs are clear; the costs of building some more capacity are well understood; and it is easy to figure out whether the investment is justified given the potential revenue from rents, tolls, or fees or the willingness of the public to pay taxes for schools and infrastructure. Society may continue along a well-defined path for long periods of time, expanding infrastructure as needed to handle growth in population and rising demand for services.

However, these periods of stability eventually are disrupted by fundamental changes in technology, institutions, political power, economic conditions, or societal norms. Technological innovation, which continually offers opportunities to improve the functioning of an existing system, from time to time makes it possible to introduce much better systems. As we saw in Chapter 2, canals took freight from toll roads in the early part of the nineteenth century; then, railroads began to take freight from the canals; and 100 years later, trucks operating over paved roads began to take freight from the railroads. And throughout this period, technological improvements for roads, canals, railroads, vehicles, and communications helped to reduce the costs of all kinds of freight transportation.

Table 15.1 indicates different stages that may be encountered in the evolution of transportation or other infrastructure-based systems. As with freight transportation, the evolution may be something that proceeds over hundreds of years, during which different infrastructure systems emerge to support new technologies. Reaching a new equilibrium may take decades; and if technological innovation continues, equilibrium may never be reached. Instead, new systems will be growing side by side with older systems that are in decline.

Section 15.2 considers each of these stages, keeping in mind that there is no predetermined sequence or timetable for passing from one stage to another. The methodologies presented earlier in this text can be useful in any of these stages of development:

- Identifying needs and objectives and generating alternatives for meeting those needs
- Specifying criteria for evaluating alternatives and conducting financial, economic, environmental, and social analysis
- Assessing the differences among alternatives using weighting schemes and an appropriate political process involving major stakeholders
- Selecting the best strategy and refining it to increase efficiency and effectiveness and to mitigate negative environmental and social impacts

Understanding the themes and issues that are likely to be encountered in each stage of development may be helpful in focusing the evaluation on the most relevant concerns and avoiding some of the most costly mistakes in evaluation. Sections 15.3 and 15.4 provide historical context for two types of infrastructure: skyscrapers and railroad lines. To gain some insight into the evolutionary process, it is useful to understand how these two infrastructure-based systems evolved over a period of decades. With skyscrapers, several technological developments of the late nineteenth century made it feasible to build much taller buildings, which meant the economics of urban real estate were markedly and forever changed. The ability to make more intensive use of land led to dramatic increases in urban development and a perhaps inevitable cycle of boom and bust. Railroads have been evolving for nearly 200 years, providing an even longer perspective on the evolution of an infrastructure-based system. Large portions of the rail system have already reached the final stages of obsolescence, and the railroads have already retrenched; operating over smaller networks and using better technology, they now serve fewer customers whose traffic is best suited to the low-cost service that railroads can provide.

Section 15.5 discusses some of the twenty-first-century challenges to be faced in moving toward more sustainable infrastructure. Many kinds of projects will be needed to meet these challenges, including rehabilitation of existing infrastructure to extend its useful life, replacement of certain elements of existing infrastructure to enhance performance, and construction of new infrastructure that embodies more sustainable technologies and designs. Examples of each of these are provided, and Section 15.6 presents a costing framework that can be used in examining when to repair and when to replace infrastructure components.

15.2 STAGES IN THE EVOLUTION OF INFRASTRUCTURE-BASED SYSTEMS

This section outlines the different stages that may be encountered in the evolution of transportation or other infrastructure-based systems. Bear in mind as you read this section that there is no predetermined sequence or timetable for passing from one stage to another.

15.2.1 Stage I: Technological Experimentation and Demonstration

From time to time, someone comes up with a new technology that seems to offer a better means of satisfying some basic need of society. The first task is to figure out how to make the technology work, which may take years of experimentation and tinkering. Eventually there will be a demonstration that the new technology works, and a few entrepreneurs or government agencies will back some initial implementations to determine whether the technology is financially feasible. Inventors figured out 200 years ago how to use a steam engine to pull a few cars along a railroad track at 10–20 mph, which was a great improvement over the speed of a horse and wagon or a canal boat. In the twenty-first century, we have seen a few demonstrations of how to use magnetic levitation for a train to reach speeds close to that of airplanes. The railroad proved to be an extremely successful technology, and within 100 years of its introduction, hundreds of thousands of miles of rail lines were constructed throughout the world. It remains to be seen whether magnetic levitation becomes a widespread alternative to air travel or remains an oddball tourist attraction for getting to and from the airport in Shanghai and perhaps a few other locations.

Relevant Methodologies

- *Measuring system performance* (Chapter 2): What kinds of performance improvements are most critical? What technologies are likely to provide such improvements in performance? How will any proposed technology affect cost, capacity, service quality, safety, and other aspects of system performance?
- *Probabilistic risk assessment* (Chapters 2 and 13): What are the risks associated with the new technology? How can those risks be minimized? Will the new technology be perceived as safe by potential users, abutters, and the general public?
- *Performance-based technology scanning* (Chapter 13): Will the anticipated improvement in performance be enough to attract users or supporters who will be willing and able to pay for the new system?

Potential Pitfalls

- Excessive investments in technologies that, even if successful, are unlikely to attract much of a market.

- Introduction of a technology without adequate safeguards, leading to highly publicized failures that cause governments or the public to reject the technology.
- Introduction of a technology without adequate consideration of externalities, leading to government or public perceptions that the technology should be rejected because of its impacts on abutters, communities, or the environment.
- Failure to consider the potential impact of new technologies, thereby delaying the introduction—and deferring potentially large benefits—of new technologies.
- Government regulations that reject or hinder new technologies that would have broad benefits to society but are opposed by special interests.

15.2.2   Stage II: Widespread, Uncoordinated Implementation

Once the technical and financial feasibility have been demonstrated, there will likely be a great many uncoordinated, fiercely competitive attempts to build up the infrastructure and related systems for the new technology. Once the automobile was shown to be feasible, dozens of manufacturers emerged, producing a great variety of styles and experimenting with steam and electrical systems for propulsion as well as with the internal combustion engine. The early automobiles had to use the existing roads, which were little better in 1900 than they had been in 1800 (and not as good as the roads that the Romans had built nearly 2000 years earlier).

Relevant Methodologies

- *Financial analysis* (Chapters 7–9): How will introduction of the new technologies be financed?
- *Brainstorming, sensitivity analysis, and scenarios* (Chapters 11 and 13): What are the options for deployment? Which factors are most likely to affect success or failure?

Potential Pitfalls

- Starting up too small, with insufficient coverage to attract enough users to support the service
- Starting up too big, resulting in too much debt and too little revenue
- Starting without adequate financial reserves, leaving the project vulnerable to a strong response from competitors or a downturn in the economy
- Waiting too long to get started, which means that competitors capture the market

15.2.3   Stage III: Development of Systems

The initial, uncoordinated efforts will provide a great deal of information and insight about what is needed for reasonably efficient implementation of the new technology. Companies involved in network-based systems (such as electricity transmission, irrigation, transportation, and telecommunications) will recognize the economies of scale and density; they will try to expand, via construction or mergers, to obtain these economies. Cities will realize that some systems, such as water delivery, need to be coordinated to avoid costly duplication of infrastructure and provide cost-effective services to customers. Companies and public agencies will decide how best to structure their systems, whether to integrate operations and maintenance of the systems, and how best to coordinate their activities with related activities. During this stage of development, attention must be given to structuring and integrating infrastructure and operations so as to handle growth.

Relevant Methodologies

- Determining the extent to which there are likely to be economies of scale, scope, or density—that is, understanding how costs and other aspects of performance vary with the structure of the system (Chapter 3)
- Identifying and assessing strategic options for expansion (Chapters 11 and 13)

Potential Pitfalls

- Creation of unregulated monopolies that in fact achieve great reductions in average cost, but that are able to charge excessive prices for their services
- Difficulties in consolidation caused by lack of **interoperability** (i.e., the failure of companies and agencies to develop compatible technologies that will facilitate coordination and cooperation)
- Competition that reduces prices to marginal costs that are below average costs, thereby limiting the ability to finance desirable increases in capacity or enhancements to facilities

15.2.4   Stage IV: Consolidation and Rationalization

The enthusiastic response to new technologies frequently leads to overbuilding, the creation of infrastructure with too little or too much capacity, and the inevitable problems in design and network structure that come from the initial uncoordinated investments. Even though the size and scope of the systems may be reasonable, there will be many opportunities for **rationalization**, which is the process of improving performance by consolidating portions of the system, eliminating bottlenecks, and abandoning obsolete or underutilized elements.

Relevant Methodologies

- Basic modeling of costs, service, competition, and financial viability
- Estimation of public benefits of consolidation and rationalization (to obtain public support and perhaps public funding for complex projects that will cause major disruptions during an extended period of construction)

Potential Pitfalls

- Difficulties in abandoning underutilized portions of a network because of public pressures to maintain services viewed as necessary by customers and local governments
- Difficulties in assembling tracts of land suitable for consolidated facilities
- Difficulties in financing the investments in a competitive environment, where prices are likely to fall quickly to new, lower marginal costs following completion of desirable projects

15.2.5   Stage V: Technological and Institutional Advancement

Managers, researchers, consultants, and regulatory agencies continually seek new technologies, management strategies, and institutional arrangements that will improve system performance. The goals during this stage are to reduce costs; enhance safety or improve service; and take advantage of better control technologies, better materials, or better management. During this stage, system improvements come from looking inward to find opportunities for doing the job better. Technological advances may result in the use of new procedures or greatly improve the performance of what were thought to be obsolete procedures.

Relevant Methodologies

- Basic modeling of costs, service, competition, and financial viability (Chapters 2, 3, and Part II)
- Performance-based technology scanning (Chapter 13)
- Project and program audits (Chapter 14)

Potential Pitfalls

- Failure to provide for adequate maintenance and rehabilitation of the system
- Failure to anticipate changes in competition, technology, or society; an overconfident management that may suddenly find that the capacity, cost, condition, or service quality of its infrastructure are woefully out of step with what is needed

EXAMPLE 15.1   New Technology Can Revive Old Methods

Recent technological advances have resulted in a major shift in the preferred methods for wastewater treatment.[1] Metal salts were widely used in England in the nineteenth century to increase the level of coagulation and flocculation in municipal wastewater treatments. While the technique was effective, it greatly increased the amount of sludge that had to be disposed, adding substantially to the overall costs of treatment. As a result, by the 1930s, the preferred method of wastewater treatment changed to unaided gravitational settling followed by a biological unit. New techniques, known as chemically enhanced primary treatments (CEPT), have recently improved the performance of primary wastewater treatment plants without generating nearly as much sludge. Since CEPT removes more of the pollutants, the demand on the biological treatments is reduced. Research efforts at the Point Loma wastewater treatment plant near San Diego, California, demonstrated the effectiveness of CEPT. Spurred by the need to meet new regulations regarding effluent, the operators came up with a CEPT method that was more efficient, produced minimal amounts of additional sludge, and increased system capacity.

Ultimately the system was so successful that Congress waived the usual requirement for secondary treatment, saving the city an estimated two billion dollars and allowing the construction of a tertiary water reclamation facility that now reuses 15% of the total wastewater flow instead of discharging it into the ocean.[2]

CEPT is now seen as an excellent, low-cost option for treating wastewater, especially in the developing world, since this method enables existing, overburdened facilities to perform better.

15.2.6   Stage VI: Responding to Competition

No system is immune to competition, and competitive threats are likely to increase as the system ages. Competition reduces demand for a system; and over time, competition could eliminate the need for a system. Several kinds of threats are possible:

- *Competition from similar systems:* Competition is based on the competitors' ability to use similar technologies to provide services to customers.
- *Competition from systems based on different technologies to provide the same service:* In this type of competition, the competitors use completely different technologies to serve the same types of needs.
- *Competition from systems based on different technologies that eliminate the need for the service:* In this type of competition, some kind of technological innovation provides a new way of doing business can potentially reduce the demand for existing services.

EXAMPLE 15.2   Competition to Airlines from Similar Systems: Low-Cost Airlines Competing with Legacy Airlines

During the late twentieth century, new airlines decided to serve selected markets from smaller airports using low-cost crews and new, efficient planes. Some of these airlines, notably Southwest Airlines in the United States and Ryan Air in Europe, were extremely successful with this strategy and diverted substantial numbers of people from the older airlines. The success of these airlines helped provide an impetus to upgrade facilities and expand capacity at airports in many smaller cities. Airports in Manchester, New Hampshire, and Providence, Rhode Island, which are about 50 miles from Logan Airport in Boston, began to compete with Logan for the same air travelers.

---

[1] This example is based on the article by Michael R. Bourke, Donald R. F. Harleman, Heidi Li, Susan E. Murcott, Gautam Narasimhan, and Irene W. Yu, "Innovative Wastewater Treatment in the Developing World," *Civil Engineering Practice* (Spring/Summer 2002): 25–34.

[2] *Ibid.*, 27.

EXAMPLE 15.3   Competition to Airlines from Systems Using Different Technologies to Provide the Same Service: High-Speed Rail Competing with Airlines

As illustrated in Chapter 13, high-speed railroads can compete effectively with airlines for trips of 200 to 500 miles. Which mode captures the largest share of the market will depend in part upon technological capabilities and in part upon public policy toward investing in transportation. In many airports, a large percentage of travelers are taking these medium-distance trips, so diverting travelers to trains can reduce the need for investment in airport capacity.

EXAMPLE 15.4   Competition to Airlines from Systems Using Technologies That Eliminate the Need for the Service: Telecommunications Competing with Airlines Travel

Videoconferencing facilities, cheap telecommunications, and the Internet allow virtual conferences that reduce the need for business travel; e-mail allows the exchange of information without requiring the use of the post office; catalog or Internet shopping allows people to purchase items without visiting stores. These changes reduce the demand for business travel, reduce the use of aircraft for mail, and increase the use of air freight to support delivery of all those items ordered online.

Relevant Methodologies

- Basic modeling of costs, service, competition, and financial viability
- Performance-based technology scanning (PBTS)
- Brainstorming and scenarios

Potential Pitfalls

- Failure to consider how current conditions might change
- Overconfidence based on past glory or past profitability
- Regulations relevant to old technologies, which limit the introduction of new technologies that may require less frequent inspection, support higher loads, or allow different methods of operation

15.2.7   Stage VII: Mitigating Social and Environmental Impacts

Social and environment impacts are likely to become more important over the life of any infrastructure-based system, for three reasons:

1. Negative impacts may no longer be acceptable.
2. Negative impacts were not initially apparent.
3. Societal norms change.

Negative Impacts May No Longer Be Acceptable

When a system is first developed, the societal needs may be so clear that any obvious negative impacts of the new system are easily accommodated. However, society eventually demands that something be done about these problems. For example, the benefits of railroads and automobiles in enhancing mobility and opportunity far outweighed the fact that both modes were originally very noisy, dirty, and risky.

After a system has been functioning, it may become apparent that its social or environmental impacts are unacceptable. As automobile accidents began to account for tens of thousands of fatalities annually, highway safety became a major issue in both vehicle design and highway design. When the public realized that drunk drivers caused a high proportion of the fatalities, they forced government agencies to impose stricter penalties and greater enforcement of laws intended to reduce driving while drunk.

Negative Impacts Were Not Initially Apparent

Implementation of new technologies often creates problems that were not anticipated and that were identified only after many years or decades of experience.

EXAMPLE 15.5   Automobile Pollution, the Clean Air Act, and Global Warming

Automobile exhaust was not a major public concern until well after there were enough automobiles to cause a noticeable degradation in air quality. Initial concerns about automobile emissions were related to air pollution and its effect on visibility, deterioration of buildings, and above all human health. By 1970, it was clear that automobile exhaust contributes heavily to smog and that air pollution is a serious health problem. About that time, governments began imposing regulations on emissions and on fuel consumption, with the result that air quality improved despite increases in automobiles and trucks. Efforts to reduce emissions focused on pollutants, including carbon monoxide, hydrocarbons, and nitrogen oxides. In the United States, emission standards for new vehicles were first imposed in 1968 and strengthened with the passage of the Clean Air Act in 1970. Within 20 years, the average emissions of carbon monoxide and hydrocarbons from new cars were only 4% of what they had been in 1970, while emissions of nitrogen oxides were only 25% of what they had been. These dramatic improvements were enabled by a new technology, the catalytic converter, which effectively prevented the release of most of these pollutants.

Although the air quality in major cities improved due to emissions regulation, a new, more difficult problem emerged—namely, the threat of global warming and climate change. As the twentieth century came to an end, scientific evidence increasingly pointed to an accelerating worldwide rise in average temperatures and a link between global warming and the emission of greenhouse gases. These gases included carbon monoxide, hydrocarbons, and nitrogen oxides; they also included carbon dioxide, which is much more difficult to control and is not normally considered a pollutant. By the early twentieth century, the threat of global warming was clearly linked to the increasing levels of carbon dioxide in the atmosphere. Today, serious efforts are under way to reduce consumption of all fossil fuels—and, in particular, to develop alternative fuels for and to increase the fuel efficiency of automobiles and trucks. In the United States, federal legislation required the use of ethanol as an additive to gasoline. Ethanol can be produced from corn, sugar cane, or other biological sources; so the use of ethanol can reduce the use of fossil fuels. This legislation led to the construction of numerous facilities for producing ethanol, along with substantial investment in rail lines that would serve the new ethanol plants. The requirement for using ethanol in gasoline created such an increase in the demand for corn that the price of corn for food increased around the world—thus creating another problem that society must deal with.

The following are some other examples of problems that influence the evolution of infrastructure.

- *Activities in one area threaten the environment in other areas:* Widespread use of irrigation and fertilizers was part of highly successful efforts to improve the yield of farms in the U.S. Midwest. Vast networks of irrigation canals and drainage systems were used to make farming possible in areas that otherwise would have been too dry or too wet. Moreover, to maximize production, lands very close to waterways and wetlands were cultivated. Because of these agricultural practices, excess fertilizers are washed into the rivers that feed into the Mississippi River and eventually flow into the Gulf of Mexico, with devastating consequences (see Box 15.1).

Box 15.1   Projects in the News: How Can the Gulf of Mexico Be Cleaned Up?

Two major approaches could help clean up the Dead Zone in the Gulf of Mexico. One is to construct better waste treatment facilities for major urban areas such as Chicago. The other, which may be much cheaper, is to convert agricultural land (back) to wetlands. However, not much progress is being made toward either approach.

The Gulf of Mexico's immense "Dead Zone"—an annually recurring phenomenon characterized by hypoxia, or low levels of dissolved oxygen—illustrates the extent to which land uses, in some cases thousands of miles away, can affect one of the world's largest bodies of water. Nutrients released from farm fields or urban areas throughout much of the United States travel via the Mississippi River downstream to the Gulf, creating a zone of hypoxia larger than some small states. Although the problem has been recognized for more than two decades, solutions have been slow in coming. Today, an awareness of the myriad efforts that will be needed to halt hypoxia in the Gulf is beginning to take shape. But will this awareness turn into action any time soon?. . . .

Because of their ability to effect denitrification, wetlands and riparian areas can act as highly effective "nitrogen sinks." Plants and other organisms in wetlands can also sequester nitrogen, while phosphorus can be curtailed via sedimentation or other physical processes common to wetlands. For these reasons, many view the restoration of wetlands as a vital step in the effort to shrink the Dead Zone. To be effective, however, this approach must be conducted on a vast scale. . . .

(continued)

(continued)

Depending upon the nutrient standard to be met, between 189,000 and 322,000 acres (76,000 and 130,000 hectares) would be needed to effect the necessary nutrient reductions, entailing a present-value cost of either $870 million or $1.4 billion. As expensive as that option sounds, the present-value cost to upgrade the treatment plants of the Metropolitan Water Reclamation District of Greater Chicago was estimated to be between $2.4 and $2.9 billion, depending upon the nutrient standard to be met.

Despite report after report addressing Gulf hypoxia, The Wetlands Initiative's Hey notes that actions to address its causes remain elusive. "It's not that we don't know how to solve the problem," he says, "We seem not to have the courage or the interest to come to grips with the problem."

Jay Landers, "Halting Hypoxia," *Civil Engineering Practices* (June 2008): 34, 64–65.

- *Concerns for energy cost influence design of buildings:* As energy costs and public concerns with global warming rise, companies are beginning to be interested in reducing the energy required in large buildings. For example, the Pearl River Tower in Guangzhou, China, was designed to be the most energy efficient of the world's tallest skyscrapers. Although the original goal of being energy neutral proved too expensive, the building is expected to consume less than half of the energy of a similar building that was built to a traditional design.[3]

  An independent, third-party rating system has been developed that certifies sustainable building practices and is used by many companies interested in being leaders in sustainable construction (see Box 15.2). LEED certification is available for different kinds of construction, including office buildings, apartment buildings, schools, and homes. A complex set of criteria was established for each type of building, and points are awarded for such things as energy efficiency, water conservation, and landscaping. Detailed information is available at the website of the U.S. Green Building Council.

Box 15.2   LEED Certification and the Design of "Green Buildings"

The Leadership in Energy and Environmental Design (LEED) Green Building Rating System™ encourages and accelerates global adoption of sustainable green building and development practices through the creation and implementation of universally understood and accepted tools and performance criteria.

LEED is a third-party certification program and the nationally accepted benchmark for the design, construction and operation of high performance green buildings. LEED gives building owners and operators the tools they need to have an immediate and measurable impact on their buildings' performance. LEED promotes a whole-building approach to sustainability by recognizing performance in five key areas of human and environmental health: sustainable site development, water savings, energy efficiency, materials selection and indoor environmental quality.

"U.S. Green Building Council, www.usgbc.org

- *Reducing the risks of earthquakes:* Structures can be built that are more likely to withstand earthquakes, and systems can be designed that are more likely to continue operations even if an earthquake destroys some of its components. New buildings and other structures in earthquake zones are today designed to withstand earthquakes. In many locations, structures are strengthened so they are more likely to withstand the lateral and vertical movements that may be caused by an earthquake. In Japan, for example, where earthquakes are common, JR East and other railroads retrofitted many of their bridges to reduce the risk that a high-speed train would derail or overturn

---

[3] Roger E. Frechette III and Russell Gilchrist, "Seeking Zero Energy," *Civil Engineering* (January 2009): 38–48.

during an earthquake. In northern California, it was decided to retrofit the Claremont Tunnel, which was constructed in 1929 to transport water to Oakland and other cities. If a portion of this tunnel were destroyed, then a half million people would be without water. To avert such a disaster, the East Bay Municipal Utility District spent $66 million to construct a bypass tunnel through the fault zone. The bypass tunnel was designed to accommodate a lateral movement of up to 8.5 feet (2.6 m).[4]

Societal Norms Change

Over time, as societies become more affluent, they tend to be less tolerant of risk, more interested in style and aesthetics, and more conscious of the factors affecting their personal well-being. As a result, systems that were acceptable when built may no longer be acceptable 20 or 50 years later.

Changes in societal expectations are perhaps most notable in the area of telecommunications. In ancient history, the Persians, Chinese, and Romans built highways that were intended in part for armies to move quickly throughout their empires. The same roads were also used to provide communications between the capital cities and the other cities and states that were part of the empire. In the mid-nineteenth century, the telegraph allowed instantaneous communications, but at a cost that only governments and corporations could routinely afford. The telephone provided direct communication for the masses; but only 50 years ago, the cost for long-distance calls was so high that each such call was a family event. Cell phones freed users from their land lines and allowed people to make calls from almost anywhere to almost anywhere. Today, e-mail, texting, the Worldwide Web, and Twitter can keep people in almost continuous contact.

EXAMPLE 15.6   Airline Terminals Emulate Shopping Malls

Travelers are no longer willing to go to the airport, check in, and sit patiently waiting to board their plane. And they can no longer show up just before flight time, because of the need to pass through security. Thus it is no longer enough for airports to provide a few seats near the departure gates. Moreover, it is clear that people are willing to spend quite a lot of money when they are forced to spend a half hour or more in a confined area. Airports today must have restaurants, shopping opportunities, first-class lounges, Internet connections, and attractive design; these features improve the comfort and quality of the traveling experience, and they provide income to help pay for the expansive and expensive terminals (Figure 15.1). For similar reasons, colleges today must provide not only adequate housing for students but also excellent sports facilities, dining options, and wireless Internet connections in every classroom and in every dormitory room. For students deciding where to apply, the quality of the living experience may be as important as the quality of the education.

Figure 15.1 A Modern Terminal at the Indianapolis Airport. This mall-like terminal opened in 2008, replacing a much smaller facility that had few passenger amenities beyond security. A new exit to Interstate 70 made the terminal much more accessible for travelers coming to or from the south or the west.

---

[4] Sarah Holtz Wilson, David F. Tsztoo, Carl R. Handford, and Kenneth Rossi, "Safeguarding a Lifeline," *Civil Engineering* (May 2008): 58–65.

15.2.8 Stage VIII: Retrenching and Obsolescence

Competition and changes in societal needs will cause some systems to shrink and perhaps disappear. If well-managed, a system can decline slowly, over a matter of decades, by continually eliminating unnecessary or unprofitable components and cutting back to profitable core operations. If poorly managed, a system may simply hold on for too long and then suddenly collapse. Today, there is an ongoing drama involving telecommunications. Are land lines obsolete? Will everyone go wireless? Will people communicate via the Internet? What will happen to all of those telephone poles?

As with the previous stages, many of the methodologies presented in this text will be useful in deciding whether or not a system or technology is obsolescing and, if so, figuring out how to age gracefully. Shortsighted management may fail to foresee problems until it is too late to do anything; bullheaded managers may decide to invest vast sums to rehabilitate a system that is actually no longer competitive. Some of the largest expenditures on rail passenger stations were made in the 1930s, just at the time when air travel and automobiles were about to put most rail passenger services out of business. It is far better to have a realistic outlook on competition than to blindly go ahead with major investments aimed at providing marginal improvements to a system that is more than marginally noncompetitive.

15.3 SKYSCRAPERS AND BUILDING BOOMS

You and your directors were well-advised in the choice of your symbol. For a tower, with its light and its belfry, has always been a source of inspiration. . . . Thus your Tower partakes of the character of the ancient towers of refuge and defense. . . . Your high tower should, therefore, be a symbol of God to you and others, standing out boldly and erect as a plea for righteousness and purity in business corporations, and as a monumental protest against the exploitation of the poor.[5]

The technologies necessary to construct buildings of almost unlimited height became available in the mid to late nineteenth century (Table 15.2). In locations where land values were high, the potential

Table 15.2 Technological Innovations Required for the Construction of Skyscrapers

Steel frames	1848	James Bogardus demonstrates how to use cast-iron posts and beams to support a building without relying upon the walls, but height is limited to 7 or 8 stories.
	1880	Using steel rather than cast iron allows framing much higher buildings because steel is more flexible, stronger, and can be joined with rivets rather than nuts and bolts. Leroy Buffington shows how tall, steel-framed buildings could be built.
	1883	William Le Baron Jenny uses Buffington's method in design for the first skyscraper, the Home Insurance Company Building in Chicago, which was built using a steel frame rather than using steel to reinforce thick masonry walls.
Elevator	1853	Graves Otis demonstrates an elevator at the Crystal Palace in London.
	1870	Equitable Building in New York City is the first building to install an elevator.
	1889	Otis installs the first electric elevator.
Utilities		Water is installed for fire protection and sanitation.
		Electricity is provided for elevators and lighting.
Fire protection		Building codes call for fire-retardant materials, fire escapes, sprinkler systems, inspections, and fire drills.
Zoning		Zoning is required to deal with effect of buildings on wind currents and sunlight.

Source: John Tauranac, *The Empire State Building* (New York: St. Martin's Griffin, 1995).

[5] This bit of hyperbole was addressed to Met Life stockholders in 1915, according to John Tauranac, *The Empire State Building* (New York: St. Martin's Griffin, 1995), 38. The president of MetLife was apparently more concerned with his pocketbook than with his view, as he is reported to have responded by saying, "The tenants will foot the bill."

benefits of building a skyscraper were immediately obvious. While the visceral appeal of towers may have appealed to certain visionaries, overly powerful executives, and politicians, the ability to replace a 5-story building with a 25-story building excited the financial instincts of developers and real estate agents. Whether buildings were designed for office space or for apartments, the ability to rent far more building space on the same site was a prospect too enticing to resist.

Skyscrapers were erected in New York City at an astonishing pace (see Tauranac, pp. 41–42). In the 1890s, several 20- to 30-story buildings were constructed in the city. By 1929, the city had 188 that were more than 20 stories tall and nearly 2500 that were more than 10 stories tall. In 1899, the 29-story Park Row Building was the tallest in the world; the world record was broken three more times in the next 14 years in Manhattan, capped in 1913 by the 55-story, 787-foot-high Woolworth Building. During the 1920s, the Chrysler Building, at 1048 feet, held the record for a while until it was surpassed by the Empire State Building, which had 102 stories and was more than 1250 feet tall.

This incredible construction boom in Manhattan was fueled by the fact that skyscrapers allowed many more people and businesses to crowd into an area with some of the highest property values in the world. The new technology made it possible to house an order of magnitude more people than previously could be housed on a site. Once developers realized the profits that could be made by rebuilding the city to a taller standard, the proposals and construction efforts multiplied—even after vacancy rates started to increase in 1926 and a report to President Coolidge concluded that the country was overbuilt. According to Tauranac:

> Developers had been desperate to develop, but by 1927 they were desperate to rent, to fill their vacancies, especially in apartment houses where the problem was critical. . . . Realtors recognized that the profession could not view the extremely heavy construction program without apprehension of a glut on the market. (p. 83)

Despite the slump and the dire warning, in both 1928 and 1929, developers filed plans for more than 700 new buildings in New York City. And some of these were quite spectacular plans. The planning for the Empire State Building and for Rockefeller Center were both initiated in this ambiguous period— still at the height of the construction boom, but just before the bottom was about to fall out of the New York City real estate market as well as the worldwide economy.

The pace of construction seen in New York City in the 1920s was later replicated in many other locations around the world, notably Hong Kong, Tokyo, Shanghai, and Dubai. Unfortunately, the instincts that cause developers to begin bigger and better projects are also perhaps too powerful to resist, even in the face of contrary economic evidence. The real estate bubble that burst in 2008 was linked to the same kinds of ''irrational exuberance'' and shaky financial dealings that led to the burst of the bubble in New York City in 1929.

Once a real estate bubble bursts, it may be many years before development resumes. In the United States and elsewhere, construction of skyscrapers stalled because of the Depression and World War II. Only after the war was it possible to resume construction of skyscrapers. As in the earlier period, technological and economic forces shaped what was built (see Picardi's reflections in the accompanying box).

I began my career in structural engineering in 1947 and was in responsible charge of structural design of many notable high-rise buildings including two of the world's highest over the next 40 years. I can attest to this period as one of great technological change in building design. Allowable stresses were increased by about a third and new high-strength steel was developed and more precise methods of computation with computers were used to solve structural problems with speed and accuracy. Architects were anxious to express new concepts of modern design and postmodern design. The light aluminum and glass exterior curtain wall was developed and the ''glass box'' building was born. ''Less is more'' was a popular theme.

(continued)

(continued)

  Thus we had as the state of the art lightweight steel frames with glass and metal curtain walls. The new designs provided for large column-free space around the central core of the buildings. The tube design was developed providing long-span truss floor systems between closely spaced exterior wall columns and a grid of columns in the building core which housed elevators, stairways, mechanical shafts and washrooms. Horizontal mechanical systems ran parallel and through the open spaces in the trusses and the structural steel was fireproofed with sprayed-on asbestos cement material. After asbestos became a dirty word, other cementatious spray-on material was developed. . . .

  Money was the primary mechanism which drove these changes. We had to build economically to offset the increasing costs of both labor and materials. Cost saving was the name of the game, and engineers were notable by their ability to carry out the ever-changing design concepts of the architects with as little tonnage of structural steel as possible. For example, in the pre–World War II era, buildings in the 25- to 50-story range had steel in the order of 20 to 40 pounds per square foot. The postwar buildings in the 40- to 100-story ranges had steel weights in the order of 30 pounds per square foot. There were no catastrophic failures of these structures from fire, impact, or explosions until 9/11.

  What we do next involves the answers to many questions and numerous decisions by participants in the design and construction process, government and concerned public. Certainly examination of fireproofing systems and structural details is a high priority, including where possible the retrofitting of existing structures. Modification of zoning ordinances limiting height and location of high buildings and changes in building codes relative to fire, explosion, and impact safety should be considered. All this involves not only technical but also political and cultural considerations. 9/11, while a terrible tragedy, should serve as a wake-up signal for a more secure building environment in this new era of terrorist war.

  Perhaps the day of megastructures in the United States is and should be over for many reasons other than safety. Concentration of thousands of people in a single structure has serious infrastructure effects on transportation systems and the environment. Time is wasted and people suffer from over-stress as they rush like ants each working day into their workplace, often from homes hours away. I, having been there, seen it, and done it, think there are better answers to housing the core operations of commerce than trying to make megastructures safe, efficient, and healthy workplaces.

  E. Alfred Picardi, P.E., ASCE, May 5, 2002

15.4 CASE STUDY: EVOLUTION OF THE U.S. RAIL SYSTEM

The railroads have experienced what is likely a typical life-cycle for infrastructure-based systems, from revolutionary to dominant to mature to declining to decommissioned. The rail industry's evolution over a period of nearly 200 years therefore provides insights into the types of problems encountered—or are likely to be encountered—by other infrastructure-based systems. An industry that began with hundreds of tiny companies in the United States consolidated into a coordinated system dominated by four mega-systems. Regulation and government support, initially required to avoid overdevelopment and monopolistic excess, eventually were required to help the industry reduce its network, recover from bankruptcies, and abandon large segments of business that were better served by other modes of transportation. Today, while railroads clearly have advantages over highway and airline competitors in terms of fuel efficiency and emissions, these advantages do not necessarily mean that railroads will capture more traffic. Technological advances have been critical in allowing the railroads to remain competitive—and even dominant—in some markets. Understanding where the industry retains clear advantages has allowed the railroads to segment their markets and focus on the most promising areas: intermodal traffic and bulk traffic. Effective public policy and continuing public support may be needed to sustain the rail system, especially for passenger transportation.

15.4.1   Introduction of Railroads

The introduction of railroads provided an incredible improvement in mobility and a dramatic reduction in the cost of moving freight.[6] Cheap transportation plus the rapid advances of the industrial revolution spurred economic growth in the early nineteenth century. In the United States, railroads were instrumental in opening up vast areas for development, in terms of both making land available for farmers and ranchers and providing access to new hotels in prime locations for rich tourists.

> *The steam engine, the railroad locomotive, the transatlantic steamship, the telegraph, and the camera together virtually completed the conquest of nature by the go-ahead age. What followed were in effect refinements. . . . Even the increasingly anticipated conquest of the air would, in essence, be no more than an added convenience. Its effect was simply to turn days into hours. It was the shortening of months into days by the railroads that was the basic revolution in transportation.*
>
> Page Smith, *The Nation Comes of Ages: A People's History of the Ante-Bellum Years* (p. 819)

To protect their investments in canals, state and local governments at first tried to hold off the railroads. Over time, the legal impediments to railroad building were struck down by courts or repealed by state legislatures; by the middle of the nineteenth century, the railroads had an irresistible momentum. Small lines constantly consolidated in an effort to increase their access to capital and improve their service. New lines were established, and as soon as they proved themselves (or went bankrupt), they were taken up by larger lines and incorporated into a "system." Boston entrepreneurs, who had no serious opportunities for building canals, financed railroads to compete with New York City's advantageous Erie Canal route to the Northwest Territories. By 1850, Boston had links as far as Ohio and throughout New England. Three thousand miles of rail line were built for $70 million, and the railroad was profitable.

The emphasis quickly shifted from preventing railroad development to promoting it. This trend was especially evident in the Mississippi Valley region, where vast expanses of lands were unsold and unsettled because of their distance from markets. Illinois gave the Central Illinois Railroad 2.6 million acres of land to help finance construction, because costs had run $10 million over budget. This land was initially worth $0.16 per acre (i.e., about $430,000). Five years later, it was worth $10/acre; 10 years after that it, was worth $30 an acre for a total of $78 million.

General economic trends were at times driven by and at times disastrous for the railroads. Construction of new lines led to a surge in demand for timber, iron, and coal. General cycles of boom and bust rapidly followed one after the other. In the latter part of the nineteenth century, two-thirds of the years were either severe recessions or depressions. Competition among railroads was fierce, and bankruptcies were common. Potential problems such as the air pollution from the coal-burning locomotives or the destruction of pristine areas and the division of cities did not impede the progress of the railroads.

15.4.2   Problems with Railroads: Accidents, Greed, and Corruption

Railroading was a dangerous occupation; many injuries resulted from collisions, derailments, and mistakes in switching cars in yards. Over time, improvements in brakes, coupling systems, signaling and communications, and track materials, and vehicle design all contributed to dramatic reductions in the risks associated with rail operations.

Rail lines that were built to open up new regions to development could not be justified by the revenues received from operations. Before the railroad was built, little or no development existed; and until development took place, there would be little or no traffic. For this reason, government funds,

---

[6] The history of the rail industry is well-documented. The facts in this subsection were obtained from Page Smith, *The Nation Comes of Age: A People's History of the Ante-Bellum Years* (New York: Penguin, 1981), 271–77.

loans, guarantees of railway bonds, and land grants were used to finance the many lines in the Midwest as well as the transcontinental rail line. Government investment was justified by the potential social and economic benefits of development and, especially for the transcontinental railroad, by the strategic need to link California with the rest of the country. The financial logic was that the railroad company could use the loans, mortgage the land, or sell the bonds to obtain the cash required to build the rail line. Once the rail line was built, the railroad could repay the loans by using profits from land sales or operations or by carrying government officials and freight at a discount. Although the government had given away large quantities of land to the railroads, the remaining land was far more valuable once the railroad constructed.

The railroads that built the transcontinental railroad received land valued at $392 million as of 1880, after the railroad was built. The government's land alongside the rail route had presumably appreciated by nearly that much. The railroads were allowed to issue government-backed bonds with a face value of $63 million, and they paid back that amount plus $105 million in interest. Thus the government's investment in the venture turned out to be very profitable.[7]

Unfortunately, rapid expansion of the rail network and access to vast amounts of public money led to various types of fraud: misuse of the funds intended for rail construction, construction of rail lines to too high or low a standard, and construction of very expensive lines along questionable routes. When the railroads in the Northeast were being developed, the profitability of railroads attracted unscrupulous investors and politicians eager to capture their share of the bonanza. Deals involving railroad barons such as the Goulds and politicians such as the Tweeds in New York City were infamous, as were the battles and stock manipulations of rival financiers. As the rail network moved into the West, the need for public support was much greater than it had been in the East. And the opportunity for corruption was much greater.

The greatest scandal of the scandalous Grant administration (1869–77) involved the construction of the transcontinental railroad. The Union Pacific (UP) and the Western Pacific railroads received construction grants from the U.S. government plus land grants along the right-of-way to help finance the first transcontinental railroad. To raise more capital for building the rail line, the two railroads were able to mortgage the land before construction based on the value of the land after construction. The UP owners did this through a French bank, Credit Mobilier, which they controlled. The bank declared as dividends the great majority of the funds deposited by the UP owners, who in effect defrauded the railroad and the government. Many U.S. officials, including the vice president, were implicated in the scandal.

Another type of fraud involved constructing lines that were unnecessarily expensive to build, thus increasing the contractors' profits. Mark Twain captured the essence of this strategy in his marvelously cynical portrayal of "The Gilded Age" (chapter 27):

> *Beautiful Road. Look at that, now. Perfectly straight line—straight all the way to the grave. And see where it leaves Hawkeye—clear out in the cold, my dear, clear out in the cold. That town's as bound to die as—well if I owned it I'd get its obituary ready, now, and notify the mourners. Polly, mark my words—in three years from this, Hawkeye'll be a howling wilderness. You'll see. And just look at that river—noblest stream that meanders over the thirsty earth!—calmest, gentlest artery that refreshes her weary bosom! Railroad goes all over it and all through it—wades along on stilts. Seventeen bridges in three miles and a half—forty-nine bridges from Hark-from-the-Tomb to Stone's Landing altogether—forty-nine bridges, and culverts enough to represent them all—but you get an idea—perfect trestle-work of bridges for seventy-two miles. Jeff Thompson and I fixed all that, you know; he's to get the contracts and I'm to put them through on the divide. Just oceans of money in those bridges. It's the only part of the railroad I'm interested in,—down along the line—and it's all I want, too. It's enough, I should judge.*

---

[7] Stephen E. Ambrose, *Nothing Like It in the World: The Men Who Built the Transcontinental Railroad 1863–1869* (New York: Simon & Schuster, 2000), 377.

15.4.3 Monopolistic Excess and Regulation of Railroads

Fraud, monopolistic excesses in pricing, fatal accidents, and cutthroat competition eventually created a situation that made it both necessary and possible to regulate the railroads.[8] So long as the railroads were unregulated, they were able to charge what the market could bear. This practice led to high prices and shipper complaints during periods of economic growth; during recessions, it led to bankruptcy for the railroads and loss of railroad jobs.

Initial attempts to regulate railroads were driven by several factors: inflation during the Civil War (1861–65), declining agricultural prices during the depression of the 1870s, rapid settlement of the West, and decline in demand for grain overseas. Between 1871 and 1874, Illinois, Iowa, Wisconsin, and Minnesota enacted the so-called Granger laws. These laws, though struck down in 1878 by the Supreme Court as unconstitutional, provided some examples that were followed in subsequent railroad regulations:

- Establish maximum rates.
- Create state railroad commissions to prescribe maximum rates.
- Require pro rata rates (proportional to mileage) for the same class of goods.
- Impose restrictions on mergers.

Regulation of the railroads was finally adopted at the federal level by the Interstate Commerce Commission (ICC) Act of 1887. The ICC was given the power to regulate rates, which were required to be "just and reasonable" and nondiscriminatory. Rates were required to be published, so that everyone could see whether they were fair. Additional legislation was needed to correct defects and omissions in the ICC Act to further protect the shipper.

The rail industry reached its financial zenith in the late nineteenth and early twentieth century. The railroads basically had a monopoly on intercity transportation, and people and companies were willing and able to pay rail fares and freight rates. Regulation had eliminated discrimination in rates, but it had not generally resulted in lower rates.

However, the rail industry was losing its appeal. Roads were being improved, and rapidly improving automobiles and trucks began cutting into the railroad's monopoly. U.S. involvement in World War I brought about a period of intensive use and very little maintenance of the rail system. To coordinate operations and maximize efficiency of the system, the federal government took control of the railroads from December 28, 1917, until March 1, 1920.[9] When the railroads returned to private operation after the war, they were in poor physical shape, and many lacked the financial resources to recuperate. It was evident that some railroads were stronger than others. The strong railroads were more efficient, had better routes, and served more customers. Competition among the strong railroads established the rate structure in many markets; weak railroads that had higher costs or second-rate routes could not compete. As competition among the railroads began driving prices down to levels where some railroads could not make any profits, the government recognized a need to help the industry. The Transport Act of 1920 provided some financial assistance to the railroads, established minimum rates as a way to avoid mutually destructive rate wars, and revised regulations related to rates and operations.

Over the next 15 years, the fortunes of the rail industry continued to decline. The industry began to face serious problems that would persist for decades. The weak railroad problem remained unsolved, and many railroads went into bankruptcy. Truck competition captured some of the industry's most profitable freight traffic, and automobiles and airlines sharply cut the railroads' passenger traffic. Although the railroad network was clearly overbuilt and had too many yards and lines, it was difficult to consolidate facilities (see the accompanying box) and the public challenged most attempts to abandon lines.

---

[8] For an excellent discussion of the economic and political factors that led to regulation, plus a detailed description of the first 75 years of rail regulation, see D. Philip Locklin, *Economics of Transportation*, 6th ed. (Homewood, IL: Irwin, 1966), which is the source of the information on regulation presented in this section.

[9] *Ibid.*, 225.

Many rail companies originally had a single route with a freight yard at each end of the line. Cities like Buffalo, Cincinnati, Chicago, Kansas City, and St. Louis became important rail hubs served by many individual railroads, each with its own yards and terminals in or near the city. When the railroads merged into much larger systems, they had great difficulty in coordinating operations through the maze of interconnecting lines and the multiplicity of small terminals. The problem of rationalizing operations within these major rail hubs persists today. Arthur Wellington, one of the first and greatest transportation economists, stated the problem most vividly:

How much of this abnormal magnitude [of railroad yard tracks in Buffalo] is the healthy and natural result of peculiar traffic conditions, and how much of it is mere fungous growth from diseases of management whose existence is universally felt, it would be useless to inquire here, because, as things are, it is all necessary, and there is no immediate evidence of any probable change.

Arthur M. Wellington, *The Economic Location of Railways* (1911)

With declining profits and bankruptcy, railroads had increasing difficulty in raising capital to upgrade equipment and facilities. Operating efficiency was hampered by work rules, many of them negotiated with the unions when the industry had been much more profitable, with different types of traffic, and more labor-intensive technology. Labor agreements made it very difficult to modify wages or work rules.

During World War II, the federal government limited the use of gasoline and rubber for private travel. Thus for a short time, the railroads once again were the dominant transport mode. However, the trends toward trucks, automobiles, and air travel continued after the war. The problems that became evident during the Depression remained unsolved, leading to pressure for further government intervention in the rail industry. The bankruptcy of Penn Central Railroad in 1969 precipitated what was known as the Northeast rail crisis. The Penn Central had been formed only a few years earlier by the merger of two of the largest railroads, the New York Central and the Pennsylvania, and even on the eve of its bankruptcy it was viewed as a financially solid company. The railroad's bankruptcy sent shock waves throughout the rail industry and the financial community. According to the Penn Central, its problems were fourfold: excessive labor costs because of union work rules and pay scales that reflected an era when railroads were dominant and operations were more labor intensive; inability to abandon unprofitable light-density lines; losses in providing passenger service; and maximum rate regulation.

In the early 1970s, further bankruptcies, spectacular accidents involving hazardous materials, equipment shortages, and disputes over light-density lines highlighted the need for federal action to preserve the industry. The federal government faced three options for dealing with the Northeast rail crisis: nationalization of the bankrupt railroads, government-funded rationalization of the system, and liquidation. Most in Congress were afraid of nationalization, due to the high cost of acquiring the railroads; the railroads and their customers feared political interference, especially with respect to consolidating lines and terminals and abandoning light-density lines. Congress, railroads, customers, and local government officials were all afraid of liquidation because nobody knew what would happen.

Congress eventually passed legislation that led to the government's acquisition of most of the lines owned by the Penn Central and several smaller bankrupt carriers. These lines were then merged into a new railroad called Conrail (Consolidated Rail Corporation). Congress was able to act because rail labor and rail shippers strongly supported providing help to the industry. The government invested several billion dollars in upgrading Conrail's equipment and facilities, in addition to paying several billion for the lines that were incorporated into Conrail.

Congress also dealt with the other problems highlighted by Penn Central by shifting passenger operations in 1971 to a new entity called Amtrak, relaxing regulation, instituting new procedures for line abandonment, and providing mechanisms to assist in gradual reduction of the labor force.

15.4.4   Deregulation of the Railroads

While Congress was still grappling with the Northeast rail crisis, the rail problems spread to the Midwest. In the mid-1970s, Congress was being urged to do something to respond to the collapse of the Granger roads—Milwaukee Road, Illinois Central Gulf, Rock Island, and other, smaller roads serving the region. Some people supported redoing the Conrail process by creating a "FARMRAIL," but this proposal was rejected as too expensive. Conrail clearly was going to cost the government more than $5 billion, and Congress was opposed to spending even more to redo it.

A labor strike on the Rock Island precipitated a crisis. Instead of intervening to keep the railroad operating, the government allowed the Rock Island to cease operation. The railroad sold its best assets to other railroads, which turned out to be quite a good way to break up a railroad. The big losers in this transaction were the union members who lost their jobs when the railroad went out of business.

The Carter administration was, in general, pushing for deregulation as a way to end protection of major industry and to lower prices for the consumer. Both shippers and local governments viewed allowing railroads to have pricing freedom as better than loss of service.

The end result was the passage of the Staggers Act in 1980, which eliminated much of the regulation of the rail industry. This act eliminated rate regulation for most types of rail shipments, allowed contract rates, and created new regulations aimed at expediting line abandonments and mergers. Deregulation, depending upon one's perspective, promoted innovative marketing by the railroads or led to discriminatory abuses such as those seen in the nineteenth century. Even railroad executives who normally claimed the former perspective sometimes admitted to the latter:

> *The best source I have for pricing ideas is in the court cases concerning pricing abuses prior to the creation of the ICC.*[10]

15.4.5   Technological Innovation

Deregulation was one of several factors that allowed the railroads to remain profitable at the end of the twentieth century. A second factor included negotiations with unions that allowed trains to be operated with 2 or 3 crew members rather than 4 or more, as well as other changes in union agreements that allowed more productive use of employees. A third major factor was technological improvement, including the remarkable advances in telecommunications and computers as well as improvements in railroad technology. Introduction of computers and advanced communications enabled the railroads to centralize their administrative activities, eliminate vast numbers of clerical positions, and increase the ability of managers to control operations over larger networks. Advances in rail technology included continuing improvements in materials and design for both track and vehicles as well as intensive development of two types of rail services:

1. **Unit trains** are used to transport large quantities of coal, ore, grain, or other bulk commodities directly from origin to destination. Unit trains provide the cheapest rail transportation because they are very efficient in terms of fuel, equipment utilization, and labor requirements. By capitalizing on what railroads do best, unit trains enable railroads to compete very effectively with barges moving on the inland waterway system and with heavy trucks operating on limited-access highways.

2. **Intermodal operations** involve the transportation of trailers or containers by rail, truck, and ocean carrier. Using a trailer or container makes it possible to transfer freight very easily from one mode to another, thus taking advantage of each mode's capabilities. Trucks are best at moving freight short distances to and from customers; railroads are best at moving trainloads of freight long distances;

---

[10] Mark Hodak, Conrail marketing executive, explaining to my class on Freight Transportation Management how he studied rail rate discrimination in the nineteenth century to understand how his railroad could benefit from the deregulation of most rail freight rates in 1980.

Figure 15.2 Double-Stack Container Train Entering Union Pacific's Bailey Yard in North Platte, Nebraska. Containers are loaded on such trains by automated lift equipment; a train carrying 250 containers can be loaded or unloaded in a few hours. Double-stack trains carry twice as many containers at little increase in the cost per train.

and ocean carriers are best at moving freight long distances across the oceans. A major breakthrough in intermodal operations came about in the 1980s, when ocean shipping lines and railroads developed the double-stack container train. This train carried twice as many containers at about half the cost per container. However, this dramatic cost reduction was possible only if intermodal terminals used expensive lift equipment at terminals, and service could be provided only on routes that had sufficient clearances for the much higher trains (Figure 15.2). The potential for lower costs sparked a great many projects to build new intermodal terminals and to increase clearances on most of the country's major rail routes. Double-stack trains enabled railroads to handle rapidly growing amounts of international freight as well as domestic freight like that moved by UPS.

15.4.6 Heavy Axle Load Research Program

In 1985, the rail industry initiated a long-term research program aimed at reducing the life-cycle costs for heavy freight operations. The program, administered by the Association of American Railroads (AAR) included laboratory testing, monitoring of track deterioration at various locations on railroads around the country, and the operation of a test track in Pueblo, Colorado, known as the Facility for Accelerated Service Testing (FAST).

The test results enabled researchers to quantify the effects of heavier loads and better suspension systems on the deterioration rates and life-cycle costs of the track structure.[11] Operating trains with heavier cars allows railroads to save money related to fuel, crews, and equipment by moving more freight in each train. The AAR's research program concluded that these operating benefits were greater than the added costs for the track infrastructure, and the industry decided to increase the maximum allowable axle loads from 33 to 36 tons. The heavier limits allow fewer trains to handle the same traffic, thus reducing operation costs as well as congestion in moving trains in and out of the most active coal mining regions—notably the Powder River Basin in Wyoming (Figure 15.3).

15.4.7 Outlook for the Twenty-First Century

The rail industry has great promise for selected services, including high-speed passenger service, heavy haul freight, and intermodal. Continuing advances in communications, control, and materials are likely to provide further improvements in productivity, thereby improving the financial stability of the industry. Increasing public interest in the environmental benefits of rail may help ensure the public support necessary to maintain and expand rail service.

---

[11] Michael B. Hargrove, Thomas S. Guins, and Carl D. Martland, *Economics of Increased Axle Loads—Fast/HAL Results*, Report No. LA-007 (Washington, DC: Association of American Railroads, October 1996).

Figure 15.3 Unit Coal Trains at Union Pacific Railroad's Bailey Terminal in North Platte, Nebraska. These coal trains originated in the Powder River Basin and are destined for power plants throughout the Midwest. Because heavier axle loads allow each train to carry more coal, more of it moves through this facility each day without the need for additional service tracks.

15.4.8   Lessons from the History of the Railroads

Some lessons that can be learned from the experience of the railroads are as follows:

- New technology may enable a proliferation of new products that will be implemented without sufficient consideration of what ultimately will be recognized as the best system design.
- Competition is likely to arise for any technology and for any system. The enterprises that prosper in a competitive environment are those that best serve their customers within the existing regulatory framework. Just because a technology is more fuel efficient, requires less land, or has a lower impact on the environment does not mean that the technology will prosper.
- Careful analysis can help determine what kinds of system improvements will be most helpful and how best to implement new technologies (i.e., performance-based technology scanning).
- Government regulation is a crucial factor for infrastructure-based systems, both because of the high fixed costs and the longevity of such systems. Government regulation of prices may be necessary to avoid destructive competition, and government funding may be necessary to allow a system with clear public benefits to be constructed or enhanced over time. Government support and regulation may also be helpful in allowing a gradual restriction of service over a slowly diminishing network rather than a collapse of service that would cause tremendous socioeconomic problems for users and communities.
- As a system ages, the facilities and services that previously were profitable may have to be abandoned to continue maintaining and operating the facilities and services that remain profitable.

15.5   TWENTY-FIRST-CENTURY CHALLENGES: SUSTAINABLE INFRASTRUCTURE

The historical reviews of skyscrapers and railroads show how technological changes can, over time, dramatically affect the way that infrastructure is designed, how it is used, and how it affects society. In both cases, technology allowed great reductions in cost; but regulation was required to curb private sector excesses related to pricing, aesthetics, overbuilding, or safety. Skyscrapers and railroads today are more energy efficient, more attuned to user needs, and more sustainable than they were 50 or 100 years ago. Presumably both types of infrastructure will continue to evolve in response to social, economic, financial, and environmental pressures (Figure 15.4).

Today, the need to consider sustainability is much more evident than it was when railroads and skyscrapers were introduced. The impetus to consider sustainability comes from various sources:

- Overdependence upon fossil fuels
- Global warming and related aspects of climate change (more severe weather patterns, rising sea levels, threats to ecosystems)

Figure 15.4 Modern Skyscrapers in Los Angeles.
Skyscrapers allow denser, more energy-efficient spaces that can be served more efficiently by local arterial streets, limited-access highways, and transit services (all shown in this photo). Even with such dense construction, planners have conserved some open space and created beautiful areas for pedestrians.

- Societal concerns with equity and social justice (the vast difference in quality of life in different countries and the vast inequality in incomes among the richest and poorest within each country)
- Instability in many societies related to disintegrating economies, internal conflicts, and lack of opportunity
- Congestion and pollution within the largest cities (Are there limits to growth? What can be done to improve quality of life in megacities, especially in developing countries?)
- Cracks in the automobile culture: recognition of the disproportionate amount of resources devoted to automobiles and highways, the enormous loss of life on highways, the dependence upon fossil fuels, the contribution to greenhouse gases, and the waste of time and resources caused by urban congestion
- Environmental degradation, including an accelerated rate of extinction of species, loss of habitat, the long-term threat of toxic chemicals to the water supply, and erosion

These factors will influence how infrastructure performance is measured, how goals are established, how problems are defined, and how project and programs are evaluated in the twenty-first century. The multi-objective framework presented in Part I of this text will remain useful, even if the relative importance of different factors come to the fore. The financial techniques presented in Part II will remain important, because an essential aspect of sustainable infrastructure will always be the ability to raise the cash to pay for construction, maintenance, and operation, and the ability to raise cash ultimately will reflect what is going on in the financial markets. The topics addressed in Part III—the ability to deal with uncertainty and risks, the ability of public and private sectors to work together, and the ability to manage projects and programs—will remain essential aspects of the development and evolution of sustainable infrastructure systems.

Enhancing the sustainability of infrastructure will be a continuing challenge for planners, engineers, and politicians in the twenty-first century. Urban problems related to pollution, congestion, disease, water supply, wastes, and natural disasters will grow—especially in megacities in developing countries, but also in all large or rapidly growing metropolitan areas. Battles over water supply will intensify, and the needs for wastewater treatment will increase. Renewable energy sources will need to be developed to lessen dependence upon fossil fuels, and safer, more efficient, more secure systems will be needed for greatly expanding the role of nuclear energy. It will be necessary to transform transportation and land use patterns that were developed over decades when oil was cheap, highways were easy to build, and cities were forsaken. Numerous small projects can be envisioned that will make cities more livable, more attractive, and more sustainable (Figure 15.5). Massive efforts will be needed to create or protect the world's green infrastructure and to reverse some of mankind's worst encroachments on harbors, waterways, wetlands, and threatened ecosystems.

Figure 15.5 St. Paul's Cathedral and Millennium Bridge, London. St. Paul's, a magnificent structure built after the Great London Fire of 1666, was once barely visible from the tourist boats plying the Thames. In 2000, a new pedestrian bridge linked the city's historical center with the rapidly redeveloping South Side. Pedestrians crossing the bridge are now treated to an unhurried view of this architectural gem. The bridge also connects well-designed walkways along the south side of the river with the nineteenth-century linear park along the north side. The Victoria Embankment was constructed in the 1860s for multiple purposes, including creation of open space along the river. It also helped to channel the river and provided space for a new sewer system and an underground railway.[12] Upgrading infrastructure, coordinating major investments affecting different kinds of infrastructure, invigorating downtown areas, creating more walkways to encourage pedestrians, and highlighting a city's architecture are all part of creating more sustainable cities.

Responses to these challenges will include efforts in many different areas:

- Managing the demands on infrastructure
- Maintaining and rehabilitating existing infrastructure
- Managing infrastructure more effectively
- Using infrastructure more effectively
- Mitigating the negative social and environmental impacts of existing infrastructure
- Building new, more sustainable infrastructure
- Retiring old, unsustainable infrastructure

In any of these areas, success will depend upon the same factors that have been emphasized throughout this text:

- Clear identification of the problems
- Development of objectives and measurable criteria
- Generating possible alternatives for dealing with the problems
- Analyzing alternatives with respect to financial, economic, social, and environmental impacts
- Selecting and refining the most promising projects and programs
- Implementing projects and programs efficiently and effectively
- Monitoring performance and taking actions as necessary to achieve project objectives

As illustrated in the case studies of skyscrapers and railroads, the evolution of infrastructure depends to some extent on leadership, innovation, technologies, and regulation. Efforts to transform infrastructure do not happen overnight, and it may take time for society to recognize either the need for changes or the need to regulate development. Once the technology for high-rise construction was available, real estate economics quickly drove buildings skyward in Chicago and New York; however, only after several decades of unfettered development were city authorities able to institute zoning regulations to preserve

---

[12] For the history of London's sewers and the construction of the embankment, see Stephen Halliday, *The Great Stink of London: Sir Joseph Bazalgette and the Cleansing of the Victorian Metropolis* (Phoenix Mill, England: Sutton Publishing, 1999).

light and air quality by requiring setbacks as the towers rose from street level. The rail industry struggled for nearly a century to transform itself from its role as the dominant transport mode of the late nineteenth century to a well-managed niche player with potential for growth at the beginning of the twenty-first.

Efforts to create more sustainable infrastructure will likewise takes decades of steady progress. While examples abound to illustrate how different systems are evolving, it is unclear which approaches will be most successful and which will ultimately be found to be too costly, too inefficient, or ineffective. The rest of this section provides some examples of recent projects that were recognized for their accomplishments with respect to sustainability. Example 15.7 describes Newark's efforts to preserve sewers; Example 15.8 describes Bank of America's efforts to create a "green" skyscraper; and Example 15.9 describes a small city's ongoing efforts to treat wastewater for a growing population within the confines of its existing facilities.

EXAMPLE 15.7   Rehabilitating Existing Infrastructure: Newark's Nineteenth-Century Brick Sewers

In 1990, the city of Newark, NJ embarked on a 20-year program to inspect and as necessary to rehabilitate its 68 miles of brick sewers that were more than a century old.[13] Old sewers are prone to fail, with consequences ranging from the high costs of emergency repairs to disruptions to residents and businesses to collapse of city streets and possible impacts on public health. The program involved six phases, as outlined in Table 15.3. Each phase involved the inspection of a portion of the system, determination of what kinds of repairs were needed, and completion of those repairs. As part of Phase VI, some of the sections that were first evaluated in Phases I and II were reevaluated.

Table 15.3   Newark's Six-Phase Program to Rehabilitate Its Brick Sewers

Phase	Years	Miles Inspected	Miles Rehabilitated	Cost ($millions)	Major Source of Funds
I	1990–92	13.2	4.2	$11.1	Low-interest loans from New Jersey
II	1993–96	12.6	5.5	$12.9	Low-interest loans from New Jersey
III and IV	1997–08	21.3	12.1	$24	Grant from U.S. EPA
V	2008	7.0	5.2	$19	Loan from NJ Dept. of Environmental Protection
VI (Planned)	2009–11	23.8	4.5	$16	Loan from NJ Dept. of Environmental Protection
Total	1990–2011	68	31.5	$83	

Inspection of the sewers produced numerous measures and photographs documenting the condition of the system, and that information was used to grade each segment of the system:[14]

- Grade 1: Condition is acceptable.
- Grade 2: Minimal potential for short-term collapse, but further degradation is probable.
- Grade 3: Collapse is unlikely in the near term, but further deterioration is likely.
- Grade 4: In some locations, collapse is likely in the near future.
- Grade 5: Collapse has already occurred or is imminent in some locations.

Priorities for repair also took into consideration the effect of a collapse on the surrounding environment. Three levels of risk were considered:[15]

1. *Critical A sewers:* The cost of failure and the impact of a failure on the surrounding environment would both be great.
2. *Critical B sewers:* The risk is less, but preventive action would still be cost effective.
3. *Critical C sewers:* Failure would have little or no effect unless many simultaneous failures occurred.

---

[13] This example is based on the article by Robert A. Pennington, Kristie A. Gersley, Anthony Gagliostro, Daniel T. Eagan, Alvin L. Zach, and John T. George, "Saving a City's Sewers," *Civil Engineering* (December 2008): 61–68.

[14] The grading technique was based on *The Sewerage Rehabilitation Manual* (Swindon, England: Water Resource Centre, 1994). Similar grading techniques have been developed for other kinds of structure, including bridges, highway pavements, and components of the railway track structure.

[15] The classification technique is described in *Existing Sewer Evaluation and Rehabilitation*, 2nd ed. (Reston, VA: ASCE Press, 1994).

Since all of the brick sewers were located under city streets, they were all categorized as critical A sewers. To manage the rehabilitation work, contracts were let for 18 separate projects, which allowed work to be done in reasonably sized pieces and staggered over a reasonable time period.

The main benefit of the program is that the city's brick sewers, once renovated, can be expected to perform their services for another 100 years, with much reduced need for future maintenance and emergency repairs, and with vastly reduced risks related to public health or collapsed roads.

Lessons to be learned from this example include the following:

- A program to upgrade existing infrastructure may require decades to complete.
- Inspection can determine where infrastructure is most in need of rehabilitation.
- Risks associated with failure will vary depending upon the nature of the surrounding areas.
- It will not be necessary to rehabilitate or replace the entire infrastructure, since many sections are likely to be in good condition or in locations where failure will not cause any significant problems.
- Priorities for rehabilitation are highest for portions of the infrastructure that are in the poorest condition and that are located in areas where failure would have the greatest consequences.

EXAMPLE 15.8   Building New, More Sustainable Infrastructure: One Bryant Park

Bank of America, working with the Durst Organization, designed its 55-story headquarters to achieve the highest level of LEED certification, thereby demonstrating the firm's commitment to sustainable development.[16] To become one of the first skyscrapers to achieve platinum certification, the new Manhattan skyscraper had to incorporate sustainable features that distinguish it from conventional office towers.

- Energy requirements:
  - The building has a 4.6-MW natural gas-fired power plant; excess heat is used to heat hot water for heating the building in the winter and cooling it in the summer via an absorption chiller.
  - A thermal storage plant in the basement produces ice at night, when energy demand is low, and helps with air conditioning during the day. This reduces the building's peak energy requirements, thereby lessening the demand for electricity from the most polluting and least efficient power plants serving the city (since these plants are used only during peak periods).
- Water requirements are half that of a conventional building:
  - Rail water is collected and used as gray water to flush toilets and supply the cooling system.
  - Waterless urinals are used in all men's restrooms, saving 3 million gallons per year.
- Interior design is healthier than that in a conventional building:
  - The interior incorporated many recycled materials and avoided materials with volatile organic compounds.
  - The heating and ventilation systems filter out 95% of the airborne particulates, compared to 35% for a conventional building; the air discharged from the building is cleaner than when it entered.
  - The glass-shrouded building lets in natural light.
- The structure uses recycled materials:
  - The 25,000 tons of structural steel had at least 75% recycled content.
  - Cement used in the building contained 46% granulated blast furnace slag (GBFS), a waste product of steel smelting that produces a stronger, denser, and more durable concrete. Use of GBFS was estimated to reduce material waste and new cement by 17,000 tons each, while also reducing the emission of carbon dioxide by nearly 16,000 tons.
  - The construction process resulted in less debris and less need for recycling.
- Integration within the city:
  - An underground pedestrian walkway eventually will connect subway stations on the north and south sides of the site, and a mid-block subway entrance eventually will be added.
  - The tower's entrances are sited to enhance access to Bryant Park.
  - The project helped in the reconstruction of Henry Miller's Theatre, a playhouse located on a corner of the site.

---

[16] This example is drawn from Andrew Mueller-Lust, "Crystal Clear," *Civil Engineering* (December 2008): 38–71. The tower is owned by One Bryant Park, LLP; it was developed by the Durst Organization; and Band of America is the largest tenant. The architect of record was Adamson Associates Consulting Engineers, P.C.

The $1.3-billion building includes 2.2 million sq ft of office space, most of it initially occupied by Bank of America beginning in 2008. In addition to its sustainability, the iconic building is noted for its beauty; it "features crisp folds that appear to be sculpted and precise vertical lines animated by the movement of the sun, the clouds, and the moon."

As this example shows, it is possible to build high-rise buildings that are much more sustainable in terms of materials requirements, energy and water needs, quality of the indoor environment, and integration into the nearby urban environment. Whether the innovations used in One Bryant Park will be cost effective in other applications is something that designers will have to determine for other locations and for other building projects.

EXAMPLE 15.9 Replacing Old Infrastructure with More Effective Systems: Wastewater Treatment in Kaukauna

The Heart of the Valley Metropolitan Sewerage District of Kaukauna, Wisconsin, provides services to the city of Kaukauna and nearby towns.[17] The district's original treatment plant was built in the 1930s, then converted into a regional treatment plant in the 1970s, and again upgraded and expanded in the 1990s. Two major factors made further enhancements necessary: population growth and stricter regulations regarding discharge. The district has a population of 42,000 that is expected to grow to 68,000 by 2029, which would cause peak demands to exceed the plant's capacity. Second, new state regulations required limiting ammonia nitrogen concentrations to 3.6 milligrams per liter (mg/L) in summer and 10 mg/L in winter, but the existing facility discharged effluent with 20 mg/L. The crucial design problem was the lack of space; the site had long since been hemmed in by a large mill on two sides and a canal and the Fox River on the other two sides. To handle the expected 100-year peak loads using conventional technology, the plant would need substantial elements to be added:

- Four primary clarifiers, each with a diameter of 115 feet and a capacity of 60 million gallons per day (mgd)
- Four final clarifiers, each with a diameter of 90 feet and a capacity of 25.4 mgd
- A nitrifying aeration basin area of 1 acre (200 ft by 200 feet)

Since the site was too small to accommodate these new facilities, conventional technology could not be used there. The possibility of acquiring a new site was considered but deemed too expensive. Instead, the district decided to use more efficient treatment processes, some of which had been used in Europe but not in the United States. To implement the new systems, the district was able to convert portions of the old facility to new uses, replace some of the old plant with portions of the new system, and convert some open space and a parking lot to other portions of the new system. The end result was a treatment plant that required a much smaller footprint, provided much higher capacity, and removed much more of the pollutants. Table 15.4 shows how the primary treatment technologies implemented as part of this project used one-tenth of the space while removing more of three key pollutants: carbonaceous biochemical oxygen demand (CBOD), total suspended solids (TSS), and phosphorous.

Table 15.4 Comparison of Conventional Primary Treatment Technologies with Those Implemented in Kaukauna

	Area of Facilities	Removal of CBOD	Removal of TSS	Removal of Phosphorous
Conventional	40,000 sq. ft.	30%	50%	0%
Kaukauna	3850 sq. ft.	50–60%	70–80%	75–85%

This example provides two major lessons. First, new technologies may make it possible to use existing sites far more effectively by reducing the footprint required for key facilities. Second, new technologies may be more effective, thereby resulting in a better level of service for the infrastructure.

---

[17] This example is drawn from Thomas E. Vik and Mark Surwillo, "Small Footprint, Big Promise," *Civil Engineering* (February 2008): 66–85.

15.6   UPGRADING INFRASTRUCTURE: A COSTING FRAMEWORK

Many projects and programs, such as those described in Examples 15.6 and 15.8, require decisions concerning the need to inspect, repair, rehabilitate, or replace infrastructure components. This section presents an analytical framework that can be used to assist in such decisions. The overall framework considers an infrastructure-based system made of up basic *elements* that are in turn made up of individual *components*. Table 15.5 shows some examples of the elements and components that make up infrastructure-based systems.

In designing the system, the question is what elements to use. In designing an element, the question is what components to use. Subsection 15.6.1 discusses the factors that influence the life-cycle cost of a component. Subsection 15.6.2 discusses the factors relating to the choice of components that make up an element of the system.

15.6.1   Life-Cycle Cost of a Component

The life-cycle cost of a component includes the cost of production, the cost of moving the component to the site, and the costs of inspection and repairs over the life of the component. The life of the component could be the time until it fails, or it could be the time until the risk of failure necessitates replacement.

The production cost of a component depends upon the size and shape of the component, the materials, and the production process. The quality of the component—however measured—is a different function of the same factors. The quality of structural components could relate to both engineering and aesthetic parameters, including strength, durability, susceptibility to wear or corrosion, fatigue life, ease of transport, ease of construction, appearance when new, and appearance over time. The production process can be important in terms of the material properties of the completed component, the quality control (i.e., probability of defects in the completed component), and the cost of production. For commonly used components, great economies of scale may be achieved in production, so that the costs of materials and transportation will dominate; for specialized components, a high cost may be related to design or specialized production techniques.

The production cost of a component is not the only cost of interest. It is also necessary to consider how long the component will last, how it might fail (and the consequences of failure), how well it will stand up to the environment and to usage, and how and why it will eventually have to be replaced. The costs of a component over the life of a project will therefore reflect maintenance, safety, and operating concerns. The life-cycle cost can be calculated as a net present value (NPV) or as an equivalent uniform annual cost (EUAC); the equivalent uniform annual cost will be easier to use for comparing infrastructure costs with operating costs and revenues.

The major cost elements that may be considered are as follows:

   1. Purchase of materials
   2. Transport of materials to production facility

Table 15.5   Examples of Elements and Components of Three Infrastructure-Based Systems

	Railroad	Office Building	Wastewater Treatment
Elements	Rail yards	Structural elements	Sewer system
	Main lines	Roof	Primary treatment
	Branch lines	Walls	Secondary treatment
	Turnouts	HVAC	Tertiary treatment
	Signals	Access/egress	
Components	Rail	Steel girders	Materials used for sewers (pipes, bricks)
	Ties	Concrete	Materials used for each type of treatment
	Ballast	Windows	
	Turnouts	Elevator	

3. Production of components
4. Transport of components to project site
5. Construction costs
6. Inspection costs
7. Maintenance costs
8. Rehabilitation costs
9. Replacement costs
10. Failure costs

The first four items comprise what might typically be thought of as the elements determining the unit price of the components, whether gravel, structural steel, or highly polished marble that is delivered to a construction site. Gravel is widely available, cheap, and transportation costs tend to dominate all other factors. Steel is manufactured at very large mills, and the quality of the steel may be critical; transportation costs are a much smaller fraction (10–20%) of the product cost, so supplies may be obtained from many different places around the world. Beautiful marble, suitable for use in the lobby of a major building, will be very expensive, require specialized workforces, and be available at only a few locations; material costs and skilled labor will therefore dominate.

Construction costs normally are not considered as a component cost, but ease of construction is certainly a factor in the selection of components. Composite materials may be favored in certain projects because they are as strong as and much lighter than steel, so they are cheaper to install.

The last six items are all interrelated. The nature and frequency of inspections is driven by the risks associated with component deterioration or failure. If these risks are negligible, then inspection may not be necessary. If it is necessary to conduct routine maintenance to avoid rapid deterioration, then periodic inspections will be part of a normal maintenance management program. If failure of a component is likely to lead to loss of life or severe disruption in operations or living conditions, then inspection at "safe" intervals will be essential. The purpose of the inspection is to determine if any of these problems exist:

- Any hazards that require immediate attention: a broken rail on a railroad line, a crack in an important bridge component, or a leak in a major pipeline
- Sufficient deterioration in components to require maintenance or replacement prior to the next inspection interval

The frequency of inspection must be sufficient to ensure an acceptable level of safety. This requires the interval between inspections to be short enough that defects that are not observable to an inspector are unlikely to have time to cause failures before the next scheduled inspection. On a railroad, it is possible to use ultrasonic inspection to identify internal defects within the rail. These defects generally are small, internal cracks in the steel rail that begin where there is some kind of contaminant or microscopic gap. The rail industry has long experience and considerable research that helps its experts estimate the amount of traffic that can move over the track before the crack grows large enough to break the rail. Theoretical and laboratory studies can be conducted to formulate models of crack initiation, crack growth, wear, corrosion, or other kinds of deterioration.

In general, some kinds of deterioration will increase as the component ages, and some kinds of potential failures become more likely as the component deteriorates. Thus some kinds of failure are increasingly probable, and eventually something must be done to preserve a safe operating environment. For railroads, the interval between ultrasonic inspections is related to the traffic volume and the nature of the traffic (because heavier cars cause more fatigue damage). The potential consequences of a failure determine what probability of failure is acceptable. A railway that handles high-speed passenger trains is at much greater risk than one that handles only coal and grain; high-speed passenger lines might therefore have daily inspections, whereas coal lines might have monthly or less frequent inspections.

Even if no safety hazards are identified, inspection may lead to maintenance or replacement of components. Maintenance may be justified by (1) its effect on future deterioration of the infrastructure or (2) its effect on infrastructure performance. It makes sense to patch cracks in highways before they grow into large potholes, and it makes sense to fix leaks in pipes to avoid wasting clean water that is moving toward cities.

Some components are "maintenance free" (i.e., they can be expected to remain in service without requiring inspection, maintenance, or replacement over the life of a project). Most components are not maintenance free, and their life-cycle costs include most of the factors listed above. The choice of the components therefore requires what may be rather complex trade-offs concerning initial cost, infrastructure performance, and operating performance. The basic approach to selecting components is to determine the initial costs (as described earlier) as well as the expense and frequency of inspections and maintenance activities over the life of the component.

Component life may be determined by failure, risk, economic factors, or availability of new technology. New components may be so much better that it is justifiable to replace existing components. This situation occurs if the EUAC for maintenance and operations over the remaining life of the existing component is higher than the EUAC for a new component (over the expected life of the project or of the new component, whichever is shorter).

The **expected life** is a key element for estimating the life-cycle cost. In general, the expected life of a component is the minumum time until one of the following events is expected to occur:

- The component fails.
- It is replaced as a safety hazard.
- It is replaced to avoid reducing facility performance.
- It is replaced to reduce annual maintenance costs.
- It is replaced to avoid reduced system performance.
- It is replaced by superior components.
- The project or facility reaches the end of its useful life.

Engineering relationships can be developed for each of these events (as functions of materials properties and usage parameters), and the most likely limits for each component can be studied in great detail. If the failure cost is tiny and replacement is easy, as is the case with light bulbs in easily accessible locations, then the expected life will be the expected life until failure. If failure cost is substantial, then a plan should exist to replace components before they are likely to fail. If the condition of a component can be monitored, then the actual decision to replace can be based on the condition of the component, the potential consequences of a failure, the time until the next inspection, and the likelihood of failure before the next inspection. If the component is replaced, then the costs over the next year can be assumed to be the EUAC of the new component. If the component is not replaced, then the costs over the next year will be the sum of the expected costs of inspection, repairs, and failure over that period. The decision rule could be stated as follows: replace the component if the EUAC of the new component over the next inspection period is less than the expected costs of the old component over that period.

EXAMPLE 15.10   Is It Worthwhile to Replace a Component?

Assume that a component is expected at intervals of 1 year. An inspector assigns one of four grades to the component:

1. Grade A: Condition is excellent.
2. Grade B: Some flaws are evident; maintenance costs expected to rise 20% but no danger of failure.
3. Grade C: Flaws are evident; maintenance and inspection costs expected to rise 30%; less than 10% chance of failure over next 12 months.
4. Grade D: Flaws are evident; maintenance and inspection costs are expected to rise 50%; chance of failure over next 12 months is at least 25%.

The annual cost of maintenance is normally $1,000 per year. The cost of a failure, which includes the extra costs of installing a new component under emergency conditions, depends upon the location:

- High-density areas where access is difficult: $20,000
- High-density areas where access is easy: $15,000

- Low-density areas where access is difficult: $12,000
- Low-density areas where access is easy: $7,000

If the component is replaced, the EUAC of the new component would be $3,000 per year. Under what circumstances should the component be replaced when its condition reaches grade B, grade C, or grade D?

The increases in maintenance costs are on the order of $200, $300, and $500 for grades B, C, and D. The expected costs of failure during the next year are the probabilities of failure multiplied by the costs of a failure, as shown in Table 15.6. The component should be replaced if the sum of the expected maintenance cost plus the expected failure costs exceed the $3,000 EUAC of the replacement component.

Table 15.6   Expected Maintenance Costs and Probabilities of Failure Increase as Condition Deteriorates

	Grade A	Grade B	Grade C	Grade D
Costs of maintenance	$1,000	$1,200	$1,300	$1,500
Probability of failure	0%	0%	10%	25%
Expected cost of failure:				
• High density, poor access	0	0	$2,000	$5,000
• High density, good access	0	0	$1,500	$3,750
• Low density, poor access	0	0	$1,200	$3,000
• Low density, good access	0	0	$ 700	$1,750

If condition is grade A or B, the component should not be replaced. If the component is grade C and in a high-density area with poor access, then the expected costs of the existing component will be $1,300 for maintenance and $2,000 for failure, for a total of $3,300. This is greater than $3,000 (which is the EUAC of the replacement component), so the component should be replaced. In other locations, components in grade C can be kept in service since the sum of the cost of maintenance and the expected costs of failure is less than $3,000. For components with grade D, the expected cost exceeds $3,000 in all types of locations, so all of these components should be replaced. The expected life of this type of component therefore is the expected time until condition deteriorates to grade C in high-density, inaccessible locations and the expected time until condition reaches grade D in all other locations.

Based on models or experience, it is possible to establish the inspection and maintenance activities over the expected life of the component. This information can then be combined with the initial price to obtain the NPV or EUAC for any component under any type of load and annual usage.

For a component with an expected 12-year life, the various costs might be incurred as shown in Table 15.7. The component is purchased, transported to the site, and installed in year 1. It is inspected every year for 12 years, and repairs of various types are made periodically until the component is replaced at the end of 12 years.

Table 15.7   Life-Cycle Costs of a Component

Activity	Year											
	1	2	3	4	5	6	7	8	9	10	11	12
Purchase	•											
Transport	•											
Install	•											
Inspect	•	•	•	•	•	•	•	•	•	•	•	•
Repair, type 1			•			•			•			
Repair, type 2		•		•		•		•		•		
Replace												•

The life-cycle costs of this component will be the NPV of the costs of all these activities listed in the table. If the discount rate is i, and if the probability of a failure is negligible, then the NPV of life-cycle cost can be calculated as follows:

$$NPV(C_{total}) = C_{install} + \sum C_{inspect}[Inspect(t)]/(1+i)^t$$
$$+ \sum \sum C_{repair,r}[Repair(r,t)]/(1+i)^t + C_{replace}/(1+i)^{12} \qquad \text{(Eq. 15.1)}$$

where:

$C_{install}$ = unit cost of component, including transportation to site and installation
$C_{inspect}$ = unit cost for inspection
$C_{repair,r}$ = unit cost for repair type r
$C_{replace}$ = unit cost to take the component out of service
$Inspect(t)$ = number of inspections during year t
$Repair(r,t)$ = number of maintenance activities of type r occurring during year t

The net present value of all inspections and maintenance activities needs to be calculated for everything that is expected to occur over the life of the facility, where the life is estimated separately, as described earlier. Given the NPV, it is then straightforward to determine the EUAC over the life of the component and to compare the EUAC for various components that might be used:

$$EUAC(C_{total}) = NPV(C_{total})(A/P, i, life) \qquad \text{(Eq. 15.2)}$$

This calculation obviously can get quite complex. However, it can be done, and it can be very helpful in designing and managing infrastructure to understand these relationships. Some useful insights can be gained from this structure:

- If a component will be subjected to a great deal of use, then the benefits of using a better component are clear, even if it has a much higher initial cost.
- If a component will be subjected to only modest use, then initial cost will dominate.
- R&D efforts will be devoted to improving those components with the highest costs or the highest risks.
- Improved inspections can reduce risks of failure and thereby extend the life of components.

Salvage values and the costs related to decommissioning are other factors that can be important for infrastructure costs and design.

15.6.2 Element Design

For a road, a railroad, an aqueduct, a dam, and many other types of infrastructure, it makes sense to look at the combination of components and elements that define the system. The terms **component** and **element** are used to suggest the hierarchical structure of infrastructure: a system is composed of various elements, and each element is composed of various components. For example, a rail system includes terminals, track, turnouts and interlockings (locations where routes cross or diverge), bridges, trestles, signal systems, and maintenance facilities. The track structure's major components are rails, fasteners, ties, ballast, and subgrade. A highway system includes intersections, bridges, rest areas, signs, and signals as well as the roadways. Each roadway has elements similar to those of the railway: a layer of pavement plus several layers of gravel and subgrade. Both highways and railroads have ditches for drainage.

The choice of components and the way they are combined determine the engineering capabilities of each infrastructure element. For example, the typical cross section of a railroad is defined by the following:

- The weight, shape, and metallurgy of the rail (which together determine the strength, wear life, and fatigue life of the rail)
- The weight, shape, and materials used for ties (which may be standard wood ties of various dimensions or ties made of concrete, plastic, or steel)
- The spacing of ties

- The fastening system (e.g., spikes and tie plates or premium fastening systems)
- The materials used for ballast
- The depth of the ballast
- The possible use of geotextiles or an asphalt layer to deal with wet or weak subgrade

Over many decades, railroads have developed stronger materials, mechanized techniques for installing track, and automated techniques for inspection and maintenance. The result has been to create a track structure that has lower life-cycle costs and can handle heavier trains.

The capacity and cost of the railroad also relate to the gauge (the width between the rails) and the clearances (both vertical and horizontal). Similar design features apply to highways, canals, pipelines, and airport runways. A wider roadbed allows larger vehicles but also requires greater cost.

Another design feature is the number of lanes for a highway, runways for an airport, or tracks for a railway. This aspect of the design has a great impact on service capabilities (e.g., speed, daily capacity, peak loads).

Since each element is made up of many components, a cost model can be developed for each element by combining cost models for the components. Once this is done, it is straightforward to use standard designs to obtain standard costs for, say, a mile of track that can handle coal trains or a four-lane divided highway in rural areas.

A building is a different type of infrastructure, but the same general approach will work. A building requires various components, and planners can determine the cost and performance of the various components and count the number of components required for an element—which might be a hallway, a room, or a floor of a multistory building or a single-family house.

As with a railroad or a highway, standard designs using standard components are used to come up with easily calculated cost estimates for buildings. The construction industry often expresses building costs based on the "cost per square foot" for the type of building, because the client probably is most interested in the size of the facility. The contractor needs to prepare cost estimates that easily relate to designs and measures the client understands. Therefore, even though costs really reflect the nature and number of the elements and components that make up a building, it is useful to calculate cost per square foot for an existing design or an existing building and use that as a rough estimate for future buildings.

Large contractors often develop specialized experience with specific types of infrastructure (e.g., electric power plants, offshore oil-drilling platforms, or transit systems). Experts in these areas may think in terms of the cost per megawatt of capacity, the cost per platform, or the cost per station and per route-mile—but at some level they will be building on their knowledge of the costs of components, trade-offs related to initial cost and durability, design of elements, and design of the system. The life-cycle costing approach described here is fundamental for any type of infrastructure.

15.6.3 Network Design

Infrastructure often takes the form of a network. A transportation network consists of a set of routes, intersections, and terminals. Similar networks are needed for energy distribution, water supply, wastewater collection and treatment, and transmitting information. Design issues for infrastructure networks include selecting the elements for each location as well as determining the layout of the network and its effect on accessibility and service capabilities. Capacity, cost, accessibility, and service capabilities are all relevant concerns. When the networks are first put in place, their main benefits will be the new services—mobility, cheaper energy, cleaner water, or cheaper communications. Once the network is established, further projects will be needed either to maintain service capabilities (maintenance and renewal), improve service capabilities, provide additional capacity, or reduce the negative impacts of the infrastructure (e.g., provide noise barriers for highways or reroute a road around environmentally sensitive areas).

The network is a collection of elements, so the cost of the network can be built up from the cost of the elements. The only complication is that new types of elements (complex interchanges or high-capacity elements) may be needed.

At a network level, system cost is easier to estimate than system benefits. The design questions will at first relate to the feasibility of constructing any portion of the network. It may be possible to start some networks with a single route (this is how the rail systems were initially built); but for others, a complex but locally constrained network may be needed (e.g., cable TV). Given a network for a system in the early stages of evolution, there will constantly be proposals to extend it to new regions, upgrade it to provide better service, or increase its capacity. In later stages of evolution, more proposals will be aimed at consolidation, reductions in service, replacement of components, and ways to maintain operations in the face of declining demand.

15.7   SUMMARY

Most infrastructure projects are completed as an extension to, an upgrade of, or a competitor to an existing system. Infrastructure projects therefore are almost always conceived and evaluated in the context of existing systems. At times, perhaps for long periods of time, it will be very clear which projects can and should be implemented; thus, there is little or no controversy and little or no need for extensive analysis. Projects that improve the performance of a successful system or projects that replicate systems in new locations may not require a great deal of specialized evaluation, because the costs and benefits associated with such projects are well understood. However, over time, changes in technology, institutions, political power, economic conditions, or societal norms will lead to new ideas for projects, new estimates of the potential costs and benefits of projects, and new criteria for evaluating projects.

Stages in the Evolution of Infrastructure-Based Systems

It is useful to consider the stages in the evolution of an infrastructure-based project, for in each stage the opportunities and problems are different. Table 15.8 lists eight stages and the types of questions encountered in each stage. These are not necessarily sequential, and different parts of a large system may be at different stages of development. For example, the infrastructure required to support cell phones and other wireless communications is still in the early stages of evolution, while the infrastructure used to support traditional land-line telephones is in the latter stages of evolution—even though the same companies may be involved in both types of communication.

Skyscrapers and Building Booms

Section 15.3 describes how changes in building technology allowed the construction of much higher buildings and thus drastically changed the economics of real estate construction in major cities. Taller buildings were not only feasible, they allowed more people to live or work in the same space; skyscrapers made it possible to increase population density and bring more people in close proximity for business or social purposes. In Manhattan, where incentives to consolidate were the strongest for economic and geographic reasons, skyscrapers proliferated in the early twentieth century. Today, the same incentives lead to the development of skyscrapers in many other large cities all over the globe.

Case Study: Evolution of the U.S. Rail System

Section 15.4 is an overview of the evolution of the rail industry in the United States. Portions of the rail system have gone through all eight stages of evolution, and the system today is far different from the one that dominated both passenger and freight transportation during the nineteenth century. Most of the smaller shipments and short-distance freight has been diverted to truck, and many once profitable rail lines have been abandoned or converted to rail trails. Many rail yards and terminals have been converted to other uses. Faced with competition from other modes, the railroads had to invest in the portions of their systems that were likely to remain competitive. Because of investment in new types of freight cars, better locomotives, improved signal systems, and more durable track structures, railroads remain the most

Table 15.8 Stages in the Evolution of Infrastructure-Based Systems

I	Technological Experimentation and Demonstration
	• Will the technology work?
	• Will investors support projects?
	• Will there be a market for the new technology?
	• Will public agencies approve the new technology?
II	Widespread, Uncoordinated Implementation
	• What will be needed to implement the new technology?
	• Who can build the fastest?
	• Whose approach will be the best?
	• What is needed to obtain public approval for new or expanded systems?
III	Development of Systems
	• How can we achieve economies of scale and density?
	• What organizational structure is best?
	• How much coordination is required among government agencies and developers or private companies?
	• What is the best way to handle growth?
	• How can we avoid overbuilding?
IV	Consolidation and Rationalization
	• How can we rationalize the system so as to become more efficient and more effective in providing services?
	• What is the optimal structure of our system?
V	Technological and Institutional Advancement
	• What is the source of competition: similar systems, different technologies that provide the same service, or systems that eliminate the need for the existing service?
	• What is the role for new technology?
	• How can the system be managed and regulated more effectively?
VI	Responding to Competition
	• What is the best role for the system?
	• Can the system survive competition?
VII	Mitigating Social and Environmental Impacts
	• How can the system respond to changes in societal views of the role of the system and the extent of its impacts on society and the environment?
	• How can negative social and environmental externalities be mitigated?
VIII	Retrenching and Obsolescence
	• When is it necessary to downsize?
	• What is the best approach to downsizing?
	• How can the system be reused, recycled, or dismantled?

economical way to move large shipments of bulk commodities such as grain or coal. Railroads also work with truckers and ocean carriers to move containers around the country and around the world. This type of transportation depends upon investment in new types of container ships and terminals where automated equipment can be used to transfer containers from one mode to another.

Twenty-First-Century Challenges: Sustainable Infrastructure

Over time, society's needs and resources change; and as they change, the roles for and perceptions of infrastructure will also change. In the nineteenth and twentieth centuries, new technologies and mammoth infrastructure projects enhanced mobility, helped make agriculture more productive, allowed more intensive urban development, provided heat and light, and shrank the world, together creating unprecedented prosperity and a global economy, Many consequences of these technologies and projects are highly beneficial, but some consequences are negative: pollution, overdependence upon nonrenewable energy sources, congestion, excessive use of water, inequalities in development, and degradation of the environment. In the twenty-first century, much of the interest in infrastructure will be aimed at resolving these problems and extending the benefits of modern technology to all regions of the world. Many projects and programs must be initiated in order to move the United States and the world toward more sustainable infrastructure.

Upgrading Infrastructure: A Costing Framework

An infrastructure-based system is made up of many elements, such as roadways and bridges, that are in turn made up of many compoenents, such as pavement and girders. Often the elements are combined to create a network that provides service over a wide area. Cost models constructed for each for component can be combined to obtain cost models for each element or for the entire network. Component costs over their expected lives can be estimated, taking into account initial purchase costs, inspection capabilities, maintenance requirements, and the risks of component failure. Detailed cost models can be developed to support decisions about maintaining, rehabilitating, upgrading, replacing or abandoning infrastructure.

ESSAY AND DISCUSSION QUESTIONS

15.1 Rising fuel prices, concerns about global warming, and the high costs of constructing more roads in urban areas are three factors affecting the shape of the transportation system and the nature of personal mobility in the twenty-first century. How do you think society will adjust to these trends? What is the potential role of having more fuel-efficient cars, better transit systems, more densely populated cities, more walkable cities, or more carefully managed highways (i.e., more toll roads, insurance based on automobile usage, or highway user fees based on actual time and location)?

15.2 What is the future of the suburban mall?

15.3 Select one of these infrastructure-based systems:

a. Railroads
b. Urban road networks
c. Water supply systems
d. Waste treatment systems
e. Housing
f. Telecommunications

What major technological, institutional, and competitive forces have influenced this system over the past 100 years? What major technological, institutional, and competitive forces are likely to influence this system over the next 100 years?

15.4 In a feature box in Section 15.3, E. Alfred Picardi summarizes the technological and economic forces that changed skyscraper design in the 40 years following World War II. Writing soon after the tragedy of 9/11, Picardi cited congestion and environmental overload along with safety as reasons to think about new ways of "housing commerce." Nevertheless, the design of skyscrapers has continued to evolve, and new buildings in Taipei, Kuala Lumpur, Dubai, Shanghai, New York, and elsewhere continue to seek records for height and awards for design. What do you foresee as the future of skyscrapers— bigger and better? Or will new approaches to urban design lead to the use of smaller buildings and more dispersed economic activity?

PROBLEMS

15.1 Let's think back to Manhattan in 1900, a time when much of the downtown area consisted of densely spaced 5- to 6-story buildings, but the new steel-framed construction technique was allowing much taller buildings to be built at a lower cost per square foot of rentable space. You own and operate several 5-story apartment buildings in a well-established residential portion of the city. Each building has 10 apartments, and all of the apartment buildings are profitable for you. Now you would like to expand your little real estate empire. Consider these options:

a. Continue to build 5-story apartment buildings. Assume you own a vacant lot that is suitable for constructing a 5-story apartment building with 2 apartments per story. Assume that the monthly lease rates for apartments in this location are

$200/month, and that your expenses associated with owning and maintaining the building would be $1,000 per month. Would it be worthwhile for you to spend $100,000 to construct a 5-story apartment building on this site? Your pretax minimum acceptable rate of return (MARR) is 8%. *NOTE*: You do not need to consider financing, depreciation, or taxes to answer this question, since you are just trying to determine whether it makes sense to build another apartment building.

b. Build a 20-story apartment building on the site, using the new steel-framed construction method. The cost for this building is estimated to be $380,000 with operating expenses of $3,000 per month. Rents would be somewhat higher on average because of the better views available. Would this be a better option than the 5-story building? Discuss.

c. Since you already own several 5-story apartment buildings, you may wonder if you should tear them down and build more 20-story buildings. Assume that it would cost $20,000 to demolish one of the existing buildings, which would increase the cost for the new 20-story building from $380,000 to $400,000. Does this option make sense from a financial perspective? If you decided to do it, how would you deal with your tenants?

15.2 During the 1960s, the U.S. railroads increased the maximum allowable weight limit of freight cars from 200,000 to 263,000 pounds. Since a freight car typically weighed about 60,000 pounds, the net weight that could be transported in a freight car was increased from 140,000 pounds (70 tons) to about 200,000 pounds (100 tons). At about the same time, a shift was occurring in the way that grain was transported. Before the 1960s, most grain was loaded into 40-foot boxcars at small grain elevators located throughout the Midwest. This approach had the merit of using standard boxcars that were cheap and could easily be shifted to other types of service; however, it took a long time to unload grain from a boxcar. Beginning in the 1960s, covered hopper cars began to be used much more extensively for grain, because these cars could be unloaded much more quickly than boxcars (the hoppers were located at the bottom of the car; cars ready to be unloaded were moved to tracks located over a grill above a conveyor belt; when the hoppers were opened, the grain simply poured out of the hoppers, through the grill, and onto the conveyor belt).

a. Suppose your railroad is expanding the capacity of its fleet of grain cars. Do you recommend buying 70-ton boxcars, 100-ton boxcars, or 100-ton covered hoppers? Assume that the purchase price of a 70-ton boxcar is $40,000, the purchase price of a 100-ton boxcar is $45,000, and the price of a covered hopper is $50,000. Assume that the cost of unloading a boxcar is at least $100, whereas the cost of unloading a covered hopper is no more than $20. Assume that cars make 30 trips per year, that they last 30–50 years, that equipment maintenance costs are about the same for all cars, and that operating and track maintenance costs are lower for the covered hoppers.

b. If the railroad can save so much by buying covered hoppers, perhaps it should scrap the fleet of 40-foot boxcars and replace them with covered hoppers. Assume that operating costs are $20.00 per 1000 gross ton-miles, no matter what kind of car is used. Assume that the average loaded trip is 1000 miles. Should the 40-foot boxcars be replaced before the end of their serviceable life? Assume that the MARR for the railroad is 10%. Discuss.

15.3 During the 1980s, the railroads and their international shipping partners introduced the double-stack container train. This innovation was driven by common sense: Since containers are much less dense than coal and many other commodities handled by rail, a train load of containers is not very heavy. If

containers could be stacked one on top of another, the same train crew could haul roughly twice as many containers. Moreover, the freight cars used to move the containers could be shorter and lighter than the 89-foot flatcars once needed to carry two containers. Operating costs for containers moving on the traditional 89-foot cars amounted to approximately $0.80 per mile. With double-stack operations, the cost per container mile dropped to approximately $0.40 per mile. These savings were so large that American President Lines, an ocean shipping line that transported containers across the Pacific, purchased double-stack equipment to move containers from West Coast ports to destinations in the U.S. Midwest and East. The railroads then faced had to decide whether to replace their existing fleets of 89-foot flatcars with the new double-stack equipment. The most cost-effective double-stack equipment consisted of an articulated set of 5 platforms that carried a total of 10 containers. The purchase price was approximately $400,000. The intermodal equipment typically made 80 trips per year with an average distance of 1000 miles per trip. Should the railroads continue to operate single-stack trains, or should they shift as soon as possible to the double-stack equipment? Assume an MARR of 12% and a 30-year life for the double-stack equipment.

15.4 The previous question considered whether railroads should replace existing intermodal flat cars with new double-stack equipment, assuming that the double-stack cars could be used over a particular route. This is actually a major assumption, since double-stack operations required overhead clearances of more than 20 feet, whereas most routes had clearances of 19 feet or less. Double-stack operations called for increasing the clearances in tunnels, under bridges, and under any other structures that crossed over rail lines. Double-stack trains were first implemented in the West, where trains could travel hundreds of miles without encountering any bridges at all. In the East, dozens of bridges with low clearances could be within 50 miles of a port. If it costs an average of $1 million to increase clearances on one bridge, would it be worth increasing the clearances for any of the following routes?

a. A 2000-mile route with 50,000 containers per year and 40 bridges (e.g., the route between a major rail intermodal terminal near a West Coast port and a rail intermodal terminal near Chicago)

b. A 20-mile route with 700,000 containers per year and 100 bridges (e.g., the route between ports and a major rail intermodal terminal, such as the rail route between the major ports of Los Angeles/Long Beach and the major rail terminal to the east of Los Angeles)

c. A 50-mile route with 10,000 containers per year and 40 bridges (e.g., the route between a small port and an inland container terminal, such as the route between the Port of Boston and the container terminals in Worcester, Massachusetts)

d. A 100-mile route with 70,000 containers per year and 40 bridges (e.g., the route between an inland terminal and the nearest rail line with double-stack clearances, such as the route from Worcester, MA, to Albany, NY).

CASE STUDY: EVOLUTION OF THE U.S. RAIL SYSTEM

Read this case, which is in Section 15.4, and be ready to discuss the following questions:

1. Compare and contrast the role of the rail system in 1850, 1900, 1950, and 2000. How and why did the role of the rail system change with respect to its influence on transportation costs, economic development, and settlement of the country?

2. To what extent did technological, managerial, and government regulatory factors influence the evolution of the rail industry? Which factors were most critical during each stage of the evolution?

3. What kinds of projects were required in each stage of development for the rail industry?

4. What lessons can be learned from this history of the rail system that could be applied to the evolution of other transportation systems or other infrastructure-based systems (e.g., urban transit systems, highway networks, electrical power generation, and telecommunications)?

5. In other countries, governments have made massive investments in high-speed passenger rail systems, and they heavily subsidize rail service for both passengers and freight. In the United States, the government has made relatively minor investments in Amtrak—on the order of $1 to 2 billion per year as opposed to $5–$20 billion per year for some of the systems in Europe and Japan. Should the federal or state governments be more aggressive in promoting rail in the United States?

Chapter 16

Final Thoughts and Further Reading

Civil engineers are problem solvers, but we need to broaden the scope of our services to include problem definition. Civil engineers must go beyond thinking in terms of project specific limits and scopes of work and become involved in systemwide, program-related decisions and policymaking to achieve long-term, sustainable solutions. We must be facilitators of collaboration among multiple agencies/owners and across jurisdictional boundaries. We must also take a leadership role in developing acceptable and sustainable methods of funding infrastructure development and asset management.

Kathy J. Caldwell, ASCE News, June 2009

CHAPTER CONCEPTS

Section 16.1 Introduction

Section 16.2 Key Lessons in Part I: Building Infrastructure to Serve the Needs of Society
  Understand the story for the project
  Consider all aspects of performance
  Enhance sustainability of projects and programs
  Model performance
  Assess performance
  Develop scenarios

Section 16.3 Key Lessons in Part II: Comparing Economic and Financial Impacts over the Life of Proposed Projects
  Discounting and the choice of a discount rate are critical
  Maximizing net present value and internal rate of return are key objectives
  Care is needed in using IRR to evaluate mutually exclusive projects
  Cost effectiveness can be used to evaluate projects with non-monetary benefits
  Depreciation and other rules of the game will affect cash flows

Section 16.4 Key Lessons in Part III: Developing Projects and Programs to Deal with Problems and Opportunities
  At first, anything is possible
  Intuition and expert judgment lead the analysis
  Public/private partnerships may be useful
  Anticipate risks and uncertainties

16.1   INTRODUCTION

Societies and civilizations advance through projects that seek to make life safer, healthier, more prosperous, or more secure. History abounds with tales of famous projects, from the pyramids of Egypt and Mexico to the temples of Greece and Cambodia, from the aqueducts and roads of Rome and the Great Wall of China to the canals, railroads, and telegraphs of the nineteenth century and the highways, telecommunication, dams, and water and wastewater systems of the twentieth. Countries have been bankrupted by bad projects, secured by practical projects, and advanced by bold projects. Individual fortunes have been made or squandered on projects because banks and financial markets have made it possible to direct vast sums of money toward massive undertakings anywhere in the world. Large projects entail large risks, and such projects are undertaken only if those promoting the projects can convince investors or governments to fund them. Highly capable, charismatic individuals motivated many great projects, including the Suez Canal, the Brooklyn Bridge, and the Empire State Building. However, it has not always been possible to separate the true visionaries from the charlatans and the deluded, especially when it comes to large outlays of public funds.

Project evaluation can be viewed narrowly as a set of procedures and methodologies for determining whether a proposal should be approved. Indeed, there are well-defined methods for assessing financial aspects of a project, and private companies and investors use these methods routinely in determining whether to begin projects or to invest in them. Moreover, governments today mandate environmental and social impact assessment before approving any significant project, and they use intricate economic models to estimate the effect of public infrastructure investments. Once a project is approved, the emphasis shifts from project evaluation to project management, a process that encompasses another set of procedures and methodologies.

This textbook does not take a narrow view of project evaluation, because that approach misses the most interesting and challenging aspects of projects and program. An agency such as the World Bank—or your local bank—may indeed evaluate many proposals for constructing new roads, buildings, or water resource systems. But someone has to come up with those proposals; and those proposals could emerge from a careful consideration of what society needs, reflect insight into opportunities offered by new technologies, or simply be pipe dreams that appeal to public emotions but have little chance of success. Much of the challenge and excitement in project evaluation deals with the earliest stages of a project: clarifying the needs of society, anticipating technological opportunities, and pulling together ideas or objectives for a possible project or program.

Sections 16.2 through 16.4 briefly outline the key lessons to be gleaned from this textbook. These sections are not intended to be a summary of the entire book. Each chapter already has its own summary, and attempting to combine these summaries would be a tedious task both for the author and for the reader. Instead, these sections select the most important concepts and insights that will be continue to be relevant to readers long after they have forgotten the details encountered in this text.

Section 16.5 provides opportunities for further reading. The list is divided into four categories: textbooks, books about infrastructure projects and programs, articles about infrastructure projects and programs, and articles about evaluation methodology.

16.2   KEY LESSONS IN PART I: BUILDING INFRASTRUCTURE TO SERVE THE NEEDS OF SOCIETY

16.2.1   Understand the Story for the Project

Every project has a story, and every story has several components. What are the context and the history of the project? What needs does the project address? How will the proposed project meet those needs? What other approaches are available, and why is the proposed approach the best? How much will it cost, and who will pay for it? What are the broader impacts on society? What is the proper role for government? Why should the public support the project? How will negative externalities be mitigated?

> *Real wealth is the ability to produce more with less—to generate a flow of goods and services without having to sacrifice something else of equal value. It is not created by taking time away from other activities and devoting it to money-making.*
> Lester C. Thurow, "Building Wealth," *The Atlantic Monthly* (June 1999): 57–69.

16.2.2   Consider All Aspects of Performance

Projects will have financial, economic, social, and environmental impacts. Any or all of these impacts could be important in evaluation.

16.2.3   Enhance Sustainability of Projects and Programs

Sustainable projects (1) have adequate financing for construction, maintenance, and operations, (2) have a fair distribution of costs and benefits to society, and (3) can continue indefinitely without significant depletion of resources or disruption of the environment. Programs initiated in one era may ultimately prove to be nonsustainable because of financial, social, or environmental problems. Changes in technology may make a system obsolete, changes in social norms may make a system inadequate, and changes in scientific knowledge may reveal unacceptable environmental problems with a system.

16.2.4   Model Performance

Engineering-based models can be developed and used to assess the impacts of proposed alternatives on various aspects of system performance: cost, capacity, service, safety, environmental impacts, and economic impacts. Models of varying complexity can be used at different stages of the process; relatively simple models can be extremely helpful when different options have markedly different impacts on performance. Models can be structured to illustrate many important economic concepts, including productivity, pricing and competition, utility, and multiplier effects.

16.2.5   Assess Performance Using a Political Process

Since there are multiple aspects of performance, it is generally impossible to create a single measure of performance that provides a readily accepted, objective basis for evaluation. Documenting the impacts of various alternatives on performance can be done objectively, but assessing the importance of each aspect of performance and deciding which alternative is best cannot be objectively done. Assessment is ultimately a political matter that, in a democratic society, requires input from all those who will be affected by the project. Workshops with stakeholders, public hearings, and input from a variety of experts will help determine which options are worth pursuing. In particular, remember that cost and financial success are always relevant, but they are never everything. Environmental impacts, social justice, aesthetics, and other societal concerns must be considered, and these concerns may dominate project evaluation and selection.

> *The need is to subordinate economic to aesthetic goals—to sacrifice efficiency, including the efficiency of organizations, to beauty. Nor must there be any nonsense about beauty paying in the long run. It need not pay. It is through the state that the society must assert the superior claims of aesthetic over economic goals and particularly of environment over cost.*
> John Kenneth Galbraith, "Liberty, Happiness and the Economy," *The Atlantic Monthly*
> (June 1967): 521–26.

16.2.6 Develop Scenarios

Infrastructure projects have a very long life—so be sure to consider the types of changes that might occur over a period of 20–50 years. Decisions based solely upon current assumptions and perspectives concerning prices, economic conditions, technology, environmental impacts, societal needs, and government regulations may soon be regretted. The past 50 years have seen major increases in energy cost, unbelievable advances in communications and computing, development of new construction materials, heightened societal concerns for environment impacts and sustainability, political restructuring of Europe, and economic resurgence of Asia. What will the next 20–50 years bring? How will governments respond to the threats of global warming, to the need to reduce dependence upon fossil fuels, to the continuing disparity between rich and poor nations? Will energy become ever more expensive, or will there be breakthroughs in renewable resources?

16.3 KEY LESSONS FOR PART II: COMPARING ECONOMIC AND FINANCIAL IMPACTS OVER THE LIFE OF PROPOSED PROJECTS

16.3.1 Discounting Future Values

In freshman physics, knowing calculus and remembering that force equals mass times acceleration $(F = MA)$ will go a very long way. Likewise, in engineering economics, you can go a long way simply by knowing how to use a spreadsheet and remembering this formula:

$$\text{Present Value} = \text{Future Value}/(1 + i)^t \qquad \text{(Eq. 16.1)}$$

16.3.2 Equivalence of Cash Flows

Given a discount rate i, it is possible to use Eq. 16.1 repeatedly to convert any expected time stream of financial or economic costs and benefits into an equivalent present value. You can also work in the other direction to find the future value of any stream of financial or economic costs and benefits. In particular, you can use trial and error to find an annuity over any desired number of periods that equals any desired present or future value. Chapter 7 develops the equivalence relationships among present value, annuities, and future values. Appendix A includes tables of the equivalence factors for selected time periods and discount rates.

16.3.3 Importance of the Discount Rate

The discount rate plays a major role in project evaluation because it determines the relative importance of present and future values. With higher discount rates, future costs and benefits become less important, and it is harder to justify large projects with long-term benefits.

16.3.4   Choice of a Discount Rate

In general, the discount rate should be equal to the minimum acceptable rate of return of the individual or organization who is evaluating the time stream of costs and benefits. The choice of a discount rate depends upon the perspective of the individual or organization. Four distinct cases are worth comment.

Banks and other potential investors in a project must determine whether their share of the project's cash flows justify their investment. From their perspective, the choice of a discount rate is a market-driven financial decision based on three factors:

1. The real return that can be obtained from a very safe investment
2. The average annual rate of inflation
3. The perceived risk associated with a project or the company or agency that is funding the project

The higher the perceived risk, the more that future cash flows will be discounted. In other words, the higher the perceived risk, the less cash that an entrepreneur or a developer will be able to raise from banks or investors for a project.

For companies and entrepreneurs, the choice of a discount rate is based not so much on market factors as on their own perceptions of the risks associated with the project and their experience and opportunities for undertaking other projects. They must consider three internal factors:

1. Their weighted average cost of capital
2. The rates of return they can obtain on other projects
3. Their assessment of risks

For governments, the discount rate should be based on a rate that reflects the opportunity costs to taxpayers. In the United States, the Government Accountability Office (GAO) establishes the discount rate to be used for federal projects (generally in the range of 7–8%).

Agencies that are created to serve specific needs, such as housing for the poor or the elderly, may raise funds by selling tax-exempt bonds at very low interest rates. Since these agencies are designed to meet specific societal goals, and since they are not using taxpayer money, they can use their cost of capital (the interest on their tax-free bonds) as a discount rate.

16.3.5   Maximizing Net Present Value

The first step in evaluating alternative projects often involves calculating the net present value (NPV) of the project's net financial benefits. If a public agency is doing the analysis, then economic benefits (including any social or environmental impacts that can be expressed in monetary terms) may be included in addition to the financial benefits directly associated with the project. There are four things to remember:

1. Any project with positive NPV is worth further consideration.
2. If projects are evaluated in terms of the equivalent future value or an equivalent annuity, the rankings will be the same as obtained in ranking by NPV. Therefore, it makes sense to use future values or annuity values if they are more convenient to work with than present values.
3. Among competing projects, the one with the highest NPV is the best, at least from a financial or economic perspective.
4. Other factors may actually determine which project is best, especially for publicly funded projects with objectives that are difficult to express in monetary terms or projects that have significant externalities.

16.3.6   Internal Rate of Return

Many private companies use the internal rate of return (IRR) to rank projects. The IRR is the discount rate for which a project's net present value is zero. If the IRR is greater than the company's hurdle rate, then the project will be approved. When evaluating independent projects, the IRR approach will identify as worthwhile the same projects that would be identified by using NPV, assuming that the hurdle rate is used as the discount rate.

16.3.7   Mutually Exclusive Projects: Justifying Incremental Costs Using IRR

For mutually exclusive projects, the IRR method cannot be used to correctly identify the best project. This is because a small project with a high IRR might actually have much less benefit than a larger project. To deal with this problem, the financial or economic evaluation should begin with the project with the least investment requirements. If that project has an acceptable IRR, then it becomes the base project. If that project does not have an acceptable IRR, then consider the other options, beginning with the project with the next-higher investment. The first project with an acceptable IRR becomes the base project. To justify a higher investment, the incremental benefits must justify the incremental cost (i.e., the incremental IRR must be above the hurdle rate). This is an important point to remember because the best of a set of mutually exclusive projects might not have the highest IRR (but it will have the highest net present value—that is why it is the best!)

16.3.8   Cost Effectiveness

In many public projects, it is either difficult or impossible to place a price on significant portions of anticipated costs or benefits. If the benefits can be measured, and if the costs can be expressed in financial terms, then it is at least possible to measure the cost effectiveness of various alternatives. The best options will have the lowest cost per unit of benefit. Whether the best option is worth pursuing will be a subjective decision to be decided by the developer, entrepreneur, board of directors of a private company, or the officials responsible for a government decision.

16.3.9   Rules of the Game and Their Effects on Cash Flows

Rules of the game include accounting rules, taxation, zoning, safety regulations, environmental restrictions, and building codes. These rules are relevant to project evaluation because they may limit what can be built, and they may change the cash flows associated with the project. Legislatures often adjust the rules in order to promote economic development or to encourage private investment that will address public goals. The financial or economic analysis can be adjusted to reflect changes in the rules of the game, so that it is possible to understand how changes in depreciation, tax credits, or the density allowed by zoning will affect the cash flows associated with a project.

16.4   KEY LESSONS FOR PART III: DEVELOPING PROJECTS AND PROGRAMS TO DEAL WITH PROBLEMS AND OPPORTUNITIES

16.4.1   At First, Anything Is Possible

The greatest challenges and the greatest opportunities occur in the earliest stages of project evaluation, when needs are poorly defined and options are poorly understood. Figuring out what the problem really is, defining the problem in a way that invites diverse solutions, and responding to the problems effectively and creatively can be extremely rewarding, in terms of the success of whatever is done as well as the intellectual satisfaction of those involved in the process.

16.4.2   Intuition and Expert Judgment Lead the Analysis

Good solutions seldom come straight from a textbook, nor do they come from using complex techniques of operations research that claim to find the optimal solution to a problem. Good solutions arise from a deep understanding of the problems or needs and a clear idea of what might be done. Experts who have developed a comprehensive conceptual framework for addressing system perform-ance often can contribute a great deal to the early stages of project evaluation, especially with regard to the kinds of technical approaches possible. Users, abutters, and members of the public can contribute their understanding of needs, identify issues that are important, and—in the aggregate if not always individually—help in applying some common sense to the discussion. Brainstorming, systematic analysis, sensitivity analysis, and scenarios can all be useful in eliciting ideas and in determining what approaches might work best in dealing with a problem.

16.4.3   Public-Private Partnerships May Be Useful

The public and private sectors have different perspectives on problems and opportunities and different skills related to project evaluation. Public-private partnerships have the potential for enhancing what can be done by either sector working alone. Here are four reasons for pursuing such partnerships:

1. The project requires the complementary strengths of the private and public sectors.
2. There are insufficient benefits for either sector to undertake a project, but the combined public and private benefits make the project worthwhile.
3. By working with the private sector, public agencies can undertake more projects.
4. Governments may contribute to projects whose financial benefits are insufficient to attract private financing, but whose economic benefits are sufficient to justify public participation.

In general, the public sector has greater ability to consider multiple criteria in assessing projects, a longer time horizon, and more interest in societal and environmental impacts. The private sector has the ability to attract capital for what may be risky projects and, in some cases, can assemble highly qualified teams to manage and construct complex projects. Seeking private participation in public projects is something that deserves careful consideration of the potential benefits; it is not something to be based on ideology. Neither sector is automatically more or less efficient, nor is either sector more or less subject to corrupting influences.

16.4.4   Anticipate Risks and Uncertainties

The success of any complex project hinges upon many unknown or uncertain factors, and many risks could disrupt a project. Research, market surveys, and coordination with or among public agencies may help in understanding the risks and reducing the uncertainties. Models of system performance can be used with various sets of assumptions to try to understand which factors are most critical to project success or failure and what kinds of user responses or technological innovations might affect the project.

Safety and security are major concerns for any infrastructure project. The possibility of personal injuries or fatalities must be factored into the design, construction process, and operating procedures for an infrastructure-based system. Though it is impossible to eliminate all risks, it is possible to understand risks and use a rational approach to reducing risks. Having a clear plan for ensuring safety that is based on logical risk assessment will be helpful to a company or agency, both in reducing risk and in responding to incidents if and when they happen.

16.4.5   Project Management Is the Key to Efficient Implementation

Effective project management will enable projects to be completed as designed, on time, and within budget. Project management tools include work breakdown structures, network diagrams, schedules, resource allocation models, cost estimation techniques, and budgets. This textbook introduces many of the terms and concepts used in project management, but it does not attempt full treatment of a topic that deserves a separate text.

16.4.6   Complex Projects Will Need Champions

The gestation period for very large projects may be measured in decades, and project management is not confined to the final stages of construction. Gaining approval for a project such as Boston's Big Dig or programs like the Interstate Highway Program may require seemingly endless political wrangling and nearly impossible coordination among local, state, and federal officials. Projects and programs therefore will need champions who are willing and able to fight the bureaucratic, political, and legal battles that must be waged. The most effective leaders will incorporate social and environmental elements into the initial design and be able to use the environmental impact assessment process as a means of enhancing projects and building public support for them. When finally approved, projects and programs will require leaders who can lay out a general strategy for implementation, employ managers who negotiate and supervise the actual construction contracts, and establish a path for the transition from construction to operation.

16.4.7   Most Projects Contribute to Existing Systems

Most proposed projects can be viewed as potential enhancements to an existing infrastructure-based system that deals with transportation, water resources, energy, or some other societal need. Effective project evaluation therefore requires understanding of how such systems are created, how they evolve to meet changing social and economic conditions, and how they eventually give way to obsolescence or to new technologies. Different types of projects and different kinds of issues are encountered in each stage of system evolution. In early stages, those involved struggle to determine how best to use new technologies, how to structure facilities or networks, and in general how to become more effective and more efficient. In later stages, there will be a need to adjust the size and structure of the systems to adjust to new technologies or new kinds of competition. At the end, the challenge may be to grow old gracefully and pass away.

16.4.8   The Future Might Not Be What You Expect

The future does not always resemble the past, and it can be very risky to expect trends to continue indefinitely over the life of a proposed project. By using scenarios and sensitivity analysis, it is possible to figure out what could happen if assumptions about demand, costs, competition, or the worldwide economy prove incorrect. There were plenty of warnings that the real estate bubble was about to burst

---

Box 16.1   Projects in the News

In 2006, the stock market was reaching new highs on a daily basis, the real estate market was booming, people were living off home equity loans and flipping condos, and real estate magnates rushed forward with ever more grandiose projects. Too bad that so few considered the possibility the bubble might burst. "Who could have known," you might have asked in September 2008, when declining housing prices nearly triggered the collapse of the world's banking system. Well, some people did know; and some like Michael Hudson even wrote about it, backing up his clear warning with 20 excellent graphs indicating why our "debt-ridden economy" was headed into "Japan-style stagnation or worse":

With the real estate boom, the great mass of Americans can take on colossal debt today and realize colossal capital gains—and the concomitant rentier life of leisure—tomorrow. If you have the wherewithal to fill out a mortgage application, you need never work again. What could be more inviting—or, for that matter, more egalitarian!

That's the pitch, anyway. The reality is that, although home ownership may be a wise choice for many people, this particular real estate bubble has been carefully engineered to lure home buyers into circumstances detrimental to their own best interests. The bait is easy money. The trap is a modern equivalent of peonage, a lifetime spent working to pay off debt on an asset dwindling in value.

Most everyone involved in the real estate bubble thus far has made at least a few dollars. But that is about to change. The bubble will burst, and when it does, the people who thought they would be living the easy life of a landlord will soon find out that what they really signed up for was the hard servitude of debt serfdom.

Michael Hudson, "The New Road to Serfdom," *Harper's Magazine* (May 2006): 40.

before it actually collapsed in 2007, leading to worldwide financial panic in 2009. Those who kept building condos, malls, and office buildings in hopes of making fantastic returns were among the biggest losers when the bottom fell out of the market.

16.4.9   Programs Need to Be Evaluated and Revised

Programs are established to simplify the approval, funding, and construction of projects that serve specified needs of society. Once a program is defined, clear criteria should be available for determining what projects can be funded by that program and how those projects should be constructed. If funding for a program is tied to particular taxes or user fees, then that program could continue for a very long time—even if the conditions and needs that led to the creation of the program change dramatically. It is best therefore to have a periodic process of program evaluation that provides an objective assessment of how well the program is working, whether the program is still needed, and whether unforeseen economic, environmental, or social consequences need to be mitigated.

16.5   FURTHER READING

"I love projects, don't you?" Luther Billis
(in South Pacific by Rogers & Hammerstein)

Opportunities for further reading are divided into the following categories:

- *Textbooks* (subsection 16.5.1): This section provides a short list of topics related to project evaluation rather than a list of the many available textbooks.
- *Books about Infrastructure Projects and Programs* (subsection 16.5.2): There are many interesting books about specific projects; they generally highlight the personalities, goals, technologies, trials and tribulations, and ultimate achievements. If read carefully, they will yield some insight into the project evaluation process.
- *Articles about Infrastructure Projects and Programs* (subsection 16.5.3): As with the books, the articles usually focus on the engineering achievements, the personalities, and the conflicts rather than the evaluation process, but they often are quite interesting.
- *Articles about Evaluation Methodology* (subsection 16.5.4): This section provides a small sample of the many articles about various aspects of methodology.

Any large project requires preparation of an environmental impact statement, and these statements (along with numerous commentaries and public reports) can be found online. The World Bank, the Asian Development Bank, and other funding agencies have many materials available online concerning specific projects and evaluation methodologies. Government agencies in the United States and other countries also have reports and materials related to project evaluation techniques and methodologies, many of them specific to certain types of projects.

16.5.1   Textbooks

Those interested in further technical information can find many textbooks that address the topics introduced in this textbook. The major methodological categories to consider are as follows:

- Engineering economics
- Microeconomics
- Macroeconomics
- City and regional planning
- Financial management
- Environmental engineering

Also available are courses and texts concerning each of the major infrastructure systems:

- Transportation systems
- Water resource systems
- Energy systems

Some schools offer subjects on narrower topics, such as "Big Projects," particular types of projects, or specific aspects of project financing and management.

16.5.2   Books about Infrastructure Projects and Programs

AASHTO. *Best Practices in Environmental Stewardship Competition*. Washington, DC: American Association of State Highway and Transportation Officials, 2004. "The intent of this competition was not only to showcase exemplary state DOT efforts at fostering environmental stewardship, but also to demonstrate the initiatives state DOTs are undertaking to institutionalize programs."

Al Naib, S. K. *London Docklands Past, Present and Future: An Illustrated Guide to History, Heritage and Regeneration*. Romford, Essex, UK: Research Books, 1994.

The docklands were originally constructed as port facilities in London for Great Britain's extensive international trade. As ships became larger, different types of facilities were required, and the docklands slid into decay. In the 1980s, a massive urban renewal effort converted the docklands into a variety of commercial and residential uses. The book combines concise history with many interesting photographs and maps.

Ambrose, Stephen E. *Nothing Like It in the World: The Men Who Built the Transcontinental Railroad 1863–69*. New York: Simon & Schuster, 2000.

A fascinating story of what it took—men, materials, and financing—to build a railroad across the deserts and mountains of the American West.

Bevis, Trevor. *Water, Water, Everywhere*. Chatteris, Cambridgeshire, UK: David J. Richards Printers and Stationers, 1992.

The 500-year story of the construction of canals and the use of windmills and pumps to drain the fens of East Anglia.

Carrels, Peter. *Uphill Against Water: The Great Dakota Water War*. Lincoln: University of Nebraska Press, 1999.

Local farmers and citizens fight to stop construction of massive irrigation project.

Clausen, Meredith L. *The Pan Am Building and the Shattering of the Modernist Dream*. Cambridge, MA: MIT Press, 2005.

The story of the issues and controversy surrounding the first of many post-WWII skyscrapers in Manhattan. "A conspicuous landmark and a testament to what many in New York felt should never have been built and should never be allowed to happen again . . . a social utopia based on the use of new industrial materials and new modes of production to generate new, efficient, clean-lined forms [was] displaced by the imperatives of a capitalist economy, and instead of the decent housing for growing urban populations modernists promised, flagship buildings for corporations were built" (pp. 386–87).

Conuel, Thomas. *Quabbin—The Accidental Wilderness* (rev. ed.). Amherst: University of Massachusetts Press, 1990.

The story of the creation of Quabbin Reservoir, which required the flooding of four towns in western Massachusetts to provide water for populations in the eastern part of the state.

DeBoer, David J. *Piggyback and Containers: A History of Rail Intermodal on America's Steel Highways*. San Marino, CA: Golden West Books, 1992.

Fredrich, A. S. *Sons of Martha—Civil Engineering Readings in Modern Literature*. Reston, VA: ASCE Press, 1989.

This book is pure fun!

Gordon, John Steele. *A Thread Across the Ocean: The Heroic Story of the Transatlantic Cable*. New York: Walker & Company, 2002.

Laying a cable across the Atlantic reduced the speed of news from weeks or months to seconds; after several failed attempts, the project was completed using the Great Eastern, the huge steam/sailing ship designed by I. K. Brunel.

Graham-Leigh, John. *London's Water Wars: The Competition for London's Water Supply in the Nineteenth Century*. London: Francis Boutle Publishers, 2000.

At the beginning of the nineteenth century, various water companies built competing systems for delivering running water to London neighborhoods. The situation led to many abuses of customers and legal battles, and eventually to recognition of the need to regulate water delivery as a public utility.

Greene, Julie. *The Canal Builders: Making America's Empire at the Panama Canal*. New York: Penguin Press, 2009.

Greene's unusual history of the Panama Canal presents the workers' view of this mammoth project.

Gutner, Tamar L. *Banking of the Environment: Multilateral Development Banks and Their Environmental Performance in Central and Eastern Europe*. Cambridge, MA: MIT Press, 2002.

An investigation into the ways that the World Bank and others balance economic, environmental, and social concerns in their attempts to promote development and reduce poverty in poor countries.

Halliday, Stephen. *The Great Stink of London: Sir Joseph Bazalgette and the Cleansing of the Victorian Metropolis*. Stroud, Gloucestershire, UK: Sutton Publishing, 1999.

The installation of sewers and the creation of the Thames embankment as a means of cleaning up the Thames in the mid-1800s: this book is as fascinating as its title!

Hughes, Thomas P. *Rescuing Prometheus: Four Monumental Projects That Changed the Modern World*. New York: Vintage Books, 1998.

The development of management systems for complex projects including the Central Artery/Tunnel—the "Big Dig"—in Boston and ARPANET, a precursor to the Internet.

Koeppel, Gerard. *Bond of Union: Building the Erie Canal and the American Empire*. Cambridge MA: Da Capo Press, 2009.

Larson, Erik. *The Devil in the White City: Murder, Magic, and Madness at the Fair That Changed America*. New York: Vintage Books, 2003.

A delightful history of mass murder at the time of the creation of the "White City" along the banks of Lake Michigan to host the World's Columbian Exposition in 1893.

Lewis, Tom. *Divided Highways: Building the Interstate Highways, Transforming American Life*. New York: Penguin Books, 1997.

McCullough, David. *The Great Bridge: The Epic Story of the Building of the Brooklyn Bridge*. New York: Simon & Schuster, 1972.

————. *The Johnstown Flood: The Incredible Story Behind One of the Most Devastating "Natural" Disasters America Has Ever Known*. New York: Simon & Schuster, 1968.

————. *The Path Between the Seas: The Creation of the Panama Canal, 1870–1914*. New York: Simon & Schuster, 1977.

McDonald, Frank, and Kathy Sheridan. *The Builders: How a Small Group of Property Developers Fuelled the Building Boom and Transformed Ireland*. Dublin: Penguin Ireland, 2008.

A portrait of the types of individuals whose decisions fueled the real estate bubble that burst in Ireland and around the world in 2007.

Newhouse, Elizabeth L., ed. *The Builders: Marvels of Engineering*. Washington, DC: National Geographic Society, 1992.

Great pictures and good overviews of major projects in all areas of civil engineering; a relatively inexpensive reference that captures the excitement of big projects, but has little detail concerning project evaluation.

Nye, David E. *Electrifying America: Social Meanings of a New Technology*. Cambridge, MA: MIT Press, 1990.

Okrent, Daniel. *Great Fortune: The Epic of Rockefeller Center*. London: Penguin Books, 2004.
How Rockefeller Center came to be developed during the depths of the Great Depression on underutilized land in central Manhattan.

Oppitz, Leslie. *Lost Railways of East Anglia*. Newbury, Berkshire, UK: Countryside Books, 2004.
Brief but detailed history of introduction first of the horse-drawn and later the electric tramway into the cities and towns of this region northeast of London. Many of the issues dealt with at that time remain central issues for modern transit operations.

Payne, Robert. *The Canal Builders: The Story of Canal Engineers Through the Ages*. New York: Macmillan, 1959.

Pellow, David Naguib. *Garbage Wars: The Struggle for Environmental Justice in Chicago*. Cambridge, MA: MIT Press, 2004.

Peters, Tom. *Building the Nineteenth Century*. Cambridge, MA: MIT Press, 1996.

Pierce, Patricia. *Old London Bridge: The Story of the Longest Inhabited Bridge in Europe*. London: Headline Book Publishing, 2001.
The 750-year history of a bridge that at one time was the retail center of London and the site of many trendy homes.

Pole, Graeme. *The Spiral Tunnels and the Big Hill: A Canadian Railway Adventure*. Vancouver: Altitude Publishing Canada Ltd., 1995.
The construction of the spiral tunnels that, when completed in 1909, reduced the ruling grade on Canadian Pacific's transcontinental line through the Rocky Mountains, enabling longer trains, faster speeds, and less expensive operations.

Portny, Stanley E., Samuel J. Mantel, Jack R. Veredith, Scott M. Shafer, Margaret M. Sutton, and Brian E. Kramer. *Project Management: Planning, Scheduling, and Controlling Projects*. New York: John Wiley & Sons, 2008.

Reisner, Marc. *Cadillac Desert: The American West and Its Disappearing Water*. New York: Penguin Books, 1993.
This book documents the struggles, by politicians and government officials in Los Angeles and elsewhere, to find and divert water for agriculture and cities.

Richmond, Peter. *Ballpark: Camden Yards and the Building of an American Dream*. New York: Simon & Schuster, 1993.
The construction of a new, old-style, urban ballpark that incorporated structures and designs from Baltimore's industrial past.

Ridgeway, James. *Powering Civilization: The Complete Energy Reader*. New York: Pantheon Books, 1982.
Ridgeway compiles readings about the various forms of energy and traces the extraction, transportation, and use of coal, oil, natural gas, nuclear power, and alternative energy sources. The readings provide compelling insights into the powerful forces affecting the exploitation of energy sources.

Rose, Mark H. *Interstate: Express Highway Politics, 1941–1956*. Lawrence: Regents Press of Kansas, 1979.
The politics that influenced the design, location and financing of the Interstate Highway System.

Sabbagh, Karl. *Skyscraper: The Making of a Building*. New York: Penguin Books, 1989.
The building of the 50-story Worldwide Plaza in New York City.

Schodek, Daniel L. *Landmarks in American Civil Engineering*. Cambridge, MA: MIT Press, 1987.
Short articles on more than 100 projects selected by the ASCE as notable achievements.

Standiford, Les. *Last Train to Paradise: Henry Flagler and the Spectacular Rise and Fall of the Railroad That Crossed an Ocean*. New York: Crown Publishers, 2002.
Construction of a railroad from Jacksonville to Key West, a spectacular feat that opened southern Florida to development and transformed Miami from a tiny port into a major resort destination.

Talese, Gay. *The Bridge*. New York: Walker & Company, 2003.
A history of the construction of the Verrazano-Narrows Bridge that focuses on the ironworkers and others who actually built it.

Tauranac, John. *The Empire State Building: The Making of a Landmark*. New York: Scribner, 1995.

Tsipis, Yanni K. *Images of America: Building the Mass Pike*. Charleston, SC: Arcadia Publishing, 2002.
One of the popular "Images of America" series, this is an annotated collection of photographs concerning the construction of the Mass Pike and its controversial extension into Boston; the author is a graduate of MIT and was both a student in and teaching assistant for Project Evaluation, the class that eventually led to this text.

Vance, James E. Jr. *The North American Railroad: Its Origin, Evolution, and Geography*. Baltimore and London: Johns Hopkins University Press, 1995.
A geographer's perspective on the development of the North American Railroad System.

Wood, F. J. *The Turnpikes of New England*. Pepperell, MA: Branch Line Press, 1997.
Reissue of 1919 classic; briefly describes every one of the nineteenth-century turnpikes that were authorized by the states, constructed by chartered companies, and financed by tolls.

Zimiles, Martha, and Murray Zimiles. *Early American Mills*. New York: Bramhall House, 1973.
A history of the construction of water-powered mills and mill towns throughout New England during the 1800s.

16.5.3   Articles about Projects

Acharya, Dharma. "Kathmandu-Hetauda in an Hour: Some Viable Alternatives." *Nepal Update 3*, no. 4 (1992).
This is a concise discussion of the potential costs and benefits of constructing a new, much less circuitous road to link Nepal and India; the paper uses rough estimates of unit costs for building roads and tunnels and for operating highway vehicles; there is a summary of financial, social, and political issues, plus some discussion concerning political issues.

Ardila, Arturo, and Gerhard Menchkhoff. "Transportation Policies in Bogota, Colombia: Building a Transportation System for the People." *Transportation Research Record* 1817 (2002): 130–36

Ball, Steven C. "Unconventional Expansion." *Civil Engineering* (April 2008).
The design, construction, and notable environmental features in the largest building to achieve LEED certification; also an example of delivering a project on time and on budget using a design/build team.

Boettner, Danita S., Don Koci, Darren L. Brown, and Bruce Allman. "Clean, Blend and Reuse." *Civil Engineering* (July 2009): 59–65, 86.
A $35 million remediation project aimed at cleaning up contaminated groundwater and to provide potable water to Hutchinson, KA.

Bourke, Michael R., Donald R. F. Harleman, Heidi Li, Susan E. Murcott, Gautam Narasimhan, and Irene W. Yu. "Innovative Wastewater Treatment in the Developing World." *Civil Engineering Practice* 17, no. 1 (2002): 25–34.

Breen, Cheryl, Jekabs Vittands, and Daniel O'Brien. "The Boston Harbor Project: History and Planning." *Civil Engineering Practice* 9, no. 1 (1994): 11–32.
Very good overview of the history, need, and options considered for the whole program.

Brocard, Dominique N., Brian J. Van Wheels, and Lawrence A. Williamson. "The New Boston Outfall." *Civil Engineering Practice* 9, no. 1 (1994): 33–48.
The engineering options for the new sewer system in Boston Harbor, with consideration of the geotechnical, water and pollution concerns.

Capano, Daniel E. "Chicago's War with Water: On Its Way to Pioneering Our Modern Sewer System, Chicago Survived Epidemics, Floods, and Countless Bad Days." *Invention & Technology* (Spring 2003): 51–58.

Curtis, Wayne. "Going with the Flow: Historic Dams Are Being Demolished or Vastly Altered to Allow Fish to Return to Their Historic Spawning Grounds. Is There Another Way?" *Preservation* (July/August 2003): 29–33.
Fish ladders are good for the fish, but they look awful next to historic dams and mills.

Deakin, Elizabeth. "Sustainable Transportation: U.S. Dilemmas and European Experiences." *Transportation Research Record* 1792 (2002): 1–11

Dornhelm, Rachel. "Beach Master: Coney Island has been world famous for 150 years, but who remembers that its beach is the revolutionary achievement of one embattled engineer?" *Invention & Technology* (Summer 2004): 43–48.

Drapeau, Raoul. "Pipe Dream: With Creative Engineering and Heroic Endurance, Freezing, Beleaguered Workers Pushed the Canol Pipeline through the Brutal Arctic Wilderness during World War II. But It Was a Project That Should Never Have Been Started." *Invention & Technology* (Winter 2002): 25–35.

Fox, Richard D., William F. Callahan, and Walter G. Armstrong. "Effective Facilities Planning Ensured a Successful Boston Harbor Cleanup." *Civil Engineering Practice 17*, no. 2 (2002): 25–34.

Griggs, Francis E.Jr. "The Panama Canal: Uniting the World for Seventy-Six Years." *Civil Engineering Practice*, 5(2), (Fall/Winter 1990): 71–90. (A 20-page synopsis of the "Path Between the Seas" that focuses on the trials and tribulations of building the canal.)

———. "Thomas W.H. Mosely and His Bridges." *Civil Engineering Practice* 12, no. 2 (1997): 19–38. One of the first to use iron for bridges, Mosely developed standard designs and worked with a prefab company to market railway and highway bridges at an advertised price per foot during the nineteenth century.

Grimm, Mike. "Floodplain Management." *Civil Engineering* (March 1998): 62–66.

This is a good, short example of a post audit. Because Fort Collins was a leader in the systems approach to flood control, the town escaped its 500-year flood with little property loss and only 5 deaths versus what likely would have been $5 million damage with nearly 100 fatalities if they had not implemented their flood control projects. (See subsection 13.3.6 of this text.)

Grunwald, Michael. "Everglades: The Nation's Storied Wetland Is the Focus of the World's Largest Environmental Restoration Project. But Will That Be Enough?" *Smithsonian* (March 2006): 46–57.

Hall, Sir Peter. "Speed Rail Comes to London." *Traffic Technology International* (Dec. 2001/Jan. 2002): 25–31.

A brief introduction to the high-speed rail link that had to be created between London and the Channel Tunnel.

Hecker, George E. "Hydraulic Engineering in China." *Civil Engineering Practice* 6, no. 1 (1991): 7–24. An interesting perspective on the magnitude of China's major water resource projects.

Heppenheimer, T. A. "Nuclear Power: Engineers Finally Made it Safe, but They Couldn't Make It Cheap." *Invention & Technology* (Fall 2002): 46–56.

Holly, H. Hobart. "The Charles River Basin." *Civil Engineering Practice* 8, no. 2 (1993): 77–80.

———. "Lowell Water Power System." *Civil Engineering Practice* 1, no. 2 (1986): 141–45.

———. "The Middlesex Canal." *Civil Engineering Practice* 7, no. 2 (1992): 104–6. "It was the Middlesex Canal that proved, through low freight rates and expanded traffic, that canal transportation in the U.S. was practical and economical."

Izaguirre, Ada Karina. "Private Infrastructure: A Review of Projects with Private Participation, 1990–2001." *Public Policy for the Private Sector* 250. Washington, DC: The World Bank Group, October 2000.

Johnson, Christopher. "The Law That Saved the Appalachians." *Appalachia* (June 2005): 88–97. The history of the Weeks Act, which led to the creation of the national forest system.

Joseph, Patrick. "The Battle of the Dams: Those Who Think Some of Our Rivers Are a Dammed Shame Argue for the Structures to Come Down." *Smithsonian* (November 1998): 48.

Kain, John F., and Zvi Liu. "Secrets of Success: Assessing the Large Increases in Transit Ridership Achieved by Houston and San Diego Transit Providers." *Transportation Research Part A* (1999): 601–24.

Kaplin, John, and Geoffrey Hughes. "Construction of Underground Facilities for the Narragansett Bay Combined Sewer Overflow Program, Phase I." *Civil Engineering Practice* (Fall/Winter 2008): 7–32.

Koeppel, Gerard. "A Struggle for Water." *Invention & Technology* (Winter 1994): 19–30.

The 70-year effort required to complete New York City's first major water system, which was authorized in 1774.

Langdon, Virgil L., Jr., Michael R. Hilliard, and Ingrid K. Busch. "Future Utilization and Optimal Investment Strategy for Inland Waterways." *Transportation Research Record* 1871 (2004): 33–41.
Optimizing investments over an entire system under a series of forecast scenarios taking into account scheduled and unscheduled closures that might affect the inland waterways.

Martland, C., R. Gakenheimer, K. Kruckemeyer, T. Lee, M. Murga, F. Salvucci, D. Shi, D. Sze, S. Gongal, G. Flood, R. Imai, and J. Won. "Linking the Delta: Bridging the Pearl River Delta." Lai Chi Kok, Hong Kong: The 2022 Foundation, 2003.
See Section 11.6 of this text.

Morrall, J. F., and T. M. McGuire. "Sustainable Highway Development in a National Park." *Transportation Research Record* 1702 (2000): 3–10.
Examples of sustainable highway development in Canada's Rocky Mountain National Parks, including fencing that directs animals to crossings constructed at intervals over the highway.

Mueller-Lust, Andrew. "Crystal Clear." *Civil Engineering* (December 2008): 38–71.
The design, construction, and notable environmental features of the Bank of America Tower at One Bryant Park, one of the first skyscrapers to achieve LEED platinum certification. (See Example 15.7 in this text.)

O'Neill, Tom. "Curse of the Black Gold: Hope and Betrayal in the Niger Delta." *National Geographic* (February 2007): 88–117.
Profits from oil production in Nigeria have not reached the people living near the oil fields; extreme poverty, destruction of fishing grounds, pollution, and general disillusionment have motivated insurgents willing to use violence and disruption of the oil flows if their call for local control of resources is not met.

Pennington, Robert A., Kristie A. Gersley, Anthony Gagliostro, Daniel T. Eagan, Alvin L. Zach, and John T. George. "Saving a City's Sewers." *Civil Engineering* (December 2008): 61–68.
Description of a 20-year effort to inspect and rehabilitate Newark's 68 miles of brick sewers that were originally constructed in the nineteenth century. (See Example 15.6 in this text).

Peters, Tom. "How Creative Engineers Think." *Civil Engineering* (March 1998): 48–51.
Peters uses historical examples including Brunel's bridges, the Crystal Palace, the Palm House at Kew Gardens, and the Thames tunnel to illustrate what he calls "technological thinking," a combination of the linear, objective, scientific method and the subjective matrix method. In every case, the project required new thinking and new technology to succeed.

Picardi, E. Alfred. "High Rise Building and 911." Unpublished article, May 5, 2002.
The structural engineer of record for some of the tallest buildings in the world questions the continued need for megastructures in the United States.

Powderham, Alan J. "Heathrow Express Cofferdam: Innovation & Delivery through the Single-Team Approach—Part I: Design and Construction." *Civil Engineering Practice* (Spring/Summer 2003): 25–50.
"Partnering, value and risk management, and technical innovation rescued this project from substantial delay and cost overruns following a major setback during construction.

Reich, Leonard S. "The Dawn of the Truck: It Caught on Much More Slowly than the Automobile, Partly Because of the Expense, Partly Because Horses Did a Good Job, and Partly Because People Had to Figure Out Just What It Was and Could Do." *Invention & Technology* (Fall 2000): 18–24.

Reid, Robert L. "Under One Green Roof." *Civil Engineering* (March 2009).
The new California Academy of Sciences building in San Francisco houses a museum, an aquarium, a planetarium, and scientific research operations in a vast structure designed for sustainability. Perhaps most notable is its 2.5-acre undulating roof, which is covered with vegetation, and its stunning use of windows and interior open space.

Rosales, M., and F. Gottemoeller. "Urban Design Considerations for the New Woodrow Wilson Memorial Bridge: Competition-Winning Design for Metropolitan Washington, DC." *Transportation Research Record* 1740 (2000): 104–7.

Scheader, Edward C. "The New York City Water Supply: Past, Present and Future." *Civil Engineering Practice* (Fall 1991): 7–20.

A very readable overview of New York City's water supply history, written by the director of the Department of Environmental Protection.

Schipper, Lee. "Sustainable Urban Transport in the Twenty-First Century." *Transportation Research Record* 1792 (2002): 12–19.

Schipper confronts the issues related to long-term problems with the automobile and what must be done to achieve sustainable transportation for the future, especially in very large urban areas in developing countries. This paper provides a clear perspective on what might be called "hard sustainability" (i.e., the basic environmental problems related to global warming, air quality, and dependence upon fossil fuel).

Schmutz, Armin. "Inside the World's Longest Tunnel." *Trains* (2004): 40–47.

A new 35-mile-long rail tunnel under the Alps improves transportation through Switzerland.

Sheridan, Thomas E. "The Big Canal: The Political Ecology of the Central Arizona Project." In *Water, Culture and Power*, edited by John M. Donahue and Barbara R. Johnston. Washington, DC: Island Press, 1998.

Shumway, Laurence W. "Making the Most of Transportation Infrastructure: MBTA's South Station Intermodal Transportation Center." *Civil Engineering Practice* (Spring/Summer 2001): 67–74.

Sipes, James L., and Ron Blakemore. "Aesthetics in the Landscape: How Nevada and Other States Are Integrating Aesthetics into Transportation Projects." *TR News* (February 2007): 3–12.

Tsipis, Yanni, "Central Corridor Highway Planning in Boston, 1900–1950: The Long Road to the Old Central Artery." *Civil Engineering Practice* (Fall/Winter 2003): 33–52.

Vic, Thomas E., and Mark Surwillo. "Small Footprint, Big Promise." *Civil Engineering* (February 2008): 66–85.

The use of new technologies to reduce the space needed for a more effective sewerage treatment plant. (See Example 15.8 in this text.)

World Bank. "Integrated Coastal Zone Management Strategy for Ghana." *World Bank Findings 113* (June 1998).

This example of the many studies carried out by the World Bank shows how a qualitative process led to the identification and prioritization of environmental concerns and recommendations for management strategies to deal with these concerns.

Zoellner, Tom. "Oil and Water: The Adventures of Getting One from Deep Beneath the Other." *Invention & Technology* (Fall 2000): 44–52.

16.5.4   Articles about Project Evaluation Methodology

Akitoby, Bernardin, Richard Hemming, and Gard Schwartz. *Public Investment and Public-Private Partnership*. Washington, DC: International Monetary Fund, 2007.

Antle, J. M. "Infrastructure and Aggregate Agricultural Productivity: International Evidence." *Economic Development and Cultural Change* 31, no. 3 (1983): 609–19. This paper addresses the economic benefits of investments in agriculture.

"The Arsenic Controversy." *Regulation* (Fall 2001): 42–54.

A series of articles with different perspectives regarding the danger posed by different concentrations of arsenic in drinking water.

Bamberger, M., and E. Hewitt. *Monitoring and Evaluating Urban Development Programs: A Handbook for Program Managers and Researchers*. World Bank Technical Paper 53. Washington, DC: The World Bank, 1986.

Belli, P., J. Anderson, H. Barnum, J. Dixon, and J.-P. Teng. *Handbook on Economic Analysis of Investment Operations*. Washington, DC: The World Bank, 1998.

Bishop, Richard C., and Michael P. Welsh. "Contingent Valuation: Incorporating Nonmarket Values." In *Better Environmental Decisions: Strategies for Governments, Businesses and Communities*, edited by K. Sexton, A. A. Marcus, K. W. Easter, and T. D. Burkhardt, chap. 9. Washington, DC: Island Press, 1999.

Burchell, Robert W., and Catherine C. Galley. "Projecting Incidence and Costs of Sprawl in the United States." *Transportation Research Record* 1831 (2003): 150–58.

Delatte, Norbert. "Learning from Failures: It Is Imperative That Engineers Not Only Learn about the History of Failures but Also Their Contexts as Well as Their Repercussions." *Civil Engineering Practice* 21, no. 2 (2006): 21–38.

Dorfman, Robert. "Why Benefit-Cost Analysis Is Widely Disregarded and What to Do about It." *Interfaces* 26, no. 5 (1996): 1–6.

Easter, K. William, Nir Becker, and Sandra O. Archibald. "Benefit-Cost Analysis and Its Use in Regulatory Decisions." In *Better Environmental Decisions: Strategies for Governments, Businesses and Communities*, edited by K. Sexton, A. A. Marcus, K. W. Easter, and T. D. Burkhardt, chap. 8. Washington, DC: Island Press, 1999.

Economics for the Environment Consultancy. *Review of Technical Guidance on Environmental Appraisal*. Report prepared for the UK Department of the Environment, Transport and the Regions, April 30, 1999.

Eschenbach, Ted G., and Alice E. Smith. "Violating the Identical Repetition Assumptions of EAC." *1990 International Industrial Engineering Conference Proceedings*. Institute of Industrial Engineers, 1990.

"Fundamentally Equivalent Annual Cost is a robust measure regardless of the alterations from the original project and its identical repetition assumption. . . . In reality, projects often do not repeat, but are rarely divested during their first life and dramatic cost change occurs only in the long run."

"Final Report of the Engineering Economy Subcommittee (Z94.5)." In *Industrial Engineering Terminology, ANSI Standard Z94.0—1982*, chap. 5. Norcross, GA: Industrial Engineering and Management Press, 1983.

Flyvbjerg, Bent, Mette Skamris Holm, and Søren Buhl. "Underestimating Costs in Public Works Projects: Error or Lie?" *Journal of the American Planning Association* 68, no. 3 (2002): 279–95.

"Based on a sample of 258 transportation infrastructure projects . . . it is found with overwhelming statistical significance that the cost estimates used to decide whether such projects should be built are highly and systematically misleading."

Houskamp, Melissa, and Nicola Tynan. "Private Infrastructure: Are the Trends in Low-Income Countries Different?" *Public Policy for the Private Sector* 216. The World Bank Group, October 2000.

Jaafari, Ali, and Kitsana Manivong. "Synthesis of a Model for Life-Cycle Project Management." *Computer-Aided Civil and Infrastructure Engineering* 15 (2000): 26–38.

Life-cycle project management shifts the focus from the cost, time, and quality of construction to broader objectives including return on investment, facility operability, and life-cycle integration.

Lee, Douglass B. "Fundamentals of Life-Cycle Cost Analysis." *Transportation Research Record* 1812 (2002): 203–10.

Litman, Todd. "Transportation Market Reforms for Sustainability." *Transportation Research Record* 1702 (2000): 11–20.

Meyer, John R., and Mahlon R. Straszheim. *Techniques of Transport Planning, Vol. 1: Pricing and Project Evaluation*. Washington, DC: Brookings Institution, 1971.

Mobasheri, Fred, Lowell H. Orren, and Fereidoon P. Sioshansi. "Scenario Planning at Southern California Edison." *Interfaces* 19, no. 5 (1989): 31–44.

In planning how best to expand capacity, the electric company considered 12 scenarios that might affect supply or demand for electricity (See Section 11.4.3 of this text.).

Morimoto, Risako, and Chris Hope. "An Extended CBA of Hydro Projects in Sri Lanka." *Working Paper* 15/2001. Cambridge, UK: The Judge Institute of Management, 2001.

This paper includes some sensitivity analysis of the NPV to various assumptions concerning discount rates. The project appears to require substantial future benefits to be justified, so the discussion of what discount rate to use is critical to the project achieving a positive NPV. The authors note that the choice of a discount rate is a political, not an economic decision—despite all the economic interest in the issue!

Mostashari, Ali, Joseph M. Sussman, and Stephen R. Connors. "Design of Robust Emission Reduction Strategies for Road-Based Public Transportation in Mexico City, Mexico: Multiattribute Trade-Off Analysis for Metropolitan Area." *Transportation Research Record* 1880 (2004).
Illustrates the use of scenarios and systems models in evaluating strategies for improving air quality.

Munasinghe, M. *Environmental Economics and Sustainable Development*. World Bank Environmental Paper 3. Washington, DC: The World Bank, Environmentally Sustainable Development Department, 1993.

Murcott, Susan. "Co-Evolutionary Design for Development: Influences Shaping Engineering Design and Implementation in Nepal and the Global Village." *Journal of International Development* 19 (2007): 123–44.

Nash, Christopher A. "A Cost-Benefit Analysis of Transport Projects." In *Efficiency in the Public Sector: The Theory and Practice of Cost-Benefit Analysis*, edited by Alan Williams and Emilio Giardina, 83–105. Aldershot, UK: Edward Elgar, 1993).

National Council on Public Works Improvement. *Fragile Foundations: A Report on America's Public Works*. Final Report to the President and Congress. Washington, DC: U.S. Government Printing Office, February 1988.

Neuman, Timothy R., Marcy Schwartz, Leofwin Clark, and James Bednar. *A Guide to Best Practices for Achieving Context Sensitive Solutions*. National Cooperative Highway Research Program Report 480. Washington, DC: Transportation Research Board of the National Academies, 2002.

Office of Management and Budget. *Guidelines and Discount Rates for Benefit-Cost Analysis of Federal Programs*. Circular No. A-94. Washington, DC: Executive Office of the President, October 29, 1992.
This is an example of a government agency issuing guidelines for conducting benefit-cost analyses.

Ozbay, Kaan, Dima Jawad, Neville A. Parker, and Sajjad Hussain. "Life-Cycle Cost Analysis: State of the Practice versus State of the Art." *Transportation Research Record* 1864 (2004): 62–70.

Schwartz, Marcy. "Technologies to Improve Consideration of Environmental Concerns in Transportation Decisions," *NCHRP Research Results Digest* 304, Transportation Research Board, June 2006.

Schwartz, Peter. *Art of the Long View: Paths to Strategic Insight for Yourself and Your Company*. New York: Doubleday, 1996.

Small, K. A. "Project Evaluation." In *Project Evaluation: Essays in Transportation Economics and Policy*, chap. 5. Washington, DC: Brookings Institution Press, 1999.

Stewart, Theodor J. "Thirsting for Consensus: Multicriteria Decision Analysis Helps Clarify Water Resources Planning in South Africa." *OR/MS Today* (April 2003): 30–34.

Transportation Research Board. *Integrating Sustainability into the Transportation Planning Process*. Conference Proceedings 37. Washington, DC: Transportation Research Board of the National Academy of Sciences, 2005.

Vanclay, Frank, and Daniel A. Bronstein, eds. *Environmental and Social Impact Assessment*. New York: John Wiley & Sons, 1995.

Winner, Langdon. *Autonomous Technology: Technics-Out-of-Control as a Theme in Political Thought*. Cambridge, MA: MIT Press, 1977.

Yoshizumi, Steven A., and F. David Freytag. "Major Investment Studies—Hit or Miss? Southern California Experience." *Transportation Research Record* 1702 (2000): 83–91.

Zogby, John. *AASHTO Strategic Highway Safety Plan—Case Studies*. National Cooperative Highway Research Plan, *Research Results Digest 265* (March 2002).
How states are establishing plans to improve highway safety.

Zurnkeller, Dirk, Bastian Chlond, and Wilko Manz. "Infrastructure Development in Germany under Stagnating Demand Conditions." *Transportation Research Record* 1864 (2004): 121–28.

Appendix A

Equivalence Tables

Summary

Discount Rate: 0.25%

N	[P/F,i%,N]	[A/P,i%,N]	[F/A,i%,N]
6	0.9851	0.1681	6.038
12	0.9705	0.0847	12.166
18	0.9561	0.0569	18.388
24	0.9418	0.0430	24.703
30	0.9278	0.0346	31.113
36	0.9140	0.0291	37.621
42	0.9004	0.0251	44.226
48	0.8871	0.0221	50.931

Discount Rate: 0.50%

N	[P/F,i%,N]	[A/P,i%,N]	[F/A,i%,N]
6	0.9705	0.1696	6.076
12	0.9419	0.0861	12.336
18	0.9141	0.0582	18.786
24	0.8872	0.0443	25.432
30	0.8610	0.0360	32.280
36	0.8356	0.0304	39.336
42	0.8110	0.0265	46.607
48	0.7871	0.0235	54.098

Discount Rate: 0.75%

N	[P/F,i%,N]	[A/P,i%,N]	[F/A,i%,N]
6	0.9562	0.1711	6.11
12	0.9142	0.0875	12.51
18	0.8742	0.0596	19.19
24	0.8358	0.0457	26.19
30	0.7992	0.0373	33.50
36	0.7641	0.0318	41.15
42	0.7306	0.0278	49.15
48	0.6986	0.0249	57.52

Discount Rate: 1.00%

N	[P/F,i%,N]	[A/P,i%,N]	[F/A,i%,N]
6	0.9420	0.1725	6.152
12	0.8874	0.0888	12.683
18	0.8360	0.0610	19.615
24	0.7876	0.0471	26.973
30	0.7419	0.0387	34.785
36	0.6989	0.0332	43.077
42	0.6584	0.0293	51.879
48	0.6203	0.0263	61.223

Discount Rate: 1.25%

N	[P/F,i%,N]	[A/P,i%,N]	[F/A,i%,N]
6	0.9282	0.1740	6.191
12	0.8615	0.0903	12.860
18	0.7996	0.0624	20.046
24	0.7422	0.0485	27.788
30	0.6889	0.0402	36.129
36	0.6394	0.0347	45.116
42	0.5935	0.0307	54.797
48	0.5509	0.0278	65.228

Discount Rate: 1.500%

N	[P/F,i%,N]	[A/P,i%,N]	[F/A,i%,N]
6	0.9145	0.1755	6.23
12	0.8364	0.0917	13.04
18	0.7649	0.0638	20.49
24	0.6995	0.0499	28.63
30	0.6398	0.0416	37.54
36	0.5851	0.0362	47.28
42	0.5351	0.0323	57.92
48	0.4894	0.0294	69.57

Discount Rate: 1.75%

N	[P/F,i%,N]	[A/P,i%,N]	[F/A,i%,N]
6	0.9011	0.1770	6.269
12	0.8121	0.0931	13.225
18	0.7318	0.0652	20.945
24	0.6594	0.0514	29.511
30	0.5942	0.0431	39.017
36	0.5355	0.0377	49.566
42	0.4826	0.0338	61.272
48	0.4349	0.0310	74.263

Discount Rate: 2.00%

N	[P/F,i%,N]	[A/P,i%,N]	[F/A,i%,N]
6	0.8880	0.1785	6.308
12	0.7885	0.0946	13.412
18	0.7002	0.0667	21.412
24	0.6217	0.0529	30.422
30	0.5521	0.0446	40.568
36	0.4902	0.0392	51.994
42	0.4353	0.0354	64.862
48	0.3865	0.0326	79.354

Discount Rate: 2.25%

N	[P/F,i%,N]	[A/P,i%,N]	[F/A,i%,N]
6	0.8750	0.1800	6.35
12	0.7657	0.0960	13.60
18	0.6700	0.0682	21.89
24	0.5862	0.0544	31.37
30	0.5130	0.0462	42.20
36	0.4489	0.0408	54.57
42	0.3928	0.0371	68.71
48	0.3437	0.0343	84.87

Discount Rate: 2%

N	[P/F;i%,N]	[A/P;i%,N]	[F/A;i%,N]
5	0.9057	0.2122	5.2040
10	0.8203	0.1113	10.9497
15	0.7430	0.0778	17.2934
20	0.6730	0.0612	24.2974
25	0.6095	0.0512	32.0303
30	0.5521	0.0446	40.5681
35	0.5000	0.0400	49.9945
40	0.4529	0.0366	60.4020

Discount Rate: 3%

N	[P/F;i%,N]	[A/P;i%,N]	[F/A;i%,N]
5	0.8626	0.2184	5.3091
10	0.7441	0.1172	11.4639
15	0.6419	0.0838	18.5989
20	0.5537	0.0672	26.8704
25	0.4776	0.0574	36.4593
30	0.4120	0.0510	47.5754
35	0.3554	0.0465	60.4621
40	0.3066	0.0433	75.4013

Discount Rate: 4%

N	[P/F;i%,N]	[A/P;i%,N]	[F/A;i%,N]
5	0.8219	0.2246	5.4163
10	0.6756	0.1233	12.0061
15	0.5553	0.0899	20.0236
20	0.4564	0.0736	29.7781
25	0.3751	0.0640	41.6459
30	0.3083	0.0578	56.0849
35	0.2534	0.0536	73.6522
40	0.2083	0.0505	95.0255

Discount Rate: 5%

N	[P/F;i%,N]	[A/P;i%,N]	[F/A;i%,N]
5	0.7835	0.2310	5.5256
10	0.6139	0.1295	12.5779
15	0.4810	0.0963	21.5786
20	0.3769	0.0802	33.0660
25	0.2953	0.0710	47.7271
30	0.2314	0.0651	66.4388
35	0.1813	0.0611	90.3203
40	0.1420	0.0583	120.7998

Discount Rate: 6%

N	[P/F;i%,N]	[A/P;i%,N]	[F/A;i%,N]
5	0.7473	0.2374	5.6371
10	0.5584	0.1359	13.1808
15	0.4173	0.1030	23.2760
20	0.3118	0.0872	36.7856
25	0.2330	0.0782	54.8645
30	0.1741	0.0726	79.0582
35	0.1301	0.0690	111.4348
40	0.0972	0.0665	154.7620

Discount Rate: 7%

N	[P/F;i%,N]	[A/P;i%,N]	[F/A;i%,N]
5	0.7130	0.2439	5.7507
10	0.5083	0.1424	13.8164
15	0.3624	0.1098	25.1290
20	0.2584	0.0944	40.9955
25	0.1842	0.0858	63.2490
30	0.1314	0.0806	94.4608
35	0.0937	0.0772	138.2369
40	0.0668	0.0750	199.6351

Discount Rate: 8%

N	[P/F;i%,N]	[A/P;i%,N]	[F/A;i%,N]
5	0.6806	0.2505	5.8666
10	0.4632	0.1490	14.4866
15	0.3152	0.1168	27.1521
20	0.2145	0.1019	45.7620
25	0.1460	0.0937	73.1059
30	0.0994	0.0888	113.2832
35	0.0676	0.0858	172.3168
40	0.0460	0.0839	259.0565

Discount Rate: 9%

N	[P/F;i%,N]	[A/P;i%,N]	[F/A;i%,N]
5	0.6499	0.2571	5.9847
10	0.4224	0.1558	15.1929
15	0.2745	0.1241	29.3609
20	0.1784	0.1095	51.1601
25	0.1160	0.1018	84.7009
30	0.0754	0.0973	136.3075
35	0.0490	0.0946	215.7108
40	0.0318	0.0930	337.8824

Discount Rate: 10%

N	[P/F;i%,N]	[A/P;i%,N]	[F/A;i%,N]
5	0.6209	0.2638	6.1051
10	0.3855	0.1627	15.9374
15	0.2394	0.1315	31.7725
20	0.1486	0.1175	57.2750
25	0.0923	0.1102	98.3471
30	0.0573	0.1061	164.4940
35	0.0356	0.1037	271.0244
40	0.0221	0.1023	442.5926

Discount Rate: 11%

N	[P/F,i%,N]	[A/P,i%,N]	[F/A,i%,N]
5	0.5935	0.2706	6.2278
10	0.3522	0.1698	16.7220
15	0.2090	0.1391	34.4054
20	0.1240	0.1256	64.2028
25	0.0736	0.1187	114.4
30	0.0437	0.1150	199.0
35	0.0259	0.1129	341.6
40	0.0154	0.1117	581.8

Discount Rate: 12%

N	[P/F,i%,N]	[A/P,i%,N]	[F/A,i%,N]
5	0.5674	0.2774	6.3528
10	0.3220	0.1770	17.5487
15	0.1827	0.1468	37.2797
20	0.1037	0.1339	72.0524
25	0.0588	0.1275	133.3
30	0.0334	0.1241	241.3
35	0.0189	0.1223	431.7
40	0.0107	0.1213	767.1

Discount Rate: 13%

N	[P/F,i%,N]	[A/P,i%,N]	[F/A,i%,N]
5	0.5428	0.2843	6.4803
10	0.2946	0.1843	18.4197
15	0.1599	0.1547	40.4175
20	0.0868	0.1424	80.9468
25	0.0471	0.1364	155.6
30	0.0256	0.1334	293.2
35	0.0139	0.1318	546.7
40	0.0075	0.1310	1013.7

Discount Rate: 14%

N	[P/F,i%,N]	[A/P,i%,N]	[F/A,i%,N]
5	0.5194	0.2913	6.6101
10	0.2697	0.1917	19.3373
15	0.1401	0.1628	43.8424
20	0.0728	0.1510	91.0249
25	0.0378	0.1455	181.9
30	0.0196	0.1428	356.8
35	0.0102	0.1414	693.6
40	0.0053	0.1407	1342.0

Discount Rate: 15%

N	[P/F,i%,N]	[A/P,i%,N]	[F/A,i%,N]
5	0.4972	0.2983	6.7424
10	0.2472	0.1993	20.3037
15	0.1229	0.1710	47.5804
20	0.0611	0.1598	102.4
25	0.0304	0.1547	212.8
30	0.0151	0.1523	434.7
35	0.0075	0.1511	881.2
40	0.0037	0.1506	1779.1

Discount Rate: 16%

N	[P/F,i%,N]	[A/P,i%,N]	[F/A,i%,N]
5	0.4761	0.3054	6.8771
10	0.2267	0.2069	21.3215
15	0.1079	0.1794	51.6595
20	0.0514	0.1687	115.4
25	0.0245	0.1640	249.2
30	0.0116	0.1619	530.3
35	0.0055	0.1609	1120.7
40	0.0026	0.1604	2360.8

Discount Rate: 18%

N	[P/F,i%,N]	[A/P,i%,N]	[F/A,i%,N]
5	0.4371	0.3198	7.1542
10	0.1911	0.2225	23.5213
15	0.0835	0.1964	60.9653
20	0.0365	0.1868	146.63
25	0.0160	0.1829	342.6
30	0.0070	0.1813	790.9
35	0.0030	0.1806	1816.7
40	0.0013	0.1802	4163.2

Discount Rate: 20%

N	[P/F,i%,N]	[A/P,i%,N]	[F/A,i%,N]
5	0.4019	0.3344	7.4416
10	0.1615	0.2385	25.9587
15	0.0649	0.2139	72.0351
20	0.0261	0.2054	186.7
25	0.0105	0.2021	472.0
30	0.0042	0.2008	1181.9
35	0.0017	0.2003	2948.3
40	0.0007	0.2001	7343.9

Discount Rate: 25%

N	[P/F,i%,N]	[A/P,i%,N]	[F/A,i%,N]
5	0.3277	0.3718	8.2070
10	0.1074	0.2801	33.2529
15	0.0352	0.2591	109.69
20	0.0115	0.2529	342.9
25	0.0038	0.2509	1054.8
30	0.0012	0.2503	3227.2
35	0.0004	0.2501	9856.8
40	0.0001	0.2500	30088.7

Discounting Future Cash Flows: [P/F, i%, N] = 1/(1+i)^N = 1/[F/P, i%, N]

N	2%	3%	4%	5%	6%	7%	8%	9%	10%	12%	15%	20%
1	0.9804	0.9709	0.9615	0.9524	0.9434	0.9346	0.9259	0.9174	0.9091	0.8929	0.8696	0.8333
2	0.9612	0.9426	0.9246	0.9070	0.8900	0.8734	0.8573	0.8417	0.8264	0.7972	0.7561	0.6944
3	0.9423	0.9151	0.8890	0.8638	0.8396	0.8163	0.7938	0.7722	0.7513	0.7118	0.6575	0.5787
4	0.9238	0.8885	0.8548	0.8227	0.7921	0.7629	0.7350	0.7084	0.6830	0.6355	0.5718	0.4823
5	0.9057	0.8626	0.8219	0.7835	0.7473	0.7130	0.6806	0.6499	0.6209	0.5674	0.4972	0.4019
6	0.8880	0.8375	0.7903	0.7462	0.7050	0.6663	0.6302	0.5963	0.5645	0.5066	0.4323	0.3349
7	0.8706	0.8131	0.7599	0.7107	0.6651	0.6227	0.5835	0.5470	0.5132	0.4523	0.3759	0.2791
8	0.8535	0.7894	0.7307	0.6768	0.6274	0.5820	0.5403	0.5019	0.4665	0.4039	0.3269	0.2326
9	0.8368	0.7664	0.7026	0.6446	0.5919	0.5439	0.5002	0.4604	0.4241	0.3606	0.2843	0.1938
10	0.8203	0.7441	0.6756	0.6139	0.5584	0.5083	0.4632	0.4224	0.3855	0.3220	0.2472	0.1615
11	0.8043	0.7224	0.6496	0.5847	0.5268	0.4751	0.4289	0.3875	0.3505	0.2875	0.2149	0.1346
12	0.7885	0.7014	0.6246	0.5568	0.4970	0.4440	0.3971	0.3555	0.3186	0.2567	0.1869	0.1122
13	0.7730	0.6810	0.6006	0.5303	0.4688	0.4150	0.3677	0.3262	0.2897	0.2292	0.1625	0.0935
14	0.7579	0.6611	0.5775	0.5051	0.4423	0.3878	0.3405	0.2992	0.2633	0.2046	0.1413	0.0779
15	0.7430	0.6419	0.5553	0.4810	0.4173	0.3624	0.3152	0.2745	0.2394	0.1827	0.1229	0.0649
16	0.7284	0.6232	0.5339	0.4581	0.3936	0.3387	0.2919	0.2519	0.2176	0.1631	0.1069	0.0541
17	0.7142	0.6050	0.5134	0.4363	0.3714	0.3166	0.2703	0.2311	0.1978	0.1456	0.0929	0.0451
18	0.7002	0.5874	0.4936	0.4155	0.3503	0.2959	0.2502	0.2120	0.1799	0.1300	0.0808	0.0376
19	0.6864	0.5703	0.4746	0.3957	0.3305	0.2765	0.2317	0.1945	0.1635	0.1161	0.0703	0.0313
20	0.6730	0.5537	0.4564	0.3769	0.3118	0.2584	0.2145	0.1784	0.1486	0.1037	0.0611	0.0261
25	0.6095	0.4776	0.3751	0.2953	0.2330	0.1842	0.1460	0.1160	0.0923	0.0588	0.0304	0.0105
30	0.5521	0.4120	0.3083	0.2314	0.1741	0.1314	0.0994	0.0754	0.0573	0.0334	0.0151	0.0042
35	0.5000	0.3554	0.2534	0.1813	0.1301	0.0937	0.0676	0.0490	0.0356	0.0189	0.0075	0.0017
40	0.4529	0.3066	0.2083	0.1420	0.0972	0.0668	0.0460	0.0318	0.0221	0.0107	0.0037	0.0007
45	0.4102	0.2644	0.1712	0.1113	0.0727	0.0476	0.0313	0.0207	0.0137	0.0061	0.0019	0.0003
50	0.3715	0.2281	0.1407	0.0872	0.0543	0.0339	0.0213	0.0134	0.0085	0.0035	0.0009	0.0001
100	0.1380	0.0520	0.0198	0.0076	0.0029	0.0012	0.0005	0.0002	0.0001	0.0000	0.0000	0.0000
200	0.0191	0.0027	0.0004	0.0001	0.0000	0.0000	0.0000	0.0000	0.0000	0.0000	0.0000	0.0000

Equivalent Uniform Annual Cost

Converting Present Value into an Annuity: $[A/P,i\%,N] = i*(1+i)^N/((1+i)^N - 1) = 1/[P/A,i\%,N]$

	2%	3%	4%	5%	6%	7%	8%	9%	10%	12%	15%	20%
1	1.0200	1.0300	1.0400	1.0500	1.0600	1.0700	1.0800	1.0900	1.1000	1.1200	1.1500	1.2000
2	0.5150	0.5226	0.5302	0.5378	0.5454	0.5531	0.5608	0.5685	0.5762	0.5917	0.6151	0.6545
3	0.3468	0.3535	0.3603	0.3672	0.3741	0.3811	0.3880	0.3951	0.4021	0.4163	0.4380	0.4747
4	0.2626	0.2690	0.2755	0.2820	0.2886	0.2952	0.3019	0.3087	0.3155	0.3292	0.3503	0.3863
5	0.2122	0.2184	0.2246	0.2310	0.2374	0.2439	0.2505	0.2571	0.2638	0.2774	0.2983	0.3344
6	0.1785	0.1846	0.1908	0.1970	0.2034	0.2098	0.2163	0.2229	0.2296	0.2432	0.2642	0.3007
7	0.1545	0.1605	0.1666	0.1728	0.1791	0.1856	0.1921	0.1987	0.2054	0.2191	0.2404	0.2774
8	0.1365	0.1425	0.1485	0.1547	0.1610	0.1675	0.1740	0.1807	0.1874	0.2013	0.2229	0.2606
9	0.1225	0.1284	0.1345	0.1407	0.1470	0.1535	0.1601	0.1668	0.1736	0.1877	0.2096	0.2481
10	0.1113	0.1172	0.1233	0.1295	0.1359	0.1424	0.1490	0.1558	0.1627	0.1770	0.1993	0.2385
11	0.1022	0.1081	0.1141	0.1204	0.1268	0.1334	0.1401	0.1469	0.1540	0.1684	0.1911	0.2311
12	0.0946	0.1005	0.1066	0.1128	0.1193	0.1259	0.1327	0.1397	0.1468	0.1614	0.1845	0.2253
13	0.0881	0.0940	0.1001	0.1065	0.1130	0.1197	0.1265	0.1336	0.1408	0.1557	0.1791	0.2206
14	0.0826	0.0885	0.0947	0.1010	0.1076	0.1143	0.1213	0.1284	0.1357	0.1509	0.1747	0.2169
15	0.0778	0.0838	0.0899	0.0963	0.1030	0.1098	0.1168	0.1241	0.1315	0.1468	0.1710	0.2139
16	0.0737	0.0796	0.0858	0.0923	0.0990	0.1059	0.1130	0.1203	0.1278	0.1434	0.1679	0.2114
17	0.0700	0.0760	0.0822	0.0887	0.0954	0.1024	0.1096	0.1170	0.1247	0.1405	0.1654	0.2094
18	0.0667	0.0727	0.0790	0.0855	0.0924	0.0994	0.1067	0.1142	0.1219	0.1379	0.1632	0.2078
19	0.0638	0.0698	0.0761	0.0827	0.0896	0.0968	0.1041	0.1117	0.1195	0.1358	0.1613	0.2065
20	0.0612	0.0672	0.0736	0.0802	0.0872	0.0944	0.1019	0.1095	0.1175	0.1339	0.1598	0.2054
25	0.0512	0.0574	0.0640	0.0710	0.0782	0.0858	0.0937	0.1018	0.1102	0.1275	0.1547	0.2021
30	0.0446	0.0510	0.0578	0.0651	0.0726	0.0806	0.0888	0.0973	0.1061	0.1241	0.1523	0.2008
35	0.0400	0.0465	0.0536	0.0611	0.0690	0.0772	0.0858	0.0946	0.1037	0.1223	0.1511	0.2003
40	0.0366	0.0433	0.0505	0.0583	0.0665	0.0750	0.0839	0.0930	0.1023	0.1213	0.1506	0.2001
45	0.0339	0.0408	0.0483	0.0563	0.0647	0.0735	0.0826	0.0919	0.1014	0.1207	0.1503	0.2001
50	0.0318	0.0389	0.0466	0.0548	0.0634	0.0725	0.0817	0.0912	0.1009	0.1204	0.1501	0.2000
100	0.0232	0.0316	0.0408	0.0504	0.0602	0.0701	0.0800	0.0900	0.1000	0.1200	0.1500	0.2000
200	0.0204	0.0301	0.0400	0.0500	0.0600	0.0700	0.0800	0.0900	0.1000	0.1200	0.1500	0.2000

Note: The annuity is assumed to be paid at the end of the period.

Uniform Series Compound Amount Factor = 1/Uniform Series Present Worth Factor
Future Value of Regular Payments: [F/A,i%,N] = ((1+i%)^N − 1)/I = 1/[A/F,i%,N]

	2%	3%	4%	5%	6%	7%	8%	9%	10%	12%	15%	20%
1	1.000	1.000	1.000	1.000	1.000	1.000	1.000	1.000	1.000	1.000	1.000	1.000
2	2.020	2.030	2.040	2.050	2.060	2.070	2.080	2.090	2.100	2.120	2.150	2.200
3	3.060	3.091	3.122	3.153	3.184	3.215	3.246	3.278	3.310	3.374	3.473	3.640
4	4.122	4.184	4.246	4.310	4.375	4.440	4.506	4.573	4.641	4.779	4.993	5.368
5	5.204	5.309	5.416	5.526	5.637	5.751	5.867	5.985	6.105	6.353	6.742	7.442
6	6.308	6.468	6.633	6.802	6.975	7.153	7.336	7.523	7.716	8.115	8.754	9.930
7	7.434	7.662	7.898	8.142	8.394	8.654	8.923	9.200	9.487	10.089	11.067	12.916
8	8.583	8.892	9.214	9.549	9.897	10.260	10.637	11.028	11.436	12.300	13.727	16.499
9	9.755	10.159	10.583	11.027	11.491	11.978	12.488	13.021	13.579	14.776	16.786	20.799
10	10.950	11.464	12.006	12.578	13.181	13.816	14.487	15.193	15.937	17.549	20.304	25.959
11	12.17	12.81	13.49	14.21	14.97	15.78	16.65	17.56	18.53	20.65	24.35	32.15
12	13.41	14.19	15.03	15.92	16.87	17.89	18.98	20.14	21.38	24.13	29.00	39.58
13	14.68	15.62	16.63	17.71	18.88	20.14	21.50	22.95	24.52	28.03	34.35	48.50
14	15.97	17.09	18.29	19.60	21.02	22.55	24.21	26.02	27.97	32.39	40.50	59.20
15	17.29	18.60	20.02	21.58	23.28	25.13	27.15	29.36	31.77	37.28	47.58	72.04
16	18.64	20.16	21.82	23.66	25.67	27.89	30.32	33.00	35.95	42.75	55.72	87.44
17	20.01	21.76	23.70	25.84	28.21	30.84	33.75	36.97	40.54	48.88	65.08	105.93
18	21.41	23.41	25.65	28.13	30.91	34.00	37.45	41.30	45.60	55.75	75.84	128.12
19	22.84	25.12	27.67	30.54	33.76	37.38	41.45	46.02	51.16	63.44	88.21	154.74
20	24.30	26.87	29.78	33.07	36.79	41.00	45.76	51.16	57.27	72.05	102.44	186.69
25	32.03	36.46	41.65	47.73	54.86	63.25	73.11	84.70	98.35	133.33	212.79	471.98
30	40.57	47.58	56.08	66.44	79.06	94.46	113.28	136.31	164.49	241.33	434.75	1181.88
35	49.99	60.46	73.65	90.32	111.43	138.24	172.32	215.71	271.02	431.66	881.17	2948.34
40	60.40	75.40	95.03	120.80	154.76	199.64	259.06	337.88	442.59	767.09	1779.09	7343.86
45	71.89	92.72	121.03	159.70	212.74	285.75	386.51	525.86	718.90	1358.23	3585.13	18281.31
50	84.58	112.80	152.67	209.35	290.34	406.53	573.77	815.08	1163.91	2400.02	7217.72	45497.19

Note: Annuity payments are assumed to be made at the end of the period, which is also the beginning of the next period.

Appendix B

Equivalence Relationships for Uniform Gradients and Geometric Sequences

B.1 THE ROLE OF EQUIVALENCE FACTORS IN EVALUATING INFRASTRUCTURE PROPOSALS: REVIEW

Chapter 7, "Equivalence of Cash Flows," presented the basic relationships of engineering economics that are repeatedly used in project evaluation and costing. Given an appropriate discount rate, these relationships can be used to convert an arbitrary stream of cash flows into an equivalent present value P, future value F, or annuity value A. Equations were derived for six equivalence factors:

1. [P/F,i,N] Present value P given the future value F, the discount rate i, and the number of time periods N. This factor can be used repeatedly to convert any stream of cash flows into a present value.
2. [F/P,i,N] Future value F at the end of time period N given the present value P and the discount rate i.
3. [A/P,i,N] *Capital recovery factor*: This factor is used to calculate the annuity A that will be received at the end of each of N periods that is equivalent to a specified present value P.
4. [A/F,i,N] *Sinking fund factor*: This factor is used to calculate how much must be set aside each period for N periods in order accumulate a future value F.
5. [P/A,i,N] *Uniform series present worth factor*: This factor can be used to calculate the present value of an annuity.
6. [F/A,I,N] *Uniform series compound amount factor*: This factor can be used to calculate the future value of an annuity after N periods.

Chapter 7 presented the equations for these factors under two sets of assumptions:

1. Discrete compounding
2. Continuous compounding

Discrete compounding is likely to be the preferred method when evaluating proposals for infrastructure investment. When evaluating such proposals, monthly or annual costs and benefits are commonly projected over periods of 20 or more years. It is therefore practical to discount these monthly or annual estimates of cash flows using the equivalence relationships based on discrete compounding. Any discrepancies introduced by the choice of compounding will be minor compared to the overall uncertainties in the analysis of any large infrastructure project. Many different engineering and economic factors will enter into the analysis, and some of these factors are apt to change in unpredictable ways over the planned life of the project. Due to the complexity and uncertainty of the analysis, it is essential to consider multiple options under various scenarios and to conduct sensitivity analyses that consider possible changes in the most important factors that can affect the potential success or failure of the project. As shown in examples and assignments throughout this text, spreadsheets can be made and used to understand how changes in assumptions will affect performance. A spreadsheet makes it readily possible to consider such things as

- Systematic increases or decreases in unit costs
- Changes in productivity for construction or operations
- Cyclical patterns in demand
- Sudden drops in demand (e.g., collapse of the real estate market)
- Sudden increases in cost (e.g., fuel cost)

Because complexities and uncertainties arise over the long life of infrastructure projects, results will not be very precise. Capturing all of the relevant inputs and considering all of the key performance measures is more important than seeking extreme precision in a few of the many variables that must be considered. Choosing a reasonable discount rate, and considering the different discount rates that may be used by owners and investors, is critical to the analysis. Whether discounting is assumed to be continuous or discrete turns out to be a minor nuance in the typical analysis. Using continuous rather than discrete compounding may turn an annual rate of 10% into an effective rate closer to 11%, but the typical project evaluation probably must deal with very imprecise estimates of costs, benefits, interest rates on bonds and mortgages, and especially discount rates for the owner and for potential investors. The owner's discount rate might be stated as a rather arbitrary range, such as 15–20%, and potential investors might be using discount rates that are even less precise.

In short, when evaluating infrastructure projects, simple equivalence relationships are normally all that is needed: what is the net present of the proposed project under various scenarios and various assumptions about discount rates, costs, productivity, and demand?

Nevertheless, as covered in Section B.2, some additional equivalence relationships may be useful in other contexts (e.g., in designing retirement annuities that have an annual increase in payments and thus offer some protection against inflation). While it is beyond the scope of this text to derive such relationships, it may be useful to understand that these relationships are available.

B.2 ADDITIONAL EQUIVALENCE RELATIONSHIPS: UNIFORM GRADIENTS AND GEOMETRIC SEQUENCES

Equivalence relationships can be derived for two situations involving annuities:

1. *Uniform gradients:* The annuity is 0 at the end of the first period and increases by a fixed amount G (known as the uniform gradient amount) at the end of each subsequent period. The payment for the second period will therefore be G, and the payment for the Nth period will be $(N - 1)G$.
2. *Geometric sequences:* The annuity is A at the end of the first period and increases at a fixed rate of f per period. The annuity payment for the second period will be $(1 + f)(A)$, and the payment for the Nth period will be $(1 + f)^{N-1}(A)$.

B.2.1 Uniform Gradients

If the uniform gradient is G, then the equivalent uniform annuity A can be calculated as

$$A = G\left[1/i - N/((1+i)^N - 1)\right] \tag{B.1}$$

Table B.1   Gradient to Uniform Series Conversion Factors [A/G,i%,N] for
Selected Discount Rates and Time Periods

N	4%	6%	8%	10%
5	1.9216	1.8836	1.8465	1.8101
10	4.1773	4.0220	3.8713	3.7255
15	6.2721	5.9260	5.5945	5.2789
20	8.2091	7.6051	7.0369	6.5081
25	9.9925	9.0722	8.2254	7.4580
30	11.6274	10.3422	9.1897	8.1762

Table B.2   Gradient to Present Equivalent Conversion Factors [P/G,i%,N] for
Selected Discount Rates and Time Periods

N	4%	6%	8%	10%
5	8.555	7.935	7.372	6.862
10	33.881	29.602	25.977	22.891
15	69.736	57.555	47.886	40.152
20	111.565	87.230	69.090	55.407
25	156.104	115.973	87.804	67.696
30	201.062	142.359	103.456	77.077

Using notation and terminology similar to that introduced in Chapter 7, Eq. B.1 can be rewritten as

$$A = G\,[A/G, i\%, N] \tag{B.2}$$

The expression [A/G,i%,N] is called the "gradient to uniform series conversion factor." Engineering economics texts often have tables showing numerical values for various discount rates i% and time periods N. Some examples are shown in Table B.1.

The equivalent present worth can be calculated as

$$P = A\,[P/A, i\%, N] = A\,[(1 + i)^N - 1)/(i(1 + i)^N)] \tag{B.3}$$

Substituting the expression for A from Eq. B.2 and rearranging the terms yields the following:

$$P = G\{1/i[(((1 + i)^N - 1)/i(1 + i)^N) - N/(1 + i)^N]\} \tag{B.4}$$

Using the usual notation and terminology, this can be written as

$$P = G\,[P/G, i\%, N] \tag{B.5}$$

The expression [A/G,i%,N] is called the "gradient to present equivalent conversion factor," which can also be found in tables in engineering economics texts. Some examples are shown in Table B.2.

EXAMPLE B.1   Uniform Gradients

A company plans to open a new fast-food restaurant at the end of year 2 and to open another restaurant every year thereafter until year 10. Each store is expected to have revenues of $2 million per year. The company's minimum acceptable rate of return (MARR) is 10%. What is the equivalent uniform annual revenue for the 10-year period? What is the present worth of the total revenue?

This revenue can be represented by a uniform gradient where G = $2 million. Recall that with a uniform gradient, the cash flow at the end of period 1 is zero and the cash flow at the end of period 2 is G. The equivalent uniform annual revenue A will be

$$A = \$2 \text{ million } [A/G, 10\%, 10 \text{ years}] \tag{B.6}$$

The uniform series conversion factor is 3.7255, as shown in the last column of Table B.1. Therefore, the equivalent uniform annual revenue is

$$A = \$2 \text{ million } (3.7255) = \$7.451 \text{ million} \tag{B.7}$$

The present worth of the revenue will be

$$P = \$2 \text{ million } [P/G, 10\%, 10 \text{ years}] \tag{B.8}$$

The gradient to present worth conversion factor is 22.891, as shown in the last column of Table B.2. Therefore, the equivalent uniform annual revenue is

$$P = \$2 \text{ million } (22.891) = \$45.78 \text{ million} \tag{B.9}$$

The present worth could also be obtained using the equivalent annuity value from (B.7) and the usual factor for converting an annuity to a present value

$$P = \$7.451 \text{ million } [P/A, 10\%, 10 \text{ years}] \tag{B.10}$$

$$P = \$7.451 \text{ million } (6.1446) = \$45.78 \text{ million} \tag{B.11}$$

B.2.2   Geometric Sequences

Geometric sequences could be useful in situations where costs or benefits are expected to rise at a constant rate over several periods. Unit costs might be expected to rise with inflation, energy costs might be expected to rise at a rate faster than inflation, and demand might be expected to grow at a constant rate each year.

Useful equivalence expressions can be found by using mathematical relationships concerning series and by defining so-called convenience rates that take the place of the usual discount rate in stating the results. If the discount rate is i% and the cash flows are increasing at a rate of f% per year, then the convenience rate i_{CR} is defined as

$$i_{CR} = ((1 + i)/(1 + f)) - 1 \tag{B.12}$$

The present value P of the geometric sequence of cash flows turns out to be a rather straightforward function of the initial cash flow A_1, f, and i_{CR}:

$$P = A_1/(1 + f) [P/F, i_{CR}, N] \tag{B.13}$$

Since i_{CR} generally is not an integer, the expression $[P/F, i_{CR}, N]$ is not included in the standard tables in engineering economics texts. However, your spreadsheet can probably handle nonintegral values, so you can easily calculate this expression.

For small values of i and f, the expression i_{CR} will be approximately equal to i − f. Note that i_{CR} can be positive or negative, depending upon whether the rate of increase f is smaller or larger than the discount rate.

B.3   SPREADSHEETS VERSUS TABLES

Before computers were widely available, there was considerable value in deriving closed-form solutions to many kinds of analytical problems and in providing tables similar to Tables B1 and B2. Someone who knows how to use the tables can quickly find the factors needed to solve a particular problem. Moreover, with a closed-form solution, it may be possible to avoid endless, repetitive calculations and tiresome trial-and-error approaches to a problem. For example, one engineering economics text points out that the cash flows in a geometric sequence could each be discounted or compounded to obtain the equivalent present or future value. However, the authors go on to state that "this becomes quite tedious for large N, so it is convenient to have a single equation instead."[1]

In my experience, trying to use an equation such as Eq. B.1 or B.4 often leads to mistakes, and it may take me several minutes to get the expression entered correctly into my spreadsheet. And sometimes I don't realize right away that I have made a mistake. Therefore, I am much happier just entering the cash flows and doing the calculations.

The spreadsheet itself simplifies the process. For example, in working with a geometric sequence of cash flows, you need to enter the initial cash flow, the annual increase f, and provide space for entering the years 1 to N and the cash flows for years 1 to N. However, it is not necessary to enter N values for the years and N more values for the cash flows, which would indeed become tedious for a monthly analysis of a project with a life of 30 years! Instead, you need to enter only two formulas that can then be copied across the desired range:

1. Year = prior year plus 1
2. Cash flow = prior cash flow (1+f)

Once this data is in the spreadsheet, it is easy to use spreadsheet functions to get the present value and the equivalent uniform annuity. Using a spreadsheet is so easy that you may never have to use Eq. B.1 or B.4.

EXAMPLE B.2   Using a Spreadsheet to Calculate Uniform Gradients and Geometric Sequences

A company has two sources of cash flows:

1. One is the new fast-food restaurant business identified in Example B.1. Revenues will be $2 million at the end of year 2, and they will increase by $2 million every year for 10 years.
2. The other business includes a chain of convenience stores that has revenues of $100 million per year; revenues from this business are expected to grow at 2% per year.

Calculate the present value of the revenues from each business and from the total business. Calculate the equivalent uniform annual revenue from each business and for the total business.

Table B.3 shows a simple spreadsheet that was constructed to answer this problem. Key parameters are entered in the control panel at the top, including the gradient G, the first-period cash flow, and the fixed rate of increase f for the geometric sequence. The values for the first period were entered manually. The values for subsequent periods were calculated using formulas:

- Gradient: revenue(N) = revenue(N − 1) + G
- Geometric sequence: revenue(N) = (1 + f) revenue(N − 1)
- NPV: NPV for revenue(N) = revenue(N)/(1 + discount rate)

The formulas were copied for 10 periods; they could as easily have been copied for 100 periods. The total NPV was obtained by summing the appropriate columns, and the equivalent uniform annuities were obtained using the PMT function on the spreadsheet. Note that the answers under the gradient column are the same as those obtained in Example B.1.

A more important reason for never having to use Eq. B.1 or B.4 is that they apply only in highly specialized situations. Using any closed-form solution requires simplifying assumptions. When computation was difficult, the trade-off was pretty clear: make the assumptions needed to use a readily available formula. However, now that computation is cheap, the trade-off tends to be resolved the other way: do all of the calculations, and avoid making unnecessary assumptions. You

---

[1] William G. Sullivan, Elin M. Wicks, and James T. Luxhoj, *Engineering Economy*, 12th ed. (Upper Saddle River, NJ: Prentice Hall, 2003), 107.

Table B.3 A Spreadsheet for Calculating Gradients and Geometric Sequences

Spreadsheet for Investigating Uniform Gradients and Geometric Sequences

First-period cash flow			$100.00			
Gradient			$ 2.00			
Fixed rate of increase			2%			
Discount rate			10%			
Period	Gradient	Geometric Sequence	Combined	NPV Gradient	NPV Sequence	NPV Combined
1	$ 0.00	$100.00	$100.00	$ 0.00	$100.00	$100.00
2	$ 2.00	$102.00	$104.00	$ 1.65	$ 84.30	$ 85.95
3	$ 4.00	$104.04	$108.04	$ 3.01	$ 78.17	$ 81.17
4	$ 6.00	$106.12	$112.12	$ 4.10	$ 72.48	$ 76.58
5	$ 8.00	$108.24	$116.24	$ 4.97	$ 67.21	$ 72.18
6	$10.00	$110.41	$120.41	$ 5.64	$ 62.32	$ 67.97
7	$12.00	$112.62	$124.62	$ 6.16	$ 57.79	$ 63.95
8	$14.00	$114.87	$128.87	$ 6.53	$ 53.59	$ 60.12
9	$16.00	$117.17	$133.17	$ 6.79	$ 49.69	$ 56.48
10	$18.00	$119.51	$137.51	$ 6.94	$ 46.08	$ 53.02
Total NPV				$45.78	$671.62	$717.40
Equivalent Annuity				$ 7.45	$109.30	$116.75

will end up with a spreadsheet that shows all of the relevant details, along with the ultimate answers, so you will be much more apt to spot errors as well as to understand the results of your analysis.

Credits

The author thanks the following authors and publishers for permission to use excerpts from their publications in this textbook:

Table 1.1, page 6, from *ASCE's Infrastructure Report Card Give Nation a D, Estimates Cost at $2.2 Trillion*, ASCE News, February 2009, Vol. 34., No. 2, used by permission from ASCE News.

Selections on pages 114 and 116 from The Interorganizational Committee on Principles and Guidelines for Social Impact Assessment, *Principles and guidlelines for social impact assessment in the USA,* **Impact Assessment and Project Appraisal**, Vol. 21, No. 3, Beech Tree Publishing, Guilford, Surry, UK, September 2003. Used with permission of Beech Tree Publishing.

Selections on pages 161, 186 and 187 from SKYSCRAPER: THE MAKING OF A BUILDING by Karl Sabbagh, copyright © 1989 by Karl Sabbagh. Used by permission of Viking Penguin, a division of Penguin Group (USA).

Selections on pages 407-408 from *Halting Hypoxia*, **Civil Engineering**, June 2008, pp. 54-56. Used with permission of Civil Engineering.

Selections on pages 411-412 from E. Alfred Picardi, *High Rise Buildings and 9/11,* May 5, 2002, used by permission of E. Alfred Picardi.

Selection on page 444 from Michael Hudson, *The New Road to Serfdom: an illustrated guide to the coming real estate collapse:* Copyright © 2006 by Harper's Magazine. All rights reserved. Reproduced from the May issue by special permission.

Selection on page 437 from *President-Elect Candidates Caldwell and Gouda Respond to Interview Questions,* ASCE News, June 2009, Vol. 34, No. 6 used by permission of ASCE News.

PHOTO CREDITS

Figure 4.4: Photo of rail line construction in Saudi Arabia by Vinay Mudholkar. Used by permission of Vinay Mudholkar.

Figure 3.7: Photo of rail line and I70 in Colorado (Figure 3.7) by Samuel J. Martland. Used by permission of Samuel J. Martland.

All other photos were taken by the author.

Index